DATE DUE

GAYLORD			PRINTED IN U.S.A.

'The development of global ozone policy was a complex, challenging and deeply educational experience for all involved – industry, government, the science community and environmental advocates. Stephen O Andersen and K Madhava Sarma deserve our sincerest appreciation for creating a comprehensive, accurate account of the events and decisions that eventually led to the landmark Montreal Protocol'

PAUL V TEBO, Vice-President, Safety, Health and Environment, DuPont

'The unprecedented convergence of science, diplomacy and world citizenry to protect the indivisible ozone layer is well articulated by two of the many makers of this history in this easy-to-read book. The book provides important lessons in environmental diplomacy and should be used as a manual for training future environmental leaders'

OMAR E EL-ARINI, Chief Officer, Multilateral Fund Secretariat of the Montreal Protocol

'Andersen and Sarma have written an excellent history of a defining moment in the relationship of humans to their environment. The complex social response that protected the ozone layer is one of the first, and best, examples of how environmental issues evolved from being considered only as an afterthought to being considered a strategic necessity for individuals, firms and society. The Montreal Protocol helped launch industry–government environmental leadership partnerships, design for environment and industrial ecology as critical tools to grapple with even more complex global challenges such as climate change'

BRAD ALLENBY, Vice President, Environment, Health & Safety, AT&T

'Montreal Protocol insiders will enjoy the clarity, accuracy and insights of this extraordinary history but its greater value will be for students and policy-makers who want a "how to" guide for protecting the global environment. They will learn how science served as the basis for policy, how partnerships between industry and government can work together to develop innovative solutions that work for both, and how developed and developing countries can find common ground. And they will learn that compliance with international treaties can be better enforced through collaboration and assistance than through coercion'

STEPHEN SEIDEL, Deputy Director, Office of Atmospheric Programs, US EPA

'The Vienna Convention and its Montreal Protocol are an excellent case study in how to deal with a global threat to human health and the environment. The Montreal Protocol, in particular, is a contract between governments to take action nationally to protect the ozone layer. More than that, it is a contract between them to help each other in taking those actions. This history is valuable in bringing together the actions and milestones, and in presenting some of the behind-the-scenes events that helped shape the process, expressed from the perspectives of those involved. Stephen Andersen and Madhava Sarma are uniquely qualified to present this history. They are part of it – initiators of some key aspects, active players in others but always involved and committed'

JOHN WHITELAW, Deputy, UNEP Chemicals Division, 1995 Chair of the Executive Committee of the MLF

Protecting the Ozone Layer

Protecting the Ozone Layer

The United Nations History

By Stephen O Andersen and
K Madhava Sarma

Edited by Lani Sinclair

Stephen O. Andersen
January 2003

UNEP

Earthscan Publications Ltd
London • Sterling, VA

First published in the UK and USA in 2002 by Earthscan Publications Ltd
for and on behalf of the United Nations Environment Programme

ISBN: 1 85383 905 1

Typesetting by PCS Mapping & DTP, Gateshead
Printed and bound in the UK by Creative Print and Design Wales, Ebbw Vale
Cover design by Danny Gillespie

For a full list of publications please contact:

Earthscan Publications Ltd
120 Pentonville Road, London, N1 9JN, UK
Tel: +44 (0)20 7278 0433
Fax: +44 (0)20 7278 1142
Email: earthinfo@earthscan.co.uk
Web: **www.earthscan.co.uk**

22883 Quicksilver Drive, Sterling, VA 20166-2012, USA

A catalogue record for this book is available from the British Library

Library of Congress Cataloging-in-Publication Data

Andersen, Stephen O.
    Protecting the ozone layer : the United Nations history / by Stephen O. Andersen
and K. Madhava Sarma ; edited by Lani Sinclair.
        p. cm.
    Includes index.
    ISBN 1-85383-905-1
    1. Ozone layer depletion—Prevention—History—20th century. I. Sarma, K.
Madhava, 1938- II. Sinclair, Lani. III. Title.

QC879.7 .A53 2002
363.738'7526—dc21

                                                          2002005805

Earthscan is an editorially independent subsidiary of Kogan Page Ltd and publishes in
association with WWF-UK and the International Institute for Environment and
Development

This book is printed on elemental-chlorine-free paper

Contents

List of plates, figures, tables and boxes

PLATES

FIGURES

TABLES

BOXES

About the authors

**Stephen O Andersen, Director of Strategic Climate Projects,
US Environmental Protection Agency**

 Stephen O Andersen began work on climate and ozone layer protection in 1974 as a member of the Climatic Impact Assessment Project on the environmental effects of supersonic aircraft. Prior to joining the US Environmental Protection Agency (EPA), he worked for environmental and consumer non-governmental organizations (NGOs) and was a professor of environmental economics. In 1986, he joined the fledgling EPA Stratospheric Protection team, working his way up from Senior Economist to Deputy Director of the Stratospheric Protection Division. He is currently EPA Director of Strategic Climate Projects. Since 1988, he has been Co-chair of the Technology and Economic Assessment Panel. He has also chaired the Solvents Technical Options Committee, the Methyl Bromide Interim Technology and Economic Assessment, and the Task Force on the Implications to the Montreal Protocol of the Inclusion of HFCs and PFCs in the Kyoto Protocol. He chaired the United Nations Environment Programme (UNEP) working group that developed the essential use process and has been a member of Mostafa Tolba's Informal Advisory Group. He pioneered EPA's voluntary approaches to ozone layer protection, including the phase-out of CFC food packaging, the recycling of CFCs from vehicle air conditioning, the halt to testing and training with halon, and the accelerated CFC solvent phase-out in electronics and aerospace. He created the EPA ozone and climate protection awards, and helped found the Industry Cooperative for Ozone Layer Protection and the Halons Alternative Research Corporation. He helped to negotiate the phase-out of CFC refrigerator manufacturing in Thailand and the corporate pledge to help Vietnam avoid increased dependence on ozone-depleting substances (ODSs). He served on the team that commercialized no-clean soldering and the team phasing out ODSs from solid rocket motors. He has helped organize dozens of international conferences, workshops, and technology demonstrations. He received the 1990 EPA Gold Medal, the 1994 ICOLP Global Achievement Award, the 1995 Fitzhugh Green Award, the 1995 UNEP Global Stratospheric Ozone Protection Award, the 1996 Sao Paulo Brazil State Ozone Award, the 1998 US EPA Stratospheric Ozone Protection Award, the 1998 Nikkan Kogyo Shimbun Stratospheric Protection Award, the 1999 Vietnam Ozone Protection Award, the 2000 Mobile Air Conditioning Society Twentieth Century Award for Environmental Leadership and the 2001 US DoD Award for Excellence. In 1998, he earned the UNEP Global 500 Roll of Honour. He has a PhD from the University of California, Berkeley.

K Madhava Sarma, formerly Executive Secretary, Secretariat for the Vienna Convention and the Montreal Protocol, United Nations Environment Programme

K Madhava Sarma is currently a consultant to UNEP on ozone issues and integration of the common aspects of global environmental treaties for greater synergy. He was Executive Secretary of the Secretariat for the Vienna Convention and the Montreal Protocol from 1991 to 2000. During his tenure as Executive Secretary, he served the Parties to the Protocol through the turbulent Meetings of the Parties in Copenhagen, Vienna, Montreal, and Beijing – including three replenishments of the Multilateral Fund for the Implementation of the Montreal Protocol. He streamlined the administration of the institutions of the Protocol, the reporting requirements, and other administrative obligations so that Parties could devote their full attention to resolving challenging political issues. Prior to being recruited to head the Secretariat, Madhava Sarma was a senior member of the Indian diplomatic team involved in the Montreal Protocol negotiations between the first and second Meetings of the Parties (1989–1991). During this time, he was often an effective spokesman for the developing country perspective and cosponsored many of the provisions of the London Amendment that satisfied developing countries while creating enforceable obligations to protect the ozone layer. He made other significant contributions as the senior Indian official looking after environmental policy, law, institutions and international cooperation, including responsibility for all global environmental issues. Prior to joining the national Government of India, he served as Head of District Administration, State Water Supply Board, and as Secretary to the Government, Irrigation and Power. During this state tenure, he was responsible for planning and implementation for many water supply, irrigation and energy projects. He earned the 1996 US EPA Stratospheric Ozone Protection Award and an award from UNEP 'For Extraordinary Contributions to Ozone Layer Protection'.

Foreword

Throughout its existence, the United Nations has been at the forefront of efforts to protect the global environment. The making of environmental law has been an essential part of that undertaking. Today, there are more than 230 environmental treaties, covering issues such as marine and air pollution, hazardous waste, biodiversity, desertification and climate change. Among the most successful of these treaties are the ozone agreements brokered by the United Nations Environment Programme: the Vienna Convention on the Protection of the Ozone Layer (1985) and its Montreal Protocol on Substances that Deplete the Ozone Layer (1987).

The treaties address one of the most ominous global environmental problems ever faced by humankind: the destruction, by synthetic chemicals, of the fragile mantle of stratospheric ozone that protects all life on earth from the sun's lethal ultraviolet rays. When UNEP first proposed controlling the chemicals that scientists had identified as responsible for the damage, there was resistance from many quarters. But compelling evidence marshalled by UNEP and the World Meteorological Organization, along with a growing public clamour about the potential consequences, ultimately persuaded governments to act. More than 180 countries are now party to the agreements, and as such have committed themselves to phasing out, on strict timetables, all ozone-depleting chemicals.

This book is an account of how many stakeholders – governments, scientists, industry, non-governmental organizations and the United Nations system – set aside their differences and came together to ward off a common, potentially catastrophic threat. Indeed, the agreements marked the first application of the 'precautionary principle' by which action is taken, even before the science is certain, to prevent an emerging problem from becoming a crisis, rather than waiting too late to avoid irreparable harm.

This book is also a timely contribution by UNEP to the World Summit on Sustainable Development, which is to take place in Johannesburg in September 2002 and which offers the international community an opportunity to act on the unfulfilled promises of the 'Earth Summit' ten years earlier in Rio de Janeiro, and to address the urgent and enormous unfinished business on the agenda of environment and development. For the sake of present and future generations alike, I commend this book to the widest possible readership.

Kofi A Annan
Secretary-General, United Nations

Preface

This is the first history of an environmental issue published by the United Nations Environment Programme (UNEP). We chose the issue of protection of the ozone layer for this venture for many reasons. The most important is that disseminating the story of how the global agreements on the ozone layer became outstanding successes can, hopefully, help the world community tackle other global environmental problems with equal accomplishment. I participated in the process of developing the Vienna Convention of 1985 and the Montreal Protocol of 1987 as the Minister for Environment for Germany, and recall the excitement the treaties generated among all the leaders. Similar excitement is needed today to make sustainable development a reality.

The threat to the ozone layer by manufactured chemicals was the most ominous global environmental problem ever faced by humanity. Although the discovery of this threat was made in 1974 by the Nobel Prize-winning work of Mario Molina and Sherwood Roland, and UNEP urged action to protect the ozone layer beginning in 1977, it took ceaseless persuasion by UNEP and many selfless individuals for ten years before governments took the first, short step in 1987.

The long labour of UNEP for ten years has yielded a spectacular result. The very process led to many innovations in the techniques of persuasion. The objective of the ozone treaties was certainly a difficult one: to persuade the entire world to give up the use of many profitable chemicals that had been praised as wonder chemicals. To be persuaded were not only governments, but also the producers of these chemicals, all major multi-national giants of industrialized countries, and thousands of industries that used these chemicals considered to be 'irreplaceable'. Behind them were the billions of consumers who wanted and needed the products that contained ozone-depleting chemicals: refrigeration, air conditioning, fire-fighting equipment and foams.

There was no measurement of the adverse effects of ozone depletion at that time, since there were none – yet. When proof of ozone depletion appeared in the stunning 1985 discovery that ozone was dramatically depleted over Antarctica in the Antarctic spring, there were some who argued that the impact of depletion was no more serious than the impact of moving from Chicago to Florida in the United States, and that human beings could adjust to ozone depletion easily by wearing hats and dark glasses. UNEP had to convince the world that once the depletion was started, there would be no place for humanity to hide. If we waited for adverse consequences to be apparent, it would be too late to reverse the consequences; hence, we had to act before the adverse impacts appeared, under the 'precautionary principle'.

It is to the credit of my predecessor, Mostafa Tolba, that he arranged an intricate minuet of scientists, technologists and industry before the diplomatic negotiators to persuade them to control these chemicals. For the first time, scientists played a direct part in diplomatic negotiations and helped the governments not only to understand the phenomenon of ozone depletion and its adverse effects, but also to give concrete policy options with each option leading to a particular impact on the ozone layer. The technologists were on hand to analyse the technical and economic feasibility of alternatives, so that governments could make up their minds after weighing all the consequences – environmental, technical and financial.

A stroke of genius in the 1987 Montreal Protocol was to take a mild first step, but to provide for stronger steps after periodic scientific and technological assessment. This allowed many cautious governments to join the Protocol as it progressed. Later, the Protocol was strengthened five times by adjustments and amendments, on the basis of assessments by scientists and technologists of the latest information on the status of the ozone layer and the necessity for, and the feasibility of, stronger control measures to restore the ozone layer.

Another inspiring provision was to recognize that developing countries had contributed little to the problem and hence were entitled to special consideration, even though all nations are responsible for protecting the ozone layer; this is the principle of 'common but differentiated responsibility'. The developing countries were, hence, given an additional ten years to implement the control measures so that when they began the phase-out of ozone-depleting chemicals, they could learn from the experience of the developed countries. Subsequently, on the urging of developing countries, the Protocol developed its own financial mechanism and a Multilateral Fund, contributed to by the developed countries, to meet the incremental costs of the phase-out in developing countries. The Fund to date has distributed, as grants, more than US$1.3 billion to more than 110 developing countries to switch to ozone-friendly chemicals. The result of such inclusiveness is that 184 governments have ratified the Protocol and are actively committed to phasing out ozone-depleting chemicals.

The implementation of the Protocol over the last 12 years has led to outstanding reductions in the consumption of ozone-depleting chemicals by more than 90 per cent. Implementation involved a large number of stakeholders. Many United Nations organizations such as the United Nations Development Programme (UNDP), UNEP, United Nations Industrial Development Organization (UNIDO), World Health Organization (WHO), World Meteorological Organization (WMO), Food and Agriculture Organization (FAO), Regional Economic and Social Commissions of the United Nations, and financial institutions such as the World Bank and the Global Environment Facility also played an invaluable part in implementation. Industry and industrial organizations eschewed their usual competitive spirit and shared technologies and techniques to phase out ozone-depleting chemicals. Environmental non-governmental organizations not only kept an alert eye on the issue and sounded the alarm when necessary, but also developed ozone-safe technologies and spread awareness about such technologies. The national governments employed many regulatory, economic and policy instruments to achieve the phase-out as planned.

What was the process that led to such a success? Can it be replicated? The success of the ozone treaties has led some people to think that it was an easy issue to resolve and that other environmental issues of today are more complex. Is this true? Before we can answer such questions, it is necessary to know exactly what happened during the last 25 years on the ozone issue. The facts are buried in the archives of UNEP and the world's governments.

I thought UNEP owed it to the world to share its account of how the ozone treaties evolved, before time took away the records and the leading personalities of the ozone issues. But the ozone protection story is by no means over. It will be over only in another 50 years when scientists assure the world that the ozone layer is restored and that there are no more threats. However, considering the human span, we will lose much information if we wait another 50 years to write the complete story. Hence, we decided to go ahead and publish the history for the years up to 2001. No doubt, my successor in the year 2051 will publish the second volume to complete the story!

UNEP decided that this history would be an appropriate presentation to the World Summit on Sustainable Development in Johannesburg in September 2002. The summit will bring together leaders of all the governments to review the success of the implementation of Agenda 21. This history will give them confidence that it is indeed possible to achieve development sustainably, if there is cooperation among all stakeholders. Hopefully, it will give them some hints on how to achieve such cooperation.

I am grateful to Madhava Sarma, who served as the Executive Secretary of the Secretariat for the Vienna Convention and the Montreal Protocol for more than nine years, and Stephen Andersen, who has been a co-chair of the Technology and Economic Assessment Panel (TEAP) since its inception 12 years ago, for agreeing to put together this history. It was a labour of love for them. In the typical Montreal Protocol style, they obtained contributions to this history from many of the people who made it a triumph. I hope this history will please all those who contributed to the success of the ozone agreements and serve as an authentic record of one of the world's great achievements.

Klaus Töpfer
Executive Director,
United Nations Environment Programme
Under Secretary-General, United Nations

Acknowledgements

Many individuals helped us to compile this book. Mostafa Tolba, former Executive Director of UNEP, inspired our focus on people and the organizations. His successor Klaus Töpfer and UNEP's Deputy Executive Director Shafqat Kakakhel gave us their unstinted support throughout the process. The Ozone Secretariat staff – Michael Graber, Nelson Sabogal, Gilbert Bankobeza, Ruth Batten, and the entire secretarial staff – enthusiastically helped us. Tore Brevik and Naomi Poulton of the UNEP Communications and Public Information Division provided us with full support. The Publications and Information Board of UNEP and the US Environmental Protection Agency (EPA) generously financed the expenses connected with the book. Duncan Brack of the Royal Institute of International Affairs, Jonathan Sinclair Wilson of Earthscan Publications and their editors Nina Behrman and Frances MacDermott guided us through the intricacies of putting together this book in a publishable form.

Kofi Annan, Secretary-General of the United Nations, and Klaus Töpfer, Executive Director of UNEP, set the stage and put the protocol in a global context with their authoritative Preface and Foreword.

We are grateful for the contribution from key participants of personal perspectives that are extracted in this book and included at length on a CD-ROM available from the Ozone Secretariat. These perspectives are contributed by: Dan Albritton, Richard E Benedick, Fatma Can, Brigitta Dahl, David Doniger, Linda Dunn, Omar E El-Arini, Mohamed El-Ashry, Yuichi Fujimoto, Maneka Gandhi, Qu Geping, Michael Graber, Joop van Haasteren, Morio Higashino, Paul Horowitz, Margaret Kerr, Vyacheslav Khattatov, Naoki Kojima, Ingrid Kökeritz, Geoffrey Lean, János (John) Maté, Yasuko Matsumoto, Alan Miller, John Miller, Ryusuke Mizukami, Lawrence Musset, Tsuneya Nakamura, Julian Newman, Tetsuo Nishide, Tsutomu Odagiri, Nelson Sabogal, Helen Tope and Hideaki Yasukawa.

We are grateful to Rajendra Shende of UNEP and his colleagues, Steve Gorman of the World Bank and his colleagues and Seniz H Yalcindag of UNIDO and her colleagues for helping us with the write-ups of their programmes. Omar E El-Arini, the Chief Officer of the Multilateral Fund Secretariat, and his colleagues were always ready with the information on the various aspects of the Fund and helped us greatly.

A talented team of scholars, writers and editors made particularly important contributions. Lani Sinclair, our experienced editor, not only ironed out our English from the opposite sides of the world, but also put together the excellent Chapter 1, on the science of ozone depletion. Don Smith and Penelope Canan contributed Chapter 8, which quantifies media coverage of science, public concern and policy action, and presents case studies of seminal events. Corinna Gilfillan joined Stephen O Andersen in writing Chapter 9 on environmental NGOs, with the considerable assistance of David Doniger, János (John) Maté and Alan Miller. Stephen O Andersen, Suely Carvalho and Sally Rand wrote the appendix on the Assessment Panels with the assistance of E Thomas Morehouse and Helen Tope. EPA's Caley Johnson authored the bibliography, list of meetings, and many of the other compilations included in the appendices and Ozone Secretariat CD-ROM. Williams College volunteer Mark Robinson authored descriptions of industry and environmental NGOs. UNEP DTIE's Samira de Gobert provided a valuable collection of original documents and added substantially to the bibliography. Gerald Mutisya and Martha Adila of the Ozone Secretariat created an extraordinary searchable database of participants in Montreal Protocol meetings and placed their talent for information technology at our disposal. We are grateful to all these people who so generously gave their time and talent.

A special thanks to the cartoonists and their publishers who granted permission for their cartoons to be reproduced where words alone cannot tell the story: Scott Willis – *San Jose Mercury News*; Brant Parker and Johnny Hart – Creators Syndicate Inc; Bill Schorr – *The Kansas City Star*; Don Wright – Tribune Media Services; Dana Fradon – Cartoon Bank; Mike Luckovich – Creators Syndicate; and Joseph Kariuki – UNEP.

We have benefited substantially from the review of our drafts by experts, but any errors or omissions are entirely our responsibility. Victor Buxton, Suely Carvalho, Omar E El-Arini, Yuichi Fujimoto, Paul Horwitz, Ingrid Kokeritz, Mack McFarland, E Thomas Morehouse, Satu Nurmi, Sally Rand, Robert Reinstein, Stephen Seidel, Rajendra Shende, Richard Stolarski and Iwona Rummel-Bulska were particularly helpful in providing reviews, organizing the topics within the chapters and coordinating the cross-references.

We are also indebted to those who provided us with documents from decades ago that are not available in electronic form or online: Steven Bernhardt, Brent Blackwelder, Larry Bohlen, Victor Buxton, Penelope Canan, Elizabeth Cook, Paul Crutzen, Halstead Harrison, John Hoffman, Ingrid Kokeritz, Mack McFarland, Alan Miller, Stephen A Montzka, E Thomas Morehouse, Paul Newman, Julien Paren, Michael Prather, Lindsey Roke, Susan Solomon and Steven Wofsy.

In addition, many people peer-reviewed individual chapters: Radhey Agarwal, Ward Atkinson, Jonathan Banks, Fernando Bejarano, Steven Bernhardt, Nick Campbell, David Catchpole, Sukumar Devotta, David Doniger, Arjun Dutta, Bryan Jacob, Mike Jeffs, Caley Johnson, Horst Kruse, János (John) Maté, Mack McFarland, Alan Miller, E Thomas Morehouse, Paul A Newman, Nancy Reichman, Lindsey Roke, Anne Schonfield, Miguel Stutzin, Gary Taylor, Helen Tope and Robert Wickham.

We thank Janet Andersen and K Ramalakshmi, our respective spouses, for their constant support and encouragement.

Stephen O Andersen and K Madhava Sarma

Introduction and reader's guide

'Where shall I begin, please your majesty? he asked. "Begin at the beginning,"
the King said, gravely, "and go on till you come to the end: then stop."'
Lewis Carroll, *Alice's Adventures in Wonderland* (1865)

Many scholars have made important contributions to the history of efforts to protect the fragile stratospheric ozone layer. We are proud to add ourselves to their list as we can claim the perspective, insider knowledge, and access to original documents from our close association with the ozone agreements almost from the inception. The United Nations Environment Programme (UNEP) conceived this project to compile a complete record of the protection of the ozone layer and to organize a publicly available collection of historical documents that would otherwise be discarded as time passes. We are the UNEP instruments for this task.

When we were asked by UNEP to complete this project, we were thrilled and, at the same time, apprehensive at the magnitude of the task. We took to heart the saying of Thomas Gray in his letter to Horace Walpole in 1768, 'Any fool may write a most valuable book by chance, if he will only tell us what he heard and saw with veracity' – but with a slight amendment. We heard and saw plenty in our long years working on the ozone issue, but this history is that of UNEP, not our personal accounts.

The history of who, what and when is presented by us in the main body of the chapters and relies entirely on published documents. Personal perspectives of selected key players complement the core history – including the why and how – and are attributed by quotes, boxed text and appended elaboration. A CD-ROM is available from the UNEP Ozone Secretariat containing elaborated perspectives, databases and scarce original documents.

We have followed the consultative process, the hallmark of the Montreal Protocol: seeking guidance from our colleagues; examining files and libraries at UNEP offices in Nairobi, Paris and Montreal; collecting and cataloguing records from companies, environmental NGOs and governments; and interviewing dozens of government and non-government participants. Various drafts of our chapters were circulated for peer review.

The story of the success of the Montreal Protocol is the story of thousands of individuals from an astonishingly diverse number of professions. Most of the credit rightly went to a former Executive Director of UNEP, Mostafa Tolba – the inspiration for the ozone treaties, other United Nations officials, scientists and diplomats. There were also many others who made essential contributions. These are the engineers, fire fighters, standards makers, medical doctors, regulators, lawyers, agriculturalists, government officials, training specialists, customs officers, refrigeration and air-conditioning technicians, industrial managers, environmental activists and military officers who contributed to the success. The story of the Protocol's success is, indeed, the success of the world community in harnessing the support of so many people of such diverse backgrounds to a common cause. The more than 500 ozone meetings show the enormous amount of effort needed to involve all the stakeholders. Any worldwide endeavour to achieve any objective within a specific time has to go through a similar inclusive process. Mere agreements by governments will lead nowhere without this participatory process. We hope we have succeeded in conveying the spirit of that process through this book.

This book is intended for any layperson interested in science, environment or international law. Readers unfamiliar with the nuances of the specialized vocabulary will want to consult the glossary, and a particularly keen reader will refer to the books listed as core reading in Appendix 8.

READER'S GUIDE

The chapters are organized in the most logical order for readers unfamiliar with the threat to the ozone layer and the progress made under the Montreal Protocol. However, we have endeavoured to make each chapter stand alone, and many readers will want to jump between chapters. Readers may first want to review the glossary of terms that are uniquely, and even peculiarly, defined under the Montreal Protocol. Near the beginning of the book, there is also a list of acronyms and abbreviations. The appendices include materials that readers may want to refer to throughout the book, such as key Montreal Protocol reference documents.

A CD-ROM is available in its latest updated form from the UNEP Ozone Secretariat. It contains some remarkable materials previously unavailable, or available only at the Ozone Secretariat in Nairobi. Among the treasures are: full-length copies of the original personal perspectives written for this book and extracted in the chapters; a bibliography of more than 5000 publications and memoranda; a list of more than 500 ozone meetings; a comprehensive timeline chronicling scientific discoveries, national actions, technical progress and NGO activities; a searchable database of names, countries, and affiliations of people attending UNEP ozone meetings; contact information for scholars who specialize in the Montreal Protocol; lists of members of the assessment panels; and scanned copies of thousands of UNEP documents. Readers may use these materials in resourceful and creative ways. Scholars may benefit from word comparisons of the many draft copies of the Montreal Protocol: what was controversial, and how was it resolved? Participants in the process can refresh their memories about the meetings that they attended.

While writing this book, the authors developed a Collection of Montreal Protocol and ozone protection documents from government, military, environmental and industry organizations, and from the private records of participants in this process. At the time of publication of this book, the

Collection amounted to approximately 50 large boxes of publications, memoranda, videotapes, posters, photographs and promotional materials. The authors found a proper home for the Collection at the Environmental Science and Public Policy Archives (ESPPA) at Harvard University, where these materials will be organized, maintained and made available for public use. Many readers of this book will have complementary materials that could be donated to this library. Historians will be particularly grateful for photographs, personal notes and previously confidential information such as records of government negotiating positions or corporate strategy. If readers have any such documents they wish to donate, please immediately contact the Harvard ESPPA or the Ozone Secretariat for instruction on how those collections can be contributed: The Ozone Secretariat, UNEP, PO Box 30552, Nairobi, Kenya; tel +254 2 623 851; fax +254 2 623 913; email ozoneinfo@unep.org. The Environmental Resources Librarian & Curator of the Environmental Science and Public Policy Archives, Cabot Science Library, Harvard University, 1 Oxford St, Cambridge, MA 02138-2901, USA; tel +1 617 496 6158; fax +1 617 495 5324; website http://hcl.harvard.edu/environment.

Stephen O Andersen and K Madhava Sarma

Chapter 1

The science of ozone depletion: From theory to certainty*

'Without a protective ozone layer in the atmosphere, animals and plants could not exist, at least upon land. It is therefore of the greatest importance to understand the processes that regulate the atmosphere's ozone content.'
The Royal Academy of Sciences, announcing the Nobel Prize for Chemistry, 1995, for Paul Crutzen, Mario Molina and F Sherwood Rowland

'I wanted to do pure science research related to natural processes and therefore I picked stratospheric ozone as my subject, without the slightest anticipation of what lay ahead.'
Paul Crutzen, Max Planck Institute for Chemistry, Mainz, Germany, 1995

'Above the Antarctic, the layer of ozone which screens all life on Earth from the harmful effects of the Sun's ultraviolet radiation is shattered.'
Joseph Farman, British Antarctic Survey, 1987

'We can now look at the Antarctic ozone hole and know that the ozone layer is not endowed with enormous resiliency, but is instead very fragile.'
F Sherwood Rowland, University of California at Irvine, 1987

'Stratospheric ozone depletion through catalytic chemistry involving man-made chlorofluorocarbons is an area of focus in the study of geophysics and one of the global environmental issues of the twentieth century.'
Susan Solomon, Aeronomy Laboratory, US National Oceanic and Atmospheric Administration, 1999

INTRODUCTION

The ozone layer forms a thin shield in the stratosphere, approximately 20–40km above the Earth's surface, protecting life below from the sun's ultraviolet (UV)

---

* This chapter was written by Lani Sinclair.[1]

radiation (Box 1.1). It absorbs the lower wavelengths (UV-C) completely and transmits only a small fraction of the middle wavelengths (UV-B). Nearly all of the higher wavelengths (UV-A) are transmitted to the Earth where they cause skin-aging and degrading of outdoor plastics and paint. Of the two types of UV radiation reaching ground level, UV-B is the most harmful to humans and other life forms.

Manufactured chemicals transported by the wind to the stratosphere are broken down by UV-B, releasing chlorine and bromine atoms which destroy ozone. As ozone is depleted, other factors remaining constant, increased transmission of UV-B radiation endangers human health and the environment, for example, by increasing skin cancer and cataracts, weakening human immune systems and damaging crops and natural ecosystems.

Notably, most ozone-depleting substances (ODSs) are also 'greenhouse gases' that contribute to climate change, causing sea level rise, intense storms and changes in precipitation and temperature.[2]

EARLY THEORIES: SCIENTISTS IDENTIFY AND NAME OZONE

The peculiar odour in the air after a lightning strike had been remarked upon for centuries, including references in The Iliad and The Odyssey, but it was not well understood or named until centuries later. In 1785, Martinus van Marum passed electric sparks through oxygen and noted a peculiar smell; he also found that the resulting gas reacted strongly with mercury. Van Marum and others attributed the odour to the electricity, calling it the 'electrical odour'.[3]

In 1840, Swiss chemist Christian Schönbein identified this gas as a component of the lower atmosphere and named it 'ozone', from the Greek word *ozein*, 'to smell'. He recognized that the odour associated with lightning was ozone, not electricity. He detailed his findings in a letter presented to the Academie des Sciences in Paris entitled, 'Research on the nature of the odour in certain chemical reactions'. According to Albert Leeds, writing in 1880:[4]

> *'The history of ozone begins with the clear apprehension, in the year 1840, by Schönbein, that in the odour given off in the electrolysis of water, and accompanying discharges of frictional electricity in air, he had to deal with a distinct and important phenomenon. Schönbein's discovery did not consist in noting the odour... but in first appreciating the importance and true meaning of the phenomenon.'*

A few years later, J L Soret of Switzerland identified ozone as an unstable form of oxygen composed of three atoms of oxygen (O_3).

Ozone in the atmosphere

Ozone in the upper atmosphere filters ultraviolet light
In 1879, Marie-Alfred Cornú of the École Polytechnique in Paris measured the sun's spectrum with newly developed techniques for ultraviolet spectroscopy

BOX 1.1 WHAT IS THE OZONE LAYER?

Ozone is a molecule made up of three oxygen atoms (O_3). Averaged over the entire atmosphere, of every 10 million molecules in the atmosphere, only about three are ozone. About 90 per cent of ozone is found in the stratosphere, between 10 and 50 kilometres above the Earth's surface. If all of the ozone in the atmosphere were compressed to sea-level pressure, it would constitute a layer only about 3 millimetres (0.1 inches) thick.

Solar radiation at the top of the atmosphere contains radiation of wavelengths shorter than visible light. This radiation, called ultraviolet radiation, is of three ranges. The shortest of these wavelengths, UV-C, is completely blocked from reaching Earth by oxygen and ozone. Wavelengths in the middle range, UV-B, are only partially absorbed by ozone. The higher wavelengths, UV-A, are minimally absorbed and mostly transmitted to the Earth's surface.

The ozone layer absorbs all but a small fraction of the UV-B radiation from the sun, shielding plants and animals from its harmful effects. Stratospheric ozone depletion: increases skin cancer, cataracts, and blindness; suppresses the human immune system; damages natural ecosystems; changes the climate; and has an adverse effect on plastics.

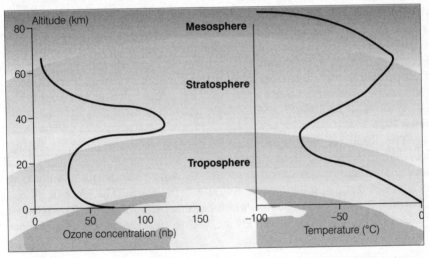

Note: The thin layer of ozone in the stratosphere is at its thickest at a height of 20–40km. It also accumulates near the ground in the troposphere, where it is a troublesome pollutant.

Source: Ozone Secretariat (2000) *Action on Ozone*, UNEP, Nairobi, p1.

Figure 1.1 *Ozone levels and temperature variation in the atmosphere*

and found that the intensity of the sun's UV radiation dropped off rapidly at wavelengths below about 300 nanometres.[5] He demonstrated that the wavelength of the 'cut-off' increased as the sun set and the light passed through more atmosphere on its path to Earth. He correctly determined that the cut-off was the result of a substance in the atmosphere absorbing light at UV wavelengths. A year later, W N Hartley of the Royal College of Science for

Ireland in Dublin concluded that this substance was ozone.[6] This conclusion was based on his laboratory studies of UV absorption by ozone. Hartley and Cornú attributed the absorption of solar radiation between wavelengths of 200 and 320 nanometres to ozone, and concluded that most of the ozone must be in the upper atmosphere.

In 1917, Alfred Fowler and Robert John Strutt, who became Lord Rayleigh, showed that a number of absorption bands could be observed near the edge of the cut-off in the solar spectrum.[7] These were consistent with the ozone absorption bands observed in the laboratory, further proving that ozone is the absorber in the atmosphere. The following year, Strutt attempted to measure the absorption by ozone from a light source located 4 miles across a valley.[8] He could detect no absorption and concluded that 'there must be much more ozone in the upper air than in the lower', and that absorption does not occur in the lower atmosphere.

Dobson discovers day-to-day and seasonal ozone variations

In 1924, Gordon M B Dobson invented a new spectrophotometer to measure the amount of ozone in the atmosphere. He discovered that there were day-to-day fluctuations in the ozone amount over Oxford, England, and that there was a regular seasonal variation.[9] He hypothesized that these variations in ozone might be related to variations in atmospheric pressure. To test this idea, he had several more spectrophotometers constructed and distributed throughout Europe. These measurements demonstrated regular variations in ozone with the passage of weather systems. One of these spectrophotometers was installed in the town of Arosa in the Swiss Alps, where measurements have been made since 1926.

The Dobson spectrophotometer splits solar radiation into light-wavelengths; because ozone absorbs only some of those wavelengths, the spectrophotometer measures the amount of ozone solar radiation interacts with as it passes through the atmosphere towards the Earth. The amount of ozone measured is expressed in Dobson units, which measure the ozone in a vertical column of the atmosphere. Worldwide, the ozone layer averages approximately 300 Dobson units.

Thomas Midgley invents CFCs

In 1928, Thomas Midgley Jr, an industrial chemist working at General Motors, invented a chlorofluorocarbon (CFC) as a non-flammable, non-toxic compound to replace the hazardous materials, such as sulphur dioxide and ammonia, then being used in home refrigerators (see Chapter 5). To prove the chemical's safety for humans, Midgley inhaled the compound and blew out candles with the inhaled vapours. By the 1950s and into the 1960s, CFCs were also used in automobile air conditioners, as propellants in aerosol sprays, in manufacturing plastics and as a solvent for electronic components.

The ozone column, the 'ozone layer' and natural balance

In 1929, F W P Götz worked with Dobson's instrument at Arosa, Switzerland, measuring the ratio of the intensity of two wavelengths at the zenith sky throughout the day. He found that the ratio of the intensities decreased as the sun set, and turned around and increased just as the sun was near the horizon; he named this the Umkehr (turnaround) effect. Thus, he invented the Umkehr method for measuring the vertical distribution of ozone and showed that the concentration of ozone reaches a maximum below an altitude of 25 kilometres.[10]

The first scientist to identify the ozone 'layer' and its full workings was Sydney Chapman, who presented his findings in 1930 in a lecture to the Royal Society of London.[11] He developed a photochemical theory of stratospheric ozone formation and destruction, based on the chemistry of pure oxygen. He explained how sunlight could generate ozone by striking molecular oxygen in the atmosphere. Chapman's findings described the chemistry this way, according to the 1993 book *Between Earth and Sky*:[12]

> '*When oxygen (O_2) in the stratosphere absorbs sunlight waves of less than 2,400 Å (angstroms), the oxygen molecule is split and two oxygen atoms are freed. Like caroming billiard balls, the two oxygen atoms go their separate ways until one free oxygen atom (O) joins a whole oxygen molecule (O_2) to create a molecule of triatomic oxygen, or ozone (O_3). Ozone (O_3), itself being highly unstable, is quickly broken up by longer-wave sunlight of 2,900 Å or by colliding with another free oxygen atom. Thus, ozone molecules are always being made and destroyed at a more or less constant rate, Chapman said, so that a relatively fixed quantity of them are always present... Chapman's comprehensive description of ozone chemistry, known thereafter as "the Chapman reactions" or "the Chapman mechanism", proved definitive, and also served to inspire the popular conception of the ozone layer as a vital atmospheric buffer protecting living organisms from deadly shortwave ultraviolet light.*'

In 1934, Götz, A R Meetham and Dobson published an interpretation of this phenomenon, pointing out that the shape of the turnaround was dependent on the shape of the altitude profile of the ozone concentration.[13] They thus provided experimental confirmation of the basic Chapman theory of ozone formation and loss.

Measurements of emissions of specific spectral lines became possible with the development of high-resolution Dobson spectrometers during the 1940s and 1950s. These instruments were pointed up towards the sky to measure the 'dayglow' and 'nightglow' of the atmosphere; they measured specific bands of molecules, such as nitric oxide (NO) and hydroxyl (OH). These measurements led to the development of a description of the chemical composition of many of the minor constituents of the atmosphere. In 1950, D R Bates and Marcel Nicolet[14] wrote their exposition on the chemistry of the hydrogen oxides in the upper atmosphere; Nicolet later described the details of the expected nitrogen oxide chemistry of the upper atmosphere.[15]

WMO global ozone observing system

In preparation for the International Geophysical Year in 1957, a worldwide network of stations was developed to measure ozone profiles and the total column abundance of ozone using a standard quantitative procedure pioneered by Dobson. The World Meteorological Organization (WMO) established the framework for ozone-observing projects, related research and publications; this network eventually became the Global Ozone Observing System, with approximately 140 monitoring stations. The British Antarctic Survey and Japanese Scientific Stations in Antarctica in 1957 installed such ozone monitors, which eventually recorded the depletion of the ozone that was later called the Antarctic ozone hole.

MODERN SCIENTISTS HYPOTHESIZE THREATS TO OZONE

Early warnings about damage to the ozone layer

Warnings about supersonic aircraft
In 1970, Paul Crutzen of The Netherlands demonstrated the importance of catalytic loss of ozone by the reaction of nitrogen oxides, and theorized that chemical processes that affect atmospheric ozone can begin on the surface of the Earth.[16] He showed that nitric oxide (NO) and nitrogen dioxide (NO_2) react in a catalytic cycle that destroys ozone, without being consumed themselves, thus lowering the steady-state amount of ozone. These nitrogen oxides are formed in the atmosphere through chemical reactions involving nitrous oxide (N_2O) which originates from microbiological transformations at the ground. Therefore, increasing atmospheric concentration of nitrous oxide that can occur through the use of agricultural fertilizers might lead to reduced ozone levels, he theorized. His hypothesis was that 'NO and NO_2 concentrations have a direct controlling effect on the ozone distributions in a large part of the stratosphere, and consequently on the atmospheric ozone production rates'.

At the same time, James Lovelock of the United Kingdom (UK) developed the electron-capture detector, a device for measuring extremely low organic gas contents in the atmosphere. Using this device in 1971 aboard a research vessel, he measured air samples in the North and South Atlantic. In 1973, he reported that he had detected CFCs in every one of his samples, 'wherever and whenever they were sought'.[17] He concluded that CFC gases had already spread globally throughout the atmosphere.

In another article published in 1970, Halstead Harrison of the Boeing Scientific Research Laboratories in the United States (USA) hypothesized that 'with added water from the exhausts of projected fleets of stratospheric aircraft, the ozone column may diminish by 3.8 percent, the transmitted solar power increase by 0.07 percent, and the surface temperature rise by 0.04 degrees K in the Northern Hemisphere'.[18] He wrote that 'several authors have expressed concern that exhausts from fleets of stratospheric aircrafts may build up to levels sufficient to perturb weather both in the stratosphere and on the surface. Indeed, calculations indicate that the quantity of added water vapor may become comparable to that naturally present'. At the time, the projected fleets of supersonic transport aircraft (SSTs) were estimated at 500 in the US, and 350 in other countries of the world.

In 1971, Harold Johnston of the US, who had carried out extensive studies of the chemistry of nitrogen compounds, showed that the nitrogen oxides produced in the high-temperature exhaust of the proposed fleet of SSTs could contribute significantly to ozone loss by releasing the nitrogen oxides directly into the stratospheric ozone layer.[19] In 1972, Crutzen elaborated on this theory with a paper that explained the process by which ozone is destroyed in the stratosphere, and presented estimates of the ozone reduction that could result from the operation of supersonic aircraft.[20] In his article, he concluded:

> *'Although it is not possible to assess at this stage the real environmental consequences of future supersonic air transport, present knowledge indicates that there exists a real possibility of serious decreases in the atmospheric ozone shield due to the catalytic action of oxides of nitrogen, emitted in the exhaust of supersonic aircraft... It is clear that the environmental problems connected with the introduction of SSTs into a region of the atmosphere which has sometimes humorously been called the 'ignorosphere' have been severely neglected... If nitrogen oxide emissions from SSTs cannot be strongly reduced, it may in the future become necessary to reach an international agreement on limitations of the world's total supersonic fleet.'*

With these studies, according to American scientist Richard Stolarski, 'the new paradigm was set: ozone production is balanced by ozone loss due to catalytic reactions of the nitrogen and hydrogen oxides, and human activities could influence this balance and affect ozone concentrations'.

In March 1971, the US House of Representatives voted not to continue funding development of the American SST. In 1973, Japan Air Lines, Pan Am, Qantas and TWA cancelled their orders for Concorde SSTs. Only British Airways and Air France were then flying Concordes across the Atlantic Ocean. Concerns about take-off and landing noise from the SST prevented them from being flown from Dulles Airport in Washington, DC, leaving New York's John F Kennedy Airport as the only US airport served by the SST.

James McDonald links ozone depletion to skin cancer

Another American, James McDonald, theorized in 1971 that even a small change in the abundance of stratospheric ozone could have significant effects in transmitting more ultraviolet radiation to the surface of the Earth, affecting the incidence of skin cancer. He testified before the US Congress that:

> *'it is my present estimate that the operation of SSTs at the now-estimated fleet levels predicted for 1980–1985 could so increase transmission of solar ultraviolet radiation as to cause something on the order of 5–10,000 additional skin cancer cases per year in just the US alone'.*

Richard Stolarski and Ralph Cicerone study chlorine exhaust from rockets

Richard Stolarski joined the scientists studying the role of chlorine in the stratosphere in 1972, when scientists at the US National Aeronautics and Space Administration (NASA) recognized that the space shuttle's solid rocket boosters would inject chlorine directly into the stratosphere in the form of hydrogen chloride. NASA awarded a contract to Stolarski and Ralph J Cicerone of the University of Michigan to examine NASA's environmental impact statement for the space shuttle, which concluded that chlorine would be spread as exhaust along the shuttle's launch trajectory. According to Stolarski:

> *'This was in 1972, before the chlorine issue had come to the forefront of the field... Another colleague asked me why we were studying the perturbation of ozone by chlorine from the shuttle since it was clearly a negligible source on a global scale. I answered that someday, someone would come up with a larger source and then maybe our work on chlorine chemistry would be significant. Little did I know that someone already had. Mario Molina and Sherry Rowland were looking into the fate of the chlorofluorocarbons which were being ubiquitously used in air conditioning, aerosol spray cans, and other applications.'*

DuPont holds manufacturers' summit to explore risk from CFCs

Responding to James Lovelock's measurements of CFCs accumulating in the atmosphere, the DuPont Company arranged a panel on The Ecology of Fluorocarbons for the world's CFC producers in 1972.[21] The invitation to the panel from Raymond McCarthy, research director of the company's Freon[tm] Products division, stated that:

> *'Fluorocarbons are intentionally or accidentally vented to the atmosphere worldwide at a rate approaching one billion pounds per year. These compounds may be either accumulating in the atmosphere or returning to the surface, land or sea, in the pure form or as decomposition products. Under any of these alternatives, it is prudent that we investigate any effects which the compounds may produce on plants or animals now or in the future.'*

As a result of that programme, 19 companies formed the Chemical Manufacturers Association's Fluorocarbon Program Panel, a research group that eventually funded at least US$20 million in research at academic and government facilities worldwide.

Two atmospheric scientists at Harvard University, Steven C Wofsy and Michael B McElroy, were also examining the effects of SSTs on ozone, and concluded in a paper published in 1974 that 'nitric oxide emitted by supersonic aircraft would lead to a significant reduction in the concentration of atmospheric ozone... A traffic model by Broderick et al for 1990 could lead to a reduction of about 2 percent in the column density of O_3'.[22]

The research of Stolarski and Cicerone, published in 1974, concluded that chlorine released in the stratosphere could deplete ozone. A single chlorine atom, through a catalytic chain reaction, could eliminate tens of thousands of ozone molecules.[23]

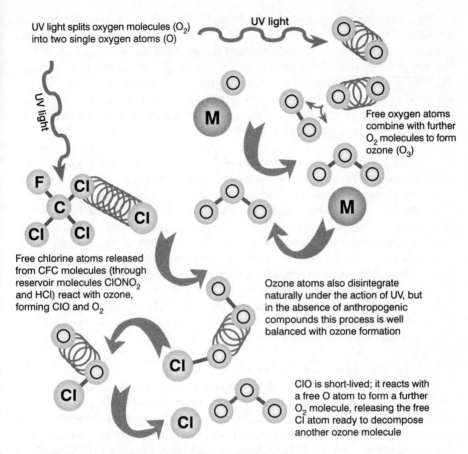

UV light splits oxygen molecules (O_2) into two single oxygen atoms (O)

UV light

Free oxygen atoms combine with further O_2 molecules to form ozone (O_3)

Free chlorine atoms released from CFC molecules (through reservoir molecules $ClONO_2$ and HCl) react with ozone, forming ClO and O_2

Ozone atoms also disintegrate naturally under the action of UV, but in the absence of anthropogenic compounds this process is well balanced with ozone formation

ClO is short-lived; it reacts with a free O atom to form a further O_2 molecule, releasing the free Cl atom ready to decompose another ozone molecule

Source: Ozone Secretariat (2000) *Action on Ozone*, UNEP, Nairobi, p4.

Figure 1.2 *The destruction of ozone molecules in the atmosphere*

Molina–Rowland hypothesis: CFCs linked to ozone depletion

Two chemists at the University of California at Irvine, Mario J Molina and F Sherwood Rowland, were the first to study CFCs (then referred to as chlorofluoromethanes, or CFMs) as a possible source of chlorine in the stratosphere. CFCs refer to all fully halogenated compounds containing chlorine, fluorine and carbon; chlorofluoromethanes contain only one carbon atom and are a subset of CFCs. CFCs turned out to be the 'larger source' of chlorine Stolarski referred to in 1972. CFCs during the 1970s had a variety of industrial uses in refrigeration, home and automobile air conditioning, aerosol propellants, the production of Styrofoam, and the manufacturing of electronic parts. US production of the two most widely used CFCs, CFC-11 and CFC-12, totalled approximately 309,000 tonnes in 1974. Total production in the rest of the world was more than 373,000 tonnes, with aerosol propellants probably accounting for about two-thirds of this, or about 249,000 tonnes. Global consumption of CFCs in 1974 was near 1 million tonnes, with about 70 per cent being used as aerosol propellants.

In a paper published in the 28 June 1974 issue of *Nature*, Molina and Rowland hypothesized that when CFCs reach the stratosphere, ultraviolet radiation causes them to decompose and release chlorine atoms, which in turn become part of a chain reaction; as a result of the chain reaction, a single chlorine atom could destroy as many as 100,000 molecules of ozone.[24] 'The chemical inertness and high volatility which make these materials suitable for technological use also mean that they remain in the atmosphere for a long time', Molina and Rowland wrote. They concluded:

> '*Chlorofluoromethanes are being added to the environment in steadily increasing amounts. These compounds are chemically inert and may remain in the atmosphere for 40–150 years, and concentrations can be expected to reach 10 to 30 times present levels. Photo-dissociation of the chlorofluoromethanes in the stratosphere produces significant amounts of chlorine atoms, and leads to the destruction of atmospheric ozone... It seems quite clear that the atmosphere has only a finite capacity for absorbing Cl atoms produced in the stratosphere, and that important consequences may result. This capacity is probably not sufficient in steady state even for the present rate of introduction of chlorofluoromethanes. More accurate estimates of this absorptive capacity need to be made in the immediate future in order to ascertain the levels of possible onset of environmental problems.*'

NGOs and the American Chemical Society sound the alarm

Molina and Rowland estimated that 'if industry continued to release a million tons of CFCs into the atmosphere each year, atmospheric ozone would eventually drop by 7 to 13 percent'. Later in 1974, they presented their findings at a meeting of the American Chemical Society and held a press conference at the encouragement of the Natural Resources Defense Council (NRDC) and the Chemical Society public affairs officer. According to Cagin and Dray, in their book *Between Earth and Sky*,[25] Rowland reported his and Molina's calculation that:

> '*if CFC production rose at the then-current rate of 10 percent a year until 1990, and then levelled off, up to 50 percent of the ozone layer would be destroyed by the year 2050. Even a 10 percent depletion, he said, could cause as many as 80,000 additional cases of skin cancer each year in the United States alone, along with genetic mutations, crop damage, and possibly even drastic changes in the world's climate*'.

Rowland and Molina called for a ban on aerosol CFCs when in September 1974 they told the American Chemical Society:

> '*if nothing was done in the next decade to prevent further release of chlorofluorocarbons, the vast reservoir of the gases that would have built up in the meantime would provide enough chlorine atoms to insure continuing destruction of the ozone layer for much of the twenty-first century. They urged that the use of the compounds as aerosol propellants be banned*'.[26]

US assessment predicts ozone depletion from CFCs, supersonic aircraft
In January 1975, the US National Academy of Sciences and Department of Transportation issued a report on 'Climate Impact Assessment Program (CIAP): Environmental Impacts of Stratospheric Flight: Biological and Climatic Effects of Aircraft Emissions in the Stratosphere', which included assessments of atmospheric science, environmental effects, and technology and economics. The report concluded that atmospheric levels of CFCs would deplete the ozone layer six times more efficiently than oxides of nitrogen from SSTs, and that ozone depletion would consequently increase the intensity of ultraviolet light at ground level.

More support for Molina–Rowland thesis
Rowland and Molina's scientific conclusions were confirmed by Wofsy, McElroy, and Nien Dak Sze in 1975, when they published a paper in *Science* that concluded: 'Freons™ [27] are a potential source of stratospheric chlorine and may indirectly cause serious reductions in the concentration of ozone... Allowing for reasonable growth in the Freon™ industry, ~ 10 percent per year, the reduction in O_3 could be 2 percent by 1980 and, if left unchecked, could grow to the disastrous level of 20 percent by the year 2000'.[28] Even if Freon[tm] use were terminated as early as 1990, 'it could leave a significant effect which might endure for several hundred years'. They also concluded that 'a fleet of 320 Concordes operating for 7 hours a day at 17 km could reduce O_3 by 1 percent. Larger fleets, such as those projected by Grobecker for 1995 to 2025, could reduce ozone by more than 20 percent in the year 2000'.

Later in February 1975, the newly created Federal Interagency Task Force on Inadvertent Modification of the Stratosphere heard testimony from McElroy, who said that bromine 'appears to be so effective at ozone depletion that it could be used as a weapon', according to a *New York Times* story.[29] Bromines and bromine compounds, including methyl bromide, were coming into increasing use in such roles as the manufacturing of plastics and fumigation of croplands (see Chapter 5).

US National Academy of Sciences confirms theories
Following the research of Wofsy and McElroy, Stolarski and Cicerone, and Rowland and Molina, the US National Academy of Sciences in March 1975 established the Panel on Atmospheric Chemistry to assess 'the extent to which man-made halocarbons, particularly CFMs [CFCs], and potential emissions from the space shuttle might inadvertently modify the stratosphere. The Panel was asked to examine critically the existing atmospheric and laboratory measurements as well as the mathematical models used to assess the impact of such pollutants on stratospheric ozone and to make recommendations on studies needed to improve understanding of the processes involved'.[30] The Academy concluded in a 1976 report:

> '*all the evidence that we examined indicates that the long-term release of CFC-11 and CFC-12 at present rates will cause an appreciable reduction in the amount of stratospheric ozone. In more specific terms, it appears that*

> *their continued release at the 1973 production rates would cause the ozone to decrease steadily until a probable reduction of about 6 to 7.5 percent is reached... The time scale of events is highly significant. This may be seen in the ozone reduction calculated for a constant CFM [CFC] release rate (1973) until 1978, when all release is halted. The ozone reduction continues to grow for a decade beyond cutoff (or cutback) and then requires an additional 65 years to recover one half of its maximum loss'.*

Noting that CFCs were produced and used around the world, the Academy advised, 'Clearly, although any action taken by the USA to regulate the production and use of CFMs [CFCs] would have a proportionate effect on the reduction in stratospheric ozone, such action must become worldwide to be effective in the long run'.

World Plan of Action, 1977
On instructions from the Governing Council of UNEP, UNEP organized a meeting of experts from many countries in Washington, DC in March 1977. This meeting resulted in a World Plan of Action on the Ozone Layer. As a part of this plan, UNEP established a Coordinating Committee on the Ozone Layer (CCOL) in which all interested countries shared the results of their studies (see Chapter 2 for full details).

Needing more precise measurements of ozone levels, NASA in October 1978 launched the Nimbus 7 satellite, and with it, two computerized instruments that started to record ozone levels: the Total Ozone Mapping Spectrometer (TOMS), which used cross-track scanning, and the Solar Backscatter Ultraviolet (SBUV). The first TOMS provided data from 1978 to 1993; the second provided data from 1991 to 1994; and the third provided data from 1996 to the present. The first SBUV operated till 1990; similar instruments have flown on several US National Oceanic and Atmospheric Administration (NOAA) satellites since then.

Further reports by the US National Academy of Sciences
In November 1978, a United Nations Environment Programme (UNEP) report to the second session of the Coordinating Committee on the Ozone Layer (CCOL) (see Chapters 2 and 3) concluded that simulations of the effects of nitrous oxide from increased fertilizer use were considerably reduced from the original 1975 estimate. Nitrous oxide is still recognized as a source of nitric oxide in the stratosphere; nitric oxide does catalytically react to deplete ozone. In 1979, the National Academy of Sciences' National Research Council followed up its earlier findings with a report by the Committee on Impacts of Stratospheric Change, and the Committee on Alternatives for the Reduction of Chlorofluorocarbon Emissions, which had been asked to study 'the effects of all substances, practices, processes, and activities which may affect the stratosphere, especially ozone in the stratosphere; the health and welfare effects of modifications of the stratosphere, especially ozone in the stratosphere; and methods of control of such substances, practices, and activities, including alternatives, costs, feasibility, and timing'.[31]

The committees concluded that:

> *'if the worldwide release of various types of CFMs [CFCs] were to continue at 1977 levels, the most probable value of eventual ozone depletion would be 16 percent (a reduction of worldwide ozone to 84 percent of which it otherwise would have been)... Despite a temporary levelling off of global CFC emissions due to the US ban on non-essential aerosol propellant use, CFC emissions will again rise and will continue to grow, unless further controls on the production and use of CFCs are initiated – in the United States as well as in the rest of the world'.*

Enumerating the effects of ozone depletion, the 1979 report concluded that:

> *'With any specified pattern of sunlight exposure in the most susceptible part of the population, skin cancer incidence rates would be higher under conditions of depleted ozone. In the United States, a 16 percent ozone depletion would result eventually in several thousand more cases of melanoma per year, of which a substantial fraction would be fatal; and several hundred thousand more cases of non-melanoma per year... crop yields from several kinds of agricultural plants are likely to be reduced as a result of a 16 percent to 30 percent ozone depletion... Larval forms of several important seafood species, as well as micro-organisms at the base of the marine food chain, would suffer appreciable killing as a result of a 16 to 30 percent ozone depletion... Climatic effects of continued CFC release at the 1977 rate would include an average warming of the Earth's surface by a few tenths of one degree Celsius before the middle of the twenty-first century'.*

With regard to the options available for ameliorative action, the report found that: 'International harmonization of attitudes and actions on the control of CFC production and use, supported by international efforts to address the principal substantive scientific questions and to achieve greater understanding of the risks posed by CFC emissions, should be essential objectives of US policy'. The report cited alternatives and substitutes for CFCs, design and construction modifications, and 'options for stimulating containment, recovery, and recycling'.

In 1979, NASA launched the Stratosphere Aerosol and Gas Experiment (SAGE), which measured ozone, water vapour, nitrogen dioxide, and aerosol extinction in the stratosphere. It operated until 1981; in 1985, SAGE II was launched on board the Earth Radiation Budget Satellite.

DISCOVERING AND MEASURING THE ANTARCTIC OZONE 'HOLE'

Recognition and proof of early findings

Researchers ignore evidence of Antarctic ozone hole
As early as October 1981, Dobson-instrument measurements from Japanese, British and other Antarctic research stations recorded a drastic 20 per cent reduction in ozone levels above Antarctica. None of the Antarctic scientists

published their results or consulted other stations to confirm their observations. Joseph Farman, head of the Geophysical Unit of the British Antarctic Survey, 'could only assume that something had gone wrong with his Halley Bay apparatus. He knew, of course, about the Molina–Rowland theory and the scientific debate over the relationship between man-made chemicals and ozone depletion, but the Dobson reading was simply too low to suggest anything but an instrument malfunction'.[32]

The next year, during the 1982 Antarctic spring in October, readings from a new Dobson instrument registered similar low ozone levels. At the same time, the ozone-measuring devices aboard the Nimbus 7 satellite had also registered low ozone levels, but the computers that logged the devices' measurements had been programmed to identify extremely low ozone measurements as erroneous, and therefore to ignore them. According to John Gribbin as reported in the 1988 book, *The Hole in the Sky*:[33]

> *'Data coming back to Goddard [Space Flight Center] from the satellite were processed automatically by computers before ever being touched by human hand (or seen by human eye), and the computers that processed the data had been programmed to reject any measurement lower than 180 Dobson units, and treat it as an anomaly. In their processing, the programs flagged the measurement as an anomaly, and reset it to 180 Dobson units for the purposes of their calculations – but fortunately, as it turns out, they also saved the original 'erroneous' measurement without processing it further... Even though very low values, 180 Dobson units, could show up in the processed data, they were flagged as erroneous, so none of the researchers took much notice of them.'*

Science and environmental effects studies maintain warnings

When the US Clean Air Act was amended in 1977, it required the US Environmental Protection Agency (EPA) to conduct studies on how human activities affect the stratosphere, and to report its findings to the US Congress. In 1981, the EPA asked the National Research Council of the National Academy of Sciences to provide an assessment of the state of knowledge on ozone depletion and its effects, to be used by the EPA in its report to the Congress. The Committee on Chemistry and Physics of Ozone Depletion and the Committee on Biological Effects of Increased Solar Ultraviolet Radiation published their report in 1982, and concluded that 'if production of two CFCs, CF_2Cl_2 [CFC-12] and $CFCl_3$ [CFC-11], were to continue into the future at the rate prevalent in 1977, the steady state reduction in total global ozone, in the absence of other perturbations, could be between 5 percent and 9 percent... The differences between current findings and those reported in 1979 are attributed to refinements in values of important reaction times'. They also reported that 'other chemicals released from human activities are understood to have the potential for affecting stratospheric ozone. Examples are methyl chloride, carbon tetrachloride, and particularly methyl chloroform', and that 'examination of the historical record of measurements of ozone does not reveal a significant trend in total ozone that can be ascribed to human activities. This observational result is consistent with those of current models, since no detectable trend would be expected on the basis of current theory'.[34]

Summarizing the biological effects of increased solar ultraviolet radiation, the report estimated that 'there will be a 2 percent to 5 percent increase in basal cell skin cancer incidence per 1 percent decrease in stratospheric ozone. The increase in squamous-cell skin cancer incidence will be about double that... A reduction in the concentration of stratospheric ozone will not create new health hazards, but will increase existing ones'. The report also made research recommendations, including maintaining 'a competent, broadly based research program that includes a long-term commitment to monitoring programs', and a global monitoring effort that 'should include both ground-based and satellite observations of total ozone and of concentrations of ozone above 35 km, where theory indicates the largest reductions might occur'.

Another National Research Council panel, the Committee on Causes and Effects of Changes in Stratospheric Ozone, in 1983 updated the information in the 1982 National Academy of Sciences report at the request of the US EPA. 'Current estimates of the steady-state reduction of total column ozone attributable to releases of CFC-11 and CFC-12 acting alone (at roughly 1980 rates) center around a value of 3 percent,' the report concluded in the first part on 'Perturbations to Stratospheric Ozone'.

> *'The calculated net column reduction is the result of a substantial decrease in ozone above 30 km, amounting to about 6 percent of the total quantity of ozone in the atmosphere, and a smaller increase in the ozone below 30 km, amounting to about 3 percent of the total ozone... [M]easurements of total ozone between 1970 and 1980 have indicated no discernible trend in the total column abundance of ozone (the net amount of ozone above a unit area of the Earth's surface)... If we consider reasonable scenarios of the recent past and potential future, model calculations suggest that the net column ozone change over the next few decades will probably be on the order of +1 percent.'*[35]

In the report's second part, 'Effects on the Biota', the National Research Council panel observed immunological changes in animals exposed to UV-B radiation and concluded that 'it has now been demonstrated that at least some of these immunological changes also occur in humans exposed to natural or artificial UV radiation. Thus, the concerns expressed in the NRC (1982) report that the immuno-suppression observed in the UV-irradiated animals might also occur in humans were well-founded.' It also found that 'most plants, including crop plants, are adversely affected by UV-B radiation. Such irradiance stunts growth, cuts down total leaf area, reduces production of dry matter, and inhibits photosynthesis in several ways.'

Japanese scientist publishes proof of ozone depletion in Antarctica
In 1984, the first published results of research on ozone depletion over Antarctica appeared when Shigeru Chubachi of the Japanese Meteorological Research Institute in Ibaraki reported his findings.[36] According to Chubachi's paper, 'In order to obtain better understanding of dynamical behaviour of atmospheric ozone in Antarctica, extensive ozone observations were carried out

BOX 1.2 WMO's ROLE IN MONITORING GLOBAL OZONE

John M Miller, former Chief of the WMO's Environment Division*

Since 1957, the World Meteorological Organization (WMO) has provided the backbone of the global ozone monitoring network. WMO assumed responsibility for the global network of ground-based ozone-measuring stations established during the International Geophysical Year in 1957. In 1960, WMO recognized the need to collect, control for quality, and make accessible the data flowing from this network. It consequently established the World Ozone Data Centre in Toronto, Canada.

This centre, operated by the Meteorological Service of Canada, now contains large data banks on both vertical profiles and total atmospheric ozone, together with ultraviolet information dating 1992. The global ozone network now operates under the umbrella of the WMO Global Atmosphere Watch programme that has, in addition to ozone monitoring, components dealing with global monitoring of greenhouse gases and regional pollution issues such as acid rain and long-range transport of pollution.

In 1975, WMO convened a group of experts to prepare a statement entitled, 'Modification of the Ozone Layer Due to Human Activities and Some Possible Geophysical Consequences'. The statement focused on the effects of supersonic aircraft and CFCs, signalled the first international warning of the potential danger of ozone decline, and recommended international action to improve understanding of the issue.

WMO launched the Global Ozone Research and Monitoring Project in 1976 to provide advice to its member countries, the United Nations, and other international bodies concerning the extent to which human activities were responsible for ozone depletion, the possible impact of ozone depletion on climate and ultraviolet radiation on the Earth's surface, and the need to strengthen long-term ozone monitoring. That same year, WMO and UNEP convened a meeting of experts from government agencies and intergovernmental and non-governmental organizations (NGOs), which adopted a 'World Plan of Action on the Ozone Layer'. WMO assumed responsibility for the part of the plan dealing with scientific and research matters.

Evidence gathered by WMO's Ozone Project from 1976 to 1982 formed the basis of a document detailing the scientific findings at that time. The document was presented to the first meeting of the Ad Hoc Group of Legal and Technical Experts for the Elaboration of a Global Framework Convention for the Protection of the Ozone Layer in 1982.

Following the measurements of the Antarctic ozone hole in 1984–1985, WMO initiated the public release of Antarctic Ozone Bulletins, which are issued every 10–14 days, beginning in mid-August. Springtime bulletins are issued for northern mid-latitudes and the Arctic regions when conditions warrant.

* John Miller is now consultant at the Air Resources Laboratory, National Oceanic and Atmospheric Administration, USA.

at Syowa Station from February 1982 to January 1983. The total amount of ozone was observed throughout the year by the standard method of extinction of sunlight in summer and of moonlight in winter, together with 49 ozonesonde soundings. The annual variation of total ozone shows two maxima, in July and November... The smallest value of total ozone since 1966 was observed in the present observation from September to October', when Chubachi's readings showed ozone levels of under 250 Dobson units.

Estimates of future worldwide ozone depletion continued to vary. Michael J Prather, McElroy and Wofsy of the Center for Earth and Planetary Physics at Harvard University in 1984 concluded that an increase in the concentration of inorganic chlorine in the stratosphere could 'cause a significant change in the chemistry of the lower stratosphere leading to a reduction potentially larger than 15 percent in the column density of ozone. This could occur, for example, by the middle of the next century, if emissions of man-made chlorocarbons were to grow at a rate of 3 percent per year.'[37]

British Antarctic Survey confirms ozone 'hole'
In May 1985, Farman, Gardiner and Shanklin of the British Antarctic Survey published their findings in *Nature*[38] confirming that ozone levels above Antarctica had been significantly depleted every Antarctic spring since at least 1981. Their paper attributed the ozone depletion to CFCs, yet scientists would not be confident in this conclusion for many more years. According to that article:

> *'Recent attempts to consolidate assessments of the effect of human activities on stratospheric ozone (O₃) using one-dimensional models for 30°N have suggested that perturbations to total O₃ will remain small for at least the next decade. Results from such models are often accepted by default as global estimates. The inadequacy of this approach is here made evident by observation that the spring values of total O₃ in Antarctica have now fallen considerably. The circulation in the lower stratosphere is apparently unchanged, and possible chemical causes must be considered... two spectrophotometers have shown October values of total O₃ to be much lower than March values, a feature entirely lacking in the 1957–73 data set. To interpret this difference as a seasonal instrumental effect would be inconsistent with the results of routine checks using standard lamps... Whatever the absolute error of the recent values may be, within the bounds quoted, the annual variation of total O₃ at Halley Bay has undergone a dramatic change... We have shown how additional chlorine might enhance O₃ destruction in the cold spring Antarctic stratosphere.'*

The phenomenon of ozone depletion over Antarctica became known as the 'ozone hole', a phrase first used in published media accounts by Rowland of the University of California, and frequently illustrated by colour slides created by NASA which depicted levels of ozone in brightly colour-coded circles around the South Pole.

Explaining high and unexpected Antarctic ozone depletion

In an article published in a June 1986 issue of *Nature*, Susan Solomon of the US NOAA Aeronomy Laboratory, Roland R Garcia, F Sherwood Rowland and Donald J Wuebbles concluded that the 'remarkable depletions in the total atmospheric ozone content in Antarctica' were 'largely confined to the region from about 10 to 20 km, during the period August to October.'[39] They suggested that chlorine compounds might react on the surfaces of polar stratospheric clouds, perturbing gas-phase chlorine in ways that could greatly accelerate ozone loss in the Antarctic lower stratosphere:

'A unique feature of the Antarctic lower stratosphere is its high frequency of polar stratospheric clouds, providing a reaction site for heterogeneous reactions. A heterogeneous reaction between HCl and $ClONO_2$ is explored as a possible mechanism to explain the ozone observations. This process produces changes in ozone that are consistent with the observations, and its implications for the behaviour of HNO_3 and NO_2 in the Antarctic stratosphere are consistent with observations of those species there, providing an important check on the proposed mechanism.'

Global scientific teams link ODS emissions to Antarctic ozone depletion

Seven international agencies teamed up to write a three-volume assessment of the state of the ozone layer in 1985: Bundesministerium für Forschung und Technologie, Commission of the European Communities, UNEP, US Federal Aviation Administration, NASA, NOAA, and WMO.[40] Approximately 150 scientists from Australia, Belgium, Brazil, Canada, the Federal Republic of Germany, France, Italy, Japan, Norway, the UK and the US contributed to the assessment, which was coordinated by NASA.

The chemicals of interest to the agencies were: nitrogen oxides from subsonic and supersonic aircraft; nitrous oxide from agricultural and combustion practices; chlorofluorocarbons used as aerosol propellants, foam-blowing agents, and refrigerants; brominated compounds, including halons used to extinguish fires and suppress explosions; carbon monoxide and carbon dioxide from combustion processes; and methane from a variety of sources, including natural and agricultural wetlands, tundra, biomass burning, and enteric fermentation in ruminants. 'It is now clear that these same gases are also important in the climate issue,' the report concluded.

Among the report's findings were that 'global trend estimates of total ozone determined from the Dobson spectrophotometer network indicate little overall support for a statistically significant trend during the 14-year period 1970–1983... Recent evidence has been presented that indicates a considerable decrease in Antarctica total ozone during the spring period since about 1968. This is presently the subject of further analysis.' Examining the predicted magnitude of ozone perturbations for a variety of emission scenarios involving a number of substances, the report concluded that 'the long-term release of chlorofluorocarbons at the 1980 rate would reduce the ozone vertical column by about 5 percent to 8 percent according to one-dimensional models ... and by a global average of about 9 percent according to two-dimensional models which involve a reduction of about 4 percent in the tropics, about 9 percent in the temperate zones, and about 14 percent in the polar regions.' In addition, the report concluded that 'One dimensional models predict that the magnitude and even the sign of the ozone column changes due to increasing CFCs depend on the future trends of CO_2, CH_4, and N_2O.'

The report found that atmospheric concentrations of CFC-11 and CFC-12 were increasing at an annual rate of about 5 per cent; methyl chloroform concentrations were increasing by 7 per cent; and carbon tetrachloride concentrations were increasing by 1 per cent. Production rates of CFC-11 and CFC-12 had increased by 16 per cent in two years, from 599 kilotons in 1982 to

694 kilotons in 1984. Production of CFC-11 had increased from slightly more than 250 kilotons in 1972 to approximately 320 kilotons in 1984; production of CFC-12 had increased from approximately 350 kilotons in 1972 to approximately 380 kilotons in 1984.

If the CFC release rate were to become twice the 1985 levels, 'the one-dimensional models predict that there will be 3 percent to 12 percent reduction of the ozone column, regardless of realistically expected increases in carbon dioxide, nitrous oxide, and methane.' The report further explained:

> *'Time-dependent scenarios were performed using one-dimensional models assuming CO_2, CH_4, and N_2O annual growth rates of 0.5 percent, 1 percent and 0.25 percent, respectively, in conjunction with CFC growth rates of 0 percent, 1.5 percent and 3 percent per year. The ozone column effects are relatively small (<3 percent over the next 70 years) for CFC increases of <1.5 percent per year, but with a CFC growth rate of 3 percent per year, the predicted ozone depletion is 10 percent after 70 years and still rapidly increasing.'*

Antarctic research expedition, 1986

In August 1986, four teams of US researchers arrived in Antarctica as part of the first National Ozone Expedition to study the ozone hole over Antarctica. Sponsors included the National Science Foundation, NASA, NOAA, the Chemical Manufacturers Fluorocarbon Program Panel, and the US Navy. The NOAA Aeronomy Laboratory team, led by Susan Solomon, made ground-based visible absorption measurements; the University of Wyoming team, led by David Hofmann, carried out balloon-based ozone and aerosol particle measurements; the State University of New York at Stony Brook team made ground-based microwave emission measurements; and the Jet Propulsion Laboratory team made ground-based solar infrared absorption measurements. All four of the teams successfully measured the formation and strengthening of the ozone hole, confirming the phenomenon. Their measurements and findings, according to NASA, strongly suggested that 'perturbed chlorine chemistry was involved. But there was still no conclusive proof that chlorine was to blame for the ozone hole, and whether the hole was a natural phenomenon having to do with changes in temperature and air circulation, or whether it was caused by chlorine compounds contributed by man-made chemicals, was still a matter of debate.'[41]

INTERNATIONAL SCIENTIFIC TEAMS LINK CFCs AND OZONE DEPLETION

Early international collaboration on ozone

Formation of an Ozone Trends Panel

Two months later, in October 1986, an International Ozone Trends Panel was formed by NASA, in collaboration with UNEP, US Federal Aviation Administration, NOAA and WMO. The Panel was a response to 'two important

BOX 1.3 HOW COMMON CONCERN FOR THE ENVIRONMENT LED TO COOPERATION IN SPACE

Vyacheslav Khattatov, Chief of Laboratory, Central Aerological Observatory

Today, it is hard to remember the polarized situation during the Cold War. For decades, the Union of Soviet Socialist Republics and the USA stood poised with nuclear and conventional weapons capable of destroying all life on Earth. It is remarkable neither attacked the other over Korea, Cuba, Vietnam, Afghanistan and a dozen other conflicts. Each nation traded insults, and the economic development of both nations suffered as military defence costs skyrocketed. There was simply no room for cooperation: this was war.

There was a significant thawing in the Cold War as a result of the global concern for stratospheric ozone protection. In September 1986, Boris Gidaspov (First Communist Party Secretariat for Leningrad, now St Petersburg) and I (Senior Scientist for Science, Soviet State Committee for Hydrometeorology and Control of the Natural Environment) participated in the UNEP Second Workshop on Demand and Technical Controls and General Control Strategies in Leesburg, Virginia, USA. We were particularly interested in a paper on regulatory equity presented by Stephen Andersen of the US EPA and asked him many questions. We had many other discussions about efforts to protect the ozone layer, and were fast becoming friends.

One discussion concerned the importance of resolving scientific uncertainty. Because the US Space Shuttle *Challenger* had exploded during launch in January 1986, America was no longer able to launch scientific satellites. NASA had a new Total Ozone Monitoring Satellite (TOMS), but no rockets; the USSR had plenty of rockets, but no TOMS. After the Leesburg meeting ended, Andersen and his wife took us to dinner. Over drinks we decided to propose the impossible: a Russian rocket would carry an American satellite to space! A meeting was quickly organized at EPA headquarters with NASA managers. NASA agreed on the importance of launching the satellite, but seemed skeptical that Boris Gidaspov and I had the authority to propose such a thing. Perhaps they were too proud to ask for assistance or lacked confidence that American politicians would allow such cooperation.

In December 1986, EPA Administrator Lee M Thomas and Yuri A Izrael, Chairman of the Soviet State Committee for Hydrometeorology and Control of the Natural Environment, agreed to pursue ozone and climate cooperation. Our proposal suddenly seemed possible to everyone.

Through the hard work and political boldness of many individuals and organizations, the dream became reality on 19 August 1991 when the SL-14 booster Meteor-3 successfully carried the TOMS into space. This was the first space cooperation of these great nations. The launch was also the first time that a team of Americans visited Plesetsk Cosmodrome, a largely secret Russian ballistic missile and space launch centre near the Arctic Circle. The TOMS data were instrumental in proving that ODSs were destroying the ozone layer and endangering life on Earth. It is ironic that space technology developed as a consequence of the Cold War would one day save humanity from destruction; it is fortunate that cooperation was possible at the exact moment it was necessary. Today the Cold War has ended, and the new Russia and the USA are full partners in the Space Station. It started with our common concern for humanity.

reports of changes in the atmospheric ozone' that occurred in 1985. 'The first report was of a large, sudden, and unanticipated decrease in the abundance of springtime Antarctic ozone over the last decade. The second report, based on satellite data, was of large global-scale decreases since 1979 in both the total column content of ozone and in its concentration near 50 km altitude.'[42]

The Ozone Trends Panel involved more than 100 scientists from around the world who would study the question of whether carefully re-evaluated ground-based and satellite data would support the findings of ozone depletion over Antarctica and on a global scale.

Further key research

Environmental scientists describe bleak future with ozone depletion
In April 1987, Margaret Kripke, a skin cancer expert at the M D Anderson Cancer Center at the University of Texas, told the US White House Domestic Policy Council that although ozone depletion was expected to increase the number of skin cancers, there were other impacts with far greater global consequences, particularly the potential impact on the global food supply and the human diseases due to the effects of ultraviolet radiation on the human immune systems.[43]

At a US Senate hearing in May 1987, Alan Teramura of the University of Maryland testified that the potential of ultraviolet radiation to damage crops and plants was indisputable.[44] 'To date the scientific community has screened roughly 200 or so species of plants and different varieties of plants, and alarmingly, they found that two out of every three of these seem to show some degree of sensitivity to ultraviolet radiation,' he testified, adding that increased ultraviolet radiation produces tropospheric ozone, which also inhibits crop growth.

Another expedition blames chlorine for ozone hole
In August 1987, the internationally sponsored Airborne Antarctic Ozone Experiment set up its base at Punta Arenas, Chile. From there, aircraft flew over Antarctica to take measurements to determine the cause of the Antarctic ozone hole. NASA's ER-2 aircraft contained 13 automated, remotely operated instruments that measured ozone, water vapour, reactive nitrogen compounds, chlorine monoxide, particles from polar stratospheric clouds, meteorological variables, and other trace gases in the lower stratosphere. A DC-8 aircraft contained another seven instruments, including a laser device which looked up into the stratosphere to measure ozone profiles. The aircraft measurements were coupled with data from three separate satellite systems and ground-based sensors in several locations on Antarctica.

The Airborne Antarctic Ozone Experiment 'determined that the cause of the Antarctic ozone hole was chlorine chemistry. Large quantities of chlorine monoxide were found which were co-located with areas of ozone depletion. The aerosol data gathered was consistent with processing on polar stratospheric clouds. But the theories which said that it was a natural phenomenon due to atmospheric dynamics were found to be inconsistent with the new data.'[45] The experiment's data showed an inverse correlation between ozone and chlorine monoxide, according to NASA. 'Because chlorine monoxide is produced by the process in which man-made chlorine destroys ozone, the large quantities observed provide strong evidence that man-made chemicals are involved in the Antarctic ozone loss process... The data obtained during the Antarctic mission show[ed] the lowest ozone levels ever recorded and directly implicate[d] man-made chemical compounds, chlorofluorocarbons, in the enormous ozone loss over this remote region in the Southern Hemisphere.'

The Montreal Protocol and the 'smoking gun', 1987

From 14–16 September 1987, a conference of the Plenipotentiaries created the Montreal Protocol on Substances that Deplete the Ozone Layer. It was signed by 24 countries and the European Economic Community (see Chapter 2).

NASA held a press conference about the Antarctic expedition's findings on 30 September 1987, at NASA's Goddard Space Flight Center in Greenbelt, Maryland. At that briefing, government scientists said they had found 'the first hard evidence that the critical environmental loss can be blamed on a man-made gas,' reported the *Washington Post*.[46] 'The scientists noted that the effects of chlorofluorocarbon (CFC) gas on the ozone layer in the stratosphere may be more severe in Antarctica than in the rest of the world because of the continent's weather patterns during its early spring. But NASA programme manager Robert Watson said that "CFCs can affect ozone globally".'

According to the *Washington Post* account:

> 'A team of 60 scientists gathered enough data to rule out theories that attributed ozone depletion to changes in the sun's output or movement of low-ozone air masses. But the researchers found ample reason to confirm the role of CFCs. Chlorine monoxide, a by-product of CFCs exposed to ultraviolet rays, was detected at levels 100 to 500 times higher than found at lower altitude, Watson said. Moreover, he said, as concentrations of ozone fell, levels of chlorine monoxide rose.'

The International Ozone Trends Panel released its report in 1988, clarifying that 'this report is different from most previous national and international scientific reviews in that the published literature was not simply reviewed, but a critical reanalysis and interpretation of nearly all ground-based and satellite data for total column and vertical profiles of ozone was performed.' Robert Watson of NASA said at a press conference announcing the panel's findings, 'Things are worse than we thought... We have strong evidence that change in the ozone is wholly or in large part due to man-made chlorine.'

At the same press conference, John Gille of the National Center for Atmospheric Research said, 'We've found more than the smoking gun. We've found the corpse.'

Among the panel's findings were:

- *'There is undisputed observational evidence that the atmospheric concentrations of source gases important in controlling stratospheric ozone*

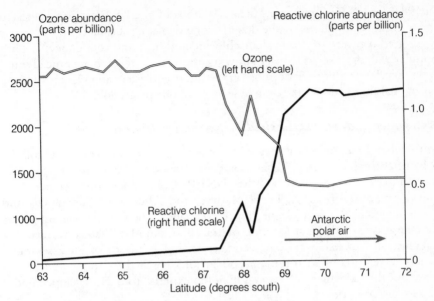

Source: Ozone Secretariat (2000) *Action on Ozone*, UNEP, Nairobi, p8.

Figure 1.3 *The 'smoking gun': Chlorine and ozone in the atmosphere*

levels (chlorofluorocarbons, halons, methane, nitrous oxide, and carbon dioxide) continue to increase on a global scale because of human activities.

- *'Analysis of data from ground-based Dobson instruments... shows measurable decreases from 1969 to 1986 in the annual average of total column ozone ranging from 1.7 to 3.0 percent, at latitudes between 30 and 64 degrees in the Northern Hemisphere. The decreases are most pronounced, and ranged from 2.3 to 6.2 percent during the winter months, averaged for December through March inclusive.*

- *'Thus, this assessment does not support the previous reports based on SBUV and TOMS data of large global decreases since 1979 in the total column of ozone (about 1 percent per year) or in the ozone concentration near 50 km altitude (about 3 percent per year).*

- *'There has been a large, sudden, and unexpected decrease in the abundance of springtime Antarctic ozone over the last decade. Ozone decreases of more than 50 percent in the total column, and 95 percent locally between 15 and 20 km altitude have been observed. The total column of ozone in the austral spring of 1987 at all latitudes south of 60°S was the lowest since measurements began 30 years ago.*

- *'The weight of evidence strongly indicates that man-made chlorine species are primarily responsible for the observed decrease in ozone within the polar vortex.*

- *'Statistically significant losses have occurred in the late winter and early spring in the Northern Hemisphere. For example, in the latitude band from 53–64°N, a loss of 6.2 percent is measured for December, January, February, and March from 1969–1986 when the data from 1965–1986 are considered.'*

In 1987, John S Hoffman and Michael J Gibbs of the US EPA first translated the scientific calculations of the dozen ozone-depleting substances into a single integrated measure of 'equivalent chlorine and bromine abundance.'[47] They popularized the use of charts showing the effect of each technically and economically feasible policy option on the future equivalent abundance, allowing policy-makers to choose the desired level of ozone protection.

Assessment panels established to synthesize findings, 1988

In October 1988, in compliance with Article 6 of the Montreal Protocol, the Scientific, Environmental, Technology, and Economic Assessment processes were informally initiated at the UNEP Conference on Science and Development, CFC Data, Legal Matters, and Alternative Substances and Technologies held in The Hague, The Netherlands. The four assessment panels were approved at the first Meeting of the Parties in May 1989, with terms of reference and timetables for completing reviews. The reports of the assessment panels were to contain the main conclusions reached by those panels and represent the judgement of several hundred experts from 21 developed and 9 developing countries.

In July 1989, Kripke of the University of Texas announced new evidence that ultraviolet radiation could weaken the body's immune system, particularly in developing countries where infectious diseases were already rampant. UV-B shuts down the part of the immune response that is governed by white blood cells, and stimulates the production of suppressor cells that halt immune defences, she concluded. Though increased UV-B would not cause epidemics among otherwise healthy people, those who already had infections would become more seriously ill, take longer to recover, or contract diseases more often.[48]

FIRST ASSESSMENT, 1989: 1987 PROTOCOL INADEQUATE, TOTAL PHASE-OUT REQUIRED

First report of the Scientific Assessment Panel

The first report of the Scientific Assessment Panel, published in November 1989, included the contributions and review of 136 scientists from 25 countries.[49] The report highlighted 'four major findings, each of which heightens the concern that chlorine- and bromine-containing chemicals can lead to a significant depletion of stratospheric ozone.' First, the Panel concluded that the weight of scientific evidence strongly indicated that manufactured chlorinated and brominated chemicals were primarily responsible for the recently discovered Antarctic ozone hole. The other three major findings were:

1   *'While at present, ozone changes over the Arctic are not comparable to those over the Antarctic, the same potentially ozone-destroying processes have been identified in the Arctic stratosphere. The degree of any future ozone depletion will probably depend on the particular meteorology of each Arctic winter and future atmospheric levels of chlorine and bromine.*

2   'The analysis of the total-column ozone data from ground-based
    instruments shows measurable downward trends from 1969 to 1988 of
    3 to 5.5 percent in the Northern Hemisphere (30 to 64°N latitudes) in
    the winter months that cannot be attributed to known natural processes.

3   'These findings have led to the recognition of major gaps in theoretical
    models used for assessment studies. Assessment models do not simulate
    adequately polar stratospheric cloud (PSC) chemistry or polar
    meteorology. The impact of these shortcomings for the prediction of ozone
    layer depletion at higher latitudes is uncertain.'

The report also concluded that the ozone depletion potentials (ODPs) of the
hydrochlorofluorocarbons (HCFCs) were significantly lower than those for the
CFCs, but that the HCFCs had much larger relative ODPs during the first 30 to
50 years after their emission into the atmosphere compared to their steady-state
ODP values.

Protocol controls inadequate to protect the ozone layer
According to the report, the panel's findings had several major implications for
public policy regarding restrictions on manufactured substances that lead to
stratospheric ozone depletion:

'Even if the control measures of the Montreal Protocol were to be implemented
by all nations, today's atmospheric abundance of chlorine (about 3 parts per
billion by volume) will at least double to triple during the next century. If the
atmospheric abundance of chlorine reaches about 9 parts per billion by volume
by about 2050, ozone depletions of 0–4 percent in the tropics and 4–12 percent
at high latitudes would be predicted, even without including the effects of
heterogeneous chemical processes known to occur in polar regions, which may
further increase the magnitude of the predicted ozone depletion.

'The surface-induced, PSC-induced chemical reactions which cause the
ozone depletion in Antarctica and also occur in the Arctic, represent additional
ozone-depleting processes that were not included in the stratospheric ozone
assessment models used to guide the Montreal Protocol. Recent laboratory
studies suggest that similar reactions involving chlorine compounds may occur
on sulfate particles present at lower latitudes, which could be particularly
important immediately after a volcanic eruption. Hence, future global ozone
layer depletions could well be larger than originally predicted.'

Complete elimination of ODS emissions needed to save the ozone layer
The Scientific Assessment Panel concluded:

'To return the Antarctic ozone layer to levels of the pre-1970s, and hence to
avoid the possible ozone dilution effect that the Antarctic ozone hole could
have at other latitudes, one of a limited number of approaches to reduce the
atmospheric abundance of chlorine and bromine is a complete elimination of
emissions of all fully halogenated CFCs, halons, carbon tetrachloride, and
methyl chloroform, as well as careful considerations of the HCFC substitutes.

Otherwise, the Antarctic ozone hole is expected to recur seasonally, provided the present meteorological conditions continue.'

First report of the Environmental Effects Assessment Panel

The first report of the Environmental Effects Panel published in November 1989 concluded that exposure to increased ultraviolet radiation can cause suppression of the body's immune system, which might lead to an increase in the occurrence or severity of infectious diseases such as herpes, leishmaniasis and malaria, and a possible decrease in the effectiveness of vaccination programmes. It also found that enhanced levels of ultraviolet radiation can lead to increased damage to the eyes, especially the cataracts, the incidence of which was expected to increase by 0.6 per cent per 1 per cent total column ozone depletion. Therefore, each 1 per cent total column ozone depletion was expected to lead to a worldwide increase of 100,000 blind persons due to ultraviolet-radiation-induced cataracts.

The panel concluded that every 1 per cent decrease of total column ozone would lead to a 3 per cent rise in the incidence of non-melanoma skin cancer, and that there was concern that an increase of the more dangerous cutaneous malignant melanoma could also occur. An increase in skin cancer would mainly affect light-skinned people with little protective pigment in their skin.

Half the plant species studied by the Environmental Effects Panel were found to be sensitive to enhanced ultraviolet radiation, resulting in plants with reduced growth and smaller leaves. These effects were applied to certain varieties of soya and wheat. 'Even small decreases in food production from UV-B effects on agriculture would significantly affect people in areas where food shortages occur even now,' the panel concluded.

Increased ultraviolet radiation was shown to have a negative influence on aquatic organisms, especially small ones such as phytoplankton, zooplankton, larval crabs and shrimp, and juvenile fish. Because many of those small organisms are at the base of the marine food web, increased ultraviolet radiation was thought to have a negative influence on the productivity of fisheries.

Other environmental effects of ozone depletion reported by the Panel included: an increased atmospheric abundance of particulates, which would aggravate urban and rural air pollution and negatively affect human health and agricultural productivity; degradation of many materials, particularly plastics that are used outdoors and particularly in tropical locations; and a contribution to anthropogenic radiative forcing of global warming.

First report of the Economic Panel

The first report of the Economic Panel (which shortly after merged with the Technology Review Panel (TRP) to become the Technology and Economic Assessment Panel (TEAP)) assessed the availability and expected market potential of alternatives and substitutes, estimated the energy efficiency and cost differences, analysed the benefits of avoiding ozone depletion, discussed the issue of technology transfer and summarized industry policies toward ozone layer protection. The report concludes:

BOX 1.4 PROGRAMMES FOR MEASURING AND ANALYSING OZONE AND ULTRAVIOLET RADIATION

Nelson Sabogal, Senior Scientific Affairs Officer, UNEP Ozone Secretariat

Several global programmes coordinate and integrate monitoring and research of ozone and ultraviolet radiation. The Global Atmosphere Watch (GAW), established by WMO in 1989, integrates WMO's atmospheric research and monitoring activities. Six World Data Centres collect, process, analyse and distribute data:

1   Ozone and UV-B radiation: World Ozone and Ultraviolet Radiation Centre, Toronto, Canada.
2   Greenhouse and other trace gases: World Data Centre for Greenhouse Gases, Tokyo, Japan.
3   Solar radiation: World Radiation Data Centre, St Petersburg, Russian Federation.
4   Aerosols: World Data Centre for Aerosols, Ispra, Italy, operated by the European Union.
5   Precipitation chemistry: World Data Center for Precipitation Chemistry, Asheville, NC, USA.
6   Surface ozone: World Data Centre for Surface Ozone (WDCSO3), Oslo, Norway.

The Network for the Detection of Stratospheric Change (NDSC), which began in January 1991, measures ozone and key ozone-related chemical compounds and parameters at remote-sounding research stations. The NDSC has the following working groups: Dobson and Brewer, Fourier Transform Infrared spectrometer (FTIR), Light Detection and Ranging (LIDAR), Microwave, Ozonesondes and Aerosol Sondes, Satellite, Spectral UV, Theory and Analysis, and UV/Visible.

In 1989, the Stratospheric Ozone Intercomparison Campaign (STOIC) was carried out at Table Mountain Observatory in California.

In 1992, the World Ozone Data Centre, created in 1960 by Environment Canada for WMO, became the World Ozone and Ultraviolet Radiation Centre (WOUDC).

In 1995, the Global Ozone Monitoring Experiment (GOME) was launched on board the European Space Agency's Earth Remote Sensing satellite (ERS-2) to measure total column amounts and vertical profiles of ozone and several trace gases such BrO, OClO, and NO_2.

Several campaigns and satellite programmes are planned for the near future, among them the Stratosphere Aerosol and Gas Experiment (SAGE III), Global Ozone Monitoring Experiment (GOME-2), the Ozone Mapping and Profiling Suite (OMPS), and the Infrared Atmospheric Sounding Interferometer (IASI).

'Notwithstanding the problems of quantifying the benefits, the basic conclusion is that the monetary value of the benefits is undoubtedly much greater than the costs of CFC and halon reductions. However, some developing nations may not have sufficient resources to make the change to new ozone-safe technologies or may have other economic, environmental, or human health concerns that are immediate and pressing than protection of the ozone layer.'

First report of the Technology Review Panel

The TRP concluded that it was technically feasible to phase down by at least 95 per cent the production and consumption of the five controlled CFCs; phase out production and consumption of carbon tetrachloride; and phase down by at least 90 per cent the production and consumption of methyl chloroform. The Parties to the Protocol considered the reports of the assessment panels when they drafted and approved the London Adjustment and Amendments of 1990.

First Synthesis Report: Policy options to protect the ozone layer

The 1989 Synthesis Report, which integrated the four assessment panels' reports, found that:

> *'ozone depletion is a global problem... While the largest ozone depletions are predicted to occur at high latitudes in both the hemispheres, enhanced levels of ultraviolet radiation will have adverse effects on people from all nations, independent of geographical position, eg, Northern or Southern Hemisphere, or economic status, eg, developed or developing. While peoples with lightly pigmented skins are most susceptible to melanoma and non-melanoma skin cancer, all peoples are susceptible to contracting eye disorders and a suppression of the immune system.'*

The cause of global ozone depletion, the Synthesis Panel wrote, is 'the current and historic use of CFCs, halons, and other chlorine and bromine-containing chemicals in the developed nations.' However, it concluded that:

> *'protection of the ozone layer will require a full partnership between developed countries that have caused the problem and those in developing countries who would now like to improve their standard of living by using these chemicals for uses such as refrigeration. The lack of technical knowledge and financial resources of developing countries inhibits the adoption of certain CFC/halon replacement technologies and the definition and implementation of the best national options for the transition to CFC-free technologies. Funding is needed to assist the transfer of technology to developing countries during the transition period because currently available resources are already strained as a result of the world debt problem and the dire economic situation of many countries.'*

The first Arctic expedition, 1989

Concern about ozone levels over the Arctic resulted in the first Airborne Arctic Stratospheric Expedition, which was staged by NASA and NOAA from Norway in 1989. During that mission, flights into the Arctic stratosphere revealed that chlorine monoxide and bromine monoxide were present at concentrations comparable to those observed over Antarctica in 1987. 'However, since the degree of ozone loss depends both on the ClO/BrO concentrations and on the duration of the elevated levels, the shorter period of cold temperatures in the Arctic diminished the impact on ozone,' NASA concluded.[50]

SECOND ASSESSMENT, 1991: QUICKER PHASE-OUT POSSIBLE, CONTROL HCFCS AND METHYL BROMIDE

At the second Meeting of the Parties to the Montreal Protocol in London in 1990, the Parties asked the Assessment Panels to submit another comprehensive assessment in November 1991; in particular, the Parties requested that the Scientific Assessment Panel include an evaluation of the ozone depletion and global warming potentials of substitutes; ozone depletion potentials of 'other halons' that might be produced in significant quantities; the impact on the ozone layer of revised control measures; and the impact on the ozone layer of engine emissions from high-altitude aircraft, rockets and space shuttles.

Second report of the Scientific Assessment Panel

Significant depletion in northern and southern hemispheres
The Scientific Assessment Panel's second report, published in 1991, concluded that ground-based and satellite observations continued to show decreases of total column ozone in winter in the northern hemisphere.[51] For the first time, there was evidence of significant decreases in spring and summer in both the northern and southern hemispheres at middle and high latitudes, as well as in the southern winter. These downward trends were larger during the 1980s than in the 1970s, and the Panel concluded that the observed ozone decreases had occurred predominantly in the lower stratosphere.

The Panel cited four other major advances since the Panel's last review in 1989:

1 *'Polar Ozone: Strong Antarctic ozone holes have continued to occur and, in 4 of the past 5 years, have been deep and extensive in area. This contrasts to the situation in the mid-1980s, where the depth and area of the ozone hole exhibited a quasi-biennial modulation. Large increases in surface ultraviolet radiation have been observed in Antarctica during periods of low ozone. While no extensive ozone losses have occurred in the Arctic comparable to those observed in the Antarctic, localized Arctic ozone losses have been observed in winter concurrent with observations of elevated levels of reactive chlorine.*

2 *'Ozone and Industrial Halocarbons: Recent laboratory research and reinterpretation of field measurements have strengthened the evidence that the Antarctic ozone hole is primarily due to chlorine- and bromine-containing chemicals. In addition, the weight of evidence suggests that the observed middle- and high-latitude ozone losses are largely due to chlorine and bromine. Therefore, as the atmospheric abundances of chlorine and bromine increase in the future, significant additional losses of ozone are expected at middle latitudes and in the Arctic.*

3 *'Ozone and Climate Relations: For the first time, the observed global lower-stratospheric ozone depletions have been used to calculate the changes in the radiative balance of the atmosphere. The results indicate that, over the last decade, the observed ozone depletions would have tended*

to cool the lower stratosphere at middle and high latitudes. Temperature data suggest that some cooling indeed has taken place there. The observed lower-stratospheric ozone changes and calculated temperature changes would have caused a decrease in the radiative forcing of the surface-troposphere system in the middle to high latitudes that is larger in magnitude than that predicted for the chlorofluorocarbon (CFC) increases over the last decade. In addition, the ozone depletion may indeed have offset a significant fraction of the radiative forcing due to increases of all greenhouse gases over the past decade.

4   *'Ozone Depletion Potentials (ODPs) and Global Warming Potentials (GWPs): A new semi-empirical, observation-based method of calculating ODPs has better quantified the role of polar processes in this index. In addition, the direct GWPs for tropospheric, well-mixed, radiatively-active species have been recalculated. However, because of the incomplete understanding of tropospheric chemical processes, the indirect GWP of methane has not, at present, been quantified reliably. Furthermore, the concept of a GWP may prove inapplicable for the very short-lived, inhomogeneously mixed gases, such as the nitrogen oxides. Hence, many of the indirect GWPs reported in 1990 by the Intergovernmental Panel on Climate Change (IPCC) are likely to be incorrect.'*

Space shuttles and rockets: Insignificant impact

More specifically, the Panel's report concluded that, in the Antarctic ozone hole: 'The low value of total-column ozone measured by TOMS in early October in 1991 was 110 Dobson units, which is a decrease of about 60 percent compared to the ozone levels prior to the late 1970s... Recent laboratory studies of heterogenous processes, re-evaluated field measurements, and modelling studies have strengthened the confidence that the cause of the Antarctic ozone hole is primarily chlorine and bromine emissions.' With regard to shuttles and rockets, 'The increase in the abundance of stratospheric chlorine from one projection of US annual launches of nine space shuttles and six Titan rockets is calculated to be less than 0.25 percent of the annual stratospheric chlorine source from halocarbons in the present-day atmosphere.'

Methyl bromide: A significant ODS

The report identified methyl bromide as an ozone-depleting substance, with an ODP of 0.6, and reported that it was the most abundant bromine source gas. The report had designated ODPs for only two halons; two years later, the 1991 Scientific Assessment designated ODPs for seven halons, ranging from 16 to 0.14. Though methyl bromide is produced naturally, the report said, 'measurements of CH_3Br made over the Atlantic Ocean show a marked interhemispheric gradient... being recorded in the Northern and Southern Hemispheres respectively. This argues for a substantial land-based source, which could well be anthropogenic.'

Second report of Environmental Effects Assessment Panel

Impacts on human health and ecosystems
The Environmental Effects Panel also published its second report in 1991, concluding that a sustained 10 per cent loss of ozone would lead to an increase in the incidence of non-melanoma skin cancers by 26 per cent, or in excess of 300,000 cases per year worldwide, and that increased ultraviolet radiation could lead to an increased incidence of melanoma.[52] The Panel also predicted that a sustained 1 per cent decrease of ozone would lead to between 100,000 and 150,000 additional cases of cataract-induced blindness worldwide.

Other findings concluded that aquatic ecosystems were already under ultraviolet radiation stress and could suffer detrimental effects; losses of phytoplankton could reduce biomass production, which is propagated throughout the whole food web, resulting in losses of biomass for human consumption. Because marine phytoplankton absorb carbon dioxide, any reduction in phytoplankton would also decrease the absorption of carbon dioxide and augment global warming. Tropospheric ozone concentrations could rise in heavily polluted areas, and other potentially harmful substances were expected to increase in all areas because of enhanced chemical reactivity.

Second report of the Technology and Economics Assessment Panel

Feasible to eliminate virtually all ODS consumption by 1997
By 1991, the Technology and Economic Assessment Panels had been merged into one panel, which concluded that worldwide CFC production had fallen 40 per cent since 1986, and halon consumption had begun to fall. 'It is technically feasible to eliminate virtually all consumption of controlled substances in developed countries by 1995–1997, if commercial quantities of transitional substances are available... As a result of rapid development of technology, the costs of eliminating controlled substances are lower than estimated in 1989 and will decline further' (see Chapter 5).

Emission scenarios conducted by the assessment panels suggested that chlorine and bromine loading could be significantly reduced through: an acceleration of the phase-out of CFCs, carbon tetrachloride, halons, and methyl chloroform; limited use of HCFCs; and a reduction in the ten-year lag for developing countries to five years. Bromine levels could be reduced if significant anthropogenic emissions of methyl bromide could be controlled. The Parties to the Protocol considered the information found in these reports when they drafted and approved the Copenhagen Adjustment and Amendment of 1992.

EXPEDITION FINDS SIGNIFICANT DEPLETION OVER THE NORTHERN HEMISPHERE

In August 1991, the second Airborne Arctic Stratospheric Expedition launched its research phase in preparation for flights into the Antarctic, Arctic, and mid-latitude stratosphere. The mission was a multi-agency effort of NASA, NOAA, the National Science Foundation, scientists from several US universities, and the chemical industry's Alternative Fluorocarbon Environmental Acceptability Study. The expedition's equipment included high altitude ER-2 aircraft, a long-range DC-8 aircraft, extensive meteorological predictions and analyses, and the TOMS on the Nimbus 7 satellite.

Mission participants concluded in their End of Mission Statement that 'during 1992, the Total Ozone Mapping Spectrometer (TOMS) satellite measurements show that the hemispheric ozone average during January, February, and most of March was lower than any previous year in the TOMS record. TOMS measurements also showed that total ozone values in the mid-latitude maximum during February were 10–15 percent lower than any previous year in the TOMS record.'

According to NASA, the mission's results 'indicate that the ozone shield of the Northern Hemisphere is increasingly vulnerable to depletion by man-made chemicals, said James Anderson of Harvard University, the study's Mission Scientist… The study also revealed important evidence supporting the conclusion that ozone decreases at mid-latitudes are associated with increased levels in chlorine and bromine in the stratosphere. Since concentrations of human-made, ozone-depleting chemicals will continue to rise during the 1990s despite current regulations, it is increasingly likely in coming years that substantial Arctic ozone losses will occur during particularly cold, protracted winters and that mid-latitude ozone trends will continue to decline.'[53]

THIRD ASSESSMENT, 1994: MOUNT PINATUBO VOLCANO DEPLETES OZONE, ARCTIC OZONE DEPLETION CONFIRMED

Third report of the Scientific Assessment Panel

In preparation for the 1995 Meeting of the Parties to the Montreal Protocol, the Scientific Assessment Panel released its third assessment in December 1994.[54] Through laboratory investigations, atmospheric observations and modelling studies, the Panel concluded that the atmospheric growth rates of several major ozone-depleting substances had slowed, demonstrating the expected impact of the Montreal Protocol and its Amendments and Adjustments. Since the stratospheric abundances of chlorine and bromine were expected to continue to grow for a few more years, increasing global ozone losses were predicted for the remainder of the decade, with gradual recovery in the 21st century. The panel's conclusions included the following:

Ozone depletion continues, aggravated by Mount Pinatubo eruption

- *'The atmospheric abundances of several of the CFC substitutes are increasing, as anticipated… Tropospheric chlorine in HCFCs increased by 5 ppt/year in 1989 and about 10 ppt/year in 1992.*
- *'Record low global ozone levels were measured over the past two years. Anomalous ozone decreases were observed in the mid-latitudes of both*

hemispheres in 1992 and 1993. The Northern Hemispheric decreases were larger than those in the Southern Hemisphere. Globally, ozone values were 1–2 percent lower than would be expected from an extrapolation of the trend prior to 1991.

- *'The stratosphere was perturbed by a major volcanic eruption. The eruption of Mt. Pinatubo in 1991 led to a large increase in sulfate aerosol in the lower stratosphere throughout the globe.*

- *'Downward trends in total-column ozone continue to be observed over much of the globe, but their magnitudes are underestimated by numerical models. Decreases in ozone abundances of about 4–5 percent per decade at mid-latitudes in the Northern and Southern Hemisphere continue to be observed by both ground-based and satellite-borne monitoring instruments.*

- *'The Antarctic ozone "holes" of 1992 and 1993 were the most severe on record... Satellite, balloon-borne, and ground-based monitoring instruments revealed that the Antarctic ozone "holes" of 1992 and 1993 were the biggest (areal extent) and deepest (minimum amounts of ozone overhead), with ozone being locally depleted by more than 99 percent between about 14–19 km in October, 1992 and 1993.*

- *'Ozone losses have been established in the Arctic winter stratosphere, and their links to halogen chemistry have been established... In the late-winter/early-spring period, additional chemical losses of ozone up to 15–20 percent at some altitudes are deduced from these observations, particularly in the winters of 1991/2 and 1992/3.*

- *'The link between a decrease in stratospheric ozone and an increase in surface ultraviolet (UV) radiation has been further strengthened. Measurements of UV radiation at the surface under clear-sky conditions show that low overhead ozone yields high UV radiation and in the amount predicted by radiative-transfer theory.*

- *'Methyl bromide continues to be viewed as a significant ozone-depleting compound... The ozone depletion potential (ODP) for methyl bromide is calculated to be about 0.6 (relative to an ODP of 1 for CFC-11).'*

Appendix 3 includes a listing of ozone depletion potentials of ODSs.

Third report of the Environmental Effects Assessment Panel

The Environmental Effects Panel's 1994 report found that a sustained 1 per cent decrease in ozone would result in approximately a 2 per cent increase in non-melanoma skin cancers, and an increase in the incidence and morbidity of eye diseases and infectious diseases.[55] 'In areas of the world where infectious diseases already pose a significant challenge to human health and in persons with impaired immune function, the added impact of UV-B-induced immune suppression could be significant,' the report concluded.

The Environmental Effects Panel also reported that ultraviolet radiation could adversely affect phytoplankton productivity and the early development stages of fish and other aquatic organisms, with one study indicating a 6–12 per cent reduction in phytoplankton production in the marginal ice zone during the

---

BOX 1.5 POPULAR RECOGNITION FOR CRUTZEN, MOLINA AND ROWLAND

When Paul Crutzen, Mario Molina, and F Sherwood Rowland received the 1995 Nobel Prize in Chemistry for their work on the science of ozone depletion, *Time* magazine (23 October, 1995) said:

> *'Rarely does Nobel-prize-winning research galvanize worldwide political action. Yet the findings that have made chemistry laureates of Sherwood Rowland of the University of California at Irvine, Mario Molina of MIT [Massachusetts Institute of Technology], and Paul Crutzen of Germany's Max Planck Institute for Chemistry did just that. Their discovery that man-made chemicals can damage the planet's protective ozone layer was instrumental in triggering the most successful global environmental treaty ever written: the 1987 Montreal Protocol limiting the use of chlorofluorocarbons, or CFCs*
> *'Crutzen's work came first, with his demonstration in 1970 that airborne chemicals called nitrogen oxides could damage the ozone gas that floats high in the Earth's stratosphere and screens out ultraviolet light, which can cause sunburn and skin cancer. The real breakthrough, though, came in 1974, when Molina and Rowland determined that CFCs are highly efficient ozone destroyers, gobbling up many times their volume in ozone molecules.*
> *'Molina's and Rowland's work didn't thrill industrialists, but it did lead to a ban on CFC-based spray cans in the US in 1978. And after an ozone hole over Antarctica was detected in 1985, industry was goaded into taking swift action.'*

The Nobel Prize 'landed like a vindication' on scientists whose 'work was long overlooked by the scientific establishment,' according to the *Los Angeles Times* (12 October, 1995). Molina said the award to environmental scientists 'shows a certain maturity in a field previously considered soft.' Rowland said he had received 'many messages from people in the environmental area saying they are doubly pleased – both by our getting the prize, but also by the fact that problems of the environment are being recognized.'

---

period of peak springtime Antarctic ozone depletion. The productivity of marine and terrestrial ecosystems could be decreased by increased ultraviolet radiation, which could also increase tropospheric ozone in heavily polluted areas.

New research on the environmental fate and impact of the hydro-fluorocarbon (HFC) and HCFC substitutes for CFCs found that some of the compounds degraded in the atmosphere to form trifluoroacetate. Trifluoroacetate is removed from the atmosphere as trifluoroacetic acid which is mildly toxic to some marine species, but would never reach toxic concentrations in the environment; trifluoroacetic acid would ultimately end up in the oceans, and the large dilutions would result in extremely low concentrations.

Third Synthesis Report

The scenarios considered by the 1994 Synthesis Panel suggested that there was only a limited number of approaches to lowering atmospheric chlorine and bromine levels beyond those already adopted by the Parties to the Montreal Protocol: elimination of methyl bromide from agricultural, structural and industrial activities; reduction of the HFCF cap and acceleration of the phase-out schedule; recovery and destruction of halons; and recovery and destruction of CFCs. The Technology and Economic Assessment Panel concluded that additional measures were technically and economically feasible.

The Parties directed future assessment work of the panels to include in March 1995 the feasibility and implications of alternatives and substitutes to HCFCs and methyl bromide, the feasibility of reducing the HCFC cap, accelerating the initial reductions, and/or speeding up the phase-out. A full range of methyl bromide control options was to be considered. The Parties to the Protocol considered the 1994 reports of the three panels when they drafted and approved the Vienna Adjustment of 1995.

Crutzen, Molina and Rowland receive Nobel Prize

In 1995, the Royal Swedish Academy of Sciences (the Royal Academy) awarded the Nobel Prize in Chemistry to Paul Crutzen, Mario Molina, and F Sherwood Rowland 'for their work in atmospheric chemistry, particularly concerning the formation and decomposition of ozone.'[56] The Royal Academy cited their work on the threat of supersonic transports and chlorofluorocarbons to the ozone layer, and the linkage of the Antarctic ozone hole to the chlorine and bromine from industrially manufactured gases.

According to the Royal Academy, Crutzen, Molina and Rowland:

> *'have all made pioneering contributions to explaining how ozone is formed and decomposes through chemical processes in the atmosphere. Most importantly, they have in this way showed how sensitive the ozone layer is to the influence of anthropogenic emissions of certain compounds. The thin ozone layer has proved to be an Achilles heel that may be seriously injured by apparently moderate changes in the composition of the atmosphere. By explaining the chemical mechanisms that affect the thickness of the ozone layer, the three researchers have contributed to our salvation from a global environmental problem that could have catastrophic consequences... Many were critical of Molina's and Rowland's calculations, but yet more were seriously concerned by the possibility of a depleted ozone layer. Today we know that they were right in all essentials. It was to turn out that they had even underestimated the risk.'*

FOURTH ASSESSMENT, 1998: MONTREAL PROTOCOL WORKING, ODSs IN THE ATMOSPHERE PEAK IN 1994

Fourth report of the Scientific Assessment Panel

At the request of the Parties to the Montreal Protocol, the Scientific Assessment Panel in 1998 provided an update of the scientific understanding of the ozone layer, involving 304 researchers from 35 countries worldwide. The major findings outlined in the Panel's 1998 report included the following.[57]

ODS in the atmosphere generally declining

- 'The total combined abundance of ozone-depleting compounds in the lower atmosphere peaked in about 1994 and is now slowly declining. Total chlorine is decreasing, but total bromine is still increasing. As forecast in the 1994 Assessment, the long period of increasing total chlorine abundances – primarily from the chlorofluorocarbons (CFCs) carbon tetrachloride (CCl_4), and methyl chloroform (CH_3CCl_3) – has ended. The declining abundance of total chlorine is due principally to reduced emissions of methyl chloroform. Chlorine from the major CFCs is still increasing slightly. The abundances of most of the halons continue to increase (for example, halon-1211, almost 6 percent per year in 1996), but the rate has slowed in recent years.

- 'The observed abundances of the substitutes for the CFCs are increasing. The abundances of the hydrochlorofluorocarbons (HCFCs) and hydrofluorocarbons (HFCs) are increasing as a result of a continuation of earlier uses and of their use as substitutes for the CFCs.

- 'The combined abundance of stratospheric chlorine and bromine is expected to peak before the year 2000. The delay in this peak in the stratosphere compared with the lower atmosphere reflects the average time required for surface emissions to reach the lower stratosphere.

- 'The role of methyl bromide as an ozone-depleting compound is now considered to be less than was estimated in the 1994 assessment, although significant uncertainties remain. The best current estimate of the ozone depletion potential (ODP) of methyl bromide is 0.4, compared to 0.6 estimated in 1994.

- 'The rate of decline in stratospheric ozone at mid-latitudes has slowed; hence, the projections of ozone loss made in the 1994 Assessment are larger than what has actually occurred. Total column ozone decreased significantly at mid-latitudes (25–60 degrees) between 1979 and 1991, with estimated linear downward trends of 4.0, 1.8, and 3.8 percent per decade, respectively, for northern mid-latitudes in winter/spring, northern mid-latitudes in summer/fall, and southern mid-latitudes year round... The observed total column ozone losses from 1979 to the period 1994–1997 are about 5.4 percent, 2.8 percent, and 5.0 percent, respectively, for northern mid-latitudes in winter/spring, northern mid-latitudes in summer/fall, and southern mid-latitudes year round, rather than the values projected in the 1994 Assessment assuming a linear trend: 7.6, 3.4, and 7.2 percent respectively.

- 'The springtime Antarctic ozone hole continues unabated.

- 'The late-winter/spring ozone values in the Arctic were unusually low in six out of the last nine years, the six being years that are characterized by unusually cold and protracted stratospheric winters.

- 'The understanding of the relation between increasing surface UV-B radiation and decreasing column ozone has been further strengthened by ground-based observations, and newly developed satellite methods show promise for establishing global trends in UV radiation. The inverse dependence of surface UV radiation and the overhead amount of ozone, *which was demonstrated in earlier Assessments, has been further demonstrated and quantified by ground-based measurements under a wide range of atmospheric conditions.'*

Fourth report of the Environmental Effects Assessment Panel

The Environmental Effects Panel in its 1998 report concluded that because of increases in ultraviolet radiation due to ozone depletion:

> *'adverse effects on the eye will affect all populations irrespective of skin colour. Adverse impacts could include: more cases of acute reactions such as "snowblindness"; increases in cataract incidence and/or severity (and thus the incidence of cataract-associated blindness); and increases in the incidence (and mortality) from ocular melanoma and squamous cell carcinoma of the eye.*
>
> *'Effects on the immune system will also affect all populations but may be both adverse and beneficial… Effects on the skin could include increases in photoaging, and skin cancer with risk increasing with fairness of skin.'*[58]

The Panel found that effects of increased ultraviolet radiation may accumulate from year to year in long-lived perennial plants and from generation to generation in annual plants, and that effects may impact terrestrial ecosystems in ways such as life-cycle timing, changes in plant form, and production of plant chemicals not directly involved in primary metabolism.

Other environmental effects cited in the report included:

- *'Solar UV-B and UV-A have adverse effects on the growth, photosynthesis, protein and pigment content, and reproduction of phytoplankton, thus affecting the food web.*
- *'Macroalgae and seagrasses show a pronounced sensitivity to solar UV-B.*
- *'Polar marine ecosystems, where ozone-related UV-B increases are the greatest, are expected to be the oceanic ecosystems most influenced by ozone depletion.*
- *'Increased UV-B has positive and negative impacts on microbial activity in aquatic ecosystems that can affect carbon and mineral nutrient cycling as well as the uptake and release of greenhouse and chemically-reactive gases.*
- *'Additional UV-B will increase the rate at which primary pollutants are removed from the troposphere.*
- *'Physical and mechanical properties of polymers are negatively affected by increased UV-B in sunlight. Increased UV-B reduces the useful lifetimes of synthetic polymer products used outdoors and of biopolymer materials such as wood, paper, wool and cotton.'*

Fourth Synthesis Report: The Montreal Protocol is working

In its 1998 report, the Synthesis Panel cited the importance of conveying to decision-makers new policy-relevant insights and information. Within this context, the Panel concluded that:

- 'The Montreal Protocol is working.
- 'The ozone layer is currently in its most vulnerable state.
- 'The ozone layer will slowly recover over the next 50 years.
- 'The issues of ozone depletion and climate change are interconnected; hence, so are the Montreal and Kyoto Protocols. Changes in ozone affect the Earth's climate, and changes in climate and meteorological conditions affect the ozone layer, because the ozone depletion and climate change phenomena share a number of common physical and chemical processes. Hence, decisions taken (or not taken) under one Protocol have an impact on the aims of the other Protocol. For example, decisions made under the Kyoto Protocol with respect to methane, nitrous oxide, and carbon dioxide will affect the rate of recovery of ozone, while decisions regarding controlling HFCs may affect decisions regarding the ability to phase out ozone-depleting substances.'

The 1998 Synthesis Panel also concluded that options to reduce the current and near-term vulnerability to ozone depletion were limited, and that few policy options were available to enhance the recovery of the ozone layer. [59] It described several scenarios that could avoid future cumulative ozone loss if certain ozone-depleting substances, including halons and methyl bromide, were eliminated in the near term.

'Failure to comply with the international agreements of the Montreal Protocol will affect the recovery of the ozone layer,' the Synthesis Panel concluded. 'For example, illegal production of 20–40 kilotons per year of CFC-12 and CFC-113 for the next 10–20 years would increase the cumulative ozone losses noted above by about 1–4 percent.'

First integrated scientific assessment of ozone depletion and climate change, 1999: Aircraft

In 1999, the Intergovernmental Panel on Climate Change's Working Groups I and III, in collaboration with the Scientific Assessment Panel, published a report on the effects of aircraft on the atmosphere. [60] The report concluded that, 'There is no experimental evidence for a large geographical effect of aircraft emissions on ozone anywhere in the troposphere.'

Susan Solomon of NOAA the same year published a review of the concepts and history of the chemistry of stratospheric ozone depletion, writing that:

'observations of ozone and of chlorine-related trace gases near 40 km provide evidence that gas phase chemistry has indeed currently depleted about 10 percent of the stratospheric ozone there as predicted, and the vertical and horizontal structures of this depletion are fingerprints for that process. More striking changes are observed each austral spring in Antarctica, where about half of the total ozone column is depleted each September, forming the Antarctic ozone hole. Measurements of large amounts of ClO, a key ozone destruction catalyst, are among the fingerprints showing that human releases of chlorofluorocarbons are the primary cause of this change.

'Enhanced ozone depletion in the Antarctic and Arctic regions is linked to heterogeneous chlorine chemistry that occurs on the surfaces of polar

BOX 1.6 WIDER LESSONS LEARNED FROM THE OZONE ISSUE

Daniel Albritton, Co-Chair, Scientific Assessment Panel

At this stage of our experience with the ozone depletion issue, it is perhaps useful to reflect on some of the valuable 'lessons learned' that might be relevant to other global environmental issues. The phenomenon of ozone layer depletion, together with the associated decision-making embodied in the Montreal Protocol, has provided a good illustration of three phases that might be common to the evolution of environmental issues.

In the first phase, the primary contribution of the scientific community is to describe the phenomenon and establish its credibility as a potential issue for consideration of the world community. The basic scientific understanding associated with the phenomenon is being built 'from the ground up' in this phase.

Once credibility is established and communities accept the notion that the issue is a 'real' one, a second phase occurs that focuses on answering the question, 'What should/can we do to address the issue?' This is a phase characterized by taking action. Scientific and technical experts focus on how to implement decisions without 'breaking the bank'.

The third and final stage might be described as 'accountability' assessment. Have the actions of phase two succeeded in addressing the environmental issue? In the case of ozone depletion, has the ozone layer recovered as expected?

We're well into phase two with ozone depletion, and perhaps transitioning into phase three. What we've learned is that the needs of the decision-making community, and the urgency of those needs, evolve as the environmental issue proceeds through its 'credibility-action-accountability' phases. But common to all three phases is the need for reliable, objective scientific information. From its inception, the Montreal Protocol invoked a means to address these information needs, via written reports or 'assessments'. The assessment process, whereby the scientific community periodically reviews 'what is known' and 'what is not known' about the underlying science of the issue, has proven to be a valuable and successful mechanism for carrying out the information transfer.

A second general lesson learned is that one of the most important properties of a molecule is its longevity in the atmosphere, gauged by scientific quantities called its atmospheric 'lifetime' or 'time constant'. Of course, the molecule must also have attributes that lead it to interact with light or with other molecules in the atmosphere. But probably the most policy-relevant property in the chemical reference handbook is how long that molecule resides in the atmosphere.

For example, in the set of compounds called perfluorocarbons, all the chlorine and bromine atoms one would find in a CFC or halon are gone, replaced by fluorine atoms. In other words, they are totally fluorinated. We have found that those compounds are probably immortal. The only way they are removed is probably destruction at the very upper part of the atmosphere by the most violent solar radiation, and perhaps in high-temperature furnaces. But other than that, when they are in the atmosphere, they are there to stay. That property and that property alone is a sobering bit of information. Compounds such as CFCs and CO_2 have lifetimes longer than our own. We have willed our CFCs to our children, and we have willed our CO_2 to our children.

We've learned that this scientific quantity, atmospheric lifetime, sets a very basic context for decision-making. A decision-maker has to decide when there is enough information to risk making a decision, balancing upsides and downsides in the process. For phenomena that have consequences for a very long time, the balance in that equation is different than for phenomena in which the consequences are short-term. The long time constant says you have discovered you are in a poker game, the deck is rigged, and you cannot quit for a very long time. A short time constant says I may wait for more information because I know that when I do make a decision, the phenomenon will reverse itself fairly rapidly. These are two very different classes of environmental issues.

We've learned other lessons from the global nature of the ozone layer issue. There are local and regional issues, such as the water quality of a river or transboundary air pollution, that can be dealt with by municipalities, a single nation, or a group of nations. But on a global issue such as ozone depletion, what we have noted from the Protocol is that there must be a global buy-in. That is a similarity between global warming and ozone depletion.

Finally, we've learned that there is tremendous value in some degree of separation of science from public policy. A scientist is perhaps most uncomfortable when he or she is asked 'What should we do?' The course of action depends not only on the science, but also on technical options, the economics of 'a' versus 'b', the politics of sovereign nations trying to decide on one global phenomenon, and considerable diplomacy, as we have seen. What Walter Lipmann recognized back in the 1920s was that it is no accident that the best service in the world is the one in which the divorce between the assembling of knowledge and the control of policy is the most perfect.

Lipmann also then points out that while the scientific information producer should not make the decisions, and the decision-makers should not be the scientific information producers, those who investigate and those who decide must have 'lunch together'. Both have knowledge to contribute to the process: science information and policy needs. The assessment process is a way of having lunch together: the research community, or some fraction of it, must interact with the decision-making process. The Montreal Protocol has set an example of that process.

What does this mean to us as researchers? I hope there are researchers who ask how they can do their research and also feel that they have made a difference. I have an answer for those researchers: scientific insights are a service. What Lipmann said was that the idea that an expert is an ineffectual person because he or she lets others make the decision is totally contrary to experience. The Montreal Protocol shows that Lipmann is right. Namely, the fact that maximum chlorine in the lower part of the atmosphere peaked at 3.2 ppb in late 1994 or early 1995 is attributable to both the experts and the decision-makers. That is a service, and that is being effective.

stratospheric clouds at cold temperatures. Observations also show that some of the same heterogeneous chemistry occurs on the surfaces of particles present at mid-latitudes as well, and the abundances of these particles are enhanced following explosive volcanic eruptions.... As human use of chlorofluorocarbons continues to decrease, these changes throughout the ozone layer are expected to gradually reverse during the twenty-first century.[61]

THE OZONE LAYER TODAY

During the 1999/2000 winter, the Third European Stratospheric Experiment on Ozone sponsored by the European Union and the NASA-sponsored SAGE III Ozone Loss and Validation Experiment obtained measurements of ozone in the Arctic stratosphere. Scientists from Europe, the USA, Canada, Russia and Japan joined forces to mount the biggest field measurement campaign of the Arctic to date. During that winter, the team measured large ozone losses inside the Arctic stratospheric polar vortex.[62]

According to the European Union's account of the campaign's findings:

> *'ozone losses of over 60 per cent have occurred in the Arctic stratosphere near 18 km altitude during one of the coldest stratospheric winters on record. These losses are likely to affect the ozone levels over Europe during spring. This is one of the most substantial ozone losses at this altitude in the Arctic. Measurements from the largest international campaign ever investigating stratospheric ozone depletion have provided more insight into the processes that control stratospheric ozone. They have also reinforced concerns that the Arctic ozone may continue to decline despite the benefits of reductions in stratospheric chlorine levels (a result of the Montreal Protocol), due to the global climate change.'*

The ozone hole in the Antarctic spring of 2000 was the largest ever recorded, covering 28.4 million square kilometres. 'Over the past several years, the annual ozone hole over Antarctica has remained about the same in both its size and in the thickness of the ozone layer,' according to a 2001 report by NASA's Goddard Space Flight Center and NOAA.[63] 'This is consistent with human-produced chlorine compounds that destroy ozone reaching their peak concentrations in the atmosphere, levelling off, and now beginning a very slow decline.' The Antarctic ozone hole in October 2001 peaked at about 26 million square kilometres, roughly the size of North America.

Chapter 2

Diplomacy: The beginning, 1974–1987

INTRODUCTION

The issue of environmental protection came to prominence in the 1960s when the adverse effects of the traditional patterns of industrial growth became pronounced. Acid rain, possible increased radiation levels from nuclear testing and problems of persistent chemicals were already apparent. Rachel Carson's *Silent Spring*, first published in 1962, detailed the impacts of the uncontrolled application of pesticides. By the beginning of the 1970s, there was a widespread sense of environmental crisis around the world. A major concern was the exploding population 'bomb' due to increasing birth rates in developing countries and decreasing mortality rates everywhere due to better health care. Growing industrialization and prosperity had led to increasing urbanization, slums, smog, traffic jams, noise, water and air pollution and waste. There were doubts about whether 'Spaceship Earth' would be able to survive.

Many argued that the present generation was passing on a huge bill to future generations. There was concern about the natural non-renewable resources of the world running out due to profligate consumption. Many countries, including India, Japan, Sweden, the UK and the USA, had taken the first steps to arrest environmental degradation in their countries. There was growing realization that the global environment and common resources of the world – the atmosphere and the oceans – might not be protected if every country looked after only its national interests. Advantages of information exchange on strategies and technologies to halt environmental degradation became obvious. At the same time, the only international organizations of that time, such as the World Bank and the United Nations organizations, had a single goal of promoting economic growth along the traditional path of industrialization in the developing countries, while occasionally articulating environmental sentiments. There was no single international focal organization to promote cooperative environmental action among nations. It was in this context that the United Nations organized the Conference on Human Environment in Stockholm, Sweden, in June 1972.

United Nations Environment Programme (UNEP), 1973

The Stockholm Conference was the culmination of a considerable process of discussion. There were three main thrusts behind the Stockholm agenda:[1]

1   The recognition that the developing nations faced massive environmental problems – especially linked to poverty – that needed to be overcome by development, while the developed countries had problems because their development had followed the wrong course.
2   The growing scientific understanding of the inter-relatedness of natural systems.
3   Growing public concern about the cumulative impacts of human activities on global environments.

In addition, there was the increasing recognition that neither individual nations, nor the North or the South acting alone, could adequately protect the global environment.

The institutional and financial arrangements set out in the Stockholm Conference report provided the basis for the establishment of the United Nations Environment Programme (UNEP) by the United Nations General Assembly. As established in 1973, UNEP had four principal components:

1   *The Governing Council:* composed of 58 member governments elected on a rotating basis for four years. The Governing Council initially met annually, but later decided to meet every two years, with additional special sessions every six years and others if necessary. It reports through the Economic and Social Council (ECOSOC) of the United Nations to the United Nations General Assembly.
2   *The Environment Secretariat:* headed by the Executive Director, who is elected by the United Nations General Assembly on the nomination of the United Nations Secretary-General, and headquartered in Nairobi, Kenya.
3   *The Environment Coordination Board (ECB):* established to ensure cooperation among all United Nations bodies having a mandate for environmental programmes. The Board was to function 'under the auspices and within the framework' of the United Nations Administrative Committee on Coordination, and under the chair of UNEP's Executive Director, reporting annually to the UNEP Governing Council.
4   *The Environment Fund:* envisioned to be a US$100 million, five-year fund.

UNEP was not to be an executing agency and did not bear the prime responsibility in the United Nations system for executing environmental projects. Project implementation was the prerogative of other agencies such as the United Nations Development Programme (UNDP) and the World Bank. However, UNEP was given the prime role in monitoring and assessment of the quality of the environment and in alerting the world to any environmental danger signals.

UNEP and the ozone layer

The issue of depletion of the stratospheric ozone layer, which surrounds the Earth and protects it from excessive radiation from the sun, was already being avidly debated in scientific circles by the time of the Stockholm Conference.

Paul Crutzen of The Netherlands theorized in 1970 that nitrogen oxides released, for example, from the application of fertilizers to soils, could lower the steady-state amount of ozone in the stratosphere. Halstead Harrison and Harold Johnston of the USA showed in 1970 and 1971 that exhausts from the proposed supersonic aircraft could deplete stratospheric ozone. Scientists showed that even a small reduction of the ozone layer could result in increased ultraviolet-B radiation reaching the Earth and consequently increasing skin cancer (see Chapter 1).

The issue of depletion of the ozone layer was covered in the preparations for the United Nations Stockholm Conference on Human Environment in 1972. The 'Pollutants' paper of the conference called for research on how human activities influenced the stratospheric transport and distribution of ozone. UNEP had been concerned with the protection of the ozone layer since its establishment in 1973. In June 1973, the address of Executive Director Maurice Strong to the first meeting of the Governing Council cited damage to the ozone layer as a possible 'outer limit' which humanity would be wise to respect. Pollution that breached that limit, he warned, may endanger the continuance of human life on the planet.

Reaction to the Molina–Rowland hypothesis, 1974

The thesis of chemists F Sherwood Rowland and Mario Molina from the University of California, Irvine, published in the journal *Nature* in 1974, linking chlorofluorocarbons (CFCs) to ozone depletion (see Chapter 1), created a stir. The significant media coverage of the press conference of Molina and Rowland at the meeting of the American Chemical Society in 1974 resulted in headlines in the US media such as 'Death to Ozone', and 'Aerosol Spray Cans May Hold Doomsday Threat'. US environmentalists were galvanized. Many consumer groups demanded a ban on the use of CFCs in aerosols, a 'frivolous use'. The Natural Resources Defense Council (NRDC), a US environmental NGO, petitioned the US Consumer Product Safety Commission to ban the use of CFCs in aerosols. In December 1974, the US House of Representatives held a hearing on a Bill to amend the Clean Air Act and on a proposal to request the National Academy of Sciences to study CFCs and to authorize the US Environmental Protection Agency (EPA) to ban CFCs if they were found harmful.

The USA was the biggest consumer and producer of CFCs at that time, accounting for about 35 per cent of the world's total. The producers in the USA were the leading chemical companies of Allied Chemical, DuPont, Kaiser, Penwalt, Racon and Union Carbide. DuPont subsidiaries also produced these chemicals in Argentina, Brazil, Canada, The Netherlands and Mexico. Allied Chemical subsidiaries produced CFCs in Canada and Mexico. Other major producers in the world included Akzo Chemie (The Netherlands); Australian Flourine Chemical and Pacific Chemical Industry (Australia); Daikin Kogyo, Mitsui, and Showa Denko (Japan); Hoechst (Germany, Spain and Brazil); Imperial Chemical Industries (UK); Kali Chemie (Germany); Montedison (Italy); and Rhone-Poulenc and Ugine Kuhlman (France and Spain).

The chemical industry in the USA was understandably defensive. The manufacturers of CFCs and the industries using CFCs started a vigorous defence of the industry. They pointed out that hundreds of thousands of jobs with a payroll of billions of dollars were dependent on the industry making and using CFCs. They termed the CFC–ozone-depletion theory highly speculative. They argued that, in any case, the USA alone could not tackle the problem without other nations cooperating as well. The industry felt that unilateral regulatory actions by the USA on CFCs would only impose a competitive disadvantage on the country without making a dent on the problem. However, the governments of Canada and the Nordic countries of Norway, Denmark, Sweden and Finland were considered to be ready for legal action. UNEP, as the environmental organization of the United Nations, was the obvious institution to promote cooperation on the issue. The UNEP Secretariat and the countries placed the ozone issue before the UNEP Governing Council for action.

In April 1975, the UNEP Governing Council backed a programme proposed by the Executive Director on the risks to the ozone layer as a part of the Outer Limits Programme.[2] The UNEP Governing Council, at its meeting in April 1976,[3] 'recognizing the potential impact that stratospheric pollution and a reduction of the ozone layer may have on mankind', requested the Executive Director 'to convene a meeting of appropriate international governmental and non-governmental organizations: to review all aspects of the ozone layer, identify related ongoing activities and future plans, and agree on a division of labour and a coordinating mechanism for: inter alia, the compilation of research activities and future plans; and the collection of related industrial and commercial information'.

THE WORLD PLAN OF ACTION, 1977

In accordance with this mandate, UNEP convened the first international meeting of experts on this issue in March 1977 in Washington, DC. Attending the meeting were experts from 32 countries, and from the United Nations headquarters, Food and Agriculture Organization (FAO), International Civil Aviation Organization (ICAO), United Nations Educational, Scientific and Cultural Organization (UNESCO), World Meteorological Organization (WMO), World Health Organization (WHO), European Economic Community (EEC), and two NGOs: the International Chamber of Commerce, representing industry, and the Standing Committee on the Problems of Environment (SCOPE) of the International Council of Scientific Unions (ICSU).

UNEP, WMO, WHO and ICSU made presentations to this meeting on the current knowledge regarding the ozone layer and the likely impacts of its depletion. Australia, Canada, Belgium, Germany, France, Italy, The Netherlands, the UK and the USA presented the status of their research on the ozone issue. A paper[4] presented by the UK discussed the implications of a possible ban on the use of CFMs (chlorofluoromethanes, a subset of CFCs). In a prophetic summary of all the debates on this issue for the next ten years, it concluded that:

'Any regulation to control CFM use would adversely affect industries producing and using CFMs. Regulatory action should not take the form of a complete moratorium on production and distribution of CFMs. Rather that, if and when regulations are introduced, selective banning of CFMs in non-essential uses (particularly aerosols, other than medical and pharmaceutical applications) should be used. Refrigeration and air conditioning systems should be required to be leak-proof, and the CFMs contained in them reclaimed (either through regulation or by using financial incentives) upon becoming obsolete. Use of CFMs as foam-blowing agents and solvents should be subject either to stringent control or, in some cases, a ban.

'The unilateral or multilateral action of certain major CFM-producing countries to control their production and distribution...would as a long-term strategy result in only a marginal impact on the problem of ozone depletion. The problem is, above all, global in nature, requiring global agreement and action.

'In the longer-term interests of stratospheric protection, it would seem highly desirable for an international regulatory body to be formed to monitor and control the whole gamut of man's activities of consequence for the stratosphere. This need is founded not only on the potentially severe effects of a reduction in stratospheric ozone, but also on the fact that longer-term protection of the stratosphere does not begin and end with the problem of CFMs.

'At the present time, the National Academy of Sciences [of the United States] Committee on the Impacts of Stratospheric Change has suggested that no regulations should be considered until a further two years' scientific investigation has been completed. On the other hand, it can be argued that the problems of irreversibility, associated with the potential depletion of stratospheric ozone by anthropogenic (human-made) CFMs, suggest a need for more immediate action, and indeed the history of depletion of other common property resources (e.g. fisheries) is replete with delays resulting in damage and loss, whilst more conclusive scientific evidence was obtained. In the final analysis, the decision on timing of regulatory action must be based on an implicit weighing-up of the potential consequences of delaying action compared with the more severe impact on industries which would occur if immediate controls were to be instituted.'

The meeting reached an agreement to initiate a World Plan of Action on the Ozone Layer.[5] The meeting did not wish to discuss actions to control emissions into the atmosphere of ozone-depleting substances. However, it considered that it would be desirable to investigate all possible modalities for such control, including the socio-economic impacts of their application. (The complete World Plan of Action may be seen in Appendix 2 below.)

The basic components of the Action Plan were numerous studies co-ordinated by international organizations:

* *Coordinated atmospheric research by WMO:* Monitoring ozone and solar radiation, measurements of selected chemicals, chemical reactions,

development of computational modelling, large-scale atmospheric transport and global constituent budgets.

*   *Studies on the impact of changes in the ozone layer on humanity and biosphere:* Monitor UV-B radiation (WMO/WHO, FAO), develop UV-B instrumentation (WMO/UNEP, WHO, FAO) and promote UV-B research (WMO, WHO, FAO).
*   *Impacts on human health (WHO):* Compile statistics on skin cancer in relation to various factors such as latitude, skin type, genetic and ethnic background; occupation and lifestyle; research on induction mechanism, skin aging and eye damage by UV-B, and possible effects of UV-B on DNA.
*   *Other biological effects (FAO):* Study responses to UV-B of selected plants, animal species, micro-organisms and aquatic plants and animals, emphasizing those related to primary productivity in the oceans; develop improved measurements of UV-B penetration into aquatic environment and study its effects on plankton.
*   *Effects on climate:* Develop computational modelling (WMO), and regional climate effects (FAO).
*   *Research on socio-economic aspects:* Production and emission data for potential ozone-depleting substances (UNEP, ICC, Organisation for Economic Co-operation and Development (OECD), International Civil Aviation Organization (ICAO); evaluate socio-economic alternatives, aircraft emissions, nitrogen fertilizers and other potential modifiers of the stratosphere (United Nations Department of Economic and Social Affairs, OECD).
*   *Institutional arrangements:* Implementation of the Action Plan: United Nations bodies; specialized agencies; international, national, intergovernmental and non-governmental organizations; and scientific institutions.

UNEP was to exercise a broad coordinating and catalytic role and establish a Coordinating Committee on the Ozone Layer composed of representatives of organizations participating in implementing the Action Plan and of countries that had major scientific programmes contributing to the Action Plan.

Much of the work included in the Action Plan was at the national level and was the financial responsibility of individual countries. Necessary coordination and advice was to be provided by the specialized agencies as indicated in the recommendations. In addition, the Executive Director of UNEP could, from time to time, convene a multidisciplinary panel of experts to provide broadly based scientific advice on the Action Plan.

In April, two weeks after the meeting of experts, the USA conducted an ad hoc intergovernmental meeting of officials from 13 countries, including all major CFC producers and UNEP, WMO, WHO, OECD and the European Commission to discuss regulations on CFCs. USA and Canada presented their proposed aerosol bans. While there was general sentiment for reduction of CFCs in non-essential aerosol bans, there was no binding agreement on any measure.[6]

COORDINATING COMMITTEE ON THE OZONE LAYER (CCOL) AND THE OZONE LAYER BULLETINS

As a result of the recommendations of the World Plan of Action, UNEP established[7] a Coordinating Committee on the Ozone Layer (CCOL) composed of representatives of all the organizations actively participating in the Action Plan and of representatives of countries that had major scientific programmes contributing to it. Initially, the members of the CCOL were 13 industrialized countries, 3 developing countries, 5 United Nations and international organizations,[8] the European Economic Community, OECD, Chemical Manufacturers Association (CMA) and International Council of Scientific Unions. Later, any interested country was allowed to participate.

The committee met first in 1977 and yearly thereafter until 1986. The information gathered by this committee was issued as UNEP *Ozone Layer Bulletins*. From 1978 to 1985 these bulletins recorded the conclusions of the world community on the science and environmental impacts of ozone depletion as they evolved, and provided the basic input to the diplomatic negotiations that were initiated in 1982. At each of its sessions, the committee examined research results on the ozone layer, its depletion and impacts of depletion and socio-economic aspects of the issue, and presented its reports to the Governing Council meetings of UNEP and the negotiating groups.

National research

Following the first session of the CCOL in Geneva in November 1977, the first *Ozone Layer Bulletin* of January 1978 reviewed the progress of various organizations and countries on the subjects allocated to them under the World Plan of Action on the Ozone Layer. That progress included findings on monitoring ozone and solar radiation, measurement of stratospheric substances including ozone, reactions of importance in the stratosphere, the one-dimensional as well as two-dimensional computer models which they were using, atmospheric transport of chemicals, their network for measurement of ultraviolet radiation (UV) by different methods, assessment of the health impacts of ozone depletion (particularly regarding skin cancer and eye cataracts), impact on plants and aquatic ecosystems, and effect of ozone on the heat and circulation regime of the atmosphere. The OECD prepared a report on its assessment of the economic impact of restrictions on CFC-11 and CFC-12 for use on aerosol propellants.

The first meeting recommended standardization and uniform calibration in the total ozone network, improvement of global coverage of ozone measurements, more extensive monitoring of solar radiation, and better monitoring of substances, particularly hydrogen, chlorine and nitrogen components of the ozone system. More study of chemical reactions, development of multi-dimensional models, and better data on distribution of skin cancer with regard to population factors such as types, lifestyles, migrations, ozone variability, and intensity of UV-B radiation were recommended. More studies on other biological effects, impacts on climate change, as well as social and economic aspects, were also recommended.

Global predictions: Ozone depletion at 15 per cent

The second *Ozone Layer Bulletin* of July 1978 reported on continuing research. The USA mentioned predictions by various modelling groups of scientists, which covered a range of ozone depletion from 10.8 per cent to 16.5 per cent from continued CFC releases at the 1975 release rate. Estimates of stratospheric ozone depletion by projected fleets of stratospheric aircraft showed less than a few tenths of a per cent of ozone depletion. US research also revealed that the systemic effect of UV radiation could be to reduce human immunity to diseases. Ozone levels for corneal damage to the eyes were established using rabbit eyes. High radiation exposure levels resulted in irreversible damage to the cornea. Through more research, data were collected on various physiological responses to increased UV-B radiation, including photosynthesis and germination. Broadband UV-B studies were conducted in a greenhouse chamber on more than 20 species of crops, including cotton, peanuts, wheat, rice, alfalfa, cucumbers and peas. The USA conducted research on various insects and found that lightly pigmented insects were much more sensitive to exposure to UV-B radiation than heavily pigmented ones such as the housefly. The Chemical Manufacturers Association arranged for CFC producers to submit production data on CFC-11, CFC-12 and hydrochlorofluorocarbons (HCFCs). The data showed that, for example, CFC-11 production, which started at very small levels in 1939, increased to 255,000 tonnes in 1976.

The *Ozone Layer Bulletin* noted the efforts already taken by Canada, the European Community, Sweden and the USA to restrict the use of the CFCs.

The second session of the CCOL was held in Bonn in November 1978. The UNEP report to this meeting noted the predicted ozone layer depletion from CFCs had climbed into the 15 per cent range, while the danger from supersonic aircraft was probably not a factor. Simulations of the effects on the ozone layer of nitrous oxide from increased fertilizer usage were considerably reduced from 1975 estimates, and such small reductions were not serious enough for concern. WMO presented the results of its symposium on the stratosphere held in Toronto in June 1978. It concluded that the long-term steady-state effect of CFC emissions at the 1977 rate would be ozone depletion of about 15 per cent, compared to the 10 per cent estimate of three years earlier, with an uncertainty range of 4 to 30 per cent. New findings from The Netherlands showed that a decline of ozone by 1 per cent would ultimately lead to a 4 per cent increase in skin cancer incidence.

The analysis of the global Dobson monitoring network did not produce any evidence of ozone depletion from CFCs by the end of 1976. Natural variations appeared to dominate the data. The Chemical Manufacturers Association was sponsoring an atmospheric lifetime experiment to determine the lifetimes of CFC-11 and CFC-12. It had been reported recently that another chlorinated compound, 1,1,1-trichloroethane (methyl chloroform), could be ozone-depleting. Carbon tetrachloride, an ozone-depleting chemical itself, was used as feedstock for producing CFCs. Its other uses were being phased out because of its toxicity.

The key questions, which formed the basis of the discussions on the protection of the ozone layer issue from then on, were:

*   Is ozone depletion actually observed and what is its extent?
*   What are the theoretical models predicting future changes in the ozone layer? How valid are the models and the assumptions of the models?
*   What should be assumed about the growth of CFCs in the future, if not controlled?
*   Are substitutes available and at what cost?
*   What scientific evidence is enough to trigger action to control the consumption of ozone-depleting chemicals?

HARMONIZING NATIONAL POLICIES, 1979–1981

The second political conference to consider harmonizing national policies for regulating production and use of CFCs was held at the invitation of the Government of the Federal Republic of Germany in Munich in December 1978 and attended by 13 industrialized countries,[9] Yugoslavia, the Commission of the European Communities (CEC), OECD and UNEP. The Soviet Union attended towards the end of the Conference. A report on the second session of the UNEP CCOL, including the 'Assessment of Ozone Depletion and its Impacts: December 1978', was presented to the conference.

The main conclusions of the conference were:

*   Studies of the technical and economic aspects of the chlorofluorocarbon question should be continued at national or regional levels, including a thorough evaluation of possible substitutes for both safety and environmental concerns. The studies should include cost-benefit analyses.
*   All governments should accept an obligation to provide annual production figures for CFC-11 and CFC-12 and their major uses to UNEP in order to obtain an assessment of total global emissions. The submission of production figures should respect commercial confidence.
*   As a precautionary measure, there should be a global reduction in the release of CFCs, and, therefore, all governments, industry and other bodies should work towards the goal of achieving a significant reduction in the release of CFCs in the next few years from 1975 levels.

This last recommendation, which dealt with the main purpose of the conference, was reached after considerable debate and negotiations. The positions of the governments concerning the need for regulations were wide apart. The USA had already issued regulations banning the manufacture of CFCs for non-essential aerosol uses, which became effective 15 December 1978; this included such items as hair sprays, deodorants and cosmetics, which accounted for over 80 per cent of all US CFC aerosol products. Exceptions were made for essential uses of CFCs such as some drugs, pesticides, lubricants and cleaners' products necessary for safe aircraft operations. Canada had achieved a 50 per cent decrease in aerosol production by voluntary agreement with industry, and had announced its intention to issue regulations in 1979. Sweden banned the manufacture and import of aerosol products containing aerosol propellants, effective 1 July 1979. The Netherlands issued regulations that required a warning label on all aerosol products containing fluorocarbons sold after 1 April 1979. The regulation further provided that a ban would be imposed on non-essential aerosol use if similar action were taken by the main producing nations. Germany was pursuing a cooperative approach with industry directed at reducing use of fluorocarbons in aerosols by 30 per cent by 1979. Norway recognized the gravity of the risks despite uncertainties, and called for international action. However, the UK, France and Italy, which produced CFCs, had no plans for regulation. They considered that the scientific evidence was not yet convincing.[10]

The third session of the CCOL took place in Paris in November 1979. The meeting noted that world production of CFC-11 and CFC-12, as reported by the CMA, fell by 17 per cent between 1974 and 1978, from 851,000 tonnes to 709,000 tonnes, largely due to the regulations by some governments. Ozone depletion was predicted at 15 per cent, at a steady rate, given the releases at 1977 levels.

A final ad hoc intergovernmental meeting to urge controls on European countries was held in Oslo, Norway, in April 1980, without the participation of the UK and France. The USA and Canada urged the Europeans to take some action but met with no response. There were no agreements at the end.

In 1980, the UNEP Governing Council discussed the reports of the CCOL[11] and appealed to governments, especially those whose consumption of CFC-11 and CFC-12 was high, to reduce their consumption and also not to increase the production capacity for these CFCs. The Council of the European Economic Community legislated in March 1980[12] a 30 per cent cutback in CFC aerosol use from 1976 levels.

At the fourth session of the CCOL in Bilthoven, The Netherlands, in November 1980, the predicted rate of ozone depletion was reduced to 10 per cent from the previous 15 per cent, due to new data on some chemical reaction rates, principally the reaction between hydroxyl radical and nitric acid. According to the models, 1 per cent depletion should have occurred already, but such an amount could not be detected with the technology then available. While the use of CFCs in aerosols declined, other uses such as in refrigeration increased. However, the total use of CFCs declined overall.

THE GOVERNING COUNCIL SETS UP A NEGOTIATING GROUP, 1981

The UNEP Governing Council discussed the issue of ozone depletion at its meeting of 1981.[13] It requested the Executive Director to initiate work to negotiate a Global Framework Convention for the Protection of the Ozone Layer and establish a working group for this purpose. The working group was to take into account the work of the meeting of an ad hoc group of legal experts in Montevideo, Uruguay, in October–November 1981. This meeting in Montevideo was held by UNEP to establish a framework and methods for development of environmental law for the enforcement of the political initiatives that followed the Stockholm conference in 1972. It selected the protection of the ozone layer as one of the three major areas on which legally binding guidelines should be developed.

The other major areas were land-based sources of marine pollution and movement of hazardous wastes. Eight other areas were also identified for action: environmental emergencies, coastal zone management, soil conservation, transboundary air pollution, international trade in potentially harmful chemicals, pollution of inland waterways, environmental impact assessment, and legal and administrative mechanisms to address pollution damage. All these issues were grouped under the title of 'Montevideo Programme'. Many of the initiatives resulting from this programme led to international and regional agreements, guidelines or principles.[14]

At the Montevideo meeting, Finland, supported by Sweden, presented a first draft of a global framework convention for the protection of the ozone layer, the elements of which were adopted as part of the future work. The elements of strategy arrived at by this meeting were: promotion of dissemination of information and public awareness; continuation, on the basis of available scientific data, of the work already initiated to arrive at the convention which would cover monitoring, scientific research, and development of technology, as well as development of policies and strategies; establishment of appropriate international machinery for the implementation and development of the convention; and development and implementation of national laws to implement the convention.

Predicted ozone depletion falls to 5–10 per cent

The fifth session of the CCOL took place in Copenhagen in October 1981, to provide an input to the UNEP ad hoc working group to arrive at a global convention for the protection of the ozone layer. Considering only CFC-11 and CFC-12, the CCOL reduced estimates of eventual ozone depletion to between 5 per cent and 10 per cent, compared to the last figure of 10 per cent. World production of CFC-11 and CFC-12, as estimated by the Chemical Manufacturers Association, fell by 18 per cent between 1974 and 1980. Most of the decrease occurred between 1974 and 1977; the decrease in 1980 was only 1 per cent. Uses in aerosols had declined, but other uses, such as foamed plastics and air conditioning, had shown an increase of CFC use. It was recognized in this meeting that, eventually, reductions in CFCs used in aerosols could be offset by growth in non-aerosol uses. The report confirmed previous findings on the link between ozone depletion and increased UV-B, and between increased UV-B and adverse effects on human health. The meeting recommended many steps to be followed by the researchers and governments to improve the predictions of ozone depletion.

AD HOC WORKING GROUP OF LEGAL AND TECHNICAL EXPERTS, 1982

The first meeting of the Ad Hoc Working Group of Legal and Technical Experts for the Elaboration of a Global Framework for the Protection of the Ozone Layer (the Working Group) was convened by UNEP in Stockholm in January 1982.[15] The CCOL presented its input to this meeting. The Environment Committee of the OECD submitted a report on several hypothetical emission scenarios and predictions of ozone depletion over time. The first three scenarios assumed a 7 per cent growth in CFC emissions, excepting carbon tetrachloride, until the year 2000, and then continued annual growth of 7 per cent, no further growth, and a 7 per cent annual decrease. In both of the first two scenarios, ozone depletion continued indefinitely with no ceiling. In the third scenario, a peak of about 6 per cent depletion occurred in 2020 and reduced gradually to 2 per cent by the end of the 21st century. The next three scenarios were based on a 3 per cent growth per year until the year 2000, and then a continued annual growth of 3 per cent; a freeze; and a 3 per cent annual decrease. Scenario 4, assuming a continued growth at 3 per cent, resulted in never-ending limitless depletion. The maximum depletion of Scenario 5, a freeze, was about 8 per cent. In the last scenario, ozone depletion rapidly decreased by the end of the century to 2 per cent. All the scenarios assumed no controls until the year 2000. The report discussed the difficulties in quantifying the effects due to uncertainties in emission rates, reaction rate constants, and the different theoretical models adopted by different scientists. The report also mentioned the hypothesis that ozone depletion would expedite climate change.

Can action wait for scientific certainty?

The report noted the argument that no action needed to be taken until scientists were more certain and the theory was proven. The main objection noted to this argument was that there was considerable natural variability in the total amount of ozone above a measuring instrument on the Earth's surface. The seasonal variation at 40° North latitude between March and October, for example, was about 13 per cent. Daily values at a station in Florida for a 650-day period in 1973–1975 varied by as much as 10 per cent between successive days and by 20 per cent for the main period. Thus, even if small stratospheric ozone depletion were detected, it would be impossible to attribute this depletion to a specific cause, to CFCs, or to human activities rather than natural causes. In any case, the statistical treatment did not show such small changes. The data available for detection of changes in vertical distribution of ozone appeared to be confused by particles from volcanic eruptions and showed disagreement between types of measurement. The global total amount of ozone did not change much until 1981, despite a significant and steady increase of up to 20 per cent of ozone in the lower atmosphere. Other contributions to variations in ozone included solar activity, the cooling effect in the upper atmosphere, increased carbon dioxide in the atmosphere, nitrogen oxides from fertilizers, changes in land use, and planned supersonic aircraft at high altitudes.

The meeting noted that consumption of CFCs was not confined predominantly to the USA and the European Economic Community. Other countries had increased consumption by 36 per cent from 1976 to 1979, offsetting much of the decrease in the use of aerosols in the USA and European Economic Community. Many countries produced and distributed CFCs, including Argentina, Australia, Belgium, Brazil, Canada, China, Czechoslovakia, the Federal Republic of Germany, France, the German Democratic Republic, Greece and India. Hence, the meeting noted that action to protect the ozone layer must not be limited to the USA and to the European countries. Not only were the uses of CFC-11 and CFC-12 increasing, but uses of other ozone-depleting chemicals such as CFC-113 and CFC-114, methyl chloroform, carbon tetrachloride and HCFC-22, were also increasing.

The complexities in replacing these chemicals had been noted by the meeting. The constraints in the substitution of other chemicals for CFC-11 and CFC-12 were higher cost and certain physical properties, such as toxicity, flammability and vapour pressure, which made them less desirable for specific purposes. The substitutes in aerosols for example, the hydrocarbons, would have an increased risk of fire or explosions. Substitute products such as roll-ons would require public acceptance. Dimethyl ether would be acceptable as a substitute, but was flammable and had not been cleared for toxic effects. CFC emissions from disposal of refrigeration systems accounted for more than half of all emissions. Recovery of these CFCs would be possible, although more expensive than the manufacture of new refrigerants. It would also be difficult to enforce recycling. Alternatives, such as methylene chloride for blowing flexible foams, would pose their own health problems and required further toxicological studies. There was considerable apprehension that control of CFCs would have adverse effects such as loss of employment and export earnings for countries with CFC industries, shifting employment within a country, greater cost of the product to users, the probability that the substitute will be less suitable, and an additional cost of monitoring and enforcement.

A menu of regulatory options

The reports of the US National Academy of Sciences and the OECD Environment Committee were presented to the meeting and contained comprehensive discussions of options for control of emissions. The broad regulatory options considered were:

- Bans on CFCs in specific applications.
- Standards for allowable emissions from specific applications.
- Quotas or ceilings on CFC production or purchase for specific uses. These quotas would be implemented by marketable or transferable permits.
- Taxes on production and use of CFCs to increase costs and reduce demand.
- Deposit refund systems to create user incentive to recycle.
- Subsidies to promote the creation of recycling and recovery systems.
- Education and labelling to promote public awareness.

It was recognized that it might be easier to reduce CFC use in the USA and Western Europe because alternative technologies and public support were available, but additional action might become increasingly difficult. Unilateral action was unlikely because of the competitive disadvantage suffered by a country that reduced CFC usage by adopting more expensive and less suitable substitutes and technologies.

FIRST DRAFT CONVENTION AND DISCUSSIONS, 1982

A first draft international convention for the protection of the stratospheric ozone layer, prepared on the basis of the deliberations in Montevideo in 1981, was officially submitted by Finland and Sweden to this meeting. This draft covered all the parameters ultimately covered by the Vienna Convention in 1985. It prescribed:

- An obligation for each country to prevent activities that can harm the ozone layer, and to cooperate and exchange information.
- A scientific and technological committee to promote research and review information, and an agreement among the contracting parties to promote the technology transfer and knowledge.
- A meeting of the parties every two years after entry into force of the convention, and a secretariat to be provided by UNEP with specific duties.
- Participation of all non-party states and the United Nations organizations as observers; any body or agency technically qualified in the protection of the ozone layer could participate in the meetings unless at least one-third of those present objected.
- The settlement of disputes through a third contracting party mediation or through arbitration.
- The amendment of the convention and the annexes, which detailed areas of research and information exchange, and adoption of new annexes through normal procedure at a meeting where it was adopted by a two-thirds majority of the contracting parties present and voting. A simplified procedure in which a contracting party could suggest an amendment by simplified procedure, would be circulated to all the governments; if no one objected, it was treated as adopted, and all parties would ratify it. The amendments for a new annex would enter into force for the contracting parties, which had accepted it 60 days after two-thirds of the parties had deposited an instrument of acceptance.
- Any state which became a party to the convention after the entry into force of amendments or new annexes being considered a party to the present convention as amended, including the new annexes, unless it objected specifically and agreed to be a party only to the unamended convention.
- The Executive Director of UNEP as the depository to the convention.
- Withdrawal of a party from the convention after five years from the date on which the convention had come into force with respect to that contracting party.

The meeting ended without any conclusions, since it was evident that scientific and technical elements needed to be integrated into this mainly legal framework.

Discussions continue on the convention, 1982–1983

> *'[The Governments] go on a strange paradox, decided only to be undecided, resolved to be irresolute, adamant for drift, solid for fluidity.'*
> Winston Churchill, 1874–1965 (speech to the
> House of Commons, 12 November 1936)

The UNEP Governing Council, at its 1982 meeting, requested the Working Group to continue its work.[16] The first part of the second session of the Working Group, held in Geneva in December 1982, continued the discussion on the draft convention.[17] The discussion brought out several important points relevant for many other global environmental issues since then, including:

• Should the convention be drafted in such general terms that any state finds it possible to ratify it?
• Should it be a rapid movement towards concrete first steps, or only provide guidelines for a world response?
• Should the group wait for more scientific evidence, or proceed on the basis of existing evidence?
• Should there be simultaneous negotiations on a protocol for specific action and a convention, or should a protocol wait for additional time?

Some delegations opposed any reference to CFCs as a potential cause of ozone depletion. The definition of the ozone layer itself was a matter of contention and, specifically, whether the ozone at or within a few kilometres of the ground level should be excluded. On the issue of obligations, some felt that specific obligations should be imposed to protect the ozone layer since the risks were already apparent. Others felt that they would not be able to fulfil obligations, which might or might not be required. A number of developing country delegates said that their obligation should be defined in light of the practical means at their disposal and in accordance with their capacities. The issues of technology transfer as a part of scientific and technological cooperation were raised. Some delegations mentioned that since much information and technology was in the private domain, unqualified transfer of such technologies might not be possible. Only a few delegations supported amendments by simplified procedure, suggested by Finland and Sweden. Meanwhile, in November 1982[18] the European Community decreed that the production capacity of CFC-11 and CFC-12 should not increase beyond the 1980 level, and contended that this action was adequate to protect the ozone layer.

Predicted ozone depletion drops to 3–5 per cent

The sixth session of the CCOL took place in Geneva in April 1983, immediately prior to the next session of the working group. At this CCOL meeting, the estimates of ozone depletion were further reduced to between 3 and 5 per cent, compared to the 5–10 per cent of the previous report. The change from the 1981 figures was due to new data on certain chemical reaction rates. The report mentioned that the range of predicted ozone reduction did not adequately represent the true uncertainty. The new changes in the concentrations of carbon oxides, nitrous oxides, ammonia and halocarbons might all have affected ozone levels. No evidence of changes in total ozone had been observed so far. The relatively large natural variability of ozone created difficulties. The reduction in production of CFC-11 and CFC-12 had halted; between 1980 and 1981, there had been a production increase of 2 per cent. The report concluded that the increase in ozone concentrations in the lower stratosphere might contribute significantly to the greenhouse effect. The submission of the Chemical Manufacturers Association to the meeting mentioned that its research showed that if CFC production remained constant, the total column ozone would decrease by 1.5 per cent in the year 2050; with CFC growth of 3 per cent per year, it would decrease by 1.8 per cent. In two other scenarios, in which other trace gases such as nitrous oxides, carbon dioxide and ammonia increased, ozone would not be depleted but, in fact, would increase.

FIRST SPECIFIC PROPOSAL TO CONTROL CFCs, 1983

The second part of the second session of the working group was held in Geneva in April 1983.[19] This meeting considered the draft from the previous meeting and a new annex to the convention from Finland, Norway and Sweden (the Nordic proposal) to end the use of CFC-11 and CFC-12 in aerosol cans, except for essential uses, with each party deciding its own target dates and essential uses. It also provided that parties should promote the development and application of the best practicable technologies to reduce emissions from non-aerosol uses. The parties were asked to assist each other and encourage the developing countries to participate in those actions. Each contracting party was to provide data regarding its production of CFCs, use of CFCs for production of aerosols, and uses it considered essential. This proposal for specific action, suggested as an annex to the convention of general principles under discussion, later matured into the Montreal Protocol.

There were many divergent views on the proposals for the convention. The result of these discussions was a large number of words as alternatives to general obligations, reflecting the different opinions of the countries. In accordance with the general tradition in international negotiations, such alternative words were placed within square brackets until consensus was achieved. Article 2b, for example, read that the parties should 'cooperate in taking appropriate legislative or administrative measures and in [harmonizing] [endeavouring to harmonize] policies to [control] [limit, reduce [and] [or] prevent] [human activities under their jurisdiction [or control]] [should it be found that these activities] [that] [have] [release of substances which cause] [or are likely to have] [significant] adverse effects [by reason of their] [resulting from] [modification of the ozone layer]'. In addition, the whole paragraph was placed in square brackets, indicating divergent opinions on whether it was at all appropriate to incorporate any paragraph on cooperation on policy measures in the convention.

There was consensus, in principle, on incorporation of two annexes to the convention: Annex 1, regarding research and systematic observations, and Annex 2, regarding the substances on which the information should be submitted, but not on the need for an additional annex on measures to control CFCs. Annex 1 encompassed a large number of substances such as carbon, nitrogen and hydrogen substances, in addition to CFCs. Regarding obligations, the developing countries expressed the view that they should be allowed to use the best practical means at their disposal to fulfil the obligations of research and information exchange, and in accordance with their capabilities. Japan doubted whether any specific regulatory action should be mentioned at all, in view of the lack of scientific evidence.

FURTHER NEGOTIATIONS, 1983–1985

In 1983, the Governing Council requested[20] the Executive Director to assist the working group further. The third session of the group took place in Geneva in October 1983,[21] and considered the comments of the countries on the Nordic proposal to control CFCs.

- Australia considered that it should not be an annex to the convention, but a separate protocol. Belgium considered that each country defining its own essential uses would cause distortions in trade between various countries.
- Canada considered that each country should be given discretion on how it would regulate its consumption and objected to giving each country's production data.
- Denmark supported the proposal.
- Japan considered that the facts of change in the ozone layer, identification of substances causing such change, and the mechanism of the destruction of the ozone layer, had not yet been scientifically established, and that there was no need to put any legal obligation on countries to regulate CFCs.
- The UK considered the Nordic proposal as unnecessary and unsound, based on the fact that the CCOL had further reduced its estimate of the depletion of the ozone layer. Further, it opposed a protocol and held that any eventual protocol should go no further than the position of the European Economic Community, which had agreed to a temporary reduction of production, subject to review; it felt there was no need for further action at present.
- The USA strongly supported the proposal but as a separate protocol of specific commitments rather than as an annex to the convention, which was only a statement of general principles. It suggested that provisions should be added on the timings of controls, reporting requirements and technical assistance for the implementation of the controls with respect to aerosols. It also suggested that reductions in emissions from other CFC uses could be allowed to compensate for any shortfalls in reductions in aerosol use.

The USA, however, opposed the proposal that regulation of non-aerosol uses should start.

The Nordic countries introduced a revised proposal, drafted as a protocol instead of an annex, which took into account the comments received from a large number of countries since the previous meeting of the Open-Ended Working Group. The revised draft was generally welcomed, but some found it premature to consider any agreement on controls of CFCs.

Broad consensus on convention but none on specific action

The next session of the meeting, the second part of the third session, was held in Vienna in January 1984.[22] At the end of this meeting, a revised draft of the convention emerged. Remarkably, despite many differences of opinion among countries on the causes and extent of ozone depletion, the convention itself was almost completely agreed upon with differences remaining on a few aspects. The convention, however, did not contain any concrete obligations to control any ozone depleting substances. The text finally agreed to on control policies (Article 2b) read: 'Cooperate in taking appropriate legislative or administrative measures and in harmonizing appropriate policies to control, limit, reduce or prevent human activities under their jurisdiction or control should it be found that these activities have or are likely to have adverse effects resulting from modification or likely modification of the ozone layer.'

One contentious issue of the meeting was the eligibility of the European Economic Community (through the Commission of the European Economic Community) to ratify a convention. Some countries, including the USA, insisted that it could do so only if a majority of states of the European Economic Community were signatories to this convention. The European Economic Community replied that since the competencies of the communities and the member states were complementary, there was no need for such a condition. Another difference was about the settlement of disputes. One alternative suggested a vague formulation, while another alternative suggested specifically negotiation, mediation, arbitration, or use of the International Court of Justice. A third difference was about whether reservations could be expressed to the convention or to any protocol thereto. The technical annexes regarding research, systematic observations and information exchange were wholly agreed upon, except that the specific mention of CFCs and HCFCs in the list of chemicals that were currently thought to have a potential to modify the ozone layer was still contentious.

There were many differing opinions on the second draft of the protocol. Three fairly unstructured alternatives emerged for controlling aerosols. The first suggested that parties take appropriate measures to phase out the use of CFCs in aerosol products within a time to be decided. Each party could decide, for itself, the essential uses that would be exempt, but should report every two years on the uses it considered essential. The second alternative suggested that at least a reduction of 30 per cent be achieved by a certain time, compared to its maximum level. The third alternative suggested that parties should, within two years, take appropriate measures to eliminate the use of CFCs in non-essential aerosol products, and that each party submit to a meeting of parties the list of uses it considered essential. Regarding non-aerosol uses, it was suggested that the parties should promote the development and application of the best practicable technologies to reduce emissions from non-aerosol uses.

A new scientific assessment, October 1984

The Governing Council discussed the convention issue again in May 1984 and requested[23] the Executive Director to convene another meeting of the working group to complete the work on convention and to work on a possible draft protocol concerning the control of CFCs. The first part of the fourth session of the working group was held in Geneva in October 1984.[24] The meeting started with a presentation by the chair of the technical working group, Robert Watson of US National Aeronautics and Space Administration (NASA), on the executive summary of the assessment of the ozone layer modification and its impacts, which had been prepared by the CCOL. Watson explained that the atmospheric concentrations of several source gases such as nitrous oxide, methane, nitrogen oxides, carbon dioxide and chlorine species, were increasing simultaneously. Their effects would be interactive. Some of them would increase the total column of ozone, whereas others would reduce it.

The total column of ozone from 1970 to 1980 showed no trend, but there was a statistically meaningful decrease in ozone between 32km and 40km during that period. The consistent feature of all the predictions was that the substantial vertical redistribution of ozone would involve a decrease above 30km, and an increase in the lower stratosphere and troposphere. This redistribution could have consequences for the tropospheric climate. Predictions of changes in the total column ozone had varied substantially during the last two years for two reasons: changes in the understanding of the chemistry, and increased recognition that the atmospheric concentration of gases was currently changing and partially offset the effects on the ozone layer. The calculated changes in ozone column content were highly dependent on the assumed scenarios for the future atmospheric concentrations of carbon dioxide, methane, nitrous oxides, chlorine oxides and other chemicals. The conclusion of the report was that:

> *'A giant experiment was being performed on the atmosphere. Humanity was perturbing the carbon, nitrogen, and chlorine cycles on a global scale and in an unprecedented manner; the consequences of this experiment for the future could not be known with any certainty.'*

It was broadly agreed at the meeting that the convention and protocol would be financed by contributions from the parties based on their United Nations scale of assessment.

Emergence of the 'Toronto Group' and further proposals for controls on CFCs

With regard to the proposed protocol to the convention, two documents were placed before the working group. In addition to the second revised draft protocol, which emerged from the previous session with all the proposals in brackets, another draft was submitted by Canada, Finland, Norway, Sweden and the USA (the Toronto Group). The new draft emerging from the Toronto Group proposal and the discussions in the working group contained four options for control of the use of CFCs to suit all countries in different stages of such control:

1   Eighty per cent reduction of total annual CFC use, including in production for exports, in six years, in three stages.
2   Ban on all domestic use of non-essential aerosols within four years and on export of non-essential aerosols in six years.
3   Seventy per cent reduction within six years in three stages, with a freeze in production capacity.
4   Twenty per cent reduction within four years.

The new draft also included cooperation to develop best practicable technologies to limit emissions from the foam, refrigeration and solvent sectors, periodic review of control measures, reporting on use of CFCs in aerosols, and listing of essential uses on national laws. It added that parties should conduct research on substitutes for CFCs in all sectors and give technical assistance to developing countries.

Opposition by European Economic Community and Japan to controls

The alternatives proposed in these two drafts of the Protocol (ie, the second revised draft and the new Toronto Group proposal) showed a clear difference of views. During discussions, there was general agreement that alternatives needed to be developed and the countries should pursue research. There was also a general agreement that there should be a review of control measures in the future. There was disagreement on the need to report the amounts of CFCs used for aerosols and other purposes, and on which uses could be considered as essential. The European Economic Community and its members argued that since the alternatives to aerosols were cheaper than CFCs, market forces alone might dictate a reduction in CFC use, rendering a protocol unnecessary. They considered the proposal to reduce the use in aerosols by 80 per cent in six years very impractical. Some even argued that there were many uncertainties about the link between CFCs and ozone depletion, and that no one knew whether CFCs could be singled out for individual regulatory action.

The European Economic Community argued that the new draft would simply lead to increase of consumption of CFCs due to increased CFC use in other sectors, whereas a ceiling on production capacity would also limit consumption automatically in all sectors. It countered with a proposal for a reduction of 30 per cent in the use of CFC-11 and CFC-12 within two years, a freeze on production capacity, relaxation for developing countries to the extent necessary, and cooperation to reduce emissions in non-aerosols.

The USA and other members of the Toronto Group countered the European Economic Community proposals for a draft protocol on many grounds. They pointed out that the effect of chlorine on ozone depletion might not be linear. Beyond a certain limit, the decline in ozone might be precipitous and might not be reversible in the short term through human action. Even a 25 per cent depletion in ozone would cause five million cases of skin cancer, 10,000 deaths, and a loss in agricultural and fisheries productivity.

If the world waited for 100 per cent certainty, the ozone layer might suffer irreversible damage. While cooperation in research as envisaged in the convention was welcome, a protocol to reduce the emissions in the short term was also needed. Many countries had already succeeded in reducing emissions from aerosols, one of the least essential uses. Their proposals also exempted controls on essential aerosols. Concerns about the flammability of CFC substitutes had also proved unwarranted. All countries could accept these proposals, the Toronto Group suggested. The European proposal of a production cap would mean a great increase in production because production capacity was under-utilized at that time. Capping present production capacity and freezing consumption was prejudicial to developing countries and many others. The Toronto Group proposals would buy time, they argued, and lead to prudent action in the short term.

The European Economic Community replied that its proposal of a production cap was a complete answer to the problem, while the new draft from the Toronto Group was a short-term solution. While production could go up under the European proposal, it was capped by the installed capacity, which could not be increased. It would not be as catastrophic as it could be under the Toronto Group proposal, the Commission argued, since the Toronto Group proposal could lead to production without any cap. Japan opposed any specific measure until further scientific information was received.

The meeting also heard from the European Chemical Manufacturers Federation, which presented the results of a model created by several scientists. Their conclusions were:

- Even with significant growth in CFCs, the threat to the ozone layer was a very distant one.
- If, at a future time, action became necessary due to sustained increase in emissions, a cap on production capacity would effectively deal with the issue, while an aerosol ban would achieve little.
- An aerosol ban alone would not be enough if CFC consumption grew in other sectors and caused more ozone depletion.

The last meeting of the working group, now called an intergovernmental group, took place in Vienna in January 1985 as a prelude to the conference of the plenipotentiaries, ie of diplomatic agents of governments having full powers to negotiate, in contrast to the meetings of experts (the working groups) with no authority to commit their governments in any way.[25] At this meeting, a few differences remained concerning the convention. The Toronto Group reiterated a multi-option approach to the draft protocol with different stages of emissions reductions of CFCs from aerosols. The European Economic Community repeated its earlier arguments that the Toronto Group's proposals would be short-term.

THE VIENNA CONVENTION FOR THE PROTECTION OF THE OZONE LAYER, 1985

The conference of the plenipotentiaries for a Convention for the Protection of the Ozone Layer took place in Vienna on 18–22 March 1985. Thirty-four countries attended the meeting, including ten developing countries. Attending, as observers, were China, industrial organizations – including two from Europe – and the International Chamber of Commerce. The United Nations Industrial Development Organization (UNIDO), WMO, the European Economic Community and OECD were also present. No environmental NGO was present.

This meeting had before it the fifth revised draft convention and the technical annexes,[26] which were provisionally agreed upon at the fourth session of the Ad Hoc Working Group of Legal and Technical Experts, which met in Geneva in October 1984 and in January 1985. Agreement had been reached on the three issues outstanding from the third session: the European Economic Community could become a party provided that at least one of its member states was a party; a text allowing various options for settlement of disputes was provisionally agreed upon; no reservations to the convention or its protocol would be allowed except when expressly stated in the convention or the protocol. However, several delegations expressed reservations against the text on settlement of disputes, and reserved their right to reopen discussion on this issue. The European Economic Community, but none of its member states, also voiced its reservation against the requirement that it could become a party only if at least one of its member states was a party. The meeting was also considering the report of the ad hoc working group on a draft protocol on chlorofluorocarbons.

The meeting elected Winfried Lang of Austria as its President, and delegates from Brazil, Egypt, the Soviet Union and The Netherlands were elected to the bureau. The USA and Sweden made proposals regarding the settlement of disputes more or less in line with the consensus of the fifth draft. The main outstanding issue was still whether the International Court of Justice would be used as a last resort for resolving disputes between parties unless a state-party explicitly had declared that it did not accept this solution ('opting out', a text which the USA could not accept). The USA had annexed to its proposal a detailed arbitration procedure. A negotiating group discussed institutional and financial arrangements, and noted the cost of a secretariat for the convention for the first two years. Both the Executive Director of UNEP and the Secretary-General of the WMO expressed their willingness to host the secretariat. The

Executive Director of UNEP also expressed his willingness to contribute towards the costs during the initial two or three years, while the WMO offered to serve as the permanent secretariat if the parties met all the costs. The meeting decided that UNEP would discharge the functions of the secretariat in the interim until the first meeting of the Conference of the Parties to the Vienna Convention, which would designate the secretariat from among those international organizations willing to carry out the secretariat functions. The two remaining issues, which were more related to general political issues than to the protection of the ozone layer, were solved only during the final diplomatic conference: the settlement of disputes in line with the USA position, and the question of membership of the European Economic Community as proposed in the Geneva agreement.

What came to be known as the Vienna Convention, which emerged after negotiations, had 21 articles and two annexes.

• Articles 1–5 dealt with matters such as: definitions; general obligations; research and systematic observations; cooperation in the legal, scientific and technical fields; and transmission of information.
• Articles 6–10 concerned institutional arrangements such as conference of parties, secretariat, adoption of the protocols under the Convention, amendments of the Convention or a protocol under the Convention, and adoption and amendment of annexes.
• Articles 11–15 dealt with legal aspects such as settlements of disputes, signature, ratification, acceptance, approval or accession, and the right to vote.
• Articles 16–21 defined the relationship between the Convention and a protocol under the Convention, entry into force, reservations, withdrawal, depository, and authentic texts being in all the six United Nations languages.
• Annex 1 concerned research and systematic observations specifying areas such as: physics and chemistry of the atmosphere; health, biological and photodegradation effects; and effects on climate. The systematic observations dealt with many sources of importance to ozone layer depletion such as carbon substances, nitrogen substances, chlorine substances including CFCs, bromine substances and hydrogen substances.
• Annex 2 dealt with exchange of scientific, technical, social, economic and commercial information between countries.

The obligations of the parties to the Convention were to cooperate in research, observations and information exchange, and to adopt policies to control human activities that might modify the ozone layer. The only mention of CFCs came in Annex 1 as one of the many substances 'thought to have the potential to modify the chemical and physical properties of the ozone layer'.

Article 8 provided for adoption of protocols to fulfil the obligations of the Convention. Despite many attempts to take at least a first step to control CFCs, the governments failed to agree on anything concrete. To keep the process going, the USA on behalf of the Toronto Group proposed a specific Resolution on a Protocol Concerning Chlorofluorocarbons. One paragraph of the preamble to this resolution dealt with the possibility that worldwide emissions and use of CFCs could significantly deplete and otherwise modify the ozone layer, leading to adverse effects. It recognized the need to assess further the modifications of the ozone layer and their potentially adverse effects. Another paragraph recognized that precautionary measures for control of CFCs had already been taken at national and regional levels, but acknowledged that such measures might not be sufficient for protecting the ozone layer. A further paragraph gave special consideration to developing countries and the relationship between the level of industrialization of the country and its responsibilities for the protection of the ozone layer, the so-called 'common but differentiated responsibility'.

UNEP required to continue work on the protocol

The resolution requested the Executive Director of UNEP to convene a working group to continue work on a protocol on the basis of the work so far conducted. The protocol should address both short- and long-term strategies to control global production, emissions and use of CFCs, and take into account the particular situation of the developing countries, as well as updated scientific and economic assessments. Another paragraph urged all interested parties to cooperate in studies on the CFC issue and to sponsor a workshop on the subject, under the patronage of UNEP; another appealed for financial support for the process. The resolution authorized the Executive Director of UNEP to convene a diplomatic conference, if possible in 1987, for the purpose of adopting such a protocol. Finally, it proposed that all states and regional economic integration organizations should endeavour to reduce their use of CFCs, especially in aerosols, to the maximum extent possible, and, in particular, the industrialized countries should reduce aerosol use by at least 50 per cent.

The resolution triggered considerable debate. In particular, the final paragraph was strongly resisted, especially by the EEC countries. Japan said that a decision on whether to continue to work on a protocol should await the results of the work of the next CCOL meeting. The resolution was eventually adopted after replacing the emphasis on use of CFCs in aerosols in the final paragraph with a more general wording on 'control of emissions, inter alia in aerosols, by any means at their disposal, including on production or use', deleting the text on 50 per cent reduction of aerosol use in industrialized countries, and replacing 'maximum extent possible' with 'maximum extent practicable'.

The Governing Council arranges for preparatory meetings for the protocol

The UNEP Governing Council discussed the Vienna Convention in May 1985. The Council's decision[27] closely followed the text of resolution. It requested the Executive Director to convene a working group to continue the work on a protocol under the Convention to control CFCs and authorized him to convene a diplomatic conference for the purpose of adopting such a protocol, if possible in 1987. A steering group consisting of representatives from UNEP, the UK, the European Economic Community, the USA, Norway, Brazil, Egypt, India, the Soviet Union and Japan was set up to prepare the workshop in two parts. All interested parties, including industry, were invited to submit contributions on the selected subjects. It was made clear that the workshops should be used not for negotiations, but for exchange of experience and discussions.

The Convention and its resolution on CFCs, while very weak in its provisions and with no regulatory controls on CFCs, had provided a momentum to the control process. The Convention and the resolution called for negotiating a protocol to protect stratospheric ozone, and were signed by more than 20 countries, including the USA, the Soviet Union and the European countries. UNEP scheduled the next negotiations on a protocol only at the end of 1986; scientific assessments and workshops, which clarified the controversies on science and the technology available for replacing CFCs, took up the entire period before the next negotiations.

FIRST COMPREHENSIVE SCIENTIFIC ASSESSMENT, 1985

'Predictability: Does the flap of a butterfly's wings in Brazil set off a tornado in Texas?'

Edward N Lorenz (title of paper given to the
American Association for the Advancement of Science,
Washington, 29 December 1979)

The first comprehensive scientific assessment of the status of the ozone layer by UNEP and WMO, assisted by the Commission of the European Communities, the Ministry of Environment of Germany, US NASA, US Federal Aviation Administration (FAA) and US National Oceanic and Atmospheric Administration (NOAA), was released at the end of 1985 (see Chapter 1). The report concluded that continued releases of CFC-11 and CFC-12 at the 1980 rate would reduce the ozone vertical column by between 5 and 9 per cent. The models with all scenarios predicted that such a release of CFCs would reduce local ozone depletion at 40 kilometres by 40 per cent or more. The report was in its final stages of preparation when the British Antarctic Survey reported its unexpected observation of the Antarctic 'ozone hole' from Halley Bay. The Survey deduced from its observations springtime losses of ozone that were much larger than any existing theory could account for. While, initially, atmospheric scientists viewed the results with shock and surprise, the results were subsequently reconfirmed by the satellite observations of US scientists. The 'ozone hole' created a sensation among scientists and laypeople alike, and brought the attention of all to the problem of ozone depletion (see Chapter 1).

The first part of the eighth session of the CCOL took place in Nairobi on 24–28 February 1986. The report by the USA showed that the release of CFC-11 and CFC-12 at 1980 rates in an otherwise unchanged atmosphere would result in ozone column decreases in the range of 4.9 to 7 per cent. With a 3 per cent growth rate of CFCs, the depletion would be 10 per cent after 70 years. The sponsored research by the Chemical Manufacturers Association found that there was no depletion trend at all in past ozone measurements.

BOX 2.1 THE 'OZONE HOLE': THE STRIKING NAME

Gordon McBean, Canada

'Lastly, let me comment from my own personal experience. One of the realities is that in order to catch political and public attention, we have to simplify problems. In the middle 1970s, I was chairman of the Environment Canada Scientific Committee in the Long Range Transport of Atmospheric Pollutants. Now, that attracted absolutely zero attention from the public and not much more from the Minister of the time. But when it was decided to call it acid rain, the interest of people shot up. It was the same issue, but we got a fancy title. I am not sure if someone thought consciously when they referred to the ozone hole, but the reality is you get attention when you get a dramatic phrase like that.'

Source: cited in P G Le Prestre, J D Reid, E T Morehouse, Jr (eds) (1998) *Protecting the Ozone Layer: Lessons, Models and Prospects*, Kluwer Academic Publishers, Boston.

ECONOMIC AND ENVIRONMENTAL WORKSHOPS, 1986

Rome workshop: Differences on scenarios for the future

The first part of the workshop under the Resolution on a Protocol on CFCs was held in Rome in May 1986. It was organized by the Government of Italy, the European Community and UNEP. Twenty-three countries, as well as the European Community, the OECD and a number of international industrial organizations[28] participated in the meeting. As background information, the meeting received: reports of CFC-11 and CFC-12 production, sales and release data until 1984 from the Chemical Manufacturers Association; production data from the European Community, Brazil, France, Sweden, Norway, Switzerland and the USA; and reports and discussion papers from 13 countries, including some developing countries. The subjects discussed included current and projected production, use and emissions of CFCs, current regulations and the costs and effects of such regulations, potential alternative technical options and estimates of production, use and emissions of other substances than the CFCs that could modify the ozone layer. The statistics showed that in 1984, the production of CFC-11 and CFC-12 increased by 7 per cent from 1983 and was almost the same figure as reported in 1977. In effect, the CFC reduction realized by the USA and other countries by banning the use of the CFCs in non-essential aerosols had been wiped out by the increase by other countries and in other sectors.

During the discussions, industry disputed that the 7 per cent increase noted in production and consumption could be taken as the basis for future projections. The meeting discussed whether three annual growth rates for CFC-11 and CFC-12 at 1.2 per cent, 2.5 per cent and 3.8 per cent could be adopted as scenarios, as suggested by a US consultant, but there was no consensus. Industry, in particular, said that no estimate would be valid beyond five years since alternatives would emerge, and developing countries may not use CFCs to the same extent as the developed countries. The ceiling on production capacity favoured by the European Economic Community supported this argument.

The meeting then discussed the technical alternatives to refrigeration, flexible and rigid foams and solvents, noting that HCFC-22 and HCFC-502 could be developed for many of the alternative uses. The environmental problems of alternatives such as methylene chloride were mentioned. Some questioned that the experience of the USA with aerosol alternatives would be applicable to other countries, and cited incidents in the hydrocarbon filling centres caused by the explosive nature of the hydrocarbons, and the resulting increased insurance rates. Participants concluded that more innovation with alternatives was needed, but conceded that the higher the price of CFCs, the more incentive there would be for innovation with alternatives.

The meeting also heard the report of the Rand Corporation of the USA (a consulting company) regarding the long-term emission profiles of five chemicals other than CFC-11 and CFC-12: CFC-113, methyl chloroform, carbon tetrachloride, halon-1211 and halon-1301. By 2075, those five chemicals could account for about 40 per cent of the total emissions contributing to potential ozone depletion. The major uses for these chemicals were as solvents and fire extinguishers. The US EPA presented a paper on the increased levels of methane, carbon dioxide and nitrous oxides, and their impact on the ozone layer and climate.

The workshop then decided that its next session would be held in Leesburg, Virginia, USA, in September 1986 to discuss possible strategies for further control of CFCs. The strategies to be considered included periodic assessment and review of control measures, increased reporting, annual limits on production/use, cumulative limits, bans on specific uses of CFCs, best practicable technologies, emission fees, and a combination of the different strategies. These alternatives should be assessed on many criteria, including: the effect of each strategy on production, use and emissions of CFCs; effects on the ozone layer; economic costs of alternative strategies; equity or fairness to both producers and non-producers; flexibility about making easy and cheap changes if they were warranted; and administrative convenience.

UNEP/US EPA workshop, Washington, DC: Scenarios of environmental impacts of ozone depletion

UNEP and the US EPA held a separate workshop in June 1986 in Washington on the health and environmental effects of ozone modification and climate change. More than 300 policy-makers and scientists from 90 countries attended the meeting. The presentations explained the potential changes to the atmosphere that would alter the global environment. Ozone depletion could result in increased skin cancers and cataracts, suppression of the human immune system, degradation of materials, increases in local and regional air pollution and damage to crops and aquatic resources. Climate change could raise sea level, affect agricultural productivity and harm human health. After this meeting, the US EPA Administrator, Lee Thomas, wrote to the Ministers of Environment of all countries in attendance on the importance of tackling the problem of ozone depletion and declared that the protection of the stratospheric ozone layer was high on his personal agenda for the coming year.

Leesburg workshop: Clarity in concepts

The second part of the workshop on the control of CFCs was held in Leesburg on 8–12 September 1986.[29] Welcoming the participants to the workshop, Ambassador Richard Benedick, US Deputy Assistant Secretary of State, said that such an international meeting would have been inconceivable a few years ago, when there was no inkling of the global experiment being inadvertently practised on the atmosphere and how it might affect humanity. He asserted that it was not enough for some countries to take the problems seriously if others did not.

Twenty-three countries, many industry organizations, the European Economic Community, OECD, UNEP, environmental organizations such as the Natural Resources Defense Council, and a number of newspapers based in Washington attended the meeting. Several countries presented 29 papers on all aspects of control strategy alternatives, their equity and their impacts. A great variety of opinions and interesting points were presented at the meeting:

- There were six primary factors for defining a control strategy: chemical coverage; global stringency; timing; method of setting requirements; allocation; and trading. The evaluation of different strategies was based on global participation, effects of substitution, and greenhouse gas growth, according to M F Gibbs, a consultant with the ICF Corporation of the USA.
- 84 per cent of the CFC-11 emitted in 1987 would still be in the atmosphere by the year 2000, and 56 per cent by the year 2030. Stabilization of the concentrations at current levels in the atmosphere would require an 85 per cent reduction in emissions. Immediate action was needed, according to John Hoffman of the US EPA.
- Only a ceiling on production of CFC-11 and CFC-12 would contribute to the protection of the ozone layer. It would prevent runaway consumption and oblige humanity to realize the potential danger it was facing in a practical way. The production levels for different states or groups of states should be controlled through ground-level monitoring of CFC concentrations. However, differentiation should be made between developed and developing countries so that only the former would be affected and would generate the technical research for substitute products or processes, according to Mr Dupuy of the Atochem Company in France.
- A global 80 per cent cut in production over five years and a full phase-out over ten years would be sufficient time for the development and deployment of alternatives (six years later, in 1992, this suggestion to phase out CFCs by 1996 was approved by the Meeting of the Parties to the Montreal Protocol). Only planning on the basis of a complete phase-out would intensify the search for alternatives. If such a search were not instituted immediately, there would be a great disruption, forcing the need to cut CFCs sharply. Economic advantage would come to those who were first in this technology race, according to Alan Miller, David Wirth and David Doniger of the Natural Resources Defense Council.
- A simulation model presented by Irving M Mintzer of the World Resources Institute estimated the chlorine build-up in the year 2075, assuming different

growth rates of production from 1 per cent. The 1 per cent growth rate would lead to 14 parts per billion volume (ppbv) of chlorine, compared to the 2ppbv of the pre-1970s when the damage to the ozone layer was not yet seen. A production limit set at the 1984 rate would lead to a chlorine level of 7ppbv. A 95 per cent phase-down would commit the atmosphere to 7ppbv. Controls on specific uses such as aerosols, as advocated by some countries, did not reduce the risk of chlorine build-up if the market could shift demand from prohibited to allowed uses. Limits on production of CFC-11 and CFC-12 could reduce the risk of chlorine build-up if the limit were set substantially below the level of current capacity. Neither the proposal of the European Economic Community for a production limit set at the level of current capacity in place nor the Toronto Group proposal of global aerosol bans was adequate to reduce the commitment to chlorine build-up substantially below the level of 14ppbv into 2075. The ozone column would be depleted by 6 to 9 per cent in 2100 if present CFC emissions were unchanged, and by more than 30 per cent if CFC emissions increased by 3 per cent per year without limitation.

• The industry fluorocarbon programme panel presented a paper with somewhat different conclusions. Limiting CFC emissions to present levels would not be expected to lead to significant changes in ozone column, it said. Limiting CFC emissions to twice the 1984 levels would be expected to result in only a small change in total column ozone of 3.5 per cent, although vertical distribution would be modified.

• In the absence of CFCs, total ozone appeared to increase by 5 per cent from 1940 to 2100, due to the effect of methane. If CFC emissions increased by 3 per cent per year, ozone column would be depleted by more than 30 per cent in 2100, according to G Brasseur and A De Rudder of the European Community.

• Ivar Isaksen of Norway presented a two-dimensional model of temperature feedback in the stratosphere, concluding that there would be a marked depletion in ozone in higher latitudes. All the models noted that ozone was very sensitive to methane in the atmosphere.

• A cap on production capacity as suggested by the European Economic Community could ultimately lead to rising prices, at which time the least expensive reduction would occur. But there was little immediate incentive at the time for reducing emissions or for research and development of chemical and product substitutes, according to Stephen O Andersen of the US EPA.

• If a stringent cap were put on the production capacity of CFC-11 and CFC-12, it would lead to hasty adoption of the presently available and costly alternatives and limit the large opportunity for cost reduction by developing and introducing new technologies. There would be a negative impact on national economies if controls were adopted, according to K Kurasowa and K Imozeki, industry representatives from Japan. They added that Japanese small aerosol manufacturers would suffer if CFCs were banned in aerosols; controls should be based on per capita use, they argued.

- It would be preferable to act through the market with measures such as taxes, marketable permits and caps on production capacity. Rising CFC prices would give producers and users an incentive to search for substitutes; CFCs would be used only in applications where it was necessary to do so. When market measures were used, consumers would be the decision-makers, according to R Valiani, an economist from Italy.
- A cap on global production capacity as a regulatory approach would be in accordance with the concept of equity. This should provide a specific clause to meet cases that might arise when member states were not able to cover their needs for CFC-11 and CFC-12, by using their existing production capacity, by importing, or by substituting for CFCs. In these cases, member states must be able to increase their production capacities to the necessary extent. There should be an obligation to control imports of CFCs from countries that did not adopt a production capacity cap, according to L Gundling of Germany.
- Where CFCs were necessary for higher standards of living by developing countries, those countries should be allowed to have access to CFCs in the immediate future, according to Stephen O Andersen of the US EPA.
- Ingrid Kökeritz of the Government of Sweden argued that in order to control emissions, it would not be enough to control production; CFC use by countries should also be controlled. She proposed a definition of net use as 'domestic production of CFCs, plus imports of CFCs, plus imports of CFCs contained in products, minus exports of CFCs, minus exports of CFCs contained in products'. On a suggestion from the meeting, the word 'net use' was changed to 'consumption'. This definition proved invaluable when the Montreal Protocol was drafted.
- Another paper suggested that non-joiners in a protocol would be affected in their trade in four ways: product leadership, market leadership, consumer response and overall innovation, according to Stephen O Andersen of the US EPA.
- The European Community approach of a cap on production capacity would lead to use by the highest value users and would simplify monitoring and controls. Use bans would be less efficient because they involved higher transition cost, did not stimulate research for substitutes in other usage fields, and were more complicated to administer because of the multiplicity of consumers. Emission fees were economically efficient in theory, but the price elasticity would be so low that the fees ought to be punitively high to be effective and would be difficult to implement on a global basis, according to David W Pearce of the UK.
- A global emission limit (GEL) could be established based on ozone-modifying substances weighted according to their ozone depletion potential, and a national emission limit (NEL) could be allocated on the basis of population and gross national product of each country. This global emission limit would be subjected to compulsory review every three to five years with a view to updating both its scope and relevance in the light of current scientific findings, according to G V Buxton, A Chisholm and S Carbonneau of the Government of Canada.

- The European Economic Community suggestion on a cap to production capacity implied that, to be equitable worldwide, all nations must have the same access to CFCs as its members, ie 1.77kg per capita. A freeze on current world capacity would not be justified because it would be unfair to developing countries and non-producers. The EEC cap would lead to a depletion of ozone by 7 per cent in 2050 and 14 per cent in 2075. For countries in northern latitudes, this would be even more. A ban on non-essential aerosols in addition to such a cap had the advantage of substantially reducing emissions in the near term, according to Stephen R Seidel of the US EPA.
- The highest priority must be given for developing chemical alternatives to CFC-11 and CFC-12, according to Boris Gidaspov of the Soviet Union.
- Incentives should be given for conversion to CFC-free products; the public must be told to accept slightly lesser quality to protect the environment, according to Herbert Eichinger of Austria.

This workshop presented options on all the essential elements of a protocol, such as the coverage of chemicals, timing, trade restrictions, concessions to developing countries, definition of consumption, and periodic review after scientific assessment. Most importantly, there had been a clarity and consensus on the state of science: the amount of chlorine in the atmosphere was increasing; ozone would be depleted if more chlorine were added to it; it was irrelevant in which chemical the chlorine originated and in which application; what was critical was the total burden of stratospheric chlorine. The different scientific models were seen to lead ultimately to the same results with slight differences.

NEGOTIATIONS ON THE PROTOCOL, 1986–1987

The second part of the eighth session of the CCOL was held in Bilthoven, The Netherlands, in November 1986. This meeting was the committee's last. Since there was an exhaustive assessment on the state of the ozone layer during 1986, the meeting concentrated on the environmental effects of ozone depletion. The report of the CCOL enumerated the many adverse effects of ozone depletion on human beings and the environment in general. The negotiations on a protocol on CFCs were resumed at the working group meeting in December 1986 in Geneva.[30]

Divergent proposals on CFC reductions

There were four proposals considered by this meeting of the working group.

European Community proposal

- Place production cap at 1986 production levels for CFC-11 and CFC-12, and a review at an unspecified time after the report of scientific experts.

• The developing countries, which did not produce the regulated substances, would be entitled to produce an amount that did not exceed their consumption in 1986.

Canadian proposal

• Place a global emission limit on ozone-modifying substances weighted according to their ozone depletion potential, initially, for instance, at 812 kilotons per annum, the 1986 production level; place a national emission limit, which was the global limit distributed to the countries, of which 25 per cent was based on their share of the world population, and 75 per cent based on their gross national product.
• Conduct a review of the production limit at least once every five years, after a scientific report on the status of the ozone layer.
• Provide technical assistance to the developing countries to facilitate implementation of the protocol.
• Place controls on CFC-11, -12, -113, -114; HCFC-22; methyl chloroform; halon-1301; halon-1211; and carbon tetrachloride.
• Aggregate consumption by multiplying the actual consumption of a substance with its ozone depletion potential.

US proposal

• Within a time limit to be specified later, reduce the consumption of fully halogenated CFCs by 95 per cent, starting with a freeze and reducing by 20 per cent and 50 per cent at intermediate levels.
• Define consumption as annual production plus imports minus exports minus quantities that have been destroyed or permanently encapsulated.
• Conduct a scientific assessment periodically, and adjust the control measures after the consumption freeze is in place, but before implementation of subsequent control measures.
• Ban exports of CFC technologies to non-parties, unless the non-parties are in full compliance with the control measures.

Soviet Union proposal

• Place controls only on CFC-11 and CFC-12 production; conduct research for alternatives; and freeze CFC production by the year 2000.

Response to the proposals

Norway, Sweden and Finland supported the US proposal, but proposed starting with a 25 per cent reduction not later than by 1991. Switzerland expressed its readiness to discuss the proposals of the USA and Canada, and said that all uses of CFCs should be covered instead of just one. Several countries, while agreeing in principle to the proposal of Canada, mentioned the practical difficulties in accurately estimating emissions. There was intense debate on the substances to be included, with some countries arguing that only fully halogenated CFCs should be included and others arguing that all ozone-modifying substances should be included in the protocol.

The representatives of 79 European and American non-governmental organizations urged the meeting to accept a total phase-out in ten years. At this stage, no delegation proposed a complete phase-out, though the USA went as far as 95 per cent. UNEP, based on the draft earlier protocol discussions, had prepared a draft protocol[31] on all the aspects except the control measures. There was a preliminary discussion on this draft protocol which did not reach any conclusion. Articles regarding the meetings of the parties and the establishment of the secretariat were generally discussed without any major divergence of opinions. An interesting sidelight of this meeting was the opposition of the Soviet Union to the Federal Republic of Germany including a representative of West Berlin in its delegation; the German representative and representatives of the Western occupying powers of Germany refuted this criticism.

Startling change in industry attitude: Controls welcomed

Industry attitudes had changed considerably by the time of the working group meeting in December 1986. Previously, producers and users of CFCs argued that further regulations were not called for until science proved the ozone depletion theory. In September 1986, the industry Alliance for Responsible CFC Policy changed its position and said that it would support 'a reasonable global limit on the future rate of growth of fully halogenated CFC production capacity'. In the same month, the leading CFC producer, DuPont, released its own statement calling for a limit on worldwide emissions of CFCs that would be negotiated under the auspices of UNEP. DuPont had about 25 per cent of the CFC world market and about US$300 million in total annual sales.

The Natural Resources Defense Council called DuPont's policy change the biggest breakthrough since the ban on the use of CFCs in aerosols, and ascribed it to the discovery of the Antarctic ozone hole. Others considered that assertions by scientists that large unrestrained growth in CFC usage might lead to future ozone depletion caused industry to fear that the growth in demand of CFCs, which had occurred in 1984, was bound to concern governments. Further, the Vienna Convention had served notice that the ozone depletion issue was being taken seriously. The negotiations under the auspices of UNEP on a protocol under the Vienna Convention further confirmed this feeling. US producers, from that point, added to the pressure for an international protocol, wanting to avoid a situation in which the US regulated CFCs domestically while the rest of the world did not. Kevin Fay of the Alliance for Responsible CFC Policy said, 'Until an international agreement is in place, US producers and users plan to develop programs to decrease emissions through voluntary action'. Chairman of the Alliance Richard Barnett suggested that another way to manage the potential problem was for producers to develop new chemical formulations and shift production to those new ozone-benign substances. According to DuPont, 'The development of alternatives is going to happen at a rate that corresponds to the amount of pressure that is applied. Right now there is no economic or regulatory incentive to look for other routes'.[32]

FOCUSING ON THE KEY QUESTIONS

The second session of the working group, in Vienna in February 1987,[33] was presided over by Winfred Lang, the Environment Minister of Austria, whose contribution to the adoption of the Vienna Convention had been significant. He constituted four working groups to deal with the issues of:

1 How a periodic review and assessment of scientific and technical issues could be organized, and what substances represented the greatest potential threat to the ozone layer.
2 Special needs of developing countries for regulatory measures.
3 Continuation of the dialogue on regulatory measures.
4 Trade issues.

Which substances and what to control?

After discussions in the subgroups, the chairman posed four questions to address the issues:

1 Which potential ozone-depleting substances should be subjected to regulatory measures? The meeting considered many options from a list that included CFC-11, -12, -113, -114 and -115; carbon tetrachloride; methyl chloroform; methylene chloride; and halon-1211 and -1301. Some suggested an immediate list of CFC-11 and CFC-12, and a further consideration of other substances on the list.
2 How could new substances be included? Canada suggested that schedules could be drawn up for chemicals with significant impacts already quantified, chemicals whose impacts should be quantified with a view to regulation, and chemicals subject to review to determine their ozone depletion potential.
3 How should production and emission levels be calculated? There was a broad consensus on the use of consumption levels, defined as production plus imports minus exports, rather than emissions.
4 What should be the base year for controls? There was near unanimity on 1986.

The next four issues discussed by the meeting

Developing countries
How could it be ensured that a protocol on CFCs would be fair to developing countries? The developing countries argued that:

• Their contribution to ozone modification was minimal; therefore, it was not important to stress controls on CFCs in developing countries.
• The alternative technologies and substitute chemicals should be given to them.
• Developing countries should be allowed additional use of the regulated chemicals consistent with their development plans.

• Adjustments to national emissions of developed countries should be made to maintain global emissions at an agreed level.

Control of trade

There was unanimity that imports from non-parties should be restricted, and movement of capital and facilities outside the protocol areas should be discouraged. The subgroup on trade issues during this meeting considered the compatibility of the measures for controlling trade between parties to the protocol, and trade between parties and non-parties, with the rules of the international trade, especially the General Agreement on Tariffs and Trade of 1947 (GATT). It concluded provisionally that, provided it was clearly demonstrated that the measures were not arbitrary or unjustifiable, any discrimination in the treatment between parties and non-parties would be permissible under the exceptions provided by Article XX, Paragraph b of GATT concerning protection of human, animal or plant life for health, and also possibly of Paragraph G concerning conservation of exhaustible natural resources. However, it was the opinion of several experts that discrimination would not arise at all. The trade restrictions regarding non-parties should not apply to non-parties that were able to demonstrate full compliance with the control measures of the protocol. Reference was also made to certain precedents, namely the Convention on International Trade in Endangered Species, and the London Dumping Convention.

Financing and administration of the protocol

All agreed that expenditures should be charged exclusively against contributions from the parties to the protocol.

Control measures

The subgroup on control measures reported that there was some flexibility of approach, even though there was no agreement as yet. The meeting ended with a sixth devised draft protocol on CFCs, with many of the provisions in brackets. The mood at the end of this meeting was optimistic; the chief US delegate, Richard Benedick, said he expected a global protocol by the end of 1987. He regretted that a number of other countries and important producers in Europe had not yet reached the stage of being able to put forth a coordinated progressive position. Robert Watson, chief scientist for NASA's Global Habitability Project, said that within the next year scientists would know a great deal more about the causes of the ozone depletion. He said a scientific expedition to Antarctica would be launched in the next two months; major experiments were then underway using high-altitude aircraft to study how chemicals at the Earth's surface reach the ozone layer. He predicted that the experiments would help scientists understand how Earth worked as a system.

Further proof of the cause of ozone depletion

Significant developments took place in the USA between February and April 1987. In a deposition before the US House of Representatives Science, Space and Technology Subcommittee in March, Susan Solomon, the leader of the 1986 National Ozone Expedition to the McMurdo Station in Antarctica, testified that the high amount of chlorine observed in the atmosphere over Antarctica was from chlorofluorocarbons; in her opinion, CFCs were causing the ozone depletion over Antarctica. F Sherwood Rowland testified before the committee that the occurrence in Antarctica had shown that the ozone layer was rather fragile and could suffer significant depletion in a very short time. Following this deposition, the Chairman of the Subcommittee on Health and Environment, Henry A Waxman, wrote to the EPA Administrator that he expected that the USA would re-evaluate its policies and instruct the negotiators to negotiate tougher controls. He sent a copy of this letter to many other countries of the OECD and to UNEP.[34]

The Environment Council of the Commission of the European Communities met in March 1987 and relaxed its position somewhat by agreeing to negotiate a maximum reduction of CFC production of 20 per cent. This reduction would be preceded by a CFC production freeze at 1986 levels and would be applicable for another four years. Scientific review could take place after four years after the protocol's entry and provide the basis for further measures. The Commission's mandate also allowed agreement on a ban of CFC imports from non-signatory countries to avoid creating 'CFC havens'.[35]

Mathematical modellers meet and agree

UNEP held a meeting in Wurzburg, Federal Republic of Germany, in April 1987, for all of the modellers to compare models generating assessments of ozone-layer change under various strategies for CFC control. This meeting came up with 12 scenarios assuming different years of beginning controls, assuming different rates of growth in halons and CFCs, and making assumptions regarding other trace gases (carbon dioxide, methane and nitrous oxides). The conclusions[36] were summarized as:

- The effects of ODSs other than CFC-11 and CFC-12 were calculated to enhance significantly the ozone depletion estimated to be caused by CFC-11 and CFC-12 alone.
- A total freeze on CFC-11 and CFC-12 would have a significant effect on calculated ozone changes, even though other halocarbons would be allowed to grow at 3 per cent a year.
- A freeze in the emissions of all fully halogenated compounds in the developed world would have a significant impact on calculated ozone changes. Future ozone levels depended strongly on the assumed level of current emissions from the developing countries.
- When a halocarbon growth rate of 3 per cent per year was assumed, the various scenarios broadly predicted ozone losses by the year 2050, ranging from 4 to 15 per cent.
- Each of the chemicals could be given an ozone depletion potential.

THE 'BREAKTHROUGH' SESSION, APRIL 1987

The third session of the negotiating group, held in Geneva on 27–30 April 1987, was addressed by Mostafa Tolba, the Executive Director of UNEP, who had been very actively working in the background to bring about a consensus.[37] He mentioned that after the Wurzburg workshop organized by UNEP, where different models of ozone depletion seemed to agree with each other, there was no excuse to delay action on the basis of scientific debate. He presented his own proposal on what UNEP wanted to achieve. He expected a protocol to be signed in summer or early autumn 1987 and enter into force in 1988. He further suggested that in 1990, two years after the entry into force of the protocol in 1987, there should be a complete freeze on CFC production and use at 1986 levels. Thereafter, there should be a 20 per cent cutback in production and use every two years, reducing it to zero by the year 2000. Every four years from the date of entry into force of the protocol, there should be a rigorous scientific review, with the first in 1992 and the second in 1996 in order to assess the need for further reduction. At the same time, the search for substitutes, research into recovery technologies, and into the physics and chemistry of the atmosphere should continue and be accelerated.

Representatives of 34 countries, OECD, the European Economic Community, industry associations, the Environmental Defense Fund, the Natural Resources Defense Council and the World Resources Institute attended the meeting. The US House of Representatives and Senate sent observers. The meeting again organized four working groups to discuss science, trade measures, control measures and developing countries. The sixth revised draft of the protocol was discussed. During the discussions in the plenary, the countries repeated their well-known positions. The USA and the Nordic countries argued for stronger CFC controls; the European Community advocated a CFC production cap and a reduction of up to 20 per cent; and Japan maintained a moderate position without specifying any particular position. The discussions in the subgroups, however, were productive, and gave rise to great optimism that a diplomatic conference could be held in September in Montreal to finalize the protocol. To reconcile the different approaches, Sweden suggested that both production and consumption (measured as production plus imports minus exports) should be controlled in parallel, a solution that was then further elaborated on by the subgroup.

Scientific Working Group illuminates the path

The Scientific Working Group had reached a consensus on four classes of ozone depleting substances: group A, fully halogenated chlorine compounds with ozone depletion potential (ODP) values near 1; group B, fully halogenated bromine compounds with ODP values greater than 1; group C, partially halogenated chlorine compounds with ODP values less than 1 that were in widespread commercial use in 1985; and group D, partially halogenated compounds not produced in commercial quantities in 1985, but which had potentially large applications in the future as substitutes for group A. The working group also concluded that it would not be enough for only CFC-11 and CFC-12 to be controlled, since the other equally ozone depleting substances would take their place. The group's further recommendations were:

- The ODP values of halon-1211 and -1301, CFC-114 and CFC-115 were not as well established as the values for other chemical compounds; the working group requested UNEP to improve the calculations of these ODP values.
- UNEP should quantify the global warming effect of CFCs.
- It would not be enough to consider only total column ozone changes in selecting the control strategy; changes in the vertical and latitudinal distribution of ozone should also be considered. Even small changes in column ozone could result in changes in the vertical distribution of ozone, which could modify the atmospheric temperature profile. Similarly, while calculated global average ozone depletion was a useful initial guide for policy considerations, ozone depletion would be much more at high latitudes, and smaller rates would occur close to the equator.
- The Scientific Working Group also recommended that the next major review be published in early 1990.

Representatives of the European CFC industries questioned the conclusions of the meeting about both the projected growth rates and the fraction of the current production assumed in developing countries.

UNEP Executive Director, Mostafa Tolba, leads the breakthrough

The subgroup on control measures was presided over by Mostafa Tolba. After discussions, he presented an informal draft of broad consensus on control measures, which was based on his informal consultations. He prepared a text on control measures without committing any party to these conclusions. This text included CFC-11, -12, -113, -114 and -115, with a freeze in two years, a 20 per cent cut in another four years, and, after another six to eight years, a possible further 30 per cent cut. Substances could be added or removed, and further reductions could be agreed either by a majority vote or a two-thirds majority of the parties. There would be a scientific review every four years.

The subgroup on trade issues reiterated its earlier views on the justification for trade measures against non-parties. A representative of the GATT Secretariat attended the meeting and said that such an article on control of trade would be in order in accordance with Article XX, Paragraph b of GATT concerning the protection of human, animal or plant life or health. He stated, however, that the judgement as to whether the action proposed satisfied Article XX rested with the GATT contracting parties, normally in the context of a complaint brought by one GATT party against another. He also mentioned that the greater the number of commodities controlled, the larger the chances would be of a challenge from some GATT members; he stressed that his view was based on practical rather than legal considerations. There was a consensus on banning imports and exports of controlled substances from non-parties, and determining the feasibility of restricting or banning import of products produced with the ozone depleting substances. There was also consensus that the parties should discourage exports of CFC technologies except for recycling. It was also agreed that these provisions would not apply to a non-party that was in full compliance with the control measures of the protocol.

The subgroup on developing countries agreed that there should be some exemption for developing countries for five or ten years based on a per capita consumption limit, and that the parties should assist developing countries in making expeditious use of environmentally safe alternative chemicals and technologies.

The meeting decided that the next meeting of the group would take place in Montreal on 8–11 September 1987, followed by the diplomatic conference on 14–16 September. It was also decided to have further negotiations on control measures under the chairmanship of Tolba in June 1987, and a meeting of the legal drafting group in July 1987, both in Brussels.

Wide welcome by the media, 1987

The results of the meeting in April 1987, based on Tolba's informal draft, rather prematurely created a stir among the news media of the world. The *New York Times* of 1 May 1987 headlined that United Nations parleys agreed to protect the ozone layer. A Reuters news story of 18 April quoted a clash of West German and Danish experts with their European Community colleagues over the pace of reducing chemical emissions damaging the world's protective layer; West Germany and Denmark wanted total elimination of CFCs within a few years, differing sharply from the 12-nation European Community proposal to cut CFC production by 20 per cent within six years of the introduction of the new protocol. On 29 April, the *Guardian* reported that Britain was being blamed for blocking an agreement on ozone since it opposed the USA and the Scandinavian countries, at the instigation of industry. It quoted the Department of Environment spokesperson as saying, 'Britain had no intentions of acceding to pressure from United States, Germany and Denmark to accept the complete phase-out of CFCs'. Reuters reported on 30 April that ozone experts agreed in principle on a formula to reduce ozone depleting chemicals.

Several European and US environmental groups said that the draft accord did not go far enough. David Wirth of the Natural Resources Defense Council said, 'Our feeling is that the timetable is too long. A strong signal should be given to industry to go and develop alternatives to the CFCs'.

A temporary hiccup in the USA

In the USA, the *Washington Post* reported in May 1987 that the Reagan Administration was reconsidering its stand on the ozone issue, prompted largely by Interior Secretary Donald P Hodel. The correspondent reported that opposition to the proposed agreement had been rising within the Administration, particularly among the Interior Department officials who believed that the accord would violate President Reagan's philosophy of minimal government regulations. The story reported that, in a meeting of the cabinet council on domestic policy, Hodel argued for an alternative programme of 'personal protection' against

BOX 2.2 THE IMPORTANCE OF INFORMAL CONSULTATIONS

Mostafa Tolba, Former Executive Director, UNEP

'This is probably the most important lesson learned. When government representatives sit around the formal negotiating table, they are not relaxed. They are either timid or aggressive. They sure want to score points. This goal does not change in the informal consultations, but representatives are more open since they are not formally committing their governments. Above all, they get to know one another better. They get to see vividly the genuine interest of others to reach common solutions. They gradually become friends working for a common cause even if it is from different angles. Informal consultations were a key ingredient in negotiating the treaties that followed Montreal – and the recipe proved successful in all cases.'

Source: cited in P G Le Prestre, J D Reid, E T Morehouse, Jr (eds) (1998) *Protecting the Ozone Layer: Lessons, Models and Prospects,* Kluwer Academic Publishers, Boston.

ultraviolet radiation, including wider use of hats, sunglasses and sunscreen lotions. It also reported that William R Graham, Jr, the White House science adviser, questioned the accord, and quoted him as saying that his concern stemmed from substantial uncertainties about the causes and the rate of ozone layer depletion. The correspondent reported that this opposition to the agreement stunned State Department and environmental protection officials.

The *New York Times* hailed the draft agreement in its 5 May 1987 story by Philip Shabecoff. It said, 'The ozone accord could signal the beginning of a new era in international cooperation in coming to grips with a wide variety of global concerns including the warming of earth's atmosphere, the destruction of tropical rain forests and the trading in toxic substances. If the Geneva agreement is signed, it would require sovereign nations to reduce the production of CFCs, a valuable economic commodity, in order to deal with the problem that science has identified as potentially very serious but which has yet to have any significant impact'. It noted that UNEP, led by Tolba, played a key role in achieving this agreement.

SEVENTH DRAFT PROTOCOL, 1987, AND
COUNTRY COMMENTS

In order to get over the technical objection to the UNEP Governing Council originally calling for control of only CFCs, the Council decided in June 1987 that the working group should consider the full range of potential ozone-depleting substances in determining what chemicals might be controlled under the protocol. A letter from a legal advisor to the Director General of GATT cautioned that a potential exemption from the ban on imports for non-parties who were fully compliant with the protocol should have no time limit.

Before the informal consultations in June 1987, UNEP requested each of the countries to comment on the draft that emerged from the April working group meeting. The following remarks were received:

- Canada: CFC-114 and CFC-115 must be included; those countries building new capacity in 1986 must not be rewarded; controls should also be on consumption; a 20 per cent cut-back should be achieved in four years; a further cut-back should be achieved in eight years; a two-thirds majority would be needed to block further cut-backs; a clear signal to industry to work on alternatives was necessary; a production freeze would come into force two years after the protocol; an import freeze would come into force at the same time as a production freeze; low-consuming countries should be defined on the basis of per capita consumption, such as 0.05kg or 0.1kg; further discussions on control of trade might delay ratification by countries; production and import controls would discourage less developed countries and others from becoming parties and provide unfair trade advantage to the current exporters; if consumption was defined as production plus imports minus exports, consumption controls would be equitable to everyone.
- Italy: Any cut beyond 20 per cent should not be automatic and should be taken only after evaluation of all aspects by an international body.
- Denmark: Regulate consumption but have regulations for production and import; include CFC-114 and -115 in regulations; have some restrictions on halons; and enact an automatic 30 per cent further reduction.
- Norway: Regulate all chemicals, including CFC-114 and CFC-115; a freeze step was unnecessary; after the first 20 per cent reduction, the further 30 per cent reduction could be semi-automatic; unless two-thirds of the parties decided otherwise, it would enter into force; production allowance to meet demand should be given to all countries, not only developing countries, since otherwise a world monopoly would be given to present producers.
- New Zealand: Inclusion of all the ozone depleting substances; a 50 per cent reduction should be allowed ten years instead of the six or eight suggested in the bracketed draft; further safeguards were necessary for ensuring supply of CFCs to non-producing or small countries; production could be transferred from one country to another if, together, they did not cause an increase in production; a freeze of halons within three years of the entry into force of the protocol.
- USA: The protocol should be called the Protocol on Ozone-Depleting Substances, rather than the Protocol on CFCs; it would be better to control consumption rather than production, since production control alone would simply transfer production to low-consuming countries; halon controls should not be postponed three years after the entry, but should take place earlier; instead of banning exports of CFCs to non-parties, the calculation of consumption in Article 3 should subtract exports to non-parties, enabling some exports to non-parties and providing an incentive for parties to export to parties rather than to non-parties, and for parties to encourage their trading partners to join; a total ban on exports would induce the non-parties to build their own production capacity; the low-consuming countries could

be supplied during the grace period from existing member capacity; the 60 per cent proposed by the Executive Director of UNEP, based on his informal consultations, as a percentage of global production required to trigger entry into force among ratifying countries, was not adequate and should be higher; the European Economic Community's proposal that a regional economic integration organization could be treated as a single unit for the purpose of their obligations was not acceptable; a new article should be added stipulating that the parties not in compliance should be treated as non-parties and therefore subject to trade restrictions; there should be a compliance committee to consider complaints submitted to the secretariat concerning non-compliance for a party, consisting of one-third of the parties, taking into account the geographical distribution; if the committee concluded that a party was not in compliance, its recommendation should go to the parties; a party determined by the meeting of the parties to be not in compliance should be treated as a non-party for a period of time at least equal in duration to a period of its non-compliance.

The sixth revised draft from the second session in February 1987, the proposals made in the third session in April, and the informal consultations by Tolba in Brussels in June and July were combined by the Legal Drafting Group into the seventh revised draft[38] of 15 July 1987 as the basis for diplomatic negotiations in Montreal on 14–16 September 1987. This text bracketed many parts of the draft as a sign of dispute on their content. Brackets remained on the definition of CFCs, inclusion of halons for controls, the majority needed for adjustment of the protocol, industrial rationalization (the transfer of a production quota to, or receiving a production quota from, another party provided the total for both parties was within the total of control limits for them), trade, control measures, developing countries, and entry into force.

What to control? Production or consumption? And other points

Canada strongly objected to a proposal by the USA that the protocol should enter into force only after all the major producers ratified. It suggested that this proposal could be applied to consumption rather than production. Canada further objected to the formulation that production and imports alone were frozen under the proposals of the seventh draft because that would give a monopoly to European countries, which were already exporting significantly. Canada also said that the proposal would lead to a shortage of supply and higher prices, and cause less developed countries not to become parties to the protocol or, even worse, decide to construct their own production facilities. Canada also suggested that industrial rationalization between all the parties and not only the parties of the European Economic Community be permitted, so that the production allowance could be shifted between the parties, and plants could be run economically.

Another submission to the forthcoming diplomatic conference was the scientific information submitted by the Chemical Manufacturers Association. According to its information:

• Changes in the ozone level, assuming a 20 per cent reduction in use by developed countries and CFC growth rates of 3 to 5 per cent for a ten-year period by developing countries followed by a freeze, would be less than the natural variability in column ozone over time periods of years to decades.
• For all the scenarios, the association's calculations showed that a turnaround would occur in which column ozone would begin to recover during the first half of the next century.
• Delays in emission control actions over the 1994–2004 period would have little effect on stratospheric chlorine concentration.

THE MONTREAL PROTOCOL ON SUBSTANCES THAT DEPLETE THE OZONE LAYER, 1987

The Conference of the Plenipotentiaries in Montreal led to a solution only after intense negotiation. Fifty-five countries and the European Economic Community attended the negotiations. Six countries, including India, attended as observers. Eleven industry organizations, six environmental NGOs and six United Nations and intergovernmental organizations participated. Ambassador Lang of Austria presided. The seventh revised draft of the protocol and the reports of the working group formed the basis of further discussions. The Final Act of the Montreal Protocol on Substances that Deplete the Ozone Layer was arrived at on 16 September. This day, 16 September, was designated by the 49th Session of the General Assembly of the United Nations in 1994 as the International Day for the Preservation of the Ozone Layer.

Control measures

As it finally emerged in Montreal, the Protocol had control measures on two baskets of ODSs, included in Annex A. The first basket, group I, consisted of five CFCs (-11, -12, -113, -114 and -115) and the second basket, group II, consisted of three halons (-1211, -1301 and -2402). Halon-2402, which was predominantly used by the Soviet Union and many countries of Eastern Europe, was included in the controls at Norway's suggestion. Each chemical had been given an ODP recommended by the scientists. The production and consumption of each basket (group) was the addition of the figure for each substance in that basket multiplied by its ODP. This definition allowed each Party to decide for itself its strategy of choosing the ODS to reduce consumption, depending on its situation and subject to the control measure on the basket. Thus, the sectoral use controls discussed previously were abandoned in favour of control of chemicals, accommodating national diversity. For example, this gave time to Japan to find alternatives to the solvent use of CFC-113 in its electronics, even while continuing its use, by reducing consumption of the other CFCs in the basket.

Article 1 provided the definitions. A controlled substance in a manufactured product (such as a refrigerator) was excluded from the definition of controlled substances to avoid double counting, since it would already have been counted when it was produced or imported. Consumption was defined as production plus imports minus exports (in bulk, ie, traded as a pure substance or as a mixture). This definition allowed a substance to be used any time from stockpiled quantities, even after a phase-out, that would have been counted when it was produced or imported. Production was defined as amounts produced minus amounts destroyed by technologies approved by the Parties. Exports to non-parties were to be subtracted from the production figures up to 1 January 1993, under Article 3.

For Group I, there would be a freeze on both production and consumption, beginning six months after the entry into force, a 20 per cent reduction beginning 1 July 1993, and a 50 per cent reduction from 1 July 1998, unless two-thirds of the Parties representing two-thirds of the consumption decided otherwise. An additional 10 per cent production was allowed for export to meet the basic domestic needs of developing countries and for industrial rationalization. After 1 July 1998, this allowance would go up to 15 per cent. Production and consumption of halons would be frozen three years after the entry into force of the Protocol with a 10 per cent increase allowed to meet the basic domestic needs of developing countries. The baseline year would be 1986.

Something for every country: The compromises

Several very specific paragraphs were drafted in order to satisfy all of the Parties. The Soviet Union raised the special issue of its CFC manufacturing plants, which were contracted for under its five-year plan and insisted that those plants should be taken into account while fixing the base year for production and consumption. Some other countries, including Japan and Luxembourg, also had problems with manufacturing plants that were under construction in 1986. To meet this concern, a special paragraph of Article 2, Paragraph 6, was introduced, allowing countries to take into account the production from the factories that were under construction in 1986 or contracted for prior to 16 September 1987 (the date of the Protocol), and provided for in national legislation prior to 1 January 1987. It was pointed out, inter alia by Canada, that industry might like to close down a plant when the production had decreased to a certain low level and instead increase production at another of its plants, which might be located in another state. A special paragraph, Article 2, Paragraph 5, was inserted to deal with this issue. It stated that, if a Party produced less than 25,000 tonnes in 1986, it could transfer its production quota or receive a production quota from any other Party, provided that the total for both Parties was within the control limits. Such transfers had to be notified to the secretariat. The European Economic Community countries were allowed to fulfil their consumption controls jointly, but production controls would have to be fulfilled individually, under Article 2, Paragraph 8.

Strengthening controls in future: An innovative provision

Any changes to the control measures on the Annex A substances or to the ODPs, which were called 'adjustments', could be decided, based on scientific assessments under Article 6, by two-thirds of the Parties, representing at least

50 per cent of the consumption; such an adjustment needed no ratification by each Party, as an amendment would, and would be binding on all Parties six months after its notification by the depository, the Secretary General of the United Nations. This was a legal innovation, praised later by many scholars. Parties could take more stringent steps than required by the control measures under Article 2.

Trade restrictions

Trade measures against non-Parties were approved under Article 4. Imports from non-Parties were banned one year from the entry into force of the Protocol, but exports to non-Parties were to be banned from 1 January 1993. The products containing the controlled substances were to be listed; imports of such products were to be banned by Parties that had not objected to such a list. Significantly, there was no ban on export of such products to non-Parties. The Parties are to examine the feasibility of preparing a list of products produced with, but not containing, controlled substances with a view to banning imports of such products from non-Parties. Electronic products cleaned with CFC-113 were an example of such products. Each Party was to discourage export of technologies for producing or utilizing controlled substances to non-Parties. Again, significantly, the word 'ban' was not used. No Party was to provide any subsidies or credits for export of any products or technology that would facilitate the production of controlled substances to non-Parties. These restrictions would not be applicable to products or technologies for recycling of controlled substances. No trade restriction would apply to a non-Party that submits data and convinces a Meeting of the Parties that it is in full compliance with the control measures.

Concessions to developing countries

Developing countries with an annual per capita consumption of less than 0.3kg, any time within ten years, were allowed to delay their implementation of the control measures by ten years. The Parties would facilitate access of developing countries to alternative substances and technologies, assist them to make use of such alternatives, and facilitate provision of aid for this purpose, under Article 5. Article 10 provided for technical assistance to the developing countries.

Science: The guide to future controls

Article 6 provided for an assessment of the control measures at least once in four years on the basis of available scientific, environmental, technical and economic information. This had been praised as an innovative measure that facilitated an immediate first step and further stringent action progressively on scientific advice and technological feasibility

Other provisions

Annual reporting on production and consumption was mandatory for each controlled substance, under Article 7. The baseline data for 1986 were to be

BOX 2.3 PUTTING PRESSURE ON INDUSTRY FOR ITS OWN GOOD

Brigitta Dahl, Former Minister for Environment, Sweden, and current Speaker, Parliament of Sweden

From an early stage of the negotiations, we had close cooperation with other progressive countries, especially the other Nordic countries, the USA and Canada. We agreed that the use of ozone depleting substances should be phased out. On the other hand, experts from EEC countries and the European Community argued – heavily – that general production limitations would be the best solution, rather than specific restrictions.

These widely divergent viewpoints made it impossible to reach an agreement on control of CFCs at the Vienna diplomatic conference in March 1985. I considered that the most important outcome of the Vienna conference was the resolution on further actions that was signed by the Parties to the Convention.

In the late afternoon of 16 September 1987, after three days of extensive deliberations, the conference was prepared to adopt the Montreal Protocol. But I was not satisfied. I knew we had to do more – by effectively using the instruments provided by the Montreal Protocol. To me, the birth of the Montreal Protocol was a milestone. For the first time, an international instrument had been created with the aim of providing timely action to avoid an environmental disaster in the future. In my opinion, there was a need to strengthen the stipulated timetable for reduction measures in the Protocol further.

Sweden was concerned about what measures should be taken to implement the Montreal Protocol, to reduce the use of some of the substances beyond the levels required by the Protocol. Industry proposed voluntary agreements. The authorities proposed differentiated import levies for different substances. It soon became obvious that the most effective route to success was to analyse and control the overall use of CFCs and halons within a limited number of main fields of application.

The plan I presented to Parliament in February 1988 implied that Swedish consumption in 1990 should be halved in relation to the consumption in 1986. In 1994, the consumption of CFCs should in principle have come to an end in all new products and in all new plants.

The Swedish plan was radically stricter than the obligations related to the Montreal Protocol. To be able to present the Swedish offensive phase-out plan as credible, intense preparatory work was necessary. Most important were the negotiations with Swedish industry and trade, which had the opportunity to develop alternative technologies and products that could also be economically successful on an international market, thus helping to accelerate the global phase-out of the use of CFCs, and at the same time, promoting Swedish products in an international market.

We met with the industry, sector by sector, to discuss their demands, possibilities and consequences. When the rumour spread about our work with the Swedish phase-out plan, this also created an interest abroad; we were happy to welcome delegations from big international chemical companies that produced CFCs. These firm targets and timetables turned out to be decisive. The challenges to industry served as instigators of pioneering technologies, which gave the companies considerable advantages on the world market. A win–win situation had been created.

The international consequences became more profound than we had dared to expect. Many nations followed our example. I will always remember these negotiations. At one point it was obvious that, in order to reach an agreement, some ministers (among whom was today's Executive Director of UNEP, Klaus Töpfer) needed to meet alone, together with Mostafa Tolba, to make the necessary political agreements. It was after this meeting that we were able to agree on a solution that was acceptable to us all.

Box 2.4 How Sweden and other small nations got global action on ozone

Ingrid Kökeritz, Swedish delegate to ozone negotiations

My first involvement in the ozone issue was in the early 1980s when I was responsible for implementing the Swedish ban on chlorofluorocarbons (CFCs) in aerosol products. A similar regulation was also in force in Canada, Norway and the USA. The European Economic Community (EEC) – of which Sweden at that time was not yet a member – had directed its member states to freeze their production capacity of CFC-11 and CFC-12 and to reduce their use in aerosols by at least 30 per cent. In April 1980, Norway hosted an informal meeting of 'like-minded countries', attended by Canada, Denmark, Germany, The Netherlands, Norway, Sweden, the USA and the European Commission, that concluded that negotiations on an international convention for the protection of the ozone layer would be appropriate. Based on a proposal from Sweden, UNEP's Governing Council then decided in May 1981 to set up an Ad Hoc Working Group of Technical and Legal Experts (LTWG) to explore the potential for an international convention.

Sweden hosted the first LTWG meeting in Stockholm in January 1982. During the years to follow, many more meetings were held with the LTWG. The discussions went around and around, on the framework structure, the preamble, confidentiality of information, institutional arrangements, etc, but with very limited progress on substantive controls. In April 1983, Sweden therefore abandoned the idea of CFC reductions in the Convention itself and instead aimed for a protocol, linked to the Convention, which could be signed by those who were ready to do so.

From about 1973, the core advocates of ozone layer protection were individuals from Canada, Denmark, Finland, Norway, Sweden and the USA. During the negotiations we became known as the 'Toronto Group' after a coordination meeting held in Toronto in 1984. The main opponents of ozone layer protection were the European Commission and the large European Commission countries. Some developing countries, Japan, and the Soviet Union (USSR) participated but remained mostly uncommitted.

By the beginning of 1985, there was a consensus on the text in the convention itself and on peripheral text in a protocol, but no agreement on any substantive commitments, or even on their basic structure. A diplomatic conference to sign the convention was scheduled for 18–22 March 1985 in Vienna.

What all parties had proposed so far was that everybody else should do what they had already done. Faced with the prospect of a non-committing convention, which would have to wait for ratification before any more work could be done on the draft protocol, the Toronto Group changed its tactics. At a strategy meeting in Washington on 4–5 March 1985, the Toronto Group drafted a resolution for approval at the same time as the convention, calling for continued negotiations on the protocol and targeting adoption by 1987. The continued negotiations should be preceded by a UNEP workshop to promote 'a more common understanding of possible scenarios for global production, emissions and use of CFCs and other substances affecting the ozone layer and the costs and effects of various control measures'. The last paragraph urged all nations to 'endeavour to control their production and reduce their use of CFCs, especially in aerosols to the maximum extent possible', for industrialized countries by at least 50 per cent compared to their peak consumption.

The idea of continued negotiations preceded by a workshop was well received during the meeting in Vienna, but the last paragraph met strong resistance, in particular from the European Commission. A compromise was reached deleting the 50 per cent specification and changing the words 'maximum extent possible' to 'maximum extent practicable'. The Resolution on a Protocol Concerning Chlorofluorocarbons was adopted together with the Vienna Convention on Substances that Deplete the Ozone Layer, on 22 March 1985.

At this time, there was not really a serious interest in the issue at high levels in most countries, including Sweden and the USA. The US Head of Delegation at the Rome workshop told me at a coffee break, 'You know, within US EPA, this issue is still only seen as John Hoffman's pet subject'. The situation was similar in Sweden. In 1985, the Swedish EPA was reorganized, and my new boss wanted me to drop the CFC issue: 'Sweden contributes just 0.5 per cent to the emissions. What can we do about it? And what can you, a lawyer, do about it? I have more important work for you to do.' After some discussions and with support from other quarters within the EPA, I was allowed to continue to work on the CFC issue – provided that I did not request any other resources than my own time and some consultancy money that had been made available in the reorganization.

And the interest was just as weak at high levels in the Swedish Ministry in charge; at that time we did not even have a Ministry of Environment – all environmental issues fell under the Ministry of Agriculture. Reluctantly, the Ministry adhered to the pressure from Parliament and requested a report from the Swedish EPA on current use, expected development and potentials for reductions in shorter and longer perspectives.

The Rome workshop was the first time that substantive issues (current and projected CFC consumption, alternatives, costs and benefits) were discussed instead of positions. And it was the first time industry was allowed to speak. Many industry representatives had participated in the meetings, influencing the positions of their governments behind the scenes, but formally participating only as observers. Now they had to come out in the open and their arguments could be evaluated and debated on their own merits. This broke the deadlock and paved the way for the more fruitful workshop in Leesburg. The workshop in Rome was also the first time the need to control halons was mentioned.

The Leesburg workshop focused on how CFCs could be controlled through an international agreement. The mere fact that this was not a negotiating event but a workshop, allowing free discussions on a great variety of possible policy strategies and their advantages and disadvantages, encouraged a more fruitful atmosphere.

The formal negotiations on the draft protocol began in December 1986. The focus on aerosol controls was dead, and so was the united front within the former Toronto Group. Each nation now negotiated on its own.

When negotiations continued in February 1987, there was a generally positive atmosphere and the new consensus on the need to control at least the CFCs, but there was still no consensus on the general control structure or the time frame. The potential requirements with regard to the developing countries were discussed seriously for the first time, but so far with no progress on details.

In the end, a halfway agreement was reached on a freeze of 'production and/or adjusted production' of CFC-11 and CFC-12 within 'one to three' years after entry into force of the protocol at 1986 level. The USSR now declared its willingness to discuss controls along the lines proposed by the USA. Japan, however, could only accept controls on production.

To put pressure on the delegates, UNEP announced that arrangements were made for a diplomatic conference to adopt the protocol in September 1987.

The next meeting took place in April 1987 in Geneva. At this time UNEP's Executive Director Mostafa Tolba personally opened the meeting. He spelled out forcefully what he expected from the delegates: 'A Protocol text agreed during this meeting and signed in early autumn 1987. The Vienna Convention as well as the Protocol entered into force by

1988. A complete freeze on production and use of CFCs at 1986 levels in effect by 1990. Twenty per cent reduction in 1992 and then another 20 per cent reduction every two years down to zero by 2000, subject to scientific review every four years.' I recall the impression his speech made in the room: now here was someone who expected something really substantial from us all!

It took a day or two until we were back in the normal nagging and haggling mood again, taking only small steps forward. However, the set date of the diplomatic conference put pressure on countries to come to an agreed text, and the European Commission now had a wider mandate to compromise, probably underpinned by a higher environmental profile that had emerged in Germany.

From this time on, Tolba took a personal interest in the negotiations, conducting informal discussion meetings and – based on what he heard – putting forward proposals of his own. His proposals were sometimes neglected as formally he was not a party in the negotiations, but mostly they pushed the process forward.

Some further steps were taken at informal meetings with key players, arranged by Tolba in June–July 1987 as a basis for his personal proposals. But the fact remained that when the delegates arrived at the preparatory meeting in Montreal on 8 September 1987 – just a week before the diplomatic conference where the protocol was supposed to be signed by the ministers – there was only an agreement on the structure. There was no agreement on the reductions to be achieved, what chemicals to control and by when, and also not on the scope, length and conditions for a grace period for developing countries – not to mention all other details such as trade sanctions against non-parties, etc.

This last preparatory meeting in Montreal was chaotic. A number of small working groups worked in parallel, which made it impossible to keep track of what was agreed on each issue.

At the end of the day, however – or rather, in the early morning of the day when the ministers would arrive – a full text was agreed, containing all the compromises it had taken to get everyone on board. It did not meet everybody's ambition but was a good start; in particular, the concept of regular review and simplified procedures to strengthen the requirements proved very important.

Ironically, while all delegations finally agreed on how to reduce the production and consumption of CFCs and halons, industry representatives worked in the corridors on a joint statement urging the delegates to avoid any short-term reductions. Such reductions would 'produce little or no environmental benefits and create unnecessary economic disruption'. Delegates should instead focus on 'timely periodic scientific, economic and technological assessment of the need for and the timing of further ozone protection measures'!

Inspired by the effectiveness of the resolution agreed in Vienna in 1985, Lindsey Roke from the New Zealand delegation and I drafted a new resolution, calling for a UNEP workshop on emission controls and alternative technologies. We went around soliciting support from Tolba and as many countries as time permitted. The Resolution on Exchange of Technical Information, co-sponsored by 14 countries, both developed and developing, was approved along with the Protocol on Substances that Deplete the Ozone Layer on 16 September 1987.

Based on this Resolution, Tolba, Stephen O Andersen and Victor Buxton organized a major workshop in The Hague in October 1988 on Substitutes and Alternatives to CFCs and halons. Both industry and government authorities contributed valuable presentations. Several of these presentations pointed to considerable possibilities for reducing use of CFCs and halons by already available means, without waiting for new chemicals to be developed.

> But even more importantly, Tolba used The Hague workshop to initiate the first assessment under Article 6 of the Protocol on the need and potential for further actions. Four assessment panels, one on atmospheric science, one on environmental effects, one economic and one technical (the latter with five technical options subcommittees) were set up. I chaired the Technical Options Committee on Aerosols, Sterilants and Miscellaneous Uses. Tolba set our deadline to June 1989. So when the Parties to the Montreal Protocol met for the first time in Helsinki in May 1989, with one of their tasks being to decide on assessment procedures and panels, the work was almost already done. The reports from the four assessment panels were synthesized in August 1989 and resulted in the accelerated and extended phase-out schedule that was agreed in London in June 1990.
>
> Almost everyone seems to agree that Sweden and other small countries were the driving force behind the Montreal Protocol and its amendments and adjustments. Delegates from small countries have more freedom to advocate strong policy measures. Perhaps it is because we are not chemical producers and have less to lose by market transformation. Perhaps it is because we have more inhabited areas close to the pole than most other countries and tend to consider outdoor activities an important part of life. An important lesson to learn – do not think that you are powerless just because you are small!

reported by each Party within three months of entry into force of the Protocol for that Party.

There was no agreement on the non-compliance procedure. Article 8 provided for the Parties to consider and approve at their first meeting a procedure and institutional mechanism for non-compliance. This allowed approval of the non-compliance procedure through a decision of the Parties.

Article 9 provided for information exchange between the Parties. The information to be provided was on alternatives to controlled substances, technologies for recycling or destruction of controlled substances, and control strategies. Parties were to promote public awareness on the environmental effects of the ozone depleting substances. Each Party should provide a report every two years on its activities under this Article.

Article 12 provided for a secretariat to service the Meetings of the Parties under Article 11, which would be held regularly under the rules of procedure to be finalized by the first Meeting of the Parties. The Parties, under rules to be finalized in the first meeting, would contribute the funds required by the Protocol.

Any country which became a Party after the Protocol entered into force would have to fulfil the obligations that applied on the date of entry into force of the Protocol for it, under Article 17. Under Article 18, no reservations may be made to the Protocol. Any Party, other than a developing country Party operating under Article 5, could withdraw from the Protocol any time after four years after assuming the control obligations, under Article 19.

Picking up on the tradition from the resolution adopted in connection with the Vienna Convention, a resolution on exchange of technical information was drafted. This resolution requested the Executive Director of UNEP to facilitate exchange of experience and appealed to interested states to sponsor and

BOX 2.5 THE IMPROBABLE TREATY: A PERSONAL REFLECTION

Richard E Benedick, Senior Advisor, Joint Global Change Research Institute of the Pacific Northwest National Laboratory (Battelle)

In January 1985, I led a small American delegation to a little-noticed international meeting in Geneva, where we and a handful of like-minded countries tried, and finally failed in the face of strong opposition from other negotiating parties, to achieve an agreement to limit CFCs; for three years, this group had struggled in vain. The event received perfunctory attention. Two months later, a mere twenty countries signed the Vienna Convention for the Protection of the Ozone Layer. We were discouraged. Very few gamblers would have wagered that in less than three years, the Montreal Protocol would initiate a process that would soon lead to a global commitment to phase out this extremely useful family of chemicals. For me, this period contained some indelible personal vignettes.

In our quest for strong controls, the USA and its allies in the 'Toronto Group' initially faced formidable opposition from the European Union, Japan and the Soviet Union, which together accounted for two thirds of the world's production of CFCs. Against this background, my colleagues at EPA and I designed a diplomatic strategy to gain support for our goals.

We established consultations with Germany, Denmark and Belgium – these governments would implement the influential (EU) 'troika' role by the time of the critical Montreal conference. US scientific missions were dispatched to overcome skepticism. In an effort to moderate the influence of Imperial Chemical Industries, I encouraged US environmentalists to enlist their British colleagues to raise pointed questions in Parliament.

While some Congressmen complained that I was not pressing hard enough, others were critical of our strong public position. At the same time, there occurred a reactionary anti-regulatory backlash from within the Reagan administration, spearheaded by Interior Secretary Don Hodel, who belatedly realized that the negotiations might be successful. US industry, more pragmatic than ideological, advocated a firm 50 per cent reduction. While the press lampooned Hodel and others for characterizing skin cancer as a 'self-inflicted' disease and recommending 'hats and sunglasses', the domestic disarray had its effect abroad, providing encouragement to those governments still opposed to strong controls. Secretary of State George Shultz sent the issue to President Reagan for decision. President Reagan, against the advice of some of his oldest friends, became the world's first head of state to personally approve a national negotiating policy on ozone protection: he totally endorsed the strong position of the State Department and EPA.

As the September date approached, a delegation of environmental groups, dissatisfied with the 50 per cent position, came to my office to request postponement of the Montreal meeting, arguing that the results of a forthcoming Antarctic scientific mission to analyze the 'ozone hole' could justify a larger reduction. However, my scientific advisors noted that there were many theories other than CFCs to explain the puzzling phenomenon. We all agreed that it was more prudent to take what we could get in September, while designing a flexible treaty that could be reopened on the basis of future science.

Our diplomatic efforts proved successful, and it seemed that the 50 per cent target could be agreed. But a critical unexpected obstacle emerged (at the Montreal negotiations) over the base year against which future reductions would be measured. We insisted that unless the basis were a prior year (1986), parties would be encouraged to build up their production, but the Soviet chief delegate persistently stressed that unless it were 1990, his country would not join. During a coffee break from the baffling stalemate, the Russian explained to us that the problem was a five-year plan, scheduled to end in 1990, which could not be changed under the Soviet constitution. After we ascertained that the plan allowed only moderate expansion in capacity, Ambassador Lang and I later drafted, on a napkin over lunch, a special paragraph limited to the Soviet situation, and the breakthrough to final agreement was achieved.

Looking back, there are undoubtedly important lessons from the ozone experience. Montreal broke new ground in dealing with scientific uncertainty and risk management. The protocol and its institutions collaborated closely with industry to promote a virtual technology revolution. The extraordinary work of independent technical and economic assessment panels provided continual guidance to the negotiators.

In setting reasonable targets, the negotiators did not let short-term politics get ahead of scientific and economic realities. The first international action to protect the ozone layer was neither a global treaty nor a target, but rather an informal agreement in the late 1970s among the USA, Canada and a few smaller nations to ban use of CFCs in aerosol sprays. This single measure reduced total global consumption by 30 per cent and paved the way for the later protocol by demonstrating to skeptics what could be done.

Finally, the critical early ozone negotiations, which produced the most innovations, were remarkably short and small by today's standards. Ultimately, it seems to me that the ozone history defies efforts of analysts to produce a connect-the-numbers guide to successful negotiation. Leadership, by countries and by individuals, remains a crucial intangible factor. In the end, diplomacy remains more of an art than a science. The unusual dedication, persuasiveness and energy of Mostafa Tolba in the critical early years, for example, played a decisive role in the protocol's success. Much depends on serendipity, and on individuals being in the right place at the right time.

participate in a workshop on exchange of information on technologies and strategies to reduce emissions of CFCs and halons. The initiative was taken by Sweden and New Zealand, but the resolution was eventually supported by a large number of countries and finally unanimously adopted. Another resolution on reporting of data called on the Executive Director to convene, within six months, a meeting to make recommendations for harmonization of data on production, imports and exports.

The Protocol would come into force on 1 January 1989, provided that at least 11 countries representing at least two-thirds of 1986 CFC consumption ratified it, under Article 16. The meeting did not include carbon tetrachloride, methyl chloroform, methylene chloride, and HCFC-22 in the list of substances to be controlled, though the issue of including them was discussed.

Media reactions, 1987

The reaction from the media throughout the world to the Montreal Protocol was ecstatic. Regardless of philosophical orientation, all Parties and media hailed it as an unprecedented breakthrough. Many of them understood that the provisions were not adequate to protect the ozone layer, but the significance of the agreement was not diminished in any way. Many others hoped that this agreement would lead to tackling of other similar problems, such as deforestation. The *Journal of Commerce* commented in its editorial of 25 September 1987, 'For a supposedly moribund organization, the United Nations is suddenly proving surprisingly dynamic'. It also commented that this burst of activity was due less to changes at the United Nations than to changes in the global political climate. When East–West ties improved, the United Nations became a diplomatic forum. The Organization may not be perfect, but it had once again proved its value, according to the editorial. Many US senators wrote to Mostafa Tolba congratulating him on his personal achievement and UNEP's leadership, and urged Tolba to take further steps, as he promised, to ensure the recovery of the ozone layer. They urged Tolba to convene an urgent meeting to consult on the issue further; Tolba promised that a scientific meeting would soon consider this action.

Chapter 3

Diplomacy: From strength to strength, 1988–1992

INTRODUCTION

The discovery of the ozone 'hole' in 1985 in the Antarctic came as a shock not only to the informed public but also to the scientists. The losses of ozone were very high at all levels of the stratosphere and no theory or mathematical model then known could explain the extent of the ozone loss. The scientists then attempted explanations. Some ascribed it to CFCs and others to atmospheric dynamics. A resolution of the debate clearly required further observations to validate any of the theories. An expedition mounted to the US Antarctic station in McMurdo in August 1986 could not resolve the debate. Scientists including Crutzen and Molina then proposed new theories, which subsequently proved right but still required validation by observation. A second, larger expedition was then led in August 1987 by NASA with aircraft reaching the 'ozone hole' for observations. These observations, along with the ground observations, provided the 'smoking gun' to implicate the CFCs prominently in the ozone depletion. Scientists found that ozone in the stratosphere was less when the chemical chlorine monoxide was high and vice versa. Chlorine monoxide was formed in the process of catalytic destruction of ozone by CFCs and had no other natural source. This convinced the scientists that the atmospheric dynamics, which some blamed for the ozone depletion, only set up the conditions to promote ozone destruction by CFCs. The CFCs were the main villains. These results came in slightly after the diplomatic conference in Montreal that arrived at the Montreal Protocol and had no effect on the diplomats who negotiated the Protocol. The Montreal Protocol, as a result, prescribed only mild controls on CFCs and halons.

Even before the ink was dry on the Protocol of September 1987, it was clear to the delegates and to many scientists that the mild controls of the Protocol would not result in the protection of the ozone layer. Only a complete phase-out of CFCs would do, and all the countries in the world, particularly all the developing countries, need to implement the phase-out too. The five years succeeding the Montreal Protocol were witness to the exciting tale of UNEP, WMO, other United Nations organizations, some governments, scientists and technologists marshalling all their strength to achieve this aim. The scientists gathered more conclusive scientific evidence and prepared assessment reports under Article 6 of the Montreal Protocol that proved the case against all CFCs and also identified many more ozone-depleting substances that needed to be phased out. The assessment reports also gave several regulatory options to governments, clearly setting out the consequences to the ozone layer of each option (see Chapter 2). UNEP and some governments mustered the political will to adjust and amend the Protocol based on these assessment reports in 1990 and 1992, strengthening the Protocol enormously. This chapter sets out this unprecedented story of scientists and technologists guiding the governments to strengthen their obligations to protect the ozone layer.

PREPARATIONS FOR THE ENTRY INTO FORCE OF THE CONVENTION AND THE PROTOCOL

Since the entry into force of the Montreal Protocol depended very much on at least 11 Parties with more than two-thirds of the total 1986 consumption ratifying the Protocol, it became essential to clarify definitions of consumption and some other terms used in the Protocol. Consistency and comparability in data reporting was also important for other reasons, including for verifying compliance. As authorized by the Resolution on Reporting of Data, UNEP convened the first meeting of an ad hoc working group of experts for the harmonization of data on production, imports and exports of ozone depleting substances (ODSs) in Nairobi in March 1988[1] to discuss the clarifications needed. This group met again in The Hague in October 1988 to complete its work.[2] These meetings and the subgroups established in the October meeting discussed and recommended many issues to prepare for the Protocol's entry into force and the first Meeting of the Parties (MOP 1).

Clarifications for facilitating data reporting

The meetings agreed on a format for submission of data. Many countries considered it necessary for each country to submit data for each chemical as mandated by the Protocol, whereas others felt that it should be enough if collective data were submitted, since control measures were prescribed collectively. Members of the European Economic Community argued that they would report data only in the aggregate for each group in order to preserve commercial confidentiality. Others argued that the Ozone Secretariat could receive chemical data, but publish it in aggregate for each country to facilitate verification of compliance and publish data for each chemical for countries as a whole to facilitate scientific assessment. The issue could not be resolved at this meeting and was postponed until MOP 1 in 1989. It was agreed that the Secretariat could not question data submitted by a Party.

The meetings agreed on a number of clarifications on the definitions of controlled substances. They clarified that controlled substances in a manufactured product, other than a container used for transportation or storage, did not count as controlled substances; this was defined to avoid double counting. Examples of such systems were an aerosol can, a refrigerator, or a fire extinguisher. When these were exported or imported, the importing or exporting country did not count the CFCs in those products since they would have been counted when they were produced or imported. The issue of the use of CFCs as intermediates for production of other chemicals was discussed; the consensus was that where CFCs were used as an intermediate product to make a non-ODS product, they should not be counted.

Definitions

Basic domestic needs

The meaning of the term 'basic domestic needs' was discussed since it was not defined in the Protocol. Developing-country Parties with consumption of ODSs less than the limits set out in Article 5, Paragraph 1 of the Montreal Protocol, are entitled to delay their implementation of control measures of Article 2, and are referred to as Article 5 Parties. Several experts felt that the phrase 'basic domestic needs' was intended to reflect the special situation of Article 5 Parties and facilitate their development. However, it was agreed that the intent of the term was not to allow production of products containing the controlled substances to expand for the purpose of supplying other countries.

Definition of developing countries

The definition of developing countries was also a matter of debate. The March meeting agreed that the Ozone Secretariat should send an updated list of the Group of 77 developing countries as an annex to the report of the meeting, and invite other countries which considered that they should be classified as developing countries to make this known to the Secretariat. This group of developing countries, which was formed in the United Nations, originally numbered 77, but had continuously increased its number with new countries joining. In 1988, there were 126 countries in the group. The session in October recommended that the developing countries in the Group of 77, Albania, People's Republic of China, Mongolia and Namibia should constitute the list of developing countries.

Other clarifications

A subgroup discussed destruction of used substances recovered from old equipment. The group agreed that they should be treated in the same manner as newly produced substances for accounting purposes. The subgroup recommended that there should be a standing technical committee to approve the methods of destruction of such substances and to ensure that such methods are environmentally appropriate. The Protocol's definition of production provided that the quantity destroyed could be deducted from production – that is, a Party could produce more than its entitlement under the Protocol depending on the extent of its destruction of CFCs. Hence it was necessary to ensure that the methods of destruction employed actually could destroy the substances.

The Protocol as approved in 1987 did not assign, pending further studies, any ozone depletion potential (ODP) number for halon-2402. The October

1988 session recommended an ozone depletion potential of 6 for this chemical. Only the former Soviet Union, and the countries to which it supplied equipment containing it, used this substance.

These clarifications helped governments to report their production and consumption data and made it possible for UNEP to verify whether the conditions for entry into force of the Protocol, ie at least 11 Parties with more than two-thirds of the estimated global consumption of CFCs and halons, were fulfilled

An Ad Hoc Legal Drafting Group met in the margins of the working group meeting in October 1988 and prepared the recommendations in legal language.

Workshop on substitutes, The Hague, 1988

As requested in the Resolution on Exchange of Technical Information, UNEP organized a workshop on substitutes and alternatives to CFCs and halons in The Hague in October 1988.[3] Many countries and private companies made presentations on alternatives to CFCs already developed or under development. DuPont presented its development and testing regarding HCFC-141b, HCFC-123, HFC-134a, and other chemicals to commercialize as substitutes, and said the process would be ready by the end of 1990. It also announced new facilities for the production of HCFC-141b and HCFC-142b in 1989 and HFC-134a by 1990. A scientific review of ozone depletion was conducted back-to-back with this meeting.

Informal establishment of assessment panels

The October session of the working group on data also considered the outcome of the scientific review of ozone depletion and the workshop on substitutes. The Executive Director of UNEP presented the conclusions of these two groups. A plan for review of the control measures as called for in Article 6 of the Protocol was prepared. The two groups established four panels to start the review of the available scientific, environmental, technical and economic information under Article 6 of the Protocol. The science panel was to be headed by Robert Watson and Daniel Albritton of the USA, the panel on environmental effects by Van der Leun of The Netherlands and Manfred Tevini of the Federal Republic of Germany, the panel of technology experts by Victor Buxton of Canada, and the panel of economic experts by Stephen O Andersen of the US Environmental Protection Agency (EPA). A high-level group of advisers to the Executive Director, consisting of six ministers and the four chairs, was also established. A timetable for preparation of the panel's assessment reports by July 1989 was prepared, so that Parties to the Protocol could propose, on the basis of the assessment reports, adjustments or amendments of the control measures in Article 2 in time for discussion and decision at MOP 2 in 1990. Even prior to MOP 1, to be held in 1989 after the entry into force of the Vienna Convention and the Montreal Protocol, the first assessment to review the control measures of the Protocol was initiated.

Requirements of developing countries

A subgroup of the meeting of the working group considered the requirements of the developing countries for information and technology transfer. It concluded that:

- Funds would be needed to establish new manufacturing plants and installations as substitutes for old ones; to purchase new substances, which could be expected to be more expensive; and to compensate for changing equipment facilities and appliances.
- Mechanisms would be needed for creating viable processes for information exchange and technology transfers.
- The developed country Parties should actively pursue programmes to assist developing countries to obtain information on new technologies.
- Environmental agencies of developed countries should consult with their governments' organizations responsible for aid to developing countries to facilitate the funding of alternative technologies and substitute products.
- International assistance and lending institutions such as the World Bank, United Nations Development Programme (UNDP) and regional development banks, should be urged to consider the needs of the developing countries to implement the Protocol.
- The Parties should maintain as a high priority the importance of technology and information transfer to developing countries for the implementation of the Protocol.

The working group considered and endorsed the need for modification of the harmonized commodity coding system by the World Customs Council to provide new numbers for each of the controlled substances and their mixtures. Parties were urged to establish identical codes for substances, as permitted by the harmonized coding system, to facilitate collection of data.

During this meeting, the USA circulated a proposal on non-compliance suggesting establishment of a compliance committee. It recommended that any country found in breach of the Protocol should be treated as a non-Party for the purpose of voting and trade for the period of time at least equal in duration to the period of its non-compliance.

Entry into force of the Convention and the Protocol, 1988–1989

The Vienna Convention entered into force on 22 September 1988, more than three years after its adoption in Vienna, following the deposit of the 20th instrument of ratification on 24 June 1988. The Montreal Protocol provided that at least 11 instruments of ratification had to be deposited, representing at least two-thirds of the1986 estimated global consumption. This condition had been fulfilled by the end of 1988. More than 70 countries had reported consumption data; UNEP estimated that global consumption was 1.14 million tonnes of CFCs and halons. By the end of December 1988, 29 countries and the European Economic Community, accounting for 83 per cent of global consumption, had ratified the Protocol. Hence, both the Vienna Convention and the Montreal Protocol entered into force by 1 January 1989, which meant that control measures would begin to be in operation by 1 July 1989.

DISSATISFACTION OF MAJOR DEVELOPING COUNTRIES

> *'Had I been present at the Creation, I would have given some useful hints to the Almighty for the better ordering of the universe.'*
> Alfonso the Wise of Castille, 1221–1284

Many developing countries echoed this view about the Montreal Protocol. Only six developing nations had ratified the Protocol by the end of 1988: Egypt, Kenya, Malta, Mexico, Nigeria and Uganda. The developing countries which had not ratified the Protocol, particularly India and China, had meanwhile firmed up their positions on the Protocol. In May 1989, China submitted a paper to a workshop suggesting that introduction of alternative technologies would entail a great amount of investment. It could not embark on the implementation of such new technologies without technological and financial assistance. A country paper presented by India at a June 1989 workshop in Tokyo called the Montreal Protocol iniquitous because the developed countries consumed large quantities while the Article 5 Parties consumed little, but the Protocol prescribed the same 50 per cent reduction for both. The technology for substitutes, recycling and equipment modification was the monopoly of a few companies in the developed world; Article 5 Parties were not equipped financially or technically to make huge investments for development of substitutes. Even though the Montreal Protocol had a provision for transfer of technology, it was not sufficiently specific. India insisted that an international fund should be established to compensate for transfer of technology on reasonable terms, to be funded by the developed world.

The 'Saving the Ozone Layer' Conference, London, 1989

Alarmed that many of the important developing countries had not ratified the Protocol, the Government of the UK and UNEP organized a meeting in March 1989 in London called the 'Saving the Ozone Layer Conference'. Representatives of 124 governments attended the three-day conference. President Daniel Arap Moi of Kenya inaugurated the meeting, calling on all countries that had not yet agreed to phase out CFCs to accede to the terms of the Montreal Protocol on production cuts. 'It is not a concern of only a few members of an exclusive club. All members of the international community have a duty to protect the ozone layer,' he said. The conference was dominated by two questions:

1   the need to move faster in abolishing the use of CFCs and to go further than the 50 per cent reduction by the end of the century as mandated in the 1987 Protocol;
2   to recruit all the developing countries to the cause.

Demands by developing countries

While the industrialized countries attending the meeting answered the first question positively, the developing countries, led by India and China, called for many changes in the Protocol before they would ratify it. Liu Ming Pu, China's Environmental Protection Minister, called for an International Ozone Layer Protection Fund, paid for by the developed world and disbursed to developing countries which agreed to limit the use of CFCs. 'There should be a free of charge transfer of technology to any signatory of the Montreal Protocol,' he said, also reporting that developing countries resented hearing rich countries 'telling them what to do and what not to do. Setting examples is infinitely more powerful.' Z R Ansari, Minister of Environment and Forests of India, said consumption of CFCs in the industrial world was 100 times greater than per capita consumption in India, and insisted that the fund must not cut into the aid developing countries received at the time. 'Any reduction... would mean that the poor of these countries would have to wait longer for the promise of freedom from hunger and poverty. Today the poor are no more prepared to wait,' he said. 'Lest someone think of this as charity, I would like to remind them of the excellent principle of "polluter pays", adopted in the developed world.' In the view of India and China, the problem, as well as the effects of present and past CFC pollution that had yet to unfold, were overwhelmingly the product of 30 years of environmental abuse in the West. They believed that the developing world would need to be compensated for the damage it would sustain and the development it would have to forego, as a result of past emissions of CFCs and future restrictions on their use.

Vladimir Zakharov, Soviet Deputy Chair for Hydrometeorology, urged caution in the elimination of CFCs. The Montreal Protocol should not be revised on an emotional basis, he said; any move to CFC substitutes would cause many difficulties involving greater investment and restructuring of industries.

UK offers support to developing countries

These interventions, particularly by the developing countries, had an impact on the participants. 'We shall put greater emphasis on environmental needs in allocating our aid programme,' British Prime Minister Margaret Thatcher said in a speech to the meeting. 'Clearly it would be intolerable for the countries which have already industrialized, and have caused the greater part of the problems we face, to expect others to pay the price in terms of their people's hopes and well-being.' She announced at the closing session that Britain would give £3 million in aid to UNEP, double the sum previously promised. Many present noted that the UK, which opposed stringent controls until 1987, had become a proponent of stringent regulations. Rumour had it that Thatcher took the scientific briefing on the ozone issue to her home one weekend as a sceptic, and returned on Monday fully convinced of the threat. Her Oxford chemistry degree helped her to understand the need for immediate action to control CFCs. Mostafa Tolba of UNEP pledged his support for developing 'an agreed plan to raise extra resources for 1990s and beyond' to help developing countries. He said, addressing Thatcher, on whom UNEP conferred the UNEP Global 500 Environment Award, 'as the leader of the first industrial nation to use coal-

---

BOX 3.1 THE BRITISH CONTRIBUTION

*'Asked at a news conference how much the British contribution (to UNEP)
now was, Prime Minister Thatcher said it was one and a half million
pounds, which would double to three million pounds a year. When
Nicholas Ridley, the British Environment Minister, whispered to her,
"Actually, Prime Minister, it's one and a quarter million," Mrs. Thatcher
replied, 'Well, then, you'd better make it one and a half million
immediately. That will teach these people to give me the wrong briefing,
won't it?"'*

Source: *New York Times*, 8 March 1989.

---

burning extensively, you may deem it appropriate for the UK to support – or
even take a lead – in an initiative of this nature.' This meeting raised hopes that,
given some incentives, the developing world was not averse to ratifying the
Protocol.

FIRST MEETING OF THE PARTIES, HELSINKI, 1989: RESOLVE TO PHASE OUT BY 2000

The first meetings of the Conference of the Parties to the Vienna Convention[4]
and of the Parties to the Montreal Protocol[5] were held in Helsinki on 26–28
April 1989 and 2–5 May 1989 respectively. The meetings were presided over by
Kaj Barlund, the Minister of Environment of Finland, and opened by M
Kovisto, the President of Finland. They tackled immediately all the most
important problems and made significant decisions on many issues based on
the recommendations of the working group.

Progress achieved in Helsinki

On five topics in particular, as follows, the meetings adopted: administrative
decisions, procedures to deal with non-compliance, an approach to the concerns
of developing countries, a declaration for achieving phase-out of all CFCs by
the year 2000, and decisions to facilitate data collection and reporting.

Administrative decisions

* The rules of the procedure for the meetings of the Protocol and the
  financial rules for the trust funds for the Vienna Convention and the
  Montreal Protocol.
* The budgets, which included assistance to experts of developing countries
  to attend Meetings of the Parties and assist the assessment process. The
  contributions of individual Parties were to be based on United Nations
  scales of contribution by each Party, even though the contributions were
  called voluntary. This removed the ad hoc nature of voluntary contributions.

These contributions are made to the Trust Funds for the Vienna Convention and Montreal Protocol created by these meetings.

• UNEP as the secretariat for both the Vienna Convention and the Montreal Protocol.

• The establishment of separate bureaux for the Vienna Convention and Montreal Protocol. A periodical meeting of Atmospheric Research Managers of the Parties was to be convened, along with a meeting of the Bureau of the Vienna Convention to give recommendations for future research and expanded cooperation between researchers in developed and developing countries.

Procedures to deal with non-compliance

• An elaborate arbitration procedure under Article 11 of the Vienna Convention for non-compliance complaints. No Party to the Convention or the Protocol has used this procedure to date. Indeed, many other global environmental conventions have a similar procedure, which has not been used by any Party to those conventions so far.

• Establishment of an Ad Hoc Working Group of Legal Experts on Non-compliance to submit proposals on non-compliance procedures under Article 8 of the Protocol to the next meeting for consideration.

Developing countries

• A decision on the list of developing countries recommended by the working group. The total number of countries on the list was 130, which covered a major part of the world. This list contained countries such as Singapore and South Korea whose per capita incomes and consumption of the controlled substances were higher than those of some European countries. No countries of Central and Eastern Europe, excepting Albania, Romania and Yugoslavia, were included in the list. Romania and Yugoslavia were on the list since they were a part of the Group of 77.

• Components in the work plans for the implementation of the Protocol required by Articles 9 and 10 of the Protocol.

• An open-ended working group to consider, in separate sessions, proposals for amendments to strengthen the Protocol (through total phase-out of CFCs and other measures) and to develop modalities for financing mechanisms, including adequate international funding mechanisms. It was decided to invite all the non-Parties to attend these meetings, thus enabling India, China and other developing countries to make their contribution to the discussions.

• A decision to recognize the urgent need to establish international financial and other mechanisms to implement Article 5 in conjunction with Articles 9 and 10, and to enable Article 5 Parties to meet the requirements of the present and future strengthened Protocol.

Helsinki Declaration: A pledge to phase out by the year 2000

All delegations recognized that the 50 per cent reduction in the production and consumption of CFCs and a freezing of halon levels were not adequate to protect the ozone layer. Most delegations indicated their support for at least 85 per cent reduction in the use and production of CFCs. Several of these delegations expressed the view that a total phase-out of CFCs before the end of the century, at the latest, would be needed.

The governments also adopted a declaration to phase out the production and consumption of CFCs as soon as possible but not later than the year 2000, to phase out halons and other ODSs as soon as feasible, and to develop funding mechanisms to facilitate the transfer of technology to Article 5 Parties. Agreed by the entire meeting, this became known as the Helsinki Declaration. The meeting also agreed to develop alternative substances and technologies.

Decisions to facilitate data collection and reporting

- An ODP value for halon-2402 of 6.
- Clarification that the Parties were required to submit the production and consumption figures for each controlled substance as specified in Article 7 of the Protocol, and establishment of procedures to secure confidentiality of data.
- Many decisions on clarification of terms and definitions, including controlled substances in bulk, production, industrial rationalization (an exchange of production quotas between Parties, provided total production is within the limit set by the Protocol), and destruction of controlled substances as recommended by the working group meeting in October 1988.
- Clarification of the term 'basic domestic needs' as not allowing production in the Article 5 Parties of products containing controlled substances to expand the purpose of supplying other countries. This clarification became necessary as such Parties were allowed to delay their implementation of control measures by ten years and, meanwhile, to produce and consume the controlled substances to meet their 'basic domestic needs'. The non-Article 5 Parties were allowed to produce additionally and export the controlled substances to Article 5 Parties, to fulfil the 'basic domestic needs'. Attempts to define the term 'basic domestic needs' precisely did not succeed.
- Clarification that exports and imports of used controlled substances should be treated as exports and imports of newly produced substances, and included in the country's consumption limits.

Assessment panels established

MOP 1 endorsed the establishment of the four assessment panels, their composition and their terms of reference. The technology review panel consisted of five subgroups for refrigerants, foams, solvents, aerosols and halons. The terms of reference of the panels contained in detail the chapter headings, and authorized the chairs of the chapters to select experts to participate in the panels. Some additional experts were added to the panels as requested by certain delegations. A timetable for the assessment process was fixed for the finalization of the first reports by July 1989. The Open-Ended Working Group of the Parties was requested to integrate the reports of the four panels into one synthesis report and make recommendations for the amendment of control measures.

Rules of procedure

The rules of procedure were virtually the same for the Convention and the Protocol. The rules prescribed the seat of the Ozone Secretariat, Nairobi, as the place for the Meetings of the Parties, unless other arrangements were made by the Secretariat in consultation with the Parties. The rules provided for preparation of agenda for meetings, representation and credentials, setting up of committees, conduct of business, and voting. The head of the organization designated as the Secretariat, at UNEP, was designated as the secretary-general of all meetings. The secretary-general was authorized to appoint an executive secretary with the requisite staff. The rules provided for ordinary meetings as well as extraordinary meetings at the request of Parties, although no extraordinary meeting has yet been held. The Conference of the Parties to the Vienna Convention was to meet every two years, back-to-back with annual Meetings of the Parties to the Montreal Protocol at the same venue; in 1991, the meeting frequency was changed to three years. It so happened that out of the 13 Meetings of the Parties to the Montreal Protocol, only two meetings, in 1991 and 1994, took place in Nairobi, and the other 11 meetings took place outside Nairobi[6] at the invitation of the Parties and approval by Meetings of Parties. The Secretariat followed the United Nations rules and collected the incremental costs incurred in holding the meetings outside the seat of the Secretariat from the host governments.

Observers

The United Nations, all its specialized agencies, the International Atomic Energy Agency and all states not party to the agreements were to be notified of all meetings by the Ozone Secretariat so that they could send observers. The Secretariat would also notify any body or agency, national or international, governmental or non-governmental, qualified in fields relating to the protection of the ozone layer so that they could be represented by observers, unless one-third or more of the Parties present at the meetings objected to their admission. The observers could participate in the meetings on invitation from the president of the meeting, without the right to vote. Every Meeting of the Parties was attended by many observers from non-Parties. They included organizations of manufacturers of ODSs or alternatives, user industries of these chemicals, environmental organizations, organizations of commodities such as strawberries which used methyl bromide, associations of asthma sufferers who use metered-dose inhalers, youth organizations, women's organizations and others. Representatives of various interests made statements on matters of interest to them, endorsing or criticizing the proposals or reports before the meeting, or urging their own point of view. They could also be seen lobbying Parties to

---

BOX 3.2 MEETINGS OF THE OZONE RESEARCH MANAGERS

The meetings of the ozone research managers, established by the first meeting of the Conference of the Parties (COP) of the Vienna Convention, took place in 1991, 1993, 1996 and 1999, the years of the subsequent meetings of the COP. These were organized by the WMO with funds from the Vienna Convention Trust Fund. WMO and UNEP published the reports of these meetings as reports of the WMO global research and monitoring project. All countries were invited to present the results of their ozone-related research to these meetings, when participants discussed the results and made recommendations for further research. These recommendations were placed before the meetings of the bureau and the Meetings of the Parties of the Vienna Convention. The meetings of the Conference of the Parties appealed to the countries to contribute to the WMO special fund for research, since the Vienna Convention Trust Fund was just adequate to conduct meetings about expenditures and little more. The Global Environment Facility funded establishment of three global ozone-observing stations in Argentina, Algeria and China in 1993, but did not fund any other efforts thereafter. There was no other funding source for research projects.

---

support their particular interests. One interesting case is that of Taiwan, which claimed to be an independent state but was recognized by the United Nations only as a province of the People's Republic of China. Taiwan could not attend meetings as a state but was keen to attend since it had significant ODS-using industry. Permitting a delegation from Taiwan as observers, with implicit consent of China, solved the problem.

Bureaux for the Convention and the Protocol

Each Bureau of the Vienna Convention and the Montreal Protocol had five Parties elected at each Meeting of the Parties for the posts of president, three vice-presidents and a rapporteur. The posts were rotated in alphabetical order among the five geographical regions: Africa, Asia and the Pacific; Latin America and the Caribbean; Eastern Europe; Western Europe and other States. The president conducted the Meetings of the Parties; the Convention Bureau was given the task of reviewing scientific information on the ozone layer, considering programmes for research on the ozone layer, and preparing an agenda for such activities for consideration by the Parties. The Bureau of the Protocol reviewed the work of the working groups established between meetings, considered the other topics on the agenda of the next Meeting of the Parties, and reviewed the documents prepared by the Secretariat for the meetings.

The Bureau of the Protocol met usually twice every year, normally at the time of the meetings of the Open-Ended Working Group or the Meeting of the Parties. The Bureau of the Convention met once with the meetings of the ozone research managers and again before the meeting of the Conference of the Parties. At these meetings, the Ozone Secretariat explained the action taken to implement the decisions of the Meetings of the Parties, the financial position and the documents prepared for the meetings. The suggestions of the members of the bureaux were followed up by the Secretariat. In a few years, following an instruction from the bureaux, the president wrote to all the Parties to ratify the

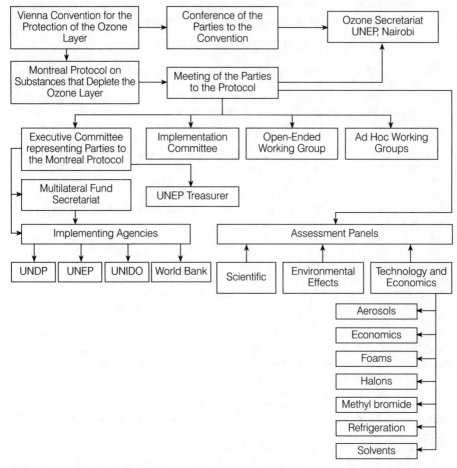

Source: Ozone Secretariat (2000) *Action on Ozone*, United Nations Environment Programme, Nairobi, p12.

Figure 3.1 *Organizational chart of the Vienna Convention and Montreal Protocol*

amendments, report the data, or pay the contributions due. In the initial years, the bureaux suggested some substantive proposals, such as amendments or adjustments to be placed before the Meetings of the Parties, non-compliance procedure, and other issues. In the later years, the bureaux played a more formal role of reviewing the progress and advising the Secretariat. The bureaux had no powers to approve expenditure.

Science panel on the status of the ozone layer

Robert Watson, Co-Chair of the Scientific Assessment Panel, presented to MOP 1 the current understanding of the state of the ozone layer and the cause of its depletion. It had now been established that CFCs and halons were a primary cause of the ozone hole over Antarctica. The majority of the approximately 3

**BOX 3.3 FINANCING THE CONVENTION AND
PROTOCOL OPERATIONS**

The process of negotiations with all governments of the world meant a sizeable expenditure for UNEP. For every meeting, arrangements had to be made for the conference halls, microphones and other equipment. The meetings had to provide for simultaneous interpretation of the discussions and for translation of the documents in all the six United Nations languages: Arabic, Chinese, English, French, Russian and Spanish. In addition, UNEP had begun financing delegates of Article 5 Parties to the ozone meetings, in order to encourage their participation. The professionals needed to process the meetings were from the Environmental Law Division of UNEP, headed first by Peter Sand and then by Iwona Rummel Bulska. The Assessment Division of UNEP processed the meetings of the Coordinating Committee on the Ozone Layer (CCOL). The resources needed came from the Environment Fund of UNEP and from the countries that were Parties to the Protocol.

At the first Meeting of the Parties in 1989, the Parties created two trust funds, the Trust Fund for the Vienna Convention and the Trust Fund for the Montreal Protocol, to receive contributions from the Parties and to meet expenditures approved by the Parties. The terms of reference for these funds were also approved. The funds would be administered by UNEP. All Parties would contribute to the Parties in the ratio of their United Nations scale of contribution. Those Parties whose contributions on the United Nations scale were less than 0.1 did not need to contribute. At the first Meeting of the Parties, and every meeting thereafter, the Parties approved budgets for the next year and for two years after on a rolling basis. The contributions were voluntary, but the decision of the Parties allocating the amount to each Party removed the ad hoc nature of voluntary contributions; most countries honoured their shares.

Of the 180 Parties, only 63 contribute to the budget and others need not contribute since their United Nations scales are less than 0.1. Twenty-seven contributing Parties are Article 5 Parties. Seven non-Article 5 Parties contribute 71 per cent of the budget: USA (22 per cent), Japan (19.7 per cent), Germany (9.6 per cent), France (6.3 per cent), UK (5.5 per cent), Italy (5 per cent) and Canada (2.5 per cent).

Almost all of the Parties fully contribute their shares, except a few Article 5 Parties. One developed country contributed a substantial, but not their full, amount. The Secretariat managed the receipts carefully to meet the expenditures. In the 12 years from 1989, the Parties contributed about US$36 million to the trust funds. About US$1 million was contributed by the UNEP Environment Fund in 1989. Some non-Article 5 Parties, Australia, Canada, the European Community, Germany, The Netherlands, Norway, Switzerland, Sweden and the UK contributed for specific purposes over and above their contribution to the trust funds.

The expenditure of the Secretariat was shared between the Vienna Convention and the Montreal Protocol Trust Funds. Of the US$28.5 million spent in the 12 years from 1989, about US$12.3 million was spent on staff and the remaining expenditure was on the organization of meetings and support to experts from Article 5 Parties to participate in the meetings. The Secretariat has five professionals supported by Secretariat staff. The Meetings of the Parties have repeatedly praised the excellent management of the Secretariat with its small number of staff.

---

parts per billion (ppb) of chlorine in the atmosphere came from human-made sources. Even with full global implementation of the regulatory measures of the Montreal Protocol, calculations indicated that atmospheric chlorine concentrations would increase to more than 6ppb during the next few decades. Under these conditions, the Antarctic ozone hole could never be replenished. In order to reduce the atmospheric levels of chlorine below current levels, it would be necessary to phase out all substances controlled under the Protocol, together with strict limitations on carbon tetrachloride and methyl chloroform. Even with a complete phase-out of CFCs and other significant ozone-depleting chemicals, it would still be decades to centuries before the chlorine amounts dropped back to levels found before the Antarctic ozone hole because of the long atmospheric residence times of these ozone-destroying substances.

PREPARATORY WORK FOR THE SECOND MEETING OF THE PARTIES

Discussions on the financial mechanism

In July 1989 UNEP convened an informal group of experts from the World Bank, UNEP, the European Economic Community, and Mackensie & Co, an independent consultant, to discuss the issue of a financial mechanism.[7] The participants recognized that only 10 out of 130 countries classified as developing countries had ratified the Protocol, and that the major producers such as Brazil, China, India and South Korea had not ratified it. For the Protocol to be fully effective, all countries must become Parties. The costs of the phase-out by Article 5 Parties could not be estimated accurately, but the meeting accepted a ballpark figure of US$400 million annually till the year 2000, taking into account that the average cost of substituting CFCs in The Netherlands so far was US$4000 per tonne, and that the Article 5 Parties' total consumption of CFCs was about 108,000 tonnes. Some experts disagreed with those figures.

The group of experts considered many ideas for the institutional arrangements for technical and financial assistance. Some experts noted that the Article 5 Parties would not accept loans from this financial mechanism because their compliance costs were much higher relative to their benefits, compared to those of non-Article 5 Parties which had managed to use CFCs for long periods with no special cost imposed on them; in addition, the magnitude of the ozone problem was largely the result of CFC consumption in the non-Article 5 Parties. Another point made was that the Article 5 Parties would not want diversion of present economic aid to solve the ozone problem, but would want the ozone aid to be additional to what was being given at the time. The experts considered different institutional arrangements such as a straightforward trust fund to which the developed countries would pledge contributions. Two ideas of the World Resources Institute were: a voluntary international environment facility, which would give adequate concessional aid in line with the economic rate of return of projects, and could persuade governments to contribute to those projects; and a conservation-oriented investment programme which was a mix of concessional loans, commercial finance, and private direct investment. The UNEP study suggested an international financial corporation, which would give loans on highly concessional rates of interest.

Ideas on non-compliance procedures

The Ad Hoc Group of Legal Experts on Non-Compliance established by the first meeting met in Geneva in July 1989.[8] After considering the written proposals by the USA, The Netherlands and Australia, consensus was reached that: the procedure should not be confrontational; it could be started by a number of Parties collectively registering concerns with the Secretariat; the procedure should not alter or weaken the comparable procedure under Article 11 of the Vienna Convention; there must be a committee to consider these issues; the decisions should be made only by the Meeting of the Parties.

A fund for developing countries: The first debate

The first session of the first meeting of the Open-Ended Working Group set up by the first Meeting of the Parties met in Nairobi in August 1989 to consider the financial mechanism and modalities.[9] UNEP presented the conclusions of the informal group, which had met earlier in Geneva. Some participants preferred existing bilateral and multilateral mechanisms. Others preferred a separate trust fund to be located with UNEP, or legally enforceable obligations of contributions by the developed countries. The meeting then established small working groups to specify the details. The Working Group on Transfer of Technology listed the activities that could be undertaken in Article 5 Parties for compliance with the Protocol. This group, of both developed and Article 5 Parties, also came up with a list of incremental costs that might be covered by such an international financial mechanism. Specific country studies would be needed regarding the total incremental costs that might have to be met by the financial mechanism. It was agreed that the beneficiaries would get funds only through the governments of the countries. The consensus was that the developed countries would contribute to these funds, and that there must be a study on the role that the existing institutions might play in the financial mechanism.

The first assessment: Analysis of scenarios for total phase-out of CFCs

The second session of the first meeting of the Open-Ended Working Group, held from 28 August to 5 September 1989,[10] considered the four reports of the assessment panels to provide a synthesis and, on the basis of the synthesis report, consider options for adjustments and amendments to the Protocol. A draft synthesis report prepared by the co-chairs of the assessment panels from their detailed reports confirmed that data from 1969 to 1988 showed a measurable downward trend of 1.8–2.7 per cent in ozone per decade in the northern hemisphere in winter months that could not be attributed to known natural processes. The weight of scientific evidence strongly indicated that chlorinated and brominated chemicals were primarily responsible for the recently discovered ozone depletion over Antarctica. Even if the 1987 Montreal Protocol were implemented by all nations, the atmospheric abundance of chlorine would at least double to triple during the next century; by 2050, ozone depletions of 0–4 per cent in the tropics and 4–12 per cent in high latitudes would be predicted. Current models underestimated ozone depletion. The Antarctic ozone hole would not disappear until the atmospheric abundance of chlorine decreased to the levels of the early 1970s. Even if consumption of ozone depleting chemicals was stopped immediately, the recovery of the ozone layer would take a long time, the reports concluded.

The Environmental Effects Panel stated that each 1 per cent of ozone depletion was predicted to lead to a worldwide increase of 100,000 cases of blindness due to ultraviolet-induced cataracts; a 1 per cent decrease of total column ozone would result in a 3 per cent rise of the incidence of non-melanoma skin cancer. In some important plant varieties, increased ultraviolet (UV-B) radiation would reduce food yields by up to 25 per cent for exposures simulating 25 per cent total column ozone depletion. Increased UV-B radiation would have a negative influence on aquatic organisms, especially small ones such as phytoplankton. Because these organisms are at the base of the marine food web, increased UV-B exposure would affect the productivity of fisheries. Tropospheric air quality would suffer due to enhanced levels of surface UV-B radiation. Increased radiation would cause degradation of many materials, particularly plastics that were used outdoors.

The Technology Review Panel (TRP) revealed that it was possible to phase down use of the five CFCs controlled under the Montreal Protocol by 95 to 98 per cent by the year 2000. The remaining demand after the year 2000 would be for maintenance of existing systems and for other minor uses such as inhalant drugs. The assumption was that hydrofluorocarbons (HFCs) and hydrochlorofluorocarbons (HCFCs) would be used as substitutes for CFCs. The HFCs were totally ozone-safe, with zero ODP, though they had a significant capacity to cause global warming, a global issue already commented upon by scientists and in the preliminary stages of consideration by the governments for a solution. The HCFCs had ODP but much less than that of CFCs. There were no substitute chemicals for halons, but many of the current halon uses could be satisfied through non-halon systems. Confining halons to essential uses and maximizing conservation and management of the banks could reduce consumption by 50 to 60 per cent by 1997. The Economic Assessment Panel cautioned that a faster phase-out would result in substantially higher costs, due to capital abandonment.

In a synthesis of these reports, the co-chairs of the panels worked out different control options for the chemicals and the impact of each of the control options on the atmospheric abundance of chlorine. These options presented by the scientists and technologists, and the model on which the calculations were based, provided a basis for decision-making by the governments. Every suggestion made by a government to vary the control measures could be analysed with respect to its impact on the ozone layer.

UNEP proposes a total phase-out

The UNEP Executive Director, on the basis of informal consultations with the governments, proposed to the meeting that:

- 1,1,1-trichloroethane (methyl chloroform) and carbon tetrachloride be included in the list of controlled substances with a schedule for phase-out;
- the production of all the controlled substances be phased out by the year 2000, a phase-out of 95 per cent of consumption by 2000, with a complete phase-out by 2005;
- any other CFCs with ODP greater than 0.1 be controlled by the Protocol over a specified timetable;
- production and consumption of halons be reduced by 50 per cent by 1995 and a target date set for phase-out;
- CFC substitution be regulated so that no substitute would have an ODP exceeding 0.02, and its greenhouse warming potential be restricted to an appropriate level to be determined by the Parties;
- all Parties provide data on all substances with a calculated ODP of greater than 0.02 whether or not controlled under the Montreal Protocol;
- export of ODS to non-Parties be counted as consumption of the exporting Party from 1 January 1991, instead of from 1 January 1993, as provided in the 1987 Protocol, to encourage more ratifications by non-Parties;
- prohibition of trade with non-Parties from 1991, instead of 1993, as provided in the Protocol of 1987.

Many Parties made proposals for adjusting and amending the Protocol. The proposals of the Executive Director and proposals made by Parties were sent to the Legal Drafting Group to be put into a single legal draft, with brackets around any wording or figures on which there was more than one suggestion. The report of the Legal Drafting Group was attached to the report of the meeting to enable governments to consider the proposals made and to make further proposals at the next meeting of the Open-Ended Working Group in November 1989.

Working plans for the future

The third session of the first meeting of the Open-Ended Working Group was held in Geneva in September 1989.[11] This meeting suggested the working plans be implemented under Articles 9 and 10 of the Montreal Protocol:

- the assessment panel reports of 1989 should be updated by 1992;
- information should be developed to fulfil the information needs of specific target groups, including the media, decision-makers, technical experts, and others;
- the Secretariat should coordinate regional workshops with detailed terms of reference;
- technical staff to gather information on technologies that reduce emissions of CFCs and halons should augment the Secretariat.
- the Secretariat should compile a list of suitable consultants who could develop a series of sector-specific case studies on alternative substances.

Proposals on control measures

The first session of the second meeting of the Open-Ended Working Group took place in Geneva in November 1989.[12] The meeting considered the final synthesis report prepared by the assessment panels, the proposed amendments and adjustments, the environmental acceptability of the substitutes, and the treatment of essential uses. A global industry representative informed the meeting that an Alternative Fluorocarbon Environmental Acceptability Study agreement had been concluded between chemical industries around the world; the industry would test the potential environmental effects of ODS alternatives and their degradation products. The meeting had before it the discussion paper on possible regulations of HCFCs and HFCs submitted by Finland, Norway and Sweden. The paper suggested that as a first step, to provide clear signals to industry, the Parties should describe the criteria to define chemicals of concern; various regulatory options should be considered, included imposing a long-term phase-out, restricting the use to special application areas or essential uses, requiring recycling and emission control measures, setting a cap on production or consumption, setting a long-term target on chlorine in the stratosphere, and defining an upper production or consumption limit. A less stringent signal to industry would be to include reporting requirements only. This could be combined with a resolution or decision setting a target for chlorine loading, limiting to essential uses the consumption of HCFCs, HFCs or others, and could also be an indication on envisaged phase-out year.

The meeting prepared a draft of the adjustments and amendments, noting the points for a decision, as reflected in the brackets around words, sentences or figures. The proposals for control measures made[13] were:

- CFCs listed in the Protocol in 1987: 50 per cent reduction by 1994 or 1995; 85 per cent by 1998 and 100 per cent by 2000; 50 per cent reduction by 1991–1992 or 1993, 85 per cent by 1995–1996 and 100 per cent by 2000; 100 per cent by 2000 with dates for intermediate reductions to be determined later; 50 per cent production reduction by 1995 and 100 per cent production and consumption phase-out by 2000, with additional production of up to 10 per cent to be allowed to meet the basic domestic needs of Parties operating under Article 5 and for industrial rationalization.
- Halons: 50 per cent reduction by 1995–1996 and 100 per cent by 2005; 50 per cent reduction by 1995 and 100 per cent by 2000 with exemptions for essential uses as decided by the Parties; 50 per cent reduction by 1995–1997 and 100 per cent as soon as feasible with exemptions for essential uses; 100 per cent production by 1995 and 100 per cent consumption by 2000; 10–50 per cent by 2000 on the basis of self-determined needs agreed to by Parties.
- Other halons: Saturated compounds that contain all, and only, carbon, fluorine, chlorine and bromine atoms. Group VII of Annex A will be the other halons with vapour pressure and/or boiling point exceeding [to be determined] limits; phase-out in the same manner as halons (Annex A, Group II). (From Finland, Norway and Sweden.)

- Other fully halogenated CFCs: Same schedule as CFCs (from Finland, Norway and Sweden); base year 1986, 1987, 1988 or 1989; could be included in Annex A, Group III, essential uses up to [5 per cent].
- Carbon tetrachloride: Include in Annex A as Group IV, base year of 1986, 1987, 1988 or 1989; 50 per cent reduction by 1991–1992 and 100 per cent by 2000; freeze six months after entry into force, same schedule as for CFCs and phase-out by 2000; 100 per cent reduction by 1995.
- Methyl chloroform: Include in Annex A as Group V, base year of 1986, 1987, 1988 and 1989, freeze by 1992–1995; further consideration of schedule in 1994, freeze by 1991–1992; 25–100 per cent by 2000, 100 per cent by 1992, 50 per cent by 1992–1993; 85 per cent by 1998; 100 per cent by 2000, up to [5 per cent] for essential uses.
- HCFCs: Ban on new equipment containing or made with by [2020–2040], phase-out by [2035–2060], use only for essential purposes, recycling and emission control measures, and phase-out by [2010] [2020] (from Finland, Norway and Sweden).
- Up to [to be determined] per cent of additional production allowed to meet the basic domestic needs of Article 5 Parties and for industrial rationalization for halons, carbon tetrachloride and methyl chloroform.

The USA suggested an amendment to ban export of CFCs by any Party (and not merely an Article 5 Party, as specified in the Protocol) from 1992 or 1993. Another suggested a ban of export of products, not only containing (as in the Protocol), but also made with, CFCs to non-Parties or those in non-compliance with the Protocol. The word 'discourage' was sought to be replaced with 'ban' in connection with exports to non-Parties of CFC technologies or subsidies for them. An additional paragraph was suggested to clarify that 'State not Party to the Protocol' include, with respect to any controlled substance, a Party that has not agreed to be bound by the control measures in effect for that substance.

A proposal was made to allow countries which produced less than 25,000 tonnes a year in 1986 to transfer or receive entitlement for production, provided the total production of the two Parties remained within the limits set by the Protocol for both Parties.

Proposals by developing countries

Proposals were made by Article 5 Parties for more specific language for technology transfer and for paying compensation to the Article 5 Parties for the incremental costs of meeting the obligations of the Protocol. Such financial assistance would be given to Article 5 Parties 'so that they may be able to become Parties to the Protocol as early as possible'. The funds for such payments would be contributed by developed countries in proportion to their consumption of CFCs in 1986. A committee of the Parties, elected every four years for a four-year term, with equal representation from both Article 5 Parties and non-Article 5 Parties, would decide on the compensation to be paid to each developing country. Another amendment suggested that Article 5 Parties could withdraw from the Protocol any time after four years after the beginning of their obligations, as non-Article 5 Parties could under the Protocol.

Another amendment by Article 5 Parties suggested that the Article 5 Party obligations could be reviewed in 1999, taking into account the resources provided and the transfer of ozone-friendly technologies so that the Article 5 Parties could leapfrog to the final ozone-friendly technologies instead of first adopting the intermediate bridge technologies. Another amendment suggested that Article 5 Parties would be informed before any product that contained, or required, ODSs was exported to them. Any Party that exported without providing such information would have to replace the product or substance without any extra cost. It was suggested that the voting procedure in the 1987 Protocol, which allowed countries with more consumption of CFCs to influence decisions, should be amended so that any decisions affecting Article 5 Parties could not be taken without the consent of all the Article 5 Parties present and voting.

A summary of other proposals

The Legal Drafting Group of the meeting summarized the other proposals:

- All known fully halogenated CFCs not already included should be added to the list of controlled substances, and technical criteria added as a way of covering all other CFCs (from Finland, Norway and Sweden).
- A definition of halons should include halons other than -1211, -1301 and -2402, which were already included (from Finland, Norway and Sweden).
- Two alternatives for HCFCs should be suggested: one specifying the HCFCs and adding a general clause to cover all others; and the other specifying that 'HCFCs which are deemed by the Technical and Scientific Assessment Panels to have an ODP equal to or greater than .03 be adopted by contracting Parties' (from Finland, Norway and Sweden).
- All chemicals with an ODP greater than 0.01 should be regulated to allow their use only in critical products and consumption areas to be approved by the Parties and placed on a list (from Finland, Norway and Sweden). The Legal Drafting Group raised the question of whether chemicals had to be specified by name for imposing controls or whether they could be regulated as a class.
- All the amendments suggested should be treated as a single amendment so that Parties either ratified them as a whole or not, thus avoiding the situation of multiple amendments and multiple categories of Parties (from the USA).
- If HCFCs were not included as controlled substances, at least the data on all the substances should be reported every year.

Another issue was that the annual control measures of the Protocol started on 1 July 1989, with the control periods from July 1 to June 30. Most industries kept their records in calendar years, and so how to convert the July–June control period to the calendar year was a point of discussion.

'A mechanical working group' was established to see which of the other provisions of the Protocol, for which no amendments were proposed, would be affected by the proposed amendments. The group made a number of suggestions about how to treat Parties that were Parties to the Protocol but not to an amendment. Since no controls were agreed to on HCFCs, a new category called transitional substances was created for them, rather than include them as controlled substances. It was also suggested that where the ODPs were not fully available for other CFCs, an ODP of 1 could be assigned since they were not consumed in great quantities.

DISCUSSIONS ON THE FINANCIAL MECHANISM, CONTROL MEASURES AND TECHNOLOGY, 1990

Financial mechanism for developing countries: The debate

The UNEP Executive Director held informal consultations in Nairobi in January 1990 with representatives of several countries on the financial mechanism. The second session of the second meeting of the Open-Ended Working Group to develop modalities for the financial mechanism was held in Geneva from 26 February to 5 March 1990. A study commissioned by UNEP calculated that the total incremental cost of the phase-out of ozone-depleting substances for the Article 5 Parties could be US$1.8 billion during about eight to ten years. These costs would be spread unevenly, with higher costs in the beginning and tapering off gradually. In his opening speech at the meeting, UNEP Executive Director Tolba mentioned the informal consultations and said that there was consensus for a financial mechanism among all the countries, though there were differences on details. He mentioned that the patents and proprietary rights for substitute chemicals and technologies, as well as the know-how required to use the patents, should be available to Article 5 Parties in an accessible and affordable manner.

Reacting quickly to the suggestions at the previous meeting in August that country studies be made, many countries, including Mexico, Egypt, Brazil, India, China, Venezuela, Kenya and Malaysia, conducted such studies. India presented its country study prepared by a British consultant in collaboration with an Indian counterpart. Mexico and Egypt presented their country studies prepared with the cooperation of the US EPA. Brazil presented similar information in oral presentations but was unable to complete a written report in time for the meeting. The India country study assessed an estimated additional cost of compliance with the Protocol to the Indian economy of US$1.2 billion, discounted at a real interest rate of 8 per cent. The bulk of this cost was to compensate for capital goods abandonment if CFCs were not available for maintenance of existing equipment. Assuming that the equipment continued to be maintained, the cost fell to US$120 million. The other case studies came to comparable conclusions, given the relative size of their economy and ODS use. The delegates to the meeting felt that, though the costs were by no means final, a rough idea of the costs was available; they agreed that lack of final costs should not impede the work on the development of a financial mechanism. Through a subgroup, the meeting further refined a list of incremental costs and also prepared guidelines for application of the list. For example, it was noted that savings could be gained in some transition processes and should be taken into account.

Another UNEP study on financing modalities which was presented to the meeting commented that a decision on financing through government contributions could be easily reached, rather than financing through levies or fees on production and consumption or emissions of controlled substances. Data gathering and research, strategy planning, establishment of country programmes, technical aid, capital investment, and programme financing all required assistance. A clearinghouse was needed for coordination, design of promotional programmes, and assistance in the transfer of technology. Additional resources mobilized for this purpose should be utilized under the supervision of the contracting Parties and received in the first instance by UNEP, which would administer the Secretariat. UNEP, UNDP and the World Bank should play a leading role in designing and implementing programmes and projects.

There was intense discussion on how contributions to the financial mechanism should be made. Many preferred mandatory contributions, while others favoured voluntary contributions on an assessed basis. Some delegates preferred voluntary contributions from the Article 5 Parties and mandatory contributions from non-Article 5 Parties. Some delegations stated that it was not important whether the contributions were mandatory or voluntary, as long as the non-Article 5 Parties considered payment obligatory. Everyone agreed that these funds should be in addition to normal development assistance, and that bilateral cooperation could be used if two countries agreed to such cooperation.

Many favoured a small executive committee to supervise the activities of such a fund; some suggested that the World Bank could fulfil this role. It was agreed that the fund should function until the requirements of the Montreal Protocol were completely met, and that it would be useful to operate with the three-year rolling plans. One delegation suggested the need to address initial funding to show goodwill and to initiate the activities of the funding mechanism. Several delegations put forward a discussion paper on the principal activities of the first three years, including the establishment of the Executive Committee, country and sector-specific studies, training courses, organizations of regional workshops, initiation of project-specific feasibility studies, and capital funding for the first projects. Preliminary estimates for country studies were given as US$18 to US$20 million and US$120 million for projects for the first three years.

Total phase-out: The debate

The first session of the third meeting of the Open-Ended Working Group took place in Geneva in March 1990[14] to discuss the proposed adjustments and amendments. In his proposals for this meeting, the UNEP Executive Director suggested that the meeting consider how much time industry would need to recover the investment in the development and production of HCFCs, and to review carefully the global warming potential of each HCFC. A point of discussion was how many tonnes the non-Article 5 Parties needed to produce for the basic needs of Article 5 Parties. The subgroup derived a figure of 100,000 tonnes, assuming a production capacity of 100,000 tonnes in Article 5 Parties and a total demand of 200,000 tonnes. Later, this figure played a part in allowing 10 per cent excess production to meet the needs of Article 5 Parties.

Technology transfer and intellectual property

The UNEP Executive Director arranged a meeting of experts in Geneva in April 1990, in cooperation with the World Intellectual Property Organization, on the role of intellectual property rights in technology transfer. The arguments on technology transfer primarily centred on how to ensure that alternative technologies, owned by private-sector companies, could be transferred to the Article 5 Parties. During the discussions, the non-Article 5 Parties argued that it was up to the Article 5 Parties to obtain the technologies by negotiating with the private sector, and that the governments of the non-Article 5 Parties had no power to compel transfer of such technologies. The developing countries wanted the governments of developed countries to assume obligations to ensure the transfer. The language for the article on technology transfer in the Protocol was therefore under dispute, with the developing countries insisting on an assurance of technology transfer, and the developed countries offering only to help, and not assure, such a transfer.

The World Intellectual Property Organization prepared a note on various kinds of licences, patents, and compulsory licensing under certain conditions. It noted that compulsory licensing provisions in the public interest existed in many countries such as the UK, Austria, Brazil, Japan, Mexico and the Nordic countries, and in the interest of public health in Canada, Colombia, India, Israel and Singapore. Such provisions were used only very rarely in extreme situations such as war; under normal circumstances, governments had no power to compel industry to market products internally or internationally. The best way of obtaining technology was through negotiations. The Organization also mentioned that it could help Article 5 Parties search for technologies that were registered with it, and noted that patents had been granted protection only in certain countries. A preliminary search by the Organization revealed that 16 alternative substances or inventions in refrigeration, propellants, foams and solvents had been registered. Twelve inventions had been granted protection only in industrialized countries, with no patents granted in any developing countries. The other four inventions had been granted patents in both developing and industrialized countries. Of these four, three were protected in Brazil, and the fourth was protected in both Brazil and China. The Organization mentioned that any person or entity might freely use any invention that was not protected by patent in that country. A product that contained the invention could not be exported to another country where a patent for the said invention existed. A search showed that Allied Signal and DuPont of the USA, Mitsubishi Oil Company, Nippon Oil Company, Matsushita Electric Industries and IC Kabushiki Kaisha of Japan had patented a variety of processes, including the manufacture of HFC-134a in 1987 and 1988.

Proposals for the structure of the financial mechanism

The second session of the third meeting to discuss the financial mechanisms was convened in Geneva in May 1990.[15] The World Bank proposed to the meeting the establishment of a global environmental facility with one billion Special Drawing Rights, a conglomerate currency of the International Monetary Fund and the World Bank, equal to about US$1.2 billion, as a minimum budget to cover action by developing countries on ozone depletion, climate change, biological diversity and international waters. The World Bank noted that this proposal was the suggestion of Germany and France, and that it was willing to serve as an agency to implement the investment programmes. The UNEP Executive Director presented a note on financial mechanisms in which the roles of the various agencies were clearly demarcated: UNEP, strategic planning and assisting developing countries in defining their needs; United Nations Development Programme, pre-investment studies and technical assistance; and the World Bank, management of external financing for investments, and programme and project implementation.

Other cooperative financing arrangements were examined for helpful features, including the Tropical Forestry Action Plan, a cooperative arrangement between the World Bank, UN Food and Agriculture Organization, UNDP, and the World Resources Institute; the 1993 programme of the World Bank and UNDP for the energy sector management action programme; the global programme on AIDS (Acquired Immune Deficiency Syndrome) started in 1990; and the 1977 International Fund for Agricultural Development. These mechanisms had succeeded in carrying out clearinghouse functions and providing technical assistance, but had not tackled the problems arising from the needs of large financial transfers, notably project design and capital funding of the type needed by the Montreal Protocol.

USA opposes additional funds or new institutions

After hearing various developing countries discuss the present status of the country studies, the discussion on financial mechanism was taken up. The head of the US delegation took the floor and said that his government would support only a mechanism within the World Bank to provide financial assistance to Article 5 Parties. He added that funds for such a mechanism would be from existing World Bank resources, and no additional funding should be required from donor countries, although voluntary contributions could be accepted. Many delegates expressed their concern and disappointment at the statement, which was contrary to the idea of a separate fund on which consensus was emerging after many months of negotiations. The Group of 77 of developing countries reacted immediately, emphasizing the importance of establishing a financial mechanism and of meeting agreed incremental costs for developing countries as a grant from a multilateral fund under the control of the Parties. The funds to be provided to developing countries would have to be additional to existing financial flows of developed aid and financing. The developing countries expressed their deep dismay over the US position, and urged the delegation to review its position and join the global consensus to save the ozone layer by shouldering its obligations.

But for this initial confrontation, the discussions went well, and the US delegation played a key role during the meeting. At the request of Mostafa Tolba, the USA prepared and presented a preliminary analysis, estimating the capital costs for CFC reductions for the first three-year funding period at US$112–215 million.[16] Anil Markandya, a UNEP consultant, prepared a budget of US$220 million for the same purpose. The UNEP Executive Director put forward a number of proposals from the suggestions made by the delegates. The proposal to constitute a multilateral fund contributed by non-Article 5 Parties to meet all the agreed incremental costs of developing countries was annexed to the report of this meeting, as well as an indicative list of incremental costs. The Executive Director suggested that the transfer of technology should be dealt with in Article 5, with an addition that the obligations of the developing countries would be subject to the financial assistance provided and to transfer of technologies.

SECOND MEETING OF THE PARTIES, LONDON, 1990: PHASE-OUT BY 2000 AND US$240 MILLION FUND APPROVED

The last and fourth meeting of the Open-Ended Working Group[17] was held in London in June 1990, back to back with the second Meeting of the Parties. Fifty-four Parties to the Montreal Protocol and 42 non-Parties, including large consumers such as Argentina, China, Columbia, India, the Philippines, the Republic of Korea, Saudi Arabia, Turkey and Zimbabwe, attended the London meeting. Industry non-governmental organizations (NGOs) attended the meeting in large numbers. Environmental NGOs were represented through Friends of the Earth, Greenpeace, Worldwatch and World Wide Fund For Nature (WWF). The critical discussions on the outstanding issues were held in small informal groups outside the plenary hall; hence, the report of the meeting did not display the flavour of the intense negotiations over many clauses.

This meeting saw a flurry of activity, with proposals emanating constantly from all the Parties and from the UNEP Executive Director on a daily basis. The Executive Director proposed further modification of the timetables for a phase-out based on his informal consultations. The proposed amendments touched many Articles of the Protocol, including the definitions, control measures, calculations of control measures, control of trade, technology transfer and financial mechanism for Article 5 Parties, reporting of data, Meetings of the Parties, and withdrawal. He suggested that, pending entry into force of the London Amendment to establish a financial mechanism, an interim financial mechanism on the same terms be established, and suggested terms of reference for the fund, the executive committee, and the inter-agency arrangements. He suggested that the United Nations scale of assessment be adopted for the fund.

A meeting of the Bureau of the Protocol, held on the sidelines of MOP 2, suggested a draft declaration by all the Parties that they would: not commercialize halons other than those controlled, report to the Secretariat on annual production and consumption of other halons, and not license production of other halons with an ODP of more than 0.1. The bureau suggested that the

---

BOX 3.4 THE IMPORTANCE OF WORDS

Except for those delegates deeply involved in the behind-the-door negotiations, most international meetings present an appearance of chaos and lack of progress to the participants and observers at the meetings. Each government proposes decisions in the language of its choice to suit its interests, and the drafting experts combine the proposals into a single text with the differences between the governments specified by placing brackets around the preferred words of each government. Each government urges the acceptance of its choice of language and rejection of the language of others. Some of the experts, including the United Nations officials, try to reach a compromise by suggesting words acceptable to all. The negotiations on the language are often carried out in small informal groups, meeting after the regular meeting hours, often late into the night. The governments tend to agree to compromise only at the last moment and, until then, each insists on its language within brackets. Frequently the disputes are about a few words held dear by the countries. Most observers at the meetings feel that the delegates are wasting everyone's time on trivial matters, ignoring the crucial issue of ozone depletion. The fact is that, behind each word, there is a country's interest.

---

use of transitional HCFCs be limited to those applications where alternatives were not available, and that the transitional substances be selected to minimize ozone loss. The proposed declaration also suggested that recovery and recycling be employed, and that transitional substances be collected and destroyed by the appropriate technologies to the end of their useful life.

Intense debate on technology transfer to developing countries

The UNEP Executive Director continued his informal consultations on transfer of technology, which eluded agreement. He submitted three drafts on technology transfer, two on 25 June and another on 27 June, each time making slight changes. While the developing countries accepted that the non-Article 5 Parties could not ensure transfer of particular technologies and that they could only take every practicable step to transfer such technologies, the developing countries questioned how they could fulfil their obligations for the reduction of consumption of ODSs, if, ultimately, the technology transfer failed to take place. The earlier drafts linked the fulfilment of the obligations by the developing countries to financial assistance and technology transfer. The draft proposed by the Executive Director initially on 25 June said that the fulfilment of obligation by the Article 5 Parties would be subject to the 'financial cooperation as provided by Article 10 and the samples of technologies transferred'. His next draft changed this provision as 'transfer of technologies as provided by this article'. In effect, the criteria for testing the implementation of this Article would not be whether transfers had taken place, but only whether each Party had taken every practicable step to ensure the transfer of such technologies. In order to compensate the disappointment felt by developing countries on this core issue, another clause was added to Article 5 that provided for a review of the operation of the financial mechanism and technology transfer. The working group ended up with a text of adjustments and amendments, which was put forward to the Meeting of the Parties, immediately following.

BOX 3.5 THE YOUTH OF AUSTRALIA: 'WE WILL INHERIT THE CONSEQUENCES OF YOUR DECISIONS'

The 'Youth of Australia' presented to the second Meeting of the Parties an eloquent declaration voicing its frustration:

> *'Over the past week of negotiations, we have been watching you. It has been at times fascinating, at times confusing, at times horrifying. We had to keep reminding each other that what is actually being debated here is the future of the ozone layer. This debate has been largely guided by short-sighted commercial gains and national self-interest. There has been more concern for semantics than for substance. There has been more self-congratulation than self-examination. We are told that countries like Britain are claiming to have led the world in international environmental cooperation, when, in reality, they have been hindering moves for faster phase-out dates. If this debate is setting the precedent for tackling the global warming problem, then we can have little hope of avoiding the greenhouse effect. Scientific imperatives were clear. Only an immediate end to the use of ozone-depleting chemicals will truly reflect the urgency of the situation. Even if we do this, it will be a further 60 years before the Antarctic ozone hole is repaired. We came to the conference with this knowledge. But it seems you did not.*
>
> *'We have a right to demand a safe future for ourselves and all generations to come. Shifting money and technology from one side of the globe to the other is not enough. What is required is a fundamental change in attitudes, values and lifestyles, particularly in the developed countries. We insist that over the next three days, you make decisions that reflect intergenerational equity and an active concern for the environment. Remember that we will inherit the consequences of your decisions. We cannot amend the Montreal Protocol; you can. You will not bear the brunt of ozone depletion; we will. We demand that you think in long term. Even the best proposals on the table now sanction an unacceptable increase in the chlorine loading of the stratosphere. Your rhetoric is not being matched by your action. Will you condemn us to a future with skin cancer, eye cataracts, immune deficiency, depleted food resources and vanishing biodiversity? At the moment we are afraid. Please do not leave our generation without hope. Our fate lies in your square brackets. You are making history. You must have the courage to save the ozone layer.'*

Note: See also Box 6.3 by Stephen O Andersen (Chapter 6).

Source: The Youth of Australia, MOP 2, London, June 1990.

USA relents on financial mechanism

After hectic negotiations behind closed doors, the USA agreed to a multilateral fund, subject to the following conditions:

- The fund must be used only to assist developing countries operating under Article 5, with incremental costs associated with adjustments made necessary by the Protocol and related exclusively to that purpose.

- The limited nature and unique nature of the fund must become especially confirmed, making clear that the fund was appropriate because there was a scientifically documented connection between the substances controlled by the Protocol and ozone depletion; the fund and the actions being financed through it could reasonably be expected to address the problem of ozone depletion; and the amount of funds needed must be limited and reasonably predictable.
- Any financial mechanism set out at the meeting should not prejudice any future arrangements the Parties might develop with respect to other environmental issues.

The USA proposed a fund secretariat with a chief, two deputy chiefs and seven professional officers. The UNEP Executive Director proposed a secretariat consisting of a secretary, a deputy secretary lawyer and three professional officers: a lawyer, administrative officer, environmental scientist). He proposed an increased budget from US$880,000 in 1990 to US$3.4 million in 1991. The budget provided specifically for assisting experts of developing countries to attend working group meetings and regional workshops, and to contribute to the assessments.

The British Prime Minister, Margaret Thatcher, urged everyone to approve the adjustments and amendments to strengthen the Protocol; she also said that countries at an early stage of industrial development had understandable concerns about adverse effects on their economic growth. It was the duty of the industrial countries to help them with substitute technologies and with additional financial costs involved, she said. Britain supported the proposal for an initial programme of action and was ready to contribute at least US$9 million, increasing to US$15 million, if other major consumers joined the Protocol.

She put pressure on the USA to agree to a Fund to assist developing countries. The British *Independent* newspaper (15 June 1990) reported that, 'Britain has been putting pressure on the US government to save an international ozone layer conference in London next week from breaking down. [Among the] key objectives, [is] to set up international funding whereby developed nations will pay poorer countries compensation for investing in alternatives. In preliminary negotiations, all the developed countries apart from the U.S. agreed this funding should be over and above the money spent on third world aid. If the U.S. cannot be persuaded to change its mind, developing nations may refuse to agree to the much stronger curbs needed to protect the ozone layer from accelerating destruction.' The *Boston Globe* (16 June 1990) reported in a front-page story, 'Bowing to pressure from allies, United Nations officials and members of Congress, the Bush Administration reversed itself yesterday and said it will support a new fund to help poor nations phase out ozone-destroying chemicals...Yesterday's announcement represented the second about-face in administration policy on the issue in the last six weeks.'

After many midnight consultations in small groups in which the UK and Tolba played a key part, the second Meeting of the Parties to the Montreal Protocol[18] approved adjustments and amendments, and took further decisions to promote the implementation of the Protocol.

London adjustments and amendment

Total phase-out of CFCs, halons and other chemicals included

The CFCs in Annex A were to be phased out gradually by the year 2000 with a 50 per cent reduction by 1995, and 85 per cent by 1997. A Meeting of the Parties would review the situation in 1992 with the objective of accelerating the reduction schedule. Halons were to be phased out by 2000, a 50 per cent reduction by 1995. The phase-out of halons was subject to essential-use exemptions, to be identified by a Meeting of the Parties by 1 January 1993. Other CFCs, carbon tetrachloride and methyl chloroform were included as controlled substances in Annex B; the base year was fixed as 1989. Other CFCs were to be phased out by 2000 with a 20 per cent reduction by 1993 and 85 per cent reduction by 1997. Carbon tetrachloride also would be phased out by 2000, with 85 per cent reduction by 1995. Methyl chloroform would be phased out by 2005 with a freeze in 1993, 30 per cent reduction in 1995, and 70 per cent reduction by 2000. There would be a review in 1992 of the feasibility of a more rapid schedule of reduction. The caution on methyl chloroform was dictated by its widespread use among many large and small industries as a solvent.

The proposals on a general definition of other CFCs, halons and HCFCs were not pursued; the chemicals to be controlled were specifically listed in the annexes by their chemical formulae. Additional production of 10 per cent of the base level, increasing to 15 per cent after the phase-out, was allowed for all the substances to meet the basic domestic needs of Article 5 Parties. New paragraphs in Article 4 provided for control of trade with non-Parties regarding substances of Annex B on the same pattern as for substances of Annex A. Industrial rationalization was allowed between Parties. A Party could transfer its production entitlement to another, provided that they fulfilled the control measures for production together. Joint reporting on consumption, as allowed for regional economic organizations such as the European Economic Community under Article 2, Paragraph 8, would be satisfied by reporting imports and exports between the organization and countries that do not belong to the organization. A controlled substance included all its isomers. HCFCs were categorized as 'transitional substances' and were placed in a new Annex C; no control measures were prescribed, but reporting under Article 7 was made mandatory.

Provisions to suit developing countries

The voting provisions in Article 2 earlier provided that apart from a two-thirds majority of those present and voting, at least 50 per cent of the countries representing total consumption must be represented in any decision. A two-thirds majority, and also a majority of both developing country Parties operating in Article 5 and of other parties not operating under Article 5 had replaced this. The grace period of Article 5 Parties for the control measures was continued. A new paragraph in Article 5 recognized that developing the capacity of the Article 5 Parties and their implementation of the control measures depended on effective implementation of the financial mechanism and transfer of technology, in accordance with the amended Articles of the Protocol. The amendments to Article 5 allowed the Article 5 Parties to notify the Ozone Secretariat that they were unable to implement their obligations due to inadequate implementation of Articles 10 and 10A; the Meetings of the Parties were authorized to make a decision on this. During the period between the notification and decision of the Meeting of the Parties, the non-compliance procedures would not apply to such Article 5 Parties. Another amendment was added, stipulating that the timetable for the Article 5 Parties would be reviewed again in 1995. The amendment allowed the Article 5 Parties to inform a Meeting of the Parties if they were unable to obtain adequate supply of controlled substances for meeting their basic domestic needs. Amendment to Article 19 allowed Article 5 Parties to withdraw from the Protocol four years after the beginning of the control obligations for them (1999) if they so chose.

The financial mechanism for developing countries approved
The meeting approved a financial mechanism, including a Multilateral Fund for the Implementation of the Montreal Protocol to be contributed by the developed countries according to the United Nations scale of contributions. The Multilateral Fund would be administered by an Executive Committee of 14 members, 7 each from the Article 5 Parties and others under the guidance of the Meeting of the Parties. The terms of reference for the Executive Committee and the fund, specifying the roles of the implementing agencies, were approved. The meeting established an interim financial mechanism pending the entry into force of the London Amendment, and adopted a budget for the Secretariat of the Multilateral Fund. An allocation of US$160 million was made for three years, to be increased to US$240 million if more Parties joined. A decision allocating the US$160 million to the non-Article 5 Parties for payment in the ratio of United Nations scales was approved. Participants accepted the offer of Canada to host the Executive Committee meetings during the interim period, and to support the participation of developing countries in those meetings.

A compromise on transfer of technology
A compromise on the concerns of developing countries regarding transfer of technology from the developed countries was eventually agreed upon as part of a broader negotiating package. The compromise text was part of the London Amendment to the Protocol in 1990 Article 10A, which read:

> '*Each Party shall take every practicable step, consistent with the programmes supported by the financial mechanism, to ensure:*
> *(a) That the best available, environmentally safe substitutes and related technologies are expeditiously transferred to Parties operating under paragraph 1 of Article 5, and*
> *(b) That the transfers referred to in subparagraph (a) occur under fair and favourable conditions.*'

The indicative list of incremental costs approved by the London Meeting of the Parties and later by the fourth Meeting of the Parties in Copenhagen as payable by the Multilateral Fund to the Article 5 Parties included 'cost of patents and designs and incremental costs of royalties' in addition to the 'capital cost of conversion' to new technologies.

BOX 3.6 A LESSON FOR HUMANITY: THE LONDON MEETING

Maneka Gandhi, Minister of State, Government of India

The events of the second Meeting of the Parties to the Montreal Protocol in London in 1990 were a turning point in the global relations between the industrialized and developing countries. Historically, the Northern countries developed the technologies, and the countries of the South took whatever products or technologies they were offered, under very iniquitous terms.

The 1987 Montreal Protocol promised 'technical assistance' without spelling out any details. At the same time, some of the companies sold CFC technologies freely to the countries of the South from 1985, while they offered these technologies only reluctantly earlier.

While some developing countries ratified this Protocol, India, China, and many other developing countries kept out of this inequitable exercise. Many also realized that any phase-out by countries of the North alone would not protect the ozone layer, if the developing countries increased their consumption of CFCs at the same time. Truly global cooperation was necessary to save the Earth!

At the London meeting, as head of the Indian delegation, I made it clear to the industrialized countries that they must not only provide the funds, but also give the ozone-friendly technologies to produce, for example, refrigerators, or India would not sign the Montreal Protocol.

The London and international press used to quiz me every day about our stand. I told them, 'Money is irrelevant if we don't have access to the knowledge. Survival is about the spread of knowledge, not money.' Asked whether I was prepared to stand out and destroy the ozone layer, I said, 'we did not destroy the ozone layer; you have done that already. Don't ask us to pay the price.' I said India was not prepared to sign an agreement that would destroy its own manufacturing base and force it to buy CFC substitutes from Britain and the USA.

One piece of news that interested the British media was my vegetarianism! The *Guardian* of 30 June 1990 said, 'Mrs Gandhi, aged 33, is a non-drinker and vegetarian. This has foxed politicians and officials of the Department of the Environment whose normal method of dealing with Third World gripes is said to be to administer a quick gin and tonic or preferably several. In almost every other respect, observers agree, it has been her conference. By the end of the week it was obvious that the woman was quite able to exploit a simple message: that as the developed countries had created the hole in the ozone layer, it was humbug to expect the developing countries to bankrupt themselves in helping to cure it.'

For two days, I was adamant. 'If you continue to clutch your patents to your chest, you may not have a world which you need patents for,' I said. 'We do not have 200 years to catch up. Maybe you should give us some of the knowledge now.' Finally, we reached a compromise that satisfied us. A clause was added saying that the developing countries would fulfil their obligations only with transfer of technology and implementation of the financial mechanism.

I hear from my friends about the same fierce debate on climate change. I hope the countries draw the correct lessons from the Montreal Protocol and find a solution to the climate-change problem soon.

Non-compliance procedure

The meeting approved, on an interim basis, a non-compliance procedure recommended by the Ad Hoc Working Group of Legal Experts and established an Implementation Committee of five members. The Working Group of Legal Experts was asked to elaborate further on procedures on non-compliance and on the terms of reference for the Implementation Committee.

Next assessment in two years

The meeting requested the panels to give their next assessment reports by November 1991, in time for consideration at MOP 4. The Scientific Assessment Panel was specifically asked to evaluate the global warming potential of chemical substitutes HCFCs and HFCs, the likely ODP of other halons that might be produced in significant quantities, and the impact of the control measures on the ozone layer. As suggested by the Soviet Union, the Scientific Assessment Panel was requested to estimate the impacts on the ozone layer of engine emissions from high-altitude aircraft, heavy rockets, and space shuttles. The Technology and Economic Assessment Panel (TEAP) was specifically asked to review the earliest feasible dates for a phase-out of methyl chloroform, evaluate the need for transitional substances in specific applications, estimate the quantity of controlled substances needed by Article 5 Parties for their basic domestic needs, estimate the likely availability of such supplies, and assess the environmental characteristics of chemical substitutes. The meeting established a technical committee to examine the issues of destruction of CFCs when they were no longer needed and to recommend methodologies for such destruction.

Working group on data reporting

The meeting established a group of experts to look into the difficulties of data submission. There was some dissatisfaction expressed with the length of time required for proposing and approving amendments; the legal experts group was requested to review Article 9 of the Vienna Convention to see whether the amendment procedure could be expedited.

London Declaration

The closure of the meeting was accompanied by one declaration and one resolution. The resolution, unanimously adopted by the governments present at the meeting and by the European Economic Community, declared that signatories would prohibit production and consumption of fully halogenated compounds containing one, two, or three carbon items and at least one atom each of bromine and fluorine not listed in the Protocol that would pose a threat to the ozone layer ('other halons'). They agreed to refrain from using 'other halons' except for those essential applications where other more environmentally suitable alternative substances or technologies were not available, and to report to the Ozone Secretariat the estimates of the annual production and consumption of such other halons. They further agreed that the use of HCFCs should be limited to those uses where alternatives were not available and should not be outside the areas of application currently used by the controlled and the transitional substances, excepting in rare cases for protection of human life or health. The transitional substances should be selected to minimize ozone depletion. Recovery and recycling should be employed to the degree possible to minimize emissions. The transitional substances should be collected to the degree possible and destroyed at the end of their final use. The countries also declared that they would review the use of transitional substances and their contribution to ozone depletion and global warming, with a view to replacement with more environmentally suitable alternatives as scientific evidence required: at present, not later than 2040 and, if possible, no later than 2020. They agreed to phase out methyl chloroform as early as possible, and urged all countries to phase out earlier than mandated by the Protocol. The declaration by many of the Western countries stated that, though the London Adjustment mandated a phase-out of CFCs by 2000, they would do it by 1997. The European Commission associated itself with this declaration.

PREPARATORY WORK FOR THE THIRD MEETING OF THE PARTIES

List of products to be banned from non-Parties

The next meeting of the Open-Ended Working Group took place in Nairobi in December 1990[19] to discuss the problems arising from the trade provisions of the Protocol, and to develop a list of products containing the controlled substances of Annex A. The list would allow the Meeting of the Parties in 1991 to annex the list to the Protocol and ban the import of such products from non-Parties by 1 January 1992, as prescribed in Article 4, Paragraph 3. The meeting concluded that there was no conflict between General Agreement on Tariffs and Trade (GATT) rules and the trade provisions of the Protocol. The working group prepared a list of products for the purpose of trade restrictions under Article 4.

Problems of data reporting

The first meeting of the Ad Hoc Group of Experts on Reporting of Data under Article 7 of the Protocol followed this meeting. Of the 62 countries that were Parties to the Protocol, only a few of the non-Article 5 Parties had reported their data. The meeting made many suggestions for facilitating collection of data. The meeting concluded that the customs classification numbers were not adequate to enable proper reporting of data. The Article 5 Parties did not have adequate know-how to collect the data. What was required was legislation requiring reporting of imports, customs classification of all the substances under the harmonized commodity system of the World Customs Organization, and licensing of all imports. A special survey by a consultant with the cooperation of industry would also be useful. The Ozone Secretariat should facilitate information exchange between countries regarding data gathering. The meeting also considered that the reporting under the Vienna Convention on a large number of substances was already being gathered by other organizations such as the World Meteorological Organization (WMO) and was more relevant to the issue of climate change than to ozone depletion.

Discussions on non-compliance procedure

The Ad Hoc Working Group of Legal Experts on Non-Compliance with the Montreal Protocol met for a second time in Geneva in April 1991.[20] One critical issue for consideration of this meeting was the link between the arbitration procedure of the Vienna Convention under Article 11 and the proceedings with the Implementation Committee established by the second Meeting of the Parties. Should the two procedures run concurrently, or one after the other? Should there be some time limit for the Parties to respond to complaints? Who should be the members of the Committee? The Implementation Committee that met immediately after this meeting also suggested a number of points for the consideration of the ad hoc group regarding the number of meetings of the committee and the right of the Implementation Committee to contact individual Parties for their clarifications.

Another question, which agitated the members, was the situation of a member of the Implementation Committee accused of non-compliance. The European Commission proposal suggested that five substitute Parties should also be elected simultaneously who would replace the accused Parties of the Implementation Committee. It also suggested different measures for different types of non-compliance. For non-compliance with data reporting requirements, the steps suggested were initial assistance by the Parties in establishing methods, and institutional mechanisms to collect and report the data. In a case of repeated negligence, the Party could be declared a non-Party for the purposes of Article 4. In a case of non-compliance with control measures, Parties might lose their right to produce with the help of industrial rationalization, and their right to implement jointly could be taken away. Where they had received assistance from the Multilateral Fund, they could be treated as non-Parties, and denied the entitlement to delay compliance with the Protocol, or have the right to receive funds from the Multilateral Fund taken away. Where a particular reduction did not take place in accordance with the provisions, the Parties might be requested to increase their reductions in the following years or might be treated as a non-Party; where there was non-compliance with trade provisions, that Party might be treated as a non-Party for the purposes of Article 4. The meeting concluded that more time was required to finalize the report and that the group would be able to report only to the fourth Meeting of the Parties in 1992.

THIRD MEETING OF THE PARTIES, NAIROBI, 1991: IMPORT OF PRODUCTS WITH CFCs BANNED FROM NON-PARTIES

The third Meeting of the Parties to the Montreal Protocol,[21] the second meeting of the Conference of the Parties to the Vienna Convention,[22] and their preparatory meetings took place in June 1991 in Nairobi. The UNEP Executive Director reported to the meeting the Protocol ratification by China, and recommended that the Multilateral Fund be increased from US$160 million to US$200 million immediately. He also noted that contributions to the Multilateral

Fund were coming in as expected. The Chairman of the Executive Committee of the Multilateral Fund, Ilkka Ristimaki of Finland, announced in his first report to the meeting that the fund was effective from 1 January 1991, with Montreal as the venue of the Multilateral Fund Secretariat and Omar El-Arini taking charge as the Chief Officer in February 1991. The Executive Committee adopted the rules of procedure, implementation guidelines, and criteria for project selection. At its fourth meeting, the Executive Committee approved work programmes by the three implementing agencies, UNDP, UNEP and the World Bank, for 1991.

The third Meeting of the Parties:

- Decided that the Ad Hoc Working Group of Legal Experts should continue its work to identify possible situations of non-compliance with the Protocol, develop an indicative list of measures that might be taken with non-compliant Parties to encourage full compliance, reflect on the role of the Implementation Committee, and reflect on the possible need for legal interpretation of the provisions of the Protocol.
- Added Turkey as a developing country for the purpose of the Montreal Protocol.
- Requested the assessment panels to work out the implications for Article 5 Parties of an earlier phase-out. They also requested the assessment panels to identify the transitional substances with the lowest potential for ozone depletion required for areas where they were essential, and to suggest a technically and economically feasible timetable, indicating associated costs, for the elimination of transitional substances. Between the second and third Meetings of the Parties, the co-chairs of the panels decided that the Economic Assessment Panel be made a Technical Options Committee and part of the Technology and Economic Assessment Panel.
- Approved, as Annex D to the Protocol, the list of products that contained controlled substances, earlier approved by the working group; it also banned the import of these products from non-Parties. The products included automobile and truck air-conditioning units, domestic and commercial equipment for refrigeration and air conditioning/heating, aerosol products except medical aerosols, portable fire extinguishers, insulation boards, panels and pipe covers, and pre-polymers. Under Article 10 on adoption and amendment of annexes, any Party could convey to the depository its inability to approve a new annex to the Convention or to a Protocol to the Convention within six months; the new annex would not be applicable to that Party. Singapore initially conveyed its objection to all items except aerosols and commercial refrigerators, but by the fourth Meeting of the Parties, had withdrawn its objections.
- Requested the Open-Ended Working Group to discuss the indicative list of incremental costs and whether any change was needed for those costs.
- Added another US$40 million to the budget of the Multilateral Fund for 1992, as a result of China ratifying the Protocol in March 1991.

BOX 3.7 TEMPORARY CLASSIFICATION OF PARTIES AS OPERATING UNDER ARTICLE 5

In the interests of expediting assistance from the Multilateral Fund to Article 5 Parties, the Secretariat, with the approval of the Implementation Committee, from time to time 'temporarily' classified many developing countries as operating under Article 5, based on information other than data from the countries, which was not forthcoming. The sixth Meeting of the Parties made a decision that if a 'temporary' Article 5 Party did not submit its 1986 data within one year of approval of its country programme and its institutional strengthening, it would lose its Article 5 status. If such a Party did not submit its data within two years of its ratification, it would lose its Article 5 status unless it asked for the assistance of the Executive Committee and the Implementation Committee. In that case, the extension would be for another two years.

This decision of the Parties helped to improve data reporting. The Executive Committee and the Multilateral Fund also insisted on periodical reports on consumption data by sector, in order to sanction projects and monitor their progress. Both the Executive Committee and the Implementation Committee reviewed the status of data reporting at each of their meetings. The Ozone Secretariat, the Fund Secretariat, and the implementing agencies continuously pressed the Parties for their data. By the eleventh Meeting of the Parties, the situation had improved significantly, with about 85 per cent of the Parties reporting. The non-reporting countries were the ones whose country programmes were under preparation.

The wording of Article 5 meant that a developing country could operate under Article 5 only after it submitted its data to prove that its consumption was below the limits set in Article 5. However, many Article 5 Parties had no idea what their use level was, and hence, had no data or resources to collect such data. They could not submit their data on time and were awaiting the assistance of the Multilateral Fund to prepare a country programme to implement the Protocol. The Meeting of the Parties approved recognition of some developing countries as operating under Article 5, even though they had not submitted their data to prove that their consumption was less than 0.3kg per capita per year, as required by Article 5.

The Meeting of the Parties requested the Open-Ended Working Group of the Parties to consider the situation in which an Article 5 Party exceeded the consumption ceiling of 0.3kg per capita. Would it then be considered a Party not operating under Article 5 and lose its ten-year grace period? What would be its base year for reduction? Would it have to follow completely the developed country control measures? What would happen if it were, at the time, a member of the Executive Committee representing the Article 5 Parties? Did it have to contribute to the Multilateral Fund?

A Co-Chair of the Scientific Assessment Panel, Robert Watson, reported to the meeting on the latest ozone observations, which showed that the rate of ozone depletion at high latitudes of both the northern and southern hemispheres was greater than previously measured. The Russian delegation objected to the findings given by Watson, doubting their veracity.

A statement by the Nordic countries, Germany and Switzerland at the end of the third Meeting of the Parties in 1991 reiterated the phase-out of CFCs by

1997 and added that, by 1995, they would limit the applications of HCFCs to uses where there were no feasible alternatives.

Second meeting of the Conference of the Parties (COP 2), 1991

The second meeting of the Conference of the Parties to the Vienna Convention (COP 2)[23] was held back to back with MOP 3. It noted that the exchange of information under Annex II of the Convention was largely fulfilled by the provision of information under Article 7 of the Protocol and the exchange of information under Article 9 of the Protocol. It requested the assessment panels of the Protocol to report on the availability of information on substances of Annex I of the Convention from other sources. It requested the Parties to assist expansion of the Global Ozone Observing System (GO3OS) Network in developing countries, and to contribute to the WMO Special Fund for Environmental Monitoring for the system. It requested the Parties to implement the recommendations of the first meeting of the ozone research managers. It decided that the Conference of the Parties of the Vienna Convention would be convened every three years after the third Meeting of the Parties in 1993, instead of every two years.

FURTHER PROGRESS IN 1991

Global Environment Facility established

Even as the Multilateral Fund was established in January 1991, a pilot phase of the Global Environment Facility was established in October 1991 with a mandate on ozone depletion, climate change, biodiversity and international waters. With headquarters in the World Bank buildings in Washington, DC, the facility was given a budget of US$1 billion for the first three years. The Facility was made permanent in 1994.

Non-compliance procedure recommended

The Ad Hoc Working Group of Legal Experts on Non-Compliance met again in Geneva on 5–8 November 1991[24] to finalize the non-compliance procedure. The Secretariat suggested that possible situations of non-compliance could arise with respect to Article 2 on control measures, Article 4 on non-compliance trade provision, Article 7 on reporting of data, Article 9 on reporting summary of activities, Article 10 on financial mechanism and non-payment of contributions, and Article 10A on failure to take every practicable step consistent with the programme supported by the financial mechanism, and non-compliance with any of the decisions of the Parties. It suggested that in a case of non-reporting, assistance could be given to a Party. It suggested the possible indicative list of measures that could be taken against non-compliant Parties: assistance, administering a caution, deprivation of certain rights such as industrial rationalization for specified periods, revised control schedules, and lastly, a declaration that the state was a non-Party for one or more articles.

The Russian Federation, a successor state to the Soviet Union, proposed that a Party could itself communicate to the Ozone Secretariat that it was unable to implement the Protocol, despite having tried its best, and that the Implementation Committee would consider such cases. Argentina proposed that the Implementation Committee maintain an exchange of information with the Executive Committee of the Multilateral Fund relating to the provision of financial assistance and technology transfer to the Article 5 Parties. This group considered a suggestion that non-governmental organizations be allowed to provide information to the Implementation Committee. Some considered that the payments to the Multilateral Fund were voluntary and not mandatory. A number of delegations expressed surprise and serious concern that anyone could interpret Article 10 not to contain an obligation to contribute to the financial mechanism. Some considered that it was better not to pursue discussion on that question because of political implications. An indicative list of measures to be applied to Parties not in compliance was approved.

The meeting also discussed whether the procedures for the amendments under Article 9 of the Vienna Convention could be expedited, and concluded that the present procedure was satisfactory.

The second assessment

The three assessment panels submitted their reports in November 1991 (see Chapter 1 for highlights of the Scientific Assessment Panel's report). The Environmental Effects Assessment Panel had gathered further research to confirm the adverse impacts of ozone depletion (see Chapter 1).

The Technology and Economic Assessment Panel noted that the bulk of ozone depleting chemicals could be phased out by 1995–1997. While halons had no replacements in some sectors, the existing bank of halon-1211 might be sufficient to maintain equipment and small quantities for most essential applications well into the next century. The bank of halon-1301 might also be adequate not only to maintain systems that remained in service for up to 45 years after production ceased, but also to supply the most essential new applications for up to 30 years after production ceased. The panel also mentioned that HCFCs were required for some applications, and that the overall ODP impact of HCFCs could be reduced if low-ODP HCFCs were selected.

In a synthesis report, the co-chairs of the panels mentioned that elimination of the Antarctic ozone hole would need a stratospheric chlorine abundance of less than 2 parts per billion by volume (ppbv). The report noted that phase-out costs were falling rapidly, and that by 1991, two years after the entry into force of the Protocol, 40 per cent of CFCs had already been phased out by the industrialized countries, much faster than anticipated by the mandated London Amendment. The synthesis report examined a number of scenarios for strengthening the Protocol, such as accelerating the control measures and reducing the grace period for Article 5 Parties by five years, with the base as the London Amendment and Adjustments. The Scientific Assessment Panel identified methyl bromide, widely used as a soil, commodity and structural fumigant, as a significantly ozone-depleting substance, and suggested controls on it.

PROPOSALS TO ACCELERATE THE PHASE-OUT

The UNEP Executive Director advocated, at the meeting of the Open-Ended Working Group in Geneva in April 1992,[25] the phase-out of CFCs by 1996 and elimination of halons much earlier, since halon banks were available for recycling. He suggested an essential-use provision, which could be refined in 1994. He also suggested control measures on HCFCs that included banning their uses in non-essential applications, halting HCFCs in new equipment by the year 2010, and allowing production of HCFCs only until the year 2025. A new set of ozone-depleting chemicals called hydrobromofluorocarbons (HBFCs) were to be included for immediate phase-out, though they were not yet widely used. Methyl bromide was to be phased out by 2010. The question of reducing the ten-year grace period to five years should also be examined. The Scientific Assessment Panel calculated that such acceleration would speed up the recovery of the ozone layer by 10 to 15 years, prevent 4.5 million additional cases of skin cancer and 350,000 cases of blindness, and result in many other benefits. Executive Director Tolba also noted that if the Article 5 Parties were to phase out faster than scheduled, they needed to be provided with much more financial assistance.

The Parties made many proposals during this meeting to strengthen the Protocol:

- The European Community: Phase out all the Annex A and B substances by 1996; place a quantitative cap on the amount of HCFCs consumed globally, as a percentage of the quantity of CFCs measured in ODP tonnes consumed in a base year plus the existing consumption of HCFCs; follow the cap with a phase-out of HCFCs and HBFCs by a date to be determined; maintain a positive or negative list use of HCFCs.
- China: The Parties should review in 1995 the implementation of financial assistance and transfer of technology to Article 5 Parties and, pending such review, only the control measures set out by London Amendment and Adjustments should be applicable to Article 5 Parties.
- Norway and Sweden: Imports and exports of recycled halons and HCFCs should be disregarded.
- Sweden, Norway, Austria and Switzerland: CFCs should be phased out by 1 January 1995, and halons by 1 January 1994; production and consumption of all HCFCs should be phased out by 1 January 2005 to 1 January 2010; installation of new equipment that used or needed HCFCs should be banned effective 1 January 2000; the Parties should draft a list of accepted uses, and each Party should ensure that after 1 January 1995, use would be restricted only to the accepted uses; each Party's consumption of HCFCs should not exceed 2 to 4 per cent of the quantity of CFCs measured in ODP tonnes; HBFCs should be phased out by 1994.
- Norway: One Party could transfer to another Party any portion of production and consumption, provided their combined levels of production or consumption did not increase.
- Canada: The Protocol should be amended to incorporate mandatory recovery of controlled substances for recycling, reclamation and/or ultimate destruction.

- USA: All the substances in Annexes A and B should be phased out by the end of 1995, subject to essential uses to be determined by 1994.
- Japan: The definition of essential uses should not be postponed to 1994, but should be determined immediately.
- Finland: CFCs should be phased out by the end of 1995, halons by the end of 1994, carbon tetrachloride by the end of 1995, and methyl chloroform by the end of 1999; regulations on HCFCs – including use of only non-recoverable HCFCs – should be banned after 2010, and a ban of all HCFCs should go into effect in 2020.

MULTILATERAL FUND OR GLOBAL ENVIRONMENT FACILITY?

The London Amendment entered into force in August 1992, thus paving the way for the Interim Multilateral Fund becoming permanent. The subject of making the financial mechanism permanent and again approving the list of its incremental costs came up before the April meeting of the Open-Ended Working Group in 1992, in preparation for approval by the fourth Meeting of the Parties to be held in December 1992 in Copenhagen. Most Article 5 Parties considered it a settled matter. The representative of the European Community, however, said that there were various concerns about the operation of the financial mechanism, and that the Parties should consider other options such as the Global Environment Facility, which had recently been established.

The European Community was concerned about the slow progress of the work of the Executive Committee and suggested that the issue of establishment of a permanent financial mechanism be further considered in late October or early November 1992. The Article 5 Parties reacted strongly to this statement: the European Community's proposal meant that there could be no further discussions before the next Meeting of the Parties in November 1992. Many Article 5 Parties, including the two largest, China and India, said they had agreed to accede to the Montreal Protocol only because of the establishment of the Multilateral Fund. They accused the non-Article 5 Parties of reneging on their promises and using the Executive Committee as a scapegoat for their failure to pay their contributions. The chair of the meeting then suggested that the Open-Ended Working Group prepare a recommendation to the fourth Meeting of the Parties to establish the Multilateral Fund along the lines of the Interim Multilateral Fund. The European Community representative emphasized that it was unable to take a position on the recommendation but, if required to, would vote against it because of its instructions. Many delegations expressed their dismay at the European Community's attitude, which boded ill for the future of the Multilateral Fund.

EARTH SUMMIT, RIO DE JANEIRO, 1992

A significant event at this time was the United Nations Conference on Environment and Development (UNCED – commonly known as the Earth

Summit), held in Rio de Janeiro, Brazil, in June 1992, 20 years after the United Nations Conference on Human Environment in 1972 in Stockholm, Sweden. The conference title was the result of the realization that environmental degradation had many socio-economic causes, and that environment and development were two sides of the same sustainable-development coin. Some 179 states attended the Earth Summit. After a hard North–South debate, the conference agreed on Agenda 21 (an agenda for the 21st century), a 700-page programme of action to achieve sustainable development.

The debate included many issues pioneered by the Montreal Protocol, such as who was responsible for the environmental degradation, who should pay for its repair, and how to establish technology transfer and financial mechanisms. Agenda 21 also contained programmes of action for social and economic dimensions such as poverty; consumption and population; conservation and management of resources such as land, atmosphere and oceans; and strengthening the role of major groups such as women, NGOs and youth. It contained guidelines for implementing financial, technology-transfer, scientific and other issues. While Agenda 21 did not solve any of the issues, it had had an effect on the debates about global environmental issues that have taken place since. The Earth Summit was also the occasion for the signing of two major agreements: the United Nations Framework Convention on Climate Change (UNFCCC), and the Convention on Biological Diversity, which followed many of the principles of the Montreal Protocol.

OPPOSITION TO METHYL BROMIDE CONTROLS

The seventh meeting of the Open-Ended Working Group took place in Geneva on 8–17 July 1992.[26] There was intense cross-examination of the chairs of both the Scientific and Technology Assessment Panels on how they calculated the ozone depletion potential for methyl bromide, how they could distinguish between natural and manufactured methyl bromide, and whether substitutes were available for all uses of methyl bromide. There was a general welcome for the proposals for a faster phase-out of Annex-A and -B substances, though there were some doubts about halons. There were differing views on HCFCs, with some arguing that different HCFCs should be dealt with differently.

There were strong comments from almost all the Article 5 Parties and some of the non-Article 5 Parties against any controls on methyl bromide, arguing that it was essential for quarantine and fumigation. Any ban on the chemical would be a serious blow to international trade in agricultural commodities. Methyl bromide had a lifetime of only two years, whereas CFCs had lifetimes of up to a century. Moreover, methyl bromide was a natural product, which was soluble in water and reactive. Information available to date did not justify the listing of methyl bromide as a controlled substance. Little was known about possible substitutes, and the economic impact and incremental cost had not been estimated. A reasonable course was to conduct a study to eliminate scientific uncertainty and, in the meantime, to reduce emissions by proper management of soil fumigation. Instead of imposing control measures, some of the delegations suggested the formation of an ad hoc working group on methyl bromide, and the involvement of other agencies such as the UN Food and Agriculture Organization in the study. Some delegations and the NGOs attending strongly supported controls on methyl bromide on the basis of the precautionary principle and also because it was a very toxic substance, already phased out by The Netherlands just for that reason.

FASTER PHASE-OUTS WELCOMED BY INDUSTRIALIZED COUNTRIES

The industrialized countries welcomed the acceleration of the phase-out schedules. The European Union favoured a much faster phase-out of HCFCs than by the year 2030 but the USA and Japan favoured the phase-out schedule extending to 2030, to allow for servicing of equipment. Article 5 Parties were unwilling to accept any acceleration of their schedules, and were not impressed by the argument that the grace period was adequate to protect their interests. The proposals for advancement of controls on Annex-A and -B substances for the industrialized countries meant that the same advancement of control measures would be applicable to Article 5 Parties, with a grace period of ten years, according to Article 5. The Article 5 Parties rejected this proposal and suggested that since there would be a review under Article 5 on the functioning of the financial mechanism and technology transfer before 1995, any change in the obligation of the Article 5 Parties should be applicable to Article 5 Parties only after that review.

The Russian Federation and some of the countries with economies in transition said they would not be able to implement the control measures beyond the London timetables for a phase-out of Annex-A and -B controlled substances. The issue on the definition of and criteria for essential uses was whether they should encompass only human health, or also the cultural and intellectual aspects of society, which arose in connection with fire protection through devices containing halons provided for aircraft, ships, oil production facilities, museums, art galleries and other facilities. Many felt that these should also be protected, particularly since they covered public safety, national security and human heritage.

INCREMENTAL COSTS

Some delegations suggested that institutional strengthening should be added to the indicative list of incremental costs approved by MOP 2 in 1990. Another suggested that research and development on substitutes, cost of research and development to adopt technology to local circumstances, and evaluation of substitutes should also be added. Some delegations, however, opposed these proposals, arguing that the list was an indicative, and not a closed, one, and that the Executive Committee had the freedom to decide on specific requests. In their view, preparing a comprehensive list might be counterproductive. The working group decided to ask for the views of the Executive Committee on these proposals.

The eighth meeting of the Open-Ended Working Group took place in Copenhagen on 17–20 November 1992,[27] just before the fourth Meeting of the Parties to the Montreal Protocol. By this time, there was agreement on the control measures for substances in Annex A and Annex B, but disagreements remained on HCFCs, methyl bromide, control measures for Article 5 Parties and a financial mechanism. A political side issue was whether the Federal Republic of Yugoslavia represented the former Yugoslavia, which had split into several countries. The countries of Western Europe considered that the Federal Republic of Yugoslavia should not be allowed to take a seat since it did not represent the former Yugoslavia; there was also no legal basis for an automatic continuation of the legal existence of the Federal Republic of Yugoslavia, which could not be considered to continue the membership of former Yugoslavia in the United Nations. Australia, Malaysia, Norway and the USA supported this view.

The Netherlands suggested an amendment to enable countries with economies in transition, including South Africa, not to contribute to the Multilateral Fund, and at the same time, not to receive money from the Fund. However, the proposal was withdrawn after opposition from many delegations. The proposal made by France and Italy on the future of the financial mechanism suggested a review of the Multilateral Fund, taking into account the experience of the Global Environmental Facility, and postponing the debate on the financial mechanism to 1995, by which time the Global Environmental Facility would be stabilized. Annual allocations to the Multilateral Fund would continue. The USA, however, opposed this proposal. Intense informal discussions, presided over by the UNEP Executive Director, took place on all the disputed issues behind closed doors, rather than in the plenary sessions.

FOURTH MEETING OF THE PARTIES, COPENHAGEN, 1992: HCFCs, METHYL BROMIDE CONTROLLED, FUND CONFIRMED

The fourth Meeting of the Parties to the Montreal Protocol[28] was presided over by His Excellency Stig Moller, Minister of Environment of Denmark; the inaugural session was attended by Her Majesty the Queen of Denmark, Margrethe II. It was revealed that this would be the last meeting of Tolba as UNEP Executive Director, and many delegates paid tribute to Tolba's contributions to the ozone agreements.

A new non-compliance procedure

After hectic negotiations, the fourth Meeting of the Parties approved a new non-compliance procedure. The strength of the Implementation Committee had been increased to ten Parties, with two members each from each geographic region. The Implementation Committee would receive, consider and report on any submission by any of the Parties or by the Ozone Secretariat on any possible non-compliance. It was authorized to request information through the Secretariat and to gather information from the territory of a Party on a specific

BOX 3.8 TRIBUTE TO MOSTAFA TOLBA

Richard E Benedick

'It was Mostafa Tolba who broke all the stereotypes of the docile United Nations servant to governments: Tolba took the risk of abandoning traditional neutrality and placing himself and the UN Environment Programme unequivocally behind strong international regulation of ozone-destroying chemicals: Tolba pleaded, provoked, cajoled, shamed, and sometimes bullied reluctant governments ever closer to the treaty provisions that he, as a scientist, knew were necessary for the world. It was an unforgettable virtuoso performance, a role that he undertook with unflagging energy and with absolutely no consideration for his own personal popularity. If the 'ozone story' can be likened to the preparation of a Michelin three-star feast, then Dr Tolba was the master chef: the rest of us were cook's apprentices – salad chefs, pastry chefs and onion peelers.'

Source: cited in W Lang (ed) (1996) *The Ozone Treaties and Their Influence on the Building of International Environmental Regimes*, Federal Ministry for Foreign Affairs, Vienna.

invitation from that Party. It was also authorized to exchange information with the Executive Committee on assistance provided to Article 5 Parties. It would report to the Meeting of the Parties, including any recommendations that it considered appropriate. The Parties would take the final decision. Any Party, not a member of the Implementation Committee but identified in a submission, would be entitled to participate in that meeting.

No Party involved in such a submission could take part in the elaboration and the adoption of recommendations. The members of the Implementation Committee or any Party involved in these deliberations should protect the confidentiality of information. The report of the Implementation Committee, which should not contain any information received in confidence, could be made available to any person upon request. All information exchanged by or with the committee related to any other recommendation of the committee should be made available by the Secretariat to any Party upon its request. That Party should protect the confidentiality of information. The procedure did not specify situations of non-compliance, leaving it to each Party to decide for itself. It included an indicative list of measures consisting of assistance, issuing cautions and suspension from the operation of the treaty.

Multilateral Fund confirmed

The debate over the Multilateral Fund was concluded successfully, with approval of the terms of reference for the permanent Multilateral Fund, the Executive Committee and contributions. India had meanwhile ratified the Vienna Convention and the Montreal Protocol; the fund for 1991–1993 was increased to US$240 million. Hence, a contribution of US$113.34 million was approved for 1993. The Parties made a commitment to a replenishment of the fund of between US$340 and US$500 million for 1994, 1995 and 1996; the total contribution for 1994 would not be less than the commitment for 1993. The meetings accepted the offer of Canada to host the Fund Secretariat on the same terms as those on which it had hosted the Interim Fund Secretariat.

The Executive Committee of the Fund was asked to submit the report of the financial mechanism and its three-year plan and budget for the next meeting. The Open-Ended Working Group was asked to make a recommendation on the level of next replenishment. The meeting also decided to evaluate and review, by 1995, the financial mechanism and to ensure its continued effectiveness, taking into account the relevant parts of Agenda 21. Hungary, Bulgaria and Poland notified the meeting that they were unable to pay their contributions for 1991, 1992 and 1993 because of their temporary economic difficulties. The meeting decided that they should be encouraged to identify other ways of making contributions, including in-kind contributions such as technologies.

The Meeting decided not to establish specific criteria for classification as a developing country, but to evaluate the case of each Party wishing to be classified individually on application.

Control measures

Faster phase-outs for CFCs, halons, carbon tetrachloride and methyl chloroform
The control measures finally adopted provided for phase-out of all Annex A and Annex B substances by 1996 by the non-Article 5 Parties, subject to the essential uses and production for the Article 5 Parties.

Controls for HCFCs: Phase-out by the year 2030
The calculated level of consumption of HCFCs in 1989 plus 3.1 per cent of the calculated level of consumption of the Annex A Group I CFCs was used as the base level; a graduated schedule of reduction, phasing out by the year 2030 of the consumption (not production) of HCFCs was adopted for the non-Article 5 Parties. The 3.1 per cent figure was arrived at after intense negotiations to suit the interests of all countries. HBFCs were to be phased out by 1996 by all countries, since no significant use was identified yet for these substances, with a provision for exemption of essential uses.

A freeze only for methyl bromide (and only for developed countries)
A freeze on methyl bromide by 1 January 1995 was prescribed for the non-Article 5 Parties; uses for quarantine and pre-shipment applications were completely exempted. The Scientific Assessment Panel and the Technology and Economic Assessment Panel were requested to make a full assessment, including the methodologies to control emissions of methyl bromide, availability of substitutes for the various current uses of methyl bromide, and the cost-effectiveness of substitutes. They were to prepare a report by the end of 1994, for consideration at the meeting in 1995. The Parties adopted a resolution that they would minimize emissions of methyl bromide, pending the further report of the panels.

No trade measures for HCFCs and methyl bromide

Trade measures were extended for HBFCs, but not for HCFCs or methyl bromide. The apprehension was that, if the non-Parties are not allowed HCFCs, they might continue to use CFCs.

Status quo for developing countries

The Article 5 Parties won their point that only the control measures of the second Meeting of the Parties would be applicable to them, with a grace period of ten years. The review in 1995 would decide whether any change in the control provisions for Annex A and Annex B would be necessary, and what control provisions for HCFCs, HBFCs and methyl bromide would apply to Article 5 Parties.

Criteria for exemptions for 'essential uses' after phase-out

The meeting decided that a use of a controlled substance should qualify as essential only: if it were necessary for health, safety or the functioning of society; if it encompassed cultural and intellectual aspects; and if there were no available technically and economically feasible alternatives or substitutes that were acceptable for the environment and human health. Exemptions on essential uses would be permitted only if all economically feasible steps had been taken to minimize essential use and emissions, and if the controlled substance was not available in sufficient quantity and quality from existing stocks of banked or recycled controlled substances. All Parties were to nominate the quantities required by them for essential uses every year. The Technology and Economic Assessment Panel was requested to review the submissions by the Parties and submit a report to the Meeting of the Parties, which would make a decision on essential uses. The meeting urged the Parties to set up international recycled halon banks.

BOX 3.9 TRADE RESTRICTIONS ON NON-PARTIES

Paragraph 8 of Article 4, which provided that trade restrictions would not apply to non-Parties that were in full compliance with the control measures of the Protocol, was utilized by some countries, pending their completion of formalities to ratify the Protocol. On the application of Colombia, a non-Party that claimed that it was in full compliance with the control measures of the Protocol, the fourth Meeting of the Parties agreed that the exception provided in Paragraph 8 of Article 4 would apply to Colombia and that no trade measures would be taken against Colombia. It also extended these exemptions to any non-Party that notified the Ozone Secretariat that it was in full compliance with the control measures and submitted data to the Secretariat to that effect.

The fifth Meeting of the Parties had before it applications from Malta, Jordan, Poland and Turkey, which, though Parties to the Protocol of 1987, were non-Parties to the London Amendment. They requested that the trade measures for substances of Annex B should not be made applicable to them since they were fully in compliance with the control measures of the Protocol for those substances. The meeting agreed that these countries could be exempted.

BOX 3.10 ISRAEL AND METHYL BROMIDE

Michael Graber, Deputy Executive Secretary, Ozone Secretariat

Each country faced its own political, financial and technical problems in doing its part to protect the ozone layer. There are many important lessons to be learned from these national struggles.

On 14 January 1988, Israel became the 46th country to sign the 1987 Montreal Protocol. However, protecting the stratospheric ozone layer was not considered to be a high priority and so steps to ratify the ozone treaties by Israel did not follow. In 1991, the Technology and Economic Assessment Panel (TEAP) and Scientific Assessment Panel (SAP) recommended that methyl bromide be added to the list of substances controlled by the Montreal Protocol. Bromine Compounds Ltd of Israel, at that time a daughter company of the Dead Sea Works Ltd, was the largest producer of methyl bromide in the world, making Israel the second largest producer in the world, after the USA. Israel's annual consumption of 2148 ODP tonnes in 1991 was by far the highest methyl bromide per capita consumption in the world. However, most of the methyl bromide produced by Bromine Compounds Ltd was exported. A further complication was that the government was part owner of Dead Sea Works.

Because the Montreal Protocol prohibited trade in ODSs with non-Parties, Bromine Compounds attempted to convince the government to ratify immediately the Vienna Convention, the Montreal Protocol and its London Amendment. However, Israel was not listed as a developing country. The London Amendment required countries not qualifying under Article 5 to contribute to the Multilateral Fund according to the United Nations scale of assessment. But government guidelines in Israel prohibited it from committing itself to international agreements if the financial resources required for this commitment were not identified. Following negotiations at the highest level between the Ministry of Finance and the Dead Sea Works Ltd, the budget required to pay the contribution to the Multilateral Fund by Israel was allocated. The necessary steps to ratify the Vienna Convention, Montreal Protocol and its London Amendment were then taken, and ratification was indeed effected on 30 June 1992.

A few months before the fourth Meeting of the Parties to the Montreal Protocol in Copenhagen in November 1992, a new government was formed under Yitzhak Rabin. He appointed Ora Namir as the Minister of Environment. Namir considered the ozone layer to be of little environmental importance, and decided that the Ministry of Environment should not represent Israel in Copenhagen. Because of pressure exerted by Bromine Compounds Ltd to participate, the Ministry of Foreign Affairs took over. The Israeli delegation to the fourth Meeting of the Parties consisted of several staff members of the Israeli Embassy in Copenhagen, and two employees of Bromine Compounds Ltd who were assisted by a consultant with knowledge of international law.

Following the fourth Meeting of the Parties in Copenhagen, Greenpeace International with Greenpeace Israel generated a news article covering half the front page of the *Ma'ariv*, a leading Israeli daily newspaper. The article accused the Israeli government and its Ministry of Environment, by then headed by Yossi Sarid, of being responsible for causing great damage to the ozone layer through its protection of the commercial interests of Bromine Compounds Ltd. The Natural Resources Defense Council, with support from like-minded corporate CEOs, also pressured the government to protect the ozone layer.

In response to this environmental activism, Prime Minister Rabin then directed Sarid, his Minister of Environment, to take appropriate action. Minister Sarid convened a meeting with representatives of Greenpeace and the Ministry of the Environment. Greenpeace was represented by Michael Affleck, head of the Greenpeace delegation;

Melanie Miller, Greenpeace advisor; John Maté, Ozone Campaign advisor; and Susy Baron, Greenpeace Mediterranean Campaign Manager. The Ministry of Environment was represented by Israel Peleg, Director General; Ruth Rotenberg, the Legal Counsellor; and myself, Head of the Air Quality Division.

Following the meeting, Minister Sarid decided that Israel would follow, to the letter, all its commitments under the Montreal Protocol, especially those related to phasing out production and consumption of methyl bromide, and that Israel would be represented only by personnel from the Ministry. Indeed, Israel was represented at the fifth Meeting of the Parties in Bangkok in 1993 by Peleg and by myself from the Ministry of Environment. It is noteworthy that Israel joined the Declaration on Methyl Bromide of 1993 requiring signatory Parties to reduce their consumption of methyl bromide voluntarily by 25 per cent at the latest by the year 2000, and to phase out totally the consumption of methyl bromide as soon as technically possible.[29]

Israel ratified the Copenhagen Amendment on 5 April 1995. By 1999, Israel had reduced methyl bromide use to 1487 ODP tonnes, well beyond the required 25 per cent reduction. The Ministry of Agriculture achieved these reductions through its extension-service programmes training farmers in agricultural methods that replaced methyl bromide with non-ODS alternatives. Regulations were also implemented by the Ministry of Environment that limited the production of methyl bromide in accordance with Israel's commitment under the Montreal Protocol. Israel now fully meets all its commitments under the Montreal Protocol.

Other decisions

At the fourth Meeting of the Parties to the Montreal Protocol, the following decisions were also made.

• There was no need to expedite the amendment procedure under Article 9 of the Vienna Convention.
• The responsibility for the legal interpretation of the Protocol rested ultimately with the Parties themselves.
• Insignificant quantities of controlled substances originating from inadvertent or coincidental production during a manufacturing process from unreacted feedstock, or from their use as process agents which were present in chemical substances as trace impurities and that were emitted during product manufacture or handling, need not be considered as covered by the definition of a controlled substance. The Parties appealed to all Parties to take steps to minimize such emissions and asked the Technology and Economic Assessment Panel to refine further its estimates of these emissions.
• If Article 5 Parties exceeded the maximum level of consumption prescribed by Article 5, and thus became ineligible for the grace period, the Implementation Committee would decide on the obligations of such Parties on a case-by-case basis.
• Five destruction technologies processes were approved.
• Recovered, reclaimed or recycled substances were exempt from being counted as controlled substances; participants urged all the Parties to recover substances as much as possible.

During the approval of the report, the representative of the Russian Federation said that although the country was committed to the phase-out of ozone-depleting substances, because of its extraordinary political, economic and social difficulties, it did not have the capacity to assume the additional obligations under the new amendments and adjustments to the Montreal Protocol.

The European Community declared that the Federal Republic of Yugoslavia could not automatically continue the membership of the former socialist Federal Republic of Yugoslavia in the United Nations, and that it did not accept representatives of the Federal Republic of Yugoslavia as representatives of Yugoslavia. The Yugoslavs again replied that such political issues did not belong at a conference on protection of the ozone layer.

A unanimous resolution at the end of the fourth Meeting of the Parties promised to reduce emissions of methyl bromide at least by 25 per cent by the end of 2000 and to phase it out as soon as technically feasible.

Chapter 4

Diplomacy: Racing towards success, 1993–2001

INTRODUCTION

The Copenhagen meeting in 1992 registered significant progress on many fronts; accelerated phase-out programme for the previously listed ODSs, a new non-compliance procedure, criteria for essential uses, a permanent Multilateral Fund and addition of HCFCs, HBFCs and methyl bromide to the list of controlled substances. At the same time, the meeting revealed new challenges. The break-up of the Soviet Union, the break-up in Yugoslavia and the instability in the new states, and the breakdown of Communism in all the states of Eastern Europe led to great social, economic and political problems in the region. Many of them announced in the Copenhagen meeting that they would not be able to contribute to the Multilateral Fund, would be unable to comply with advanced schedules of phase-out agreed to by the Copenhagen meeting and would need financial assistance even to implement the more leisurely timetables for phase-out agreed to by the second MOP in London in 1990. The Copenhagen Meeting also saw the developing countries refusing to accept any acceleration of their schedules before a review of the financial mechanism and technology transfer was completed in 1995 as provided in the Protocol, and refusing to accept controls on methyl bromide without a further study of alternatives. There was agreement of phase-out of HCFCs by the year 2030, but the Europeans felt that the time given was too long and that the phase-out could be achieved by the year 2015.

Over the next years, the phase-out by the industrialized countries by 1996 proceeded smoothly but there were many applications for exemptions for essential uses. The Technology and Economic Assessment Panel in its recommendations on the applications and the Meetings of the Parties in their approvals had to be scrupulously fair and firm even while being liberal with the exemptions for health applications of CFCs for metered-dose inhalers for asthma patients. The non-compliance by the countries in transition, the East Europeans, became real by 1997 and had to be dealt with firmly by the Implementation Committee and the MOP, even while encouraging assistance to these countries by the Global Environment Facility. The developing countries had to be persuaded to accept phase-out of methyl bromide. The Multilateral Fund performed very well. However, the many suggestions for improvement of its functioning, which came out of a review, had to be implemented to the satisfaction of all the countries. The reimbursements of the Fund every three years, in 1993, 1996 and 1999 had to be agreed to by all the Parties.

The increasing use of HCFCs caused concern to some. The European Union continued to attempt to tighten the restrictions on the controls on HCFCs by proposing adjustments to the Protocol in many meetings but succeeded only to a small extent, as all other countries were largely resistant. The financing of projects in developing countries by the Multilateral Fund to replace CFCs with HCFCs was criticized by the European Union as well as by NGOs. They contended there are ozone-safe alternatives for all applications. Other countries, however, disagreed and felt that use of HCFCs is necessary to phase out CFCs in some sectors. The use of HFC-134a, a greenhouse gas, as an alternative to CFCs was also criticized by the European Union and the NGOs. The last years of the 20th century saw the emergence of new chemicals with ozone-depletion potential, though small, in the market. These could not be controlled under the Montreal Protocol and many expressed concern that new uncontrolled ozone-depleting chemicals could emerge. The Parties discussed this issue but found no resolution.

On the whole, the consumption of the ozone-depleting chemicals came down by 85 per cent, thanks to the phase-out by the industrialized countries. The developing countries too have begun their phase-out thanks to the assistance of the Multilateral Fund. However, in the year 2000, several small countries, unable to control their imports of CFCs, exceeded their permitted quantities and were in non-compliance. They blamed illegal trade and excessive exports to their countries of CFC-using equipment for their non-compliance. The Parties are looking for solutions to their problems. This chapter details the efforts of governments to face the issues, and the considerable success achieved to date.

FIFTH MEETING OF THE PARTIES, BANGKOK, 1993: SECOND REPLENISHMENT OF THE FUND BY US$455 MILLION

No exemptions for 'essential uses' of halons

The meetings of the Open-Ended Working Group and the fifth Meeting of the Parties in 1993[1] assumed importance because of the impending phase-out of halons by the end of 1993 and the applications by some Parties for essential-use exemptions that had to be decided by the fifth Meeting of the Parties. A number of countries, including Austria, Belgium, Finland, Germany, Italy, Japan, Malta, Poland and the UK, had written to the Secretariat regarding their continued need for halons in critical uses such as civil and military aircraft. The Technology and Economic Assessment Panel (TEAP), which considered the submissions, noted that new substitutes existed for fixed fire suppression applications, and that existing halon stocks amply covered the total quantity of halons for which exemptions had been sought. It noted that when production of halons closed by the end of 1993, the feasibility of recycled halons would improve. Thus no exemptions were recommended by TEAP for halons.

The fifth Meeting of the Parties was held in Bangkok in November 1993, and opened by the Deputy Prime Minister of Thailand, Banyat Bantadtan. The meeting agreed that no production or consumption exemption was necessary for halons after the phase-out in 1994. Thus the production of halons in the non-Article 5 Parties came to a complete stop by the end of 1993; these Parties agreed that they would manage their critical needs through recycled halons. Among the Article 5 Parties, only China produced significant quantities of halons, and India produced small quantities. The other Article 5 Parties expressed their concerns that they might not get halons for their critical uses after the phase-out by the non-Article 5 Parties. The TEAP was requested to study this issue. The meeting noted that UNEP, with the assistance of the Multilateral Fund, was functioning as the clearinghouse for information on international halon bank management, and requested UNEP to hold the details of all such banks with recycled halon for sale.

Second replenishment, 1994–1996

The second replenishment of the Multilateral Fund for the years 1994–1996 was to be decided in 1993. The chair of the Executive Committee of the Multilateral Fund, Eileen Claussen of the USA, reported to the fifth Meeting of the Parties that 1991 had been a year of infrastructure development. The real work of the committee had begun in 1992, when nine country programmes were approved, and new procedures were adopted to ensure implementing agency coordination and to avoid duplication of effort. The United Nations Industrial Development Organization (UNIDO) had been added as the fourth implementing agency. A large increase in project and programme approvals had occurred in 1993. Claussen reported that the fund was clearly operational; early executive support for the institutional strengthening in Article 5 Parties would facilitate more expenditure and project development. She noted that the outstanding contributions for 1991 and 1992 stood at 19 per cent, three-quarters of which was from countries with economies in transition. She also presented a report on the replenishment needs for the next three years and proposed a budget of US$510 million for 1994, 1995 and 1996. The proposed budget of US$510 million would reduce consumption of ozone-depleting substances (ODSs) by the Article 5 Parties by 46,600 tonnes and production by 17,000 tonnes.

With very little controversy, the meeting approved the provision of US$510 million for 1994–1996, including the US$55 million that was unallocated from 1991–1994. It also approved the contribution of each non-Article 5 Party in accordance with the United Nations scales of contribution for 1994–1996. Hungary, Bulgaria, Poland and other countries with their economies in transition said they would not be able to contribute to the Multilateral Fund and requested the Executive Committee to consider the various possibilities for addressing this situation. The meeting requested the Executive Committee to prepare a report on the functioning of the Multilateral Fund so that the obligations of the Article 5 Parties for control measures could be reviewed in 1995, as mandated by Article 5, Paragraph 8, and to make efforts to get in-kind contributions for the countries with economies in transition which were unable to contribute. It authorized the Executive Committee to provide funding for a limited number of methyl bromide information projects for Article 5 Parties, although there were no control measures as yet for these countries for methyl bromide.

The meeting noted that Cyprus, Kuwait, the Republic of Korea, Saudi Arabia, Singapore and the United Arab Emirates, though classified as developing countries, consumed more than 0.3 kilograms per capita of CFCs and hence were not eligible to operate under Article 5.

Products made with, but not containing, CFCs

The meeting considered the feasibility of controlling imports from non-Parties of products made with, but not containing, substances of Group I of Annex A to the Protocol; these included products such as computers which were cleaned with CFC-113, but did not contain any ODSs. The TEAP reported that it would be very difficult and expensive to identify such products since the detection of traces of ODSs left in such products required significant expertise and resources. The meeting decided that it was not feasible to impose a ban or restriction on the import of such products at that stage.

Eastern Europe seeks special status

At the end of the fifth Meeting of the Parties, Belarus, Bulgaria, Romania, the Russian Federation and Ukraine – which were changing from Communism to a market economy – explained their social and economic difficulties and declared that the sixth Meeting of the Parties should decide on giving the countries with economies in transition a special status to provide for concessions and certain flexibility in fulfilment of their obligations.

Bangkok Declarations

Austria, Germany, Liechtenstein and Switzerland declared that CFCs could be substituted in many areas without using HCFCs, and that the phase-out schedule of HCFCs should start immediately, be completed by 2015, and not be prolonged until 2030 as permitted by the Copenhagen Amendment in 1992. Another declaration by many countries[2] also agreed with this schedule, and declared their resolve to use HCFCs only for necessary applications and to phase out HCFCs not later than 2015. Many countries[3] issued another declaration stating their resolve to reduce their consumption of methyl bromide by at least 25 per cent by the year 2000 and to phase it out entirely as soon as technically possible.

The end of separate reporting under the Vienna Convention

The third Meeting of the Conference of the Parties to the Vienna Convention (COP 3)[4] met back to back with the fifth Meeting of the Parties to the Protocol. It decided, based on the report of the Scientific Assessment Panel, that there was no need to get separate data reports under the Vienna Convention for substances specified in the Convention, and that reporting under the Protocol would suffice. It considered the recommendation of the ozone research managers from the second Meeting of the Parties that reporting under Annex I should include HFCs, which are used as substitutes for CFCs, in view of their high global warming potential, and deferred a decision on such reporting pending the decision of the Parties to the Climate Change Convention. The meeting reiterated its appeal to the Parties to contribute to the World Meteorological Organization (WMO) Special Fund to assist developing countries for ozone research.

SIXTH MEETING OF THE PARTIES, NAIROBI, 1994: RUSSIAN FEDERATION GIVES NOTICE OF NON-COMPLIANCE

Applications for essential-use exemptions for CFCs

The meetings of the Open-Ended Working Group and the Parties in 1994[5] focused, inter alia, on essential-use exemption applications for CFCs, other controlled substances for 1996 and beyond, and other issues on which the TEAP reported. The TEAP recommended an exemption for production and consumption of CFCs for metered-dose inhalers used by asthma patients, and for laboratory and analytical uses of controlled substances for all Parties globally. The applications for essential-use exemptions by Belgium, Denmark, Ireland, Italy and the UK for the use of CFC-113 for fingerprints were not accepted because there were alternatives. The Russian Federation said that halon-2402, which it alone used, could not be recycled; hence, it needed to continue production but did not apply for an exemption in time. Its request was, therefore, not considered. The total quantity of essential-use nominations recommended by the TEAP and approved by the meeting for exemption for 1996 was about 13,000 tonnes of CFCs, mostly for medical aerosols, and 56.8 tonnes of methyl chloroform. Nominations for another 4000 tonnes were not recommended by the panel and hence not considered by the meeting. The consumption of nearly 800,000 tonnes of these substances by the non-Article 5 Parties in 1986 would, by this decision, decrease to 13,000 tonnes by 1996.

France, Italy and The Netherlands nominated for exemption the use of carbon tetrachloride in manufacturing certain products. The TEAP observed that its use as a process agent was widespread in many countries, and recommended that the use need not be counted as consumption, as emissions were negligible. The issue of ozone-depleting substances as process agents assumed importance in this meeting. Some Parties might have interpreted their use in process agents as feedstock applications eligible for exemption, and other Parties, as subject to phase-out. The TEAP was unable to recommend exemption under the essential-use criteria to Parties submitting applications for such uses nominated in 1994. The meeting requested the TEAP to identify the uses of ODSs as chemical process agents, to estimate emissions in such uses, to evaluate alternatives, and to submit a report by March 1995. Meanwhile, it was decided that for 1996 only, chemical process agents would be treated as feedstock.

Are HCFCs and HFCs needed?

The TEAP reported that HCFCs were technically and economically necessary for the transition in a number of refrigeration and air-conditioning applications, the manufacture of insulating foams, selected and limited solvent applications and certain fire protection applications where space constraints existed. More environmentally acceptable alternatives to HCFCs were available for most fire-fighting applications, non-insulating foams and a majority of solvent applications, aerosols and sterilization. The Executive Committee presented a similar report listing the applications where HCFCs could be avoided and other applications where HCFCs were needed. During the debate, representatives of the European Economic Community, Austria, Finland, Norway and Sweden wanted the TEAP to investigate an earlier phase-out of HCFCs. Other countries, including the USA, Japan and several Article 5 Parties, questioned the justification for the early phase-out of HCFCs and its additional costs. One representative noted that the Multilateral Fund should encourage hydrocarbon technologies rather than HFC-134a technology for refrigeration since HFC-134a has a high global warming potential. A co-chair of the TEAP replied that hydrocarbon technology required higher investments, particularly for safety, but this could change in the future. The meeting requested the TEAP to prepare a report on alternatives to HCFCs and methyl bromide.

Products containing Annex B substances: Not worth identification

The meeting considered the issue of banning the import of products containing Annex B substances (other CFCs, carbon tetrachloride and methyl chloroform) from non-Parties. The TEAP reported that since the Protocol had already been ratified by a large number of Parties and since these substances were to be phased out by the end of 1995, it was not worthwhile to identify such products and impose import restrictions. The Meeting decided, based on the TEAP report, not to elaborate the list of such products.

Recycling

The Scientific Assessment Panel, as requested by the fifth Meeting of the Parties, studied the impact of the continued use of recycled controlled substances, and reported that recycling would benefit the ozone layer since emissions to the stratosphere would be delayed. One Party expressed apprehension about whether the exemption of recycled substances might have created a loophole. France suggested that the list of reclamation centres should be prepared and circulated to all the Parties, so that the origin of the recycled substances could be checked. Some mentioned the misuse of the trade in recycled substances. One representative said that UNEP should verify the claims concerning ozone-safe substitutes and alternatives, and advise the countries with information on such substitutes. Some participants argued that there should also be a legal regime for product safety and liability.

Conditions for exporting CFCs to developing countries

The meeting decided that the Article 5 Parties importing virgin controlled substances should send to the supplying Party, within 60 days of imports, a letter specifying the quantity of those substances, and stating that the imports were for meeting basic domestic needs. The supplying Party should send an annual summary of these details to the Ozone Secretariat. This decision, intended to curb excessive production and consumption of ODSs, was, in a later meeting, modified for the Parties to provide only data on their imports, exports and the origin/source of such trade. The meeting requested that all Parties exporting used substances take steps to ensure that the substances were labelled correctly, and make their best efforts to require their companies to ensure that proper documentation accompanied such exports. It also requested the Ozone Secretariat to study and report on this trade.

Russian Federation gives notice of non-compliance

A representative of the Russian Federation made a detailed statement to the meeting on behalf of itself and some other countries. The representative informed the meeting that the countries' phase-out of Annex A and Annex B substances would be delayed at least until 1998. The countries would require financial assistance to fulfil their obligations; they would be unable to contribute to the Multilateral Fund even though they wanted to do so; and they could contribute in kind. They suggested that special status should be given to the countries with economies in transition, a proposal that Poland fully supported.

Global Environment Facility assistance plan

The Global Environment Facility, in fulfilment of its mandate on ozone depletion, decided in 1994 to assist all countries that were eligible for assistance under its charter but not eligible for the assistance of the Multilateral Fund. The Multilateral Fund had the mandate to assist all developing-country Parties operating under Article 5, ie those whose ODS consumption was below the limit specified in Article 5. Almost all the countries of Eastern and Central Europe and the countries of the former Soviet Union were not classified as developing countries, and hence not eligible for the assistance of the Multilateral Fund. All these countries, however, became eligible for assistance from the Global Environment Facility because their per-capita incomes were below the threshold fixed by the Facility for such assistance. The Czech Republic became the first country with its economy in transition to receive assistance at the end of 1994. An operational strategy adopted in October 1995 made the ratification of the London Amendment and approval of the Implementation Committee necessary for the support of the Global Environment Facility.

A panel to supervise review of the financial mechanism

The Meeting of the Parties agreed that a steering panel consisting of representatives from Canada, France, the USA, Mexico, India and Mauritius would supervise the process for the review of the financial mechanism, as called for by a decision of MOP 4 in 1992. The meeting approved the terms of reference for the study of the financial mechanism. The NGOs present requested that they should be represented on the committee, but this request was not accepted by the meeting.

Developing countries consuming more than the limit

The meetings of the Open-Ended Working Group in 1994 discussed the cases of certain Article 5 Parties, such as South Korea and Singapore, which were classified as not operating under Article 5 because their consumption of CFCs exceeded 0.3 kilograms per capita per year. The issue was their contributions to the Multilateral Fund for the years when they operated as non-Article 5 Parties and their eligibility for assistance from the Multilateral Fund after they converted themselves into Article 5 Parties by reducing their consumption below 0.3kg per capita in a future year. The meeting also discussed the cases of Article 5 Parties temporarily classified by the Secretariat as operating under Article 5, pending full consumption-data reporting by these Parties. Developing countries can be classified as operating under Article 5 only when they report their consumption data and demonstrate that their consumption is below the limits set by Article 5. Thirty-four Article 5 Parties, though they did not report their data by the time of the 1994 meetings, were temporarily classified in order not to hamper the work of the Executive Committee of the Multilateral Fund in assisting these countries.

The sixth Meeting of the Parties agreed on these principles for such developing countries:

1   A country would be classified temporarily as operating under Article 5 only for a period of two years after this decision and, after that, the Article 5 status could not be extended without data reporting unless the country sought the assistance of the Executive Committee and Implementation Committee (for building up its capacity to report).
2   A developing country temporarily classified as operating under Article 5 would lose the status if it did not report consumption data within one year of the approval of its country programme and its institutional strengthening by the Executive Committee.
3   The projects approved for a country temporarily classified would continue to be funded even if the countries were subsequently classified as not operating under Article 5.
4   No new project would be sanctioned during the period in which the country was classified as not operating under Article 5.
5   Parties could correct their data in the interest of accuracy, but no change of classification would be permitted for that year. An explanatory note should accompany any correction.
6   In the case of developing-country Parties initially classified as not operating under Article 5 and then reclassified as operating under Article 5, any outstanding contribution to the Multilateral Fund would be disregarded only for the years in which they were reclassified as operating under Article 5.

7   Any Party reclassified as operating under Article 5 could use the remainder of the ten-year grace period though encouraged not to do so. Such Parties should not be requested to contribute to the Multilateral Fund and would not be eligible for assistance from the fund.

Declaration by developing countries

At the end of the meeting, many Article 5 Parties[6] presented a declaration urging non-Article 5 Parties to honour their pledges fully to the Multilateral Fund, and for their incremental operating costs to be paid for up to four years. All projects to phase out ODSs should be financed. The Fund should fully support the companies in Article 5 Parties that exported ODS-free products. The Parties should 'consider collectively and in the most democratic manner the need to halt the tendency to selectivity and restrictiveness of the Multilateral Fund'. The proponents of the declaration explained that the declaration was aimed at some of the decisions and the functioning of the Executive Committee.

THIRD REPORTS OF THE ASSESSMENT PANELS, 1994

The assessment panels published their third comprehensive assessment reports under Article 6 of the Montreal Protocol in December 1994 (see Chapter 1). A synthesis of the assessment reports by the co-chairs of the panels detailed further options for control measures for the Parties:

- Reduction of up to 90 per cent in the use of methyl bromide was technically and economically feasible for non-Article 5 countries.
- Limits to future growth in HCFC and methyl bromide use in Article 5 countries were feasible.
- Reducing the size of the HCFC cap was technically but perhaps not economically feasible in every non-Article 5 country. It was technically but perhaps not economically feasible to phase out HCFCs completely by 2015 because HCFCs would be necessary to service equipment with a useful lifetime beyond 2015.
- The TEAP reported the financial consequences of different scenarios for phasing out methyl bromide for Article 5 Parties. A freeze in 1998 would cost US$9.5–78.5 million; a 25 per cent reduction by 2000 would cost US$48.6–232 million; a 25 per cent reduction by 2005 and a phase-out by 2011 would cost US$86–326 million. A phase-out by 2001 was not considered feasible.
- The costs of an HCFC freeze by 2000 for Article 5 Parties were estimated at US$85–500 million; for the Copenhagen schedule plus ten years (ie freeze in 2006, 35 per cent reduction in 2014, 65 per cent reduction in 2020, 90 per cent reduction in 2025, 99.5 per cent reduction in 2030 and complete phase-out by 2040), US$115–235 million; and for a freeze by 2011 and a phase-out by 2040, US$80–160 million.
- With respect to the trade in Annex A and Annex B substances, if Article 5 producer countries were to be restricted to supplying their domestic markets

only, the non-producing Article 5 Parties would have to import 75,000 tonnes from non-Article 5 sources.
• About 40,000 tonnes of ODSs were used as process agents annually; those uses were expected to increase. A total of 10,400 tonnes were emitted. Only about 230 tonnes were from Japan, Western Europe and North America; 660 tonnes were from facilities in countries with economies in transition; and the rest were from Article 5 Parties. By use of the best available technologies, the emissions of Article 5 Parties could be eliminated with a total investment cost of US$240–730 million.
• Twelve nominations had been received for essential uses; the TEAP recommended essential-use exemptions for CFCs for metered-dose inhalers to treat asthma and chronic obstructive pulmonary disease, and in solvents for rocket manufacture; and for halon-2402 for fire protection in the countries of the former Soviet Union. Poland presented a nomination for CFC-12 for servicing domestic refrigeration, which was not recommended by the TEAP because it was not in accordance with approved criteria.

REVIEW OF CONTROL MEASURES AND FINANCIAL MECHANISM FOR DEVELOPING COUNTRIES, 1995

Review of control measures for Article 5 Parties

Paragraph 8 of Article 5, introduced as a part of the London Amendment in 1990, mandated that 'a Meeting of the Parties shall review, not later than 1995, the situation of the Parties operating under Paragraph 1 (of Article 5) including the effective implementation of the financial cooperation and transfer of technology to them, and adopt such revisions that may be deemed necessary regarding the schedule of control measures applicable to those parties.'

The Executive Committee of the Multilateral Fund prepared a study by consultants in 1995 to assist the review. The scope of the study was to analyse the phase-out schedules for Article 5 Parties and examine various options to alter the schedules. Consultants carried out this study after conducting extensive interviews and after visiting many Article 5 countries, including India and China. The study examined options, including an immediate phase-out in 1996, and options up to the year 2010 with four more years to service equipment. It examined variations, from two to four years, of meeting the incremental recurring costs. For all the scenarios, it examined the technical feasibility and costs to the Multilateral Fund, as well as the real costs to the Article 5 Parties. With a discount rate of 10 per cent, the least costly option was the phase-out in the year 2010, with a tail (permission to consume ODSs) for servicing, which cost US$855 million. The costliest option was for the phase-out in 2000, costing US$3.8 billion dollars. A phase-out in 2006 would cost US$1.8 billion. The fastest phase-out by sector, rather than by the countries, would cost US$1.357 billion. The phase-outs with tails for servicing would cost much less. The impact of the different scenarios on the chlorine loading in the atmosphere was also calculated. The year of the return of chlorine in the atmosphere to 2 parts per billion, which was the pre-1970 level, ranged from 2082 to 2087, varying very little for different options. The study concluded that, given the extraordinary technical difficulties of Article 5 countries, the earliest feasible phase-out year would be 2006.

Results of review of financial mechanism

A consultant for UNEP prepared a review of the financial mechanism. The consultant's study team visited five Article 5 Parties and interviewed many of those concerned, both Article 5 and non-Article 5 Parties. Its conclusions were:

- In the first four years, the projects sanctioned would result in the phase-out of 51,500 tonnes of ODSs, or about 25 per cent of Article 5 Party consumption. The actual phase-out would be about 12,000 tonnes due to delays in implementation.
- The institutions of the financial mechanism had approached clarification of project approval and cost definition criteria through a case-by-case approach in which the consideration of specific project proposals had resulted in decisions on policy issues.
- The policy development had been less of a priority, which had led to frustrations at the level of enterprises, Article 5 governments, and implementing agencies because individual projects had become subjects of outstanding policy resolutions.
- The Executive Committee had dealt with some policy issues quickly, whereas others had been cautiously dealt with.
- The Fund Secretariat had injected an element of discipline into the project process, avoiding the provision of fund support for ineligible costs.
- No implementing agency had shown any interest in a concessional lending programme. The institutional framework produced an increasing flow of approved projects. Implementation of the projects in a timely manner was the real issue. There had been significant delay in implementation of the projects for a variety of reasons.
- None of the implementing agencies had integrated the Montreal Protocol priorities into their ongoing development-policy dialogues with Article 5 countries or had influenced these countries to take regulatory, legislative or other actions as a result of this dialogue.
- Some of the bilateral projects had contributed to the introduction of newer ODS-free technologies and use of innovative project concepts.
- UNEP had responded effectively to the information needs of Article 5 Parties.
- The Fund Secretariat developed an inventory of approved equipment and the negative indicative lists of costs, which improved understanding.
- The Executive Committee funded the institutional strengthening activities, even though they were not specifically mentioned by the indicative list. The study confirmed that more emphasis on institutional strengthening would yield both faster project implementation and overall ODS phase-out.
- The country programmes' preparation had sensitized governments of the Article 5 Parties to the ODS issue. But the policies and strategies that had

been produced had been declarations of intent, rather than commitments to actions, particularly in the area of legislation and regulation.

- There should be more delegation from the Executive Committee to the implementing agencies, subject to the condition that the policy issues had been resolved, decisions on eligible equipment items were made, and guidelines were developed.
- The World Bank should concentrate its operations in large countries, while the United Nations Development Programme (UNDP) and UNIDO should concentrate on other countries. The World Bank should examine its system of local financial intermediaries and speed up its process.
- The implementing agencies should develop plans for integrating Montreal Protocol matters and mobilizing non-Multilateral-Fund resources in support of the Protocol objectives into their ongoing development programming discussion with Article 5 Parties.
- Priority should be given to cost-effective sectors so that the phase-out could be faster with the same resources. A system for lump-sum compensation could be considered for enterprises that cross the cost-effectiveness boundaries.

PROPOSALS FOR ADJUSTMENTS BEFORE WORKING GROUP MEETINGS, 1995

On the basis of the reports of the assessment panels, the report of the Executive Committee and the review of the financial mechanism, the Parties made many proposals to the meetings of the Open-Ended Working Group in 1995[7] for adjusting and amending the Protocol:

- India recommended that the additional production allowed to the non-Article 5 Parties to meet the basic domestic needs of Article 5 Parties could be reduced to 10 per cent or 0 per cent from the present 15 per cent.
- The European Community, Sweden and others recommended a faster phase-out of HCFCs, a reduction of the ceiling, and a positive list of applications for use of HCFCs.
- Australia, Canada, the European Community, Japan, Malawi, The Netherlands, New Zealand, Norway, South Africa and the USA recommended many amendments for the phase-out of methyl bromide, ranging from the year 2001 to 2011, and for gradually reducing the exemption given for quarantine and pre-shipment applications, without specifying the year of the phase-out.
- Australia, China, Canada, the European Community, India, Norway, Sri Lanka and the USA made different proposals for changing the control measures for the Article 5 Parties: the present timetable of a phase-out for 2010 should be extended for another 10 or 20 years for servicing the existing refrigeration equipment; Article 5 Parties should stop using CFCs for refrigeration equipment after 1999 or 2006; the grace period for Article 5 Parties should be reduced; five or ten years should be added to the present

grace period; and the phase-out dates for methyl bromide should be 2001, 2006 or 2011.

- Italy and others recommended that from 1 January 1997, no new fire equipment depending on halons should be installed unless it was for a critical use approved by the Parties.
- Sweden and others recommended production controls on HCFCs.
- Australia, the European Community, Canada, Norway, New Zealand, Switzerland and the USA recommended that exports of CFCs be permitted only if there was a shortfall in supplies of controlled substances, and if the importing Party certified that the CFCs were needed for basic domestic needs. The production of controlled substances for export should utilize only existing production capacity and should not exceed 10 per cent or 15 per cent of the calculated level of production of the exporting Party for 1994. The Meeting of the Parties could permit a higher export level.
- France, New Zealand and Switzerland argued that process-agent use was not the same as feedstock use and that, from 1998 on, use of controlled substances as chemical process agents should be exempted under the essential-use procedure on application. The exemption for process agents should apply only for 1997.

Another attempt to define 'basic domestic needs'

During a discussion on the definition of basic domestic needs, many non-Article 5 Parties suggested an amendment to the Protocol to address the issues and facilitate export by Article 5 Parties only to the extent necessary, without undermining the phase-out of other Article 5 Parties through excessive supplies. They felt that the producing Article 5 Parties should produce only to satisfy their own needs, rather than to export to other Article 5 Parties, since that was the intent of the term 'basic domestic needs' in the Protocol. The Article 5 Parties felt that the term should include the export and import needs of controlled substances, as well as products containing or made with ODSs, and trade among those Parties during grace and phase-out periods. They opposed the amendment, stating that it would limit the supply of ODSs to Article 5 countries by disallowing ODS supplies from Article 5 countries, and would have far-reaching adverse consequences. A small informal group established to discuss this issue concluded that no agreement was possible.

Possible scenarios for developing countries

A subgroup on control scenarios for Article 5 Parties concluded that only the scenarios of a phase-out by 2006 with a service tail or phase-out by 2010 could be considered for Article 5 Parties, taking into account environmental benefits and technical feasibility. A representative of the Group of 77 said that not only were limited funds a constraint, but also technology transfer, which was not occurring as it should. The Group of 77 was of the opinion that it could not accept any more speeding up of its schedules. It would not accept new controls without more financial provisions being made at the same time. One representative of the Organisation for Economic Co-operation and

Development (OECD) countries replied that it was fully committed to the Multilateral Fund, but could not commit to specific replenishment since the next replenishment for the triennium 1997–1999 would be finalized only by the Meeting of Parties in 1996.

Kenya opposed any further controls on methyl bromide until the remaining scientific uncertainties were resolved and demonstrations of methyl bromide alternatives, funded by the Multilateral Fund, were finished. Many, but not all, Article 5 Parties expressed support for that position. Their reason was that key export products, on which their economies depended, required the use of methyl bromide, for which there were no substitutes available. Many non-Article 5 Parties replied that they had committed themselves to phasing out methyl bromide by the year 2001, and that there were many alternatives to methyl bromide.

Continuing use of halons

The Halons Technical Options Committee and the TEAP said, with reference to Italy's proposal regarding a ban on use of halons in fire-fighting equipment unless the use was approved as critical, that there was no evidence that recovered halons were used for non-critical uses. Forcing HCFCs, which were considered alternatives to halons, into the wrong fire-fighting applications, would result in frivolous, unnecessary uses of HCFCs without benefit or protection to life or property. A limitation in the use of halons could backfire if equipment owners released halons in the atmosphere because of the anticipated liability to recover or destroy halons. Italy then withdrew its proposal and submitted a draft decision in its place.

SEVENTH MEETING OF THE PARTIES, VIENNA, 1995: FURTHER STRENGTHENING OF THE CONTROL MEASURES

Tenth anniversary of the Vienna Convention

The seventh Meeting of the Parties to the Montreal Protocol took place in Vienna on 5–7 December 1995.[8] Martin Bartenstein, the Minister for Environment of Austria, opened the meeting. During the meeting, Austria celebrated the tenth anniversary of the Vienna Convention with a symposium.[9] The occasion had added significance because three scientists who had made significant contributions to the ozone-depletion theory, F Sherwood Rowland, Mario Molina and Paul Crutzen, were awarded the Nobel Prize for Chemistry for 1995 for their discoveries of the causes of ozone depletion in 1970–1974. UNEP instituted ozone awards for the first time, presenting them to 20 men and women who had made extraordinary contributions to the protection of the ozone layer.

Tightening of controls on HCFCs and methyl bromide and other measures

Based on intense informal negotiations during the meeting, the seventh Meeting of the Parties adopted the following adjustments:

*   A reduction of the ceiling of 3.1 per cent of CFC plus HCFC consumption in 1989 (adopted in 1992) to 2.8 per cent of CFC plus HCFC consumption.
*   Control measures for HCFCs for Article 5 Parties of a freeze in the year 2016 at the level of 2015, and a phase-out by the year 2040.
*   Control measures for methyl bromide for non-Article 5 Parties of a freeze in 1995 at the 1991 level and a phase-out by the year 2010; for developing countries operating under Article 5, a freeze in the year 2002 at the level of average consumption of 1995–1998; and an exemption for quarantine and pre-shipment applications for all parties. The ODP of methyl bromide was reduced from 0.7 to 0.6.
*   There was no change in the control measures for Article 5 Parties for substances of Annex A and Annex B. The obligation of Article 5 Parties continued to be the London Amendment and Adjustment schedule for non-Article 5 Parties plus ten years.

Related decisions included:

*   The Meeting of the Parties would consider, by the year 2000, the need for further adjustments to the HCFC schedule for Article 5 Parties.
*   The Parties would reduce methyl bromide emissions of quarantine and pre-shipment applications through improved application techniques.
*   The TEAP report on trade measures and further control measures on methyl bromide would be considered in 1997.
*   Although no definition of basic domestic needs had been determined, the Parties would report to the Ozone Secretariat their exports to Article 5 Parties. The Article 5 Parties would be bound by the same restrictions on production for supplying the basic domestic needs as applicable to the non-Article 5 Parties in Article 2.
*   From 7 December 1995, no Party would build a new factory for supplying CFCs; the Meeting of the Parties in 1997 would provide for a licensing system, including a ban on unlicensed imports and exports. The Parties should consider a mechanism under which trade in controlled substances could take place only between Parties that comply with the Protocol. This had implications for trade by the Russian Federation, which was in non-compliance. The Russian Federation was allowed to export only to states of the former Soviet Union after ensuring that those states would not make any re-exports.
*   Process agents would be treated as similar to feedstocks for 1996 and 1997, and further decisions would be made in 1997.
*   Parties were urged to restrict the use of halons on a voluntary basis for only critical uses, and to recover halons effectively.

- Full assistance from the Global Environment Facility was recommended to the Russian Federation and all countries with economies in transition to enable implementation of control measures by these Parties.
- The TEAP was asked to provide criteria for critical agricultural uses to be exempted for methyl bromide
- The assessment panels were requested to make their next full assessment by November 1998. The Scientific Assessment Panel was requested to report on the impact of aircraft emissions on the atmosphere, in collaboration with the International Civil Aviation Organization (ICAO) and the Intergovernmental Panel on Climate Change (IPCC). The TEAP was requested to report on a number of issues relating to metered-dose inhalers.
- To satisfy the provisions of the Basel Convention on Transboundary Movements of Hazardous Wastes, the meeting decided that transfers of used substances should take place only if the recipient country had recycling facilities. The Basel Convention classified the used controlled substances of the Montreal Protocol as hazardous wastes that could be transferred to other countries only under certain conditions.

Essential uses

The TEAP recommended essential uses only for metered-dose inhalers and halon-2402 for the Russian Federation until 1999. The reason given for recommending the exemption requested by Russia was that there was no global supply of halon-2402, which is used only by Russia. Hence, it needed the exemption until it established a banking system of its own to meet critical needs. There was a debate on whether the essential-use exemption for medical aerosols was still required. Some held the view that since alternative CFC-free medical aerosols had appeared on the market, an essential-use exemption should not be sanctioned for CFC medical aerosols. Others felt that doctor and patient education was needed to gain acceptance of the alternatives. As this was a sensitive health issue, each country should devise its own strategy for phasing out CFCs in medical aerosols. Many stressed that a detailed transition strategy should be proposed by the meeting for all the countries.

Actions to improve the financial mechanism

Based on the report of a subgroup on the review of the financial mechanism, for which France was the convenor, the meeting approved a list of 21 actions for the Executive Committee to take to improve the financial mechanism. The Executive Committee was requested to report periodically to the Parties on the implementation of the actions. The actions included:

- Completion of a systematic approach to policy development, monitoring and evaluation guidelines, delegation and dissemination of policies among national ozone units and consultants.
- Completion of incremental cost guidelines for the production of CFC substitutes.

BOX 4.1 TRADE IN USED PRODUCTS CONTAINING CFCS

With the phase-out of CFCs in the non-Article 5 Parties and introduction of new models of products, such as refrigerators, using alternatives, the products that relied on CFCs were discarded. There were no buyers for these used CFC products in the non-Article 5 Parties and they were increasingly exported to Article 5 Parties. The Article 5 Parties were concerned that these products would increase their demand for CFCs and make it impossible for them to reduce their consumption of CFCs to comply with the Protocol.

Mauritius proposed at the seventh Meeting of the Parties in 1995 that Parties should take all measures, including labelling the products, to regulate export and import of products containing ODSs or equipment used in the manufacturing of such products. Parties should avert any dumping of such products in Article 5 Party countries. The meeting decided to recommend that each Party adopt measures, including labelling of products and equipment, to regulate imports and exports of products containing ODSs or technologies used in manufacturing such products, and report to the Secretariat on the steps taken.

The Secretariat informed the working group meeting in 1997 that ten Parties reported on the measures taken to regulate imports and exports of products that contained or used CFCs. The ninth Meeting of the Parties in 1997 recommended that non-Article 5 Parties specifically take measures to control the export of used products which used the controlled substances of Annexes A and B. The Article 5 Parties demanded in the meetings of 1998 that all Parties should ban the export of used CFC products. A subgroup formed to discuss this issue arrived at a compromise that if a Party itself did not manufacture a CFC product and had regulations to ban the import of such products, it could inform the Ozone Secretariat, which would communicate the list of such products to all the Parties. It was expected that all the Parties would then take measures to honour the regulations of the importing Parties.

The tenth Meeting of the Parties in 1998 agreed to the recommendation of the subgroup. Since then, the Secretariat has communicated to all the Parties such communications received regarding a Party's wish not to import particular used products. The practical impact of these communications seemed to be negligible, since the complaints about the dumping of these products continued to be voiced by many countries at the Meetings of the Parties every year from 1998 on.

- Selection of a lead agency for preparing a framework for policy dialogue with Article 5 Parties to enhance regulatory support for an ODS phase-out.
- Completion of a study on concessional lending.
- Integration of the Protocol issues in the regular development dialogue of the implementing agencies with Article 5 Parties.
- Attention to training needs, particularly of small ODS users.
- Completion of a study on technology transfer.

The Executive Committee presented a special report covering its three-year rolling business plan for the fund, which projected a cash inflow of US$132 million for 1996 and gave a detailed plan on what the resources were required for. The meeting requested the TEAP, for the first time, to determine the monies needed by the Multilateral Fund for 1997–1999.

BOX 4.2 BACKLASH: A SMALL NUMBER OF VOCAL SCEPTICS
DOUBT THE LINK BETWEEN CFCS AND OZONE DEPLETION

While scientists were getting more confirmation of the role of CFCs in ozone depletion, a small segment of people, particularly in the USA and France, published newsletters and books in an unsuccessful bid to convince everyone that CFCs were beneficial and not harmful. One of these publications, the 21st Century Science Associates' 1992 book, *The Holes in the Ozone Scare: The Scientific Evidence that the Sky is Not Falling*, argued that much more chlorine reached the atmosphere annually from sea water, volcanoes, biomass burning and ocean biota than from CFCs. CFCs had nothing to do with ozone depletion, the book argued.

Ozone-depletion sceptics claimed that the major multinational corporations such as DuPont and ICI had a common front to market their HFCs and other patented replacement chemicals for CFCs. They claimed that these corporations funded environmental organizations through their foundations to propagate the theory of ozone depletion from CFCs. They asked: what was the motive of the scientists, policy-makers, corporations and environmental movements? They all wanted to solve the population problem, which was responsible for environmental ills, by killing most of the people in the developing countries, the sceptics claimed. How was this to be achieved? Without CFCs, there would be no refrigeration, food storage would be affected, and people would die. They claimed that by scaring the general population with stories of imminent catastrophe, these policy-makers intended to justify adoption of stringent measures that would curtail economic growth and population sizes. The ozone hole was just one of several such scare stories; other such stories were global warming and PCBs, they claimed.

Business Week on 19 June 1995 published a viewpoint by Paul Craig Roberts, a fellow at the Cato Institute in Washington, DC, who wrote, 'quietly, now let us rethink the ozone', raising questions about the scientific basis of the ozone-depletion theory and blaming the ozone scare on media hype.

US Representative Tom Delay also introduced a resolution in 1995 in the US Congress to repeal the entire section of the Clean Air Act that provided for the authority to phase out ozone-depleting chemicals. The proposal did not go through.

Science magazine, in its 11 June 1993 issue, analysed this phenomenon, quoting F Sherwood Rowland as describing the 'ozone scare' book as 'a good job of collecting all the bad papers in the field in one place'. It also quoted scientific sceptic S Fred Singer as admitting that the arguments about volcanoes being the main source of ozone depletion were not correct, and that CFCs were the main source of ozone depletion.

To counter such propaganda, the Scientific Assessment Panel included in its report of 1994 a list of 'frequently asked questions' and answers to these questions to clarify doubts. These may be seen on the Ozone Secretariat website at www.unep.org/ozone.

Technology transfer: Some developing countries complain

The issue of technology transfer again came up before the seventh Meeting of the Parties in connection with the review of the obligations of the Article 5 Parties. In its report on technology transfer, the Executive Committee detailed various actions taken and reported that no impediments had been brought to its notice so far. The Article 5 Parties were not satisfied with the report and mentioned that when the technology transfers took place, they were not under 'fair and favourable terms'. India and South Korea mentioned that they could

---

Box 4.3 The Ozone Secretariat and differing data sources

The Ozone Secretariat, the Implementation Committee and the Meeting of the Parties never questioned the accuracy of data submitted by a Party since there was no provision in the Protocol for verification of data. The Secretariat would occasionally question the data on the basis of some other information.

The issue of the extent to which the Secretariat could question data submitted by a Party came up before the seventh Meeting of the Parties in 1995. The representative of Lebanon protested that the Ozone Secretariat accepted the population figures for Lebanon given by the United Nations and not the figures given by Lebanon. The lower figures of the United Nations resulted in the per capita consumption of CFCs (Annex I, Group 1 substances) of Lebanon for a particular year being higher than 0.3kg; hence, the Secretariat classified the country as not operating under Article 5. This made it ineligible for assistance from the Multilateral Fund and for the ten-year grace period. The higher population figure given by Lebanon, citing changed political conditions in the Middle East and the return of many Lebanese refugees back to their homeland, reduced its per capita consumption and made it eligible for classification as operating under Article 5. The meeting decided that, while the Ozone Secretariat could ask for clarifications of the data, the data given by the countries should be accepted finally. Lebanon was back on the list of Article 5 Parties.

---

not get the technologies for making HFC-134a on reasonable terms. The Executive Committee was requested by the seventh Meeting of the Parties to study this issue again and provide a final report to the eighth Meeting of the Parties in 1996.

Vienna Declarations

A number of countries[10] declared at the end of the seventh meeting that HCFCs were not necessary for the substitution of some uses of CFCs, and stressed the need to strengthen the control measures for both developing and developed countries. Another declaration on methyl bromide by some countries[11] promised appropriate measures to limit methyl bromide and phase it out as soon as possible.

MEETINGS IN 1996: ILLEGAL TRADE DISCUSSED, REPLENISHMENT OF THE FUND BY US$466 MILLION IN SAN JOSE, COSTA RICA[12]

Meetings of the working group and Parties in 1996 dealt with the issues of CFC-free metered-dose inhalers, illegal trade in CFCs and replenishment of the Multilateral Fund. They also covered technology transfer and reviewed terms of reference for the Technology and Economic Assessment Panel (TEAP). The eighth Meeting of the Parties took place in San Jose, Costa Rica, in December 1996, and was opened by the President of Costa Rica, Jose Maria Figueres.

Essential-use exemptions: Further study

By 1996, many non-CFC metered-dose inhalers had appeared on the market, raising questions as to why an essential-use exemption should be granted for CFC inhalers. There was a consensus that each country should follow its own strategy in this regard. Participants at the eighth Meeting of the Parties and its preparatory meeting urged all non-Article 5 Parties to take steps to pursue alternative technologies for metered-dose inhalers, while approving the essential-use nominations for them. A format was prescribed for each country to report on the use of approved essential-use exemptions. The non-Article 5 Parties were requested to prepare national strategies on transitioning to non-CFC metered-dose inhalers and give copies to the TEAP, which would consult with others and give another report at the tenth Meeting of the Parties. The TEAP was also asked to study criteria for critical agricultural uses of methyl bromide and report to the next meeting, and to study the availability of halons for critical uses.

Illegal trade

The Ozone Secretariat presented its report on illegal trade. Some shipments, which came to European ports for re-export to developing countries, might have found a way to the market within the European Community. The ships in transit sometimes unloaded CFCs with consignments incorrectly labelled. The Alliance for Responsible Atmospheric Policy, an industry group of the USA, reported that between 10,000 and 20,000 metric tonnes of CFCs illegally entered the USA each year in 1994 and 1995. The USA had taken steps to curb such illegal trade, forming an investigative group composed of the EPA, Customs Service, Internal Revenue Service and the Commerce and Justice Departments. Customs authorities, trained and provided with the requisite equipment to identify CFCs, had seized over 500 tonnes of illegal CFCs. The country had convicted 12 individuals for smuggling CFCs into the USA, with judges imposing penalties of up to US$3.5 million and imprisonment for up to five years. In addition to criminal penalties, civil penalties of about US$25,000 per kilogram had been levied. The EPA adopted a licensing process for importing used CFCs; industry and government together generated much publicity on this issue through conferences and other means.

The US Government and the Alliance for Responsible Atmospheric Policy recommended that every country have a good licensing system similar to that of the EPA for importing used substances, and that every Party report all imports and exports. All Parties should also report to the Ozone Secretariat their facilities for recycling and reclamation. Money from the Multilateral Fund should be given to developing countries to enable them to increase border patrolling to enforce the law on tracking production and exports. Funds should also be available for the purchase of pressure gauges for use at major entry ports to identify contents of containers, which could be falsely labelled. The non-Article 5 Parties should be requested to patrol their borders, enforce laws, and institute more severe criminal penalties for illegal trading of CFCs. The Parties should institute a registration number for each production facility, which should be applied to every container of virgin CFCs leaving the facility. Parties not in compliance with production controls should not be allowed to export or import CFCs. The Ozone Secretariat suggested that in addition to these steps, each Party should report to the Secretariat the destination of its exports and origins of its imports. The suggestions were widely welcomed.

Third replenishment, 1997–1999

The TEAP recommended that US$436.5 million would be required for the Multilateral Fund for 1997–1999 to enable developing countries to comply with ODS reductions according to the Protocol. An additional US$60 million would be required to maintain the accelerated reduction programmes underway in some developing countries. The TEAP had developed a computer model using the data from official sources to calculate these figures. The model took into account all the available data, such as the projects already sanctioned, the average time taken to complete projects, the average cost-effectiveness in various sectors, the control measures that needed to be followed by countries, and the phase-out status of the countries in each of the categories. The TEAP took into account not only the freeze in 1999, but also the subsequent reduction steps which the developing countries were obliged to implement.

The eighth Meeting of the Parties approved a replenishment of US$466 million for 1997–1999, and decided that US$10 million of this would be used to assist Parties starting implementation of any measure that might arise from the 1997 Meeting of the Parties. It was thought that the methyl bromide obligations for developing countries were likely to change in 1997. The meeting agreed that Parties that did not ratify the London Amendment were not obliged to contribute to the Multilateral Fund.

Technology transfer: an informal group to advise the Executive Committee

The issue of technology transfer was discussed by the Executive Committee for preparation of a report to the eighth Meeting of the Parties, as requested by the seventh Meeting of the Parties. There was a difference of opinion in the Executive Committee about how to prepare a report on the issue and how to interpret 'fair and favourable terms' as stated in Article 10A. The eighth Meeting of the Parties appointed an informal advisory group of Parties – Australia, Italy, The Netherlands, the USA, China, Colombia, Ghana and India – to advise the Executive Committee on the issue.

Terms of reference for the TEAP

The eighth Meeting of the Parties also reviewed the terms of reference for the Technology and Economic Assessment Panel (TEAP). The sixth Meeting of the Parties in 1994 had seen some criticism of the TEAP by the Nordic countries, Austria and Switzerland. Some Parties particularly disagreed with the findings of the TEAP that HCFCs and HFCs were needed in some applications to phase out CFCs. Based on a proposal by Switzerland, the seventh Meeting of

---

**BOX 4.4 THE COURSE OF THE INFORMAL ADVISORY GROUP
ON TECHNOLOGY TRANSFER**

The Informal Advisory Group met many times in 1997 and 1998 to discuss technology transfer, but because there were differences between members of this advisory group, the ninth and tenth Meetings of the Parties requested the group to continue its efforts. The group finally reached a consensus in 1999 on the issue of identifying potential impediments to technology transfer; while there was no consensus on those impediments, it discussed steps that could be taken to overcome them.

The group recommended the following seven steps for consideration by the Executive Committee:

1   A database should be prepared to contain a description and characteristics of the available ODS substitution technologies and the terms under which such technologies would be available for transfer.

2   The Article 5 Parties and their enterprises could select appropriate technologies with the assistance of the database. Appropriate representatives in Article 2 countries might also be a source of information to facilitate the technology transfer. The implementing agencies might be asked by an Article 5 Party to help in this process, but the final decision on technologies would be, consistent with Executive Committee guidelines, with the concerned Article 5 Party.

3   When an Article 5 Party found impediments in the transfer of particular technologies, consistent with Article 10A of the Protocol, it could notify the government of the country where the owners of those technologies were located. While the governments could not force the technology transfer, they might be able to assist and catalyse it.

4   The Article 5 Parties should continue to be empowered and enabled to make informed decisions regarding technologies. The implementing agencies must be requested to carry out the Article 5 Parties' wishes and, consistent with Executive Committee guidelines, not impose their own views on Article 5 Parties.

5   Article 2 Parties' governments were encouraged to develop incentives to facilitate the transfer of ODS substitution technologies to Article 5 countries, in addition to the database.

6   The companies of the Article 5 countries were encouraged to use domestic technologies for conversion or substitution when such technologies were available domestically and competitive in terms of technical effectiveness and price.

7   The transfer of technologies among the Article 5 countries should be encouraged.

---

the Parties in 1995 appointed an informal advisory group on the organization and functioning of the TEAP. This group suggested that the number of experts from developing countries and countries with economies in transition on the panels and the TEAP's technical options committee should be increased to 50 per cent. It also suggested detailed terms of reference for the TEAP and its technical options committees, which the eighth Meeting of the Parties approved:

•   Members of the TEAP would be appointed by the Meetings of the Parties, while the members of the technical options committees would be appointed by the co-chairs of these committees in consultation with the TEAP.

- A detailed procedure was drawn up whereby a 50 per cent representation of developing countries and countries with economies in transition was urged by the co-chairs of the TEAP.
- Each technical options committee was entitled to submit its reports without any modification by the TEAP, while the TEAP could comment on those reports.
- A code of conduct was prescribed for the members to ensure that they were, and were seen to be, unbiased in discharging their duties.

Fourth COP of the Vienna Convention, 1996

The fourth meeting of the Conference of the Parties to the Vienna Convention was held in 1996 in San Jose, Costa Rica, back to back with the eighth Meeting of the Parties to the Protocol. The meeting urged the Global Environment Facility to support monitoring of ozone and UV-B radiation in developing countries. It endorsed the recommendations of the third meeting of the ozone research managers on areas of ozone research, and appealed to the Parties to help the WMO and the developing countries to implement the recommendations.

TENTH ANNIVERSARY, MONTREAL, 1997: CONTROL MEASURES ON METHYL BROMIDE TIGHTENED

In 1997, the Open-Ended Working Group met in Nairobi and in Montreal, and the ninth Meeting of the Parties to the Montreal Protocol was also held in Montreal.

Proposals on Article 5 production baseline, HCFCs and methyl bromide and trade licensing

The meetings of the Open-Ended Working Group and the ninth Meeting of the Parties[13] considered a number of adjustments and amendments proposed by many countries:

- India would set the baseline for the Article 5 Parties for their production phase-out of Annex A substances at the average production for 1995–1997, and for Annex B substances, the average production for 1998–2000.
- The USA included control of trade in methyl bromide, a mandatory licensing system for trade in ozone-depleting substances by all countries, and an advanced methyl bromide phase-out to the year 2001 for developed and Article 5 Parties.
- The European Union suggested, without mentioning dates, adopting a phase-out schedule for methyl bromide for Article 5 Parties, reducing the HCFC cap from 2.8 to 2.0 per cent; advancing the HCFC phase-out from 2030 to 2015, having a phase-out schedule for the production of HCFCs, extending trade measures to methyl bromide and HCFCs, and adopting a licensing system for trade in all controlled substances.

- Australia suggested: adopting a licensing system; ensuring that Parties which were unable to implement control measures and continued to produce virgin ODSs could not export used or recycled substances, but use them themselves and reduce their production; and including, in control measures on carbon tetrachloride, interim measures of a freeze by 2000, a 50 per cent reduction by 2002, and a phase-out by 2006 by all Article 5 Parties.
- Switzerland also proposed that HCFC production should be controlled. Canada suggested that the non-Article 5 Parties should phase out methyl bromide by the year 2001, and the Article 5 Parties by 2011.
- Canada suggested that the implementation procedure be reviewed to address many additional angles, such as equitable treatment of non-compliant Parties and different types of non-compliance.

Diverse opinions on further controls

Many countries supported a methyl bromide phase-out by 2001 by the non-Article 5 Parties and by 2011 by the Article 5 Parties. Many Article 5 Parties repeated the argument that any further control measures should wait for demonstration of alternatives. The Group of 77 opposed further control measures on methyl bromide for Article 5 Parties. All countries, except the European Community and Switzerland, opposed further HCFC control measures on the grounds that further controls on HCFCs would bring only marginal benefits to the ozone layer. They argued that the key to the success of the Montreal Protocol had been a constructive partnership with industry, which could be undermined by the introduction of further controls that would not allow sufficient time to recoup the investment in the new HCFC technologies introduced to phase out CFCs quickly. If the phase-out schedule were restricted further, industries would be punished for their bold early decisions to phase out CFCs and to convert to HCFCs with lower ODPs. This would discourage industry from responding to governments in future. Many representatives wanted stability for the Protocol regime and mentioned that, for the Article 5 Parties, a review was due in the year 2000.

The tenth anniversary of the Montreal Protocol

The ninth Meeting of the Parties took place in Montreal on 9–17 September 1997. Canada celebrated the tenth anniversary of the Montreal Protocol with a workshop on various aspects of the Protocol and with a grand and solemn opening ceremony for the meeting. The meeting was opened by Christine Stewart, Minister of Environment of Canada, and was addressed by, among others, G O P Obasi, Secretary General of the World Meteorological Organization. At the workshop, a number of experts including Mario Molina and Mostafa Tolba analysed the genesis and the development of the Protocol.[14]

Further strengthening of the Protocol

The meeting, after further discussions, approved the following adjustments, amendments and decisions:

- The baseline of production for Article 5 Parties was set as the average of 1995–1997 for Annex A and Annex B substances.
- A methyl bromide phase-out by 2005 was approved for the non-Article 5 Parties and by 2015 for Article 5 Parties.
- The trade restrictions for methyl bromide were approved.
- The Multilateral Fund would meet the incremental costs of phasing out methyl bromide irrespective of the cost-effectiveness. It would provide at least US$25 million per year in 1998 and 1999. Future replenishment would take into account the new requirements. There would be transfer of alternatives and technologies for alternatives.
- In light of the TEAP assessment in the year 2002, the Meeting of the Parties would decide in 2003 on further specific interim reductions of methyl bromide for Article 5 Parties beyond 2005.
- Criteria for critical-use exemptions were approved. Uses would be considered critical if the nominating Party considered that a lack of availability of methyl bromide for that use would result in significant market disruption, that there were no feasible alternatives that were acceptable by environmental and health standards, and that alternatives were suitable to the crops and circumstances of the nominations. The critical use would be approved only if, inter alia, it was demonstrated that effort was being made to evaluate, commercialize, and get regulatory approval of alternatives.
- Any Party had been allowed, on notification to the Ozone Secretariat, to use up to 20 tonnes of methyl bromide in emergencies. The Secretariat and TEAP would assess the use according to the critical methyl bromide criteria and present the information to the next Meeting of the Parties.
- The TEAP reported the development of non-CFC metered-dose inhalers on the market and gave broad principles about a transition framework. The essential uses were recommended for inhalers and halon-2402 for 1999 for the Russian Federation. The TEAP reported on the feasibility of decommissioning halon systems, and said that up to 80 per cent of all halon-1211 applications could be taken out of service. No experience was available regarding destruction. The global supply and demand of halon-1301 were balanced and would remain so over the next 30 years.
- The meeting requested an ad hoc working group of legal experts to report in 1998 on any amendments necessary for the non-compliance procedure.

Control of trade between Parties

This meeting introduced a new article regarding control of trade with Parties. The Parties agreed to the Australian proposal for banning exports of recycled substances by Parties that continued their production in non-compliance with the Protocol. In another article, a licensing system for trade in new, used or recycled substances was made mandatory by 1 January 2000 for all Parties, with an extension of time for Article 5 Parties for HCFCs until 2005, and for methyl bromide until 2002. Extensions would be granted if countries were unable to introduce the licensing system before that date.

BOX 4.5 GOALS OF THE MONTREAL PROTOCOL TRADE PROVISIONS

Duncan Brack, Royal Institute of International Affairs

The negotiators of the Montreal Protocol had two goals in drawing up trade provisions. One goal was to maximize participation in the Protocol by shutting off non-signatories from supplies of CFCs and providing a significant incentive to join. If completely effective, this would in practice render the trade provisions redundant, as there would be no non-Parties against which to apply them.

The other goal, should participation not prove total, was to prevent industries from migrating to non-signatory countries to escape the phase-out schedules. In the absence of the trade restrictions, not only could this reaction fatally undermine the control measures, it would also help non-signatory countries to gain a competitive advantage over signatories, as the progressive phase-outs raised industrial production costs. If trade were forbidden, however, non-signatories would not only be unable to export ODSs, they would also be unable to enjoy fully the potential gains from cheaper production as exports of products containing, and eventually made with, ODSs would also be restricted. (In fact, as industrial innovation proceeded far more quickly that expected, many of the CFC substitutes proved significantly cheaper than the original ODSs – but this could not have been foreseen in 1987.)

Although it is difficult to determine states' precise motivations for joining – there are a variety of reasons, including the availability of financial support for Article 5 Parties – the trade restrictions do appear to have provided a powerful incentive, and at least some countries have cited them as the major justification for their participation. The evidence suggests that the trade provisions achieved their objectives. All CFC-producing countries and all but a handful of consuming nations have adhered to the treaty.

Source: cited in P G Le Prestre, J D Reid, E T Morehouse, Jr (eds) (1998) *Protecting the Ozone Layer: Lessons, Models and Prospects*, Kluwer Academic Publishers, Boston, p101.

New ODSs emerge in the market

The Ozone Secretariat reported on two new ozone-depleting substances, bromochloromethane and n-propyl bromide, which had entered the market, as reported by two countries. The US EPA had proposed banning bromochloromethane as a substitute for CFC solvents. Based on the information received from the Parties about these substances, which were not controlled by the Montreal Protocol, the Secretariat referred the issue to the Scientific Assessment Panel, which confirmed the ODP of bromochloromethane as between 0.11 and 0.13. The residence time of n-propyl bromide in the atmosphere was only about ten days; it was to be studied whether the short residence time implied that the calculation of its ODP might not be fully valid. The Parties expressed concern at the emergence of these new, uncontrolled ODSs and urged the Parties to report on any such substances, the Scientific Assessment Panel to assess the ODP of such substances, and the TEAP to assess the extent of use of such substances. The Parties were also urged to discourage the development and promotion of such substances.

Montreal Declarations

Some countries[15] declared at the end of the ninth Meeting of the Parties that they were concerned about the lack of results during the meeting regarding HCFCs and that the eleventh Meeting of the Parties in 1999 should make a decision on them. Some countries[16] declared that further action was necessary to phase out methyl bromide as soon as possible, and that they would promote sustainable alternatives to methyl bromide in their own countries.

MEETINGS IN 1998: 1998 ASSESSMENT CONFIRMS PROTOCOL WORKING, TENTH MEETING OF THE PARTIES IN CAIRO DISCUSSES LINK BETWEEN OZONE DEPLETION AND CLIMATE CHANGE, NON-COMPLIANCE[17]

The main issues before the 1998 meetings were the increasing abundance of halons in the atmosphere despite stopping of new production by the non-Article 5 Parties in 1994; new ozone-depleting substances appearing in the market; process agents; relationship with the Kyoto Protocol of 1997 on greenhouse gases; and the terms of reference for the study by TEAP on replenishment of the Multilateral Fund for the period 2000–2002. The assessment panels presented their fourth comprehensive assessment reports in December 1998.

The tenth Meeting of the Parties took place in Cairo in November 1998 and was opened by the Deputy Prime Minister of Egypt, Youssef Wally. On this occasion, Suzanne Mubarak, the First Lady of Egypt, and Klaus Töpfer, Executive Director of UNEP, presented prizes to the winners of the worldwide children's art competition on ozone-layer themes. The first prize went to a young artist in Egypt; other prizes went to children in Egypt, Indonesia, Iran and Niger. Jury prizes were also awarded to two children from China and Romania. Ten other entries were awarded merit certificates. UNEP gave Tolba a Global Ozone Award in celebration of his outstanding services to the protection of the ozone layer. The Minister for Environment of Egypt, Nadia Makram Ebid, helped to organize the meeting.

Concern about the increase of halons in the atmosphere

The meetings were alarmed by the finding of the Scientific Assessment Panel that the atmospheric abundance of halons was increasing, despite the phase-out in 1994 by the non-Article 5 Parties, due to emissions from existing uses and from consumption in Article 5 Parties. Australian scientists reported an unexplained abundance of halon-1202, an ozone-depleting substance not controlled by the Montreal Protocol since it was thought that no one used it. The emissions of the controlled halons came from many unnecessary uses and from critical uses, such as in aircraft, where no alternatives were available. The critical uses differed from country to country. Compulsory decommissioning from non-critical uses might lead to surreptitious leaking, since destruction might be costly. The meeting appealed to all of the Parties to develop a strategy for management of halons, including emissions reduction and the ultimate elimination of their use, and requested all Parties, including the non-Article 5 Parties, to submit their strategies by the end of July 2000.

The meeting requested the Scientific and Technology and Economic Assessment Panels to assess whether substances such as n-propyl bromide, with very short atmospheric lifetimes (less than one month), were a threat to the ozone layer. The meeting also requested the panels to identify the sources and availability of halon-1202, which was not a controlled substance. Participants decided that the Parties should discourage production and marketing of bromochloromethane, and that if any new substance posed a significant threat to the ozone layer, the Parties would take appropriate action under the Protocol. They requested the legal drafting group to consider the options under the Protocol to control new ODSs.

Use of CFCs as process agents

The meeting decided that process agents would be treated as feedstocks for non-Article 5 Parties until December 2001. Quantities used in plants in operation before 1 January 1999 would not be taken into account from 1 January 2002 if, in the case of non-Article 5 Parties, the emissions from these processes were insignificant and, in the case of Article 5 Parties, the emissions had been reduced to levels agreed by the Executive Committee to be reasonably achieved. A list of uses of controlled substances as process agents had been developed, and Parties were requested to inform the Ozone Secretariat if there were more uses. It was also agreed that no new facility with ODS process agents should be installed after 30 June 1999 except as an essential use. The Multilateral Fund was requested to develop funding guidelines for Article 5 Parties to consider project proposals.

Ozone depletion and climate change: Montreal and Kyoto Protocols

A new development in 1997 was the Kyoto Protocol arrived at by the Parties to the United Nations Framework Convention on Climate Change (UNFCCC) in December. This Protocol required the developed-country Parties to reduce their emissions of six greenhouse gases – carbon dioxide, methane, nitrous oxide, hydrofluorocarbons (HFCs), perfluorocarbons (PFCs), and sulphur hexa-fluoride (SF_6) – from 2008 to 2012. The HFCs and PFCs, which had high global warming potentials but zero ozone-depletion potentials, entered the market as substitutes for CFCs. They became very popular in many countries, including Article 5 Parties where the Multilateral Fund financed projects to phase out CFCs by using HFCs. This raised concerns about whether the problem of ozone depletion was being solved at the expense of the problem of global warming.

The Article 5 Parties were apprehensive about whether they would have to change their technologies once again to stop using HFCs, under the mandate of Kyoto Protocol. The fourth meeting of the UNFCCC was held in 1998 just prior to the tenth Meeting of the Parties of the Montreal Protocol. It decided to request all the Parties to the UNFCCC (almost all of whom were also Parties to the Montreal Protocol) to bear in mind the relationship of the two Protocols and develop information on minimizing emissions of HFCs and PFCs. The tenth Meeting of the Parties responded to this decision by requesting the Parties to submit data on HFCs and PFCs to the Secretariat of the UNFCCC and all the Montreal Protocol bodies to cooperate with the UNFCCC on this issue. It also requested that a joint workshop be held with the Intergovernmental Panel on Climate Change on ways to minimize the emissions of HFCs and PFCs.

Study for the fourth replenishment

The meeting requested the TEAP to assess the funds needed by the Multilateral Fund for 2000–2002. On the initiative of some non-Article 5 Parties, the meeting requested the treasurer to study the proposal, similar to the one adopted by the Global Environmental Facility, for a system of a fixed currency exchange rate mechanism for the replenishment of the Multilateral Fund. Under this system, each country would pay in its own convertible currency, rather than in US dollars. The amount due from each of the Parties would be fixed once in three years in the currency of that Party.

Minor changes in non-compliance procedure

The ad hoc group of legal experts reviewed the non-compliance procedure and recommended only minor changes, which were accepted by the tenth Meeting of the Parties. An approved improvement was to set a time limit of three months for Parties accused of non-compliance to reply, and a limit of six months for the Ozone Secretariat to place a complaint before the Implementation Committee.

Cairo Declarations

Many countries[18] made a declaration at the end of the tenth Meeting of the Parties urging all the bodies of the Montreal Protocol not to support use of HCFCs where alternatives were available. They also urged all the Parties to consider the ODS replacement technologies, taking into account their global warming potential.

Fourth reports of the assessment panels, 1998

The assessment panels presented their fourth comprehensive assessment reports in December 1998. Their conclusions were that the overall rate of decline of ozone at mid-latitudes had slowed, but it was not clear whether that decline heralded the recovery of the ozone layer (see Chapter 1). The Environmental Assessment Panel report quantified the number of skin cancers and cataracts that had been avoided due to the Montreal Protocol (see Chapter 1).

The TEAP reported that the halon-1202 observed was the result of inadvertent production and release during halon-1211 production in Article 5 Parties. There were alternatives for more than 95 per cent of the current uses of methyl bromide, of which 19–23 per cent was for quarantine and pre-shipment uses. There were many alternatives available for these uses; the Parties could consider rescinding the blanket exemption and placing a cap on this consumption so that the alternatives were promoted. The TEAP also reported on its work regarding minimization of emissions of HFCs and HCFCs.

BOX 4.6 A PIONEERING APPROACH TO INTERNATIONAL
CONTROL STRATEGIES

Patrick Szell, Chairman, Legal Drafting Group

The purpose of the Montreal Protocol was to promote protection of the ozone layer. In pursuit of this goal, the traditional control strategies were either too weak to have real impact or too strong to be appropriate. What was needed was a new approach, somewhere between the two: a regime that was non-confrontational, conciliatory and cooperative, that would cajole, encourage or otherwise help Parties that were in breach of their obligation to achieve full compliance. There was a strong feeling that if Parties felt they were being subjected to some kind of judicial process, they would become defensive and turn in on themselves, with the result that the ozone layer would be the loser.

With a more constructive approach based on a recognition that non-compliance is frequently the consequence not of malice or greed, but rather of technical, administrative or economic problems, a regime that worked with, rather than against, Parties in difficulty was sought. In theory, the regime devised is weaker for having to be approved as a decision than if it had formed an integral and binding part of the Protocol as originally adopted. Given, however, that the non-compliance regime is intended to be cooperative, its lack of binding force is somewhat academic.

We should not allow an understandable desire for quick results and strong recommendations from the Implementation Committee to blind us to the fact that the committee already has – contrary to what most people would have believed possible – made considerable strides forward already. The global environment's first comprehensive non-compliance regime is functioning regularly and efficiently in what is a delicate area of inter-Party relations. It has already prompted valuable improvements in Parties' observance of the Protocol's obligations: this is no mean achievement for a proposal which, it will be recalled, was originally tabled merely as a negotiation ploy and which few, if any, who were in Montreal imagined would make a major impact on the day-to-day working of the Protocol. The initial task of establishing the Committee as an essential feature of the Protocol has now been completed, and the Committee has started to tackle more sensitive issues. It has approached these new challenges with skill and imagination. We have good reason to look forward to its future achievements with confidence, albeit a confidence that will have to be tempered with patience.

Source: cited in W Lang (ed) (1996) *The Ozone Treaties and Their Influence on the Building of International Environmental Regimes*, Federal Ministry for Foreign Affairs, Vienna.

MEETINGS IN 1999: BEIJING AMENDMENT, FREEZE IN PRODUCTION OF HCFCS AND TRADE RESTRICTIONS, REPLENISHMENT OF THE FUND BY US$440 MILLION

European proposals on HCFCs and methyl bromide

The assessment reports came up for discussion in the Open-Ended Working Group meeting[19] preceding the eleventh Meeting of the Parties.[20] During this meeting, the European Community proposed adjustments and amendments to controls on HCFCs, production controls, reduction of the HCFC cap to include only 2 per cent of CFC consumption in 1989 and not 2.8 per cent as it was then, trade controls, and the rationalization of the measures for Article 5 Parties. It suggested that the quantities of methyl bromide used for quarantine and pre-shipment use be frozen at 1995–1998 levels, and that reporting of the consumption for such uses be mandatory. It also suggested reduction of the additional production allowed for non-Article 5 Parties for meeting the basic domestic needs of Article 5 Parties, in step with the reduction of consumption by the Article 5 Parties. Regarding new ozone-depleting substances, it suggested that a lighter procedure be adopted instead of the present cumbersome procedure of amendments. While there was welcome for the reduction of CFC production to meet the needs of the Article 5 Parties, the proposals on HCFCs and methyl bromide met with stiff resistance from many countries. There was no consensus on the procedure for new substances.

Fourth replenishment, 2000–2002

The TEAP reported on the replenishment of the Multilateral Fund for 2000–2002 and explained that the growth rate of CFC consumption by Article 5 Parties had been much less than estimated in the previous TEAP report of 1996. Therefore, for strict compliance with the Protocol, the funds required for 2000–2002 were reduced to US$306 million, compared to the previous replenishment of US$465 million. However, the 85 per cent reduction target in 2007 meant that the replenishment for 2003–2005 would have to be much higher, nearly US$700 million. The TEAP proposed that advance funding for 2000–2002 might be given, and that replenishment be fixed at US$500 million to utilize the capacity built up in Article 5 Parties and the implementing agencies, and to benefit the ozone layer by faster reductions. The advance funding was justified on several grounds: environmental (benefit to the ozone layer), business (inefficiency in sharply varying approvals every three years), administrative (avoiding waste of higher built-up implementation capacity), and to benefit small countries (which could be neglected if allocations were merely adequate).

During the discussion, many stressed the need for non-investment activities such as training, provision and dissemination of information, and awareness-raising, particularly for small countries and the service sector in all countries. Many countries, while agreeing to advance funding, stressed that funding be given by the contributors as concessional loans and other forms of innovative funding. The Article 5 Parties opposed concessional loans and preferred only grants. The Meeting of the Parties finally approved a replenishment of US$440 million for 2000–2002. No concessional loans were mentioned.

In his presentation to the meeting, the representative of the Global Environmental Facility noted that the Facility had assisted the phase-out of ozone-depleting substances in 19 countries with economies in transition, with total funding of US$140 million. In these countries, consumption dropped from about 190,000 tonnes in 1990 to less than 15,000 tonnes in 1997. Production by the Russian Federation would close down by the year 2000. The Facility considered that it had fulfilled its mandate.

Listing new ODSs in the Protocol: No agreement on lighter procedure

The Legal Drafting Group discussed the proposal for a lighter procedure for amending the Protocol to control new ODSs emerging in the market. The group asked the Open-Ended Working Group first for a policy decision regarding many aspects to enable drafting the legal text. What was the basis for selection of new substances to be controlled? How much of a majority would be needed to include new substances? Would the addition of the new substances be carried out in one or two steps – would the substances be identified first and control measures later? What would be the conditions for entry into force? How could Parties, if they so wished, opt out of the application of these decisions? The Legal Drafting Group noted that an expedited procedure for adding new substances did not obviate the need for amendments if the Parties wanted to add other related obligations, such as reporting or control of trade with non-Parties for those substances. The chair of the Legal Drafting Group refrained from drafting the legal text since these policy issues needed to be decided upon first.

The eleventh Meeting of the Parties took place in Beijing in December 1999, with all the Parties present in a very optimistic mood on the eve of the new millennium. The President of China, Jiang Zeming, graced the occasion. Xie Zhenhua, the Minister of Environment of China, played a key role in the meeting.

Last tightening and Beijing Declaration

In its final decisions, the eleventh Meeting of the Parties approved:

- A reduction in the allowance for production for meeting the needs of Article 5 Parties, along with a consumption-reduction schedule of the Article 5 Parties.
- A freeze in HCFC production from 1 January 2004 for non-Article 5 Parties and from 2016 for Article 5 Parties; trade restrictions on HCFCs from 2004. The Article 5 Parties requested the TEAP to review by 30 April 2000 whether HCFCs would be adequately available to them after the freeze in 2004.
- Adding the new ODS, bromochloromethane, to the list of controlled substances, with a phase-out by January 2002.
- Each Party providing data on methyl bromide used for quarantine and pre-shipment applications. The pre-shipment applications were specified as those occurring within 21 days prior to export to meet the official requirements of the importing country or existing official requirements of the exporting country. The TEAP was requested to report on alternatives for quarantine pre-shipment at the 2003 meeting.
- A replenishment of US$440 million for the Multilateral Fund for 1997–1999, noting the carry-over of US$35.7 million and arrears from the countries with economies in transition of US$34.7 million.
- The basis of contribution to the Fund changing from dollars to a fixed exchange rate mechanism. Through this mechanism, Parties could pay in

BOX 4.7 ENTRY INTO FORCE: THE GAP BETWEEN
AGREEMENT AND COMMITMENT

While crafting the Vienna Convention, the Montreal Protocol and the Amendments to the Protocol, governments approved a process for the entry into force of these legal agreements for states approving these agreements. Under this process, each government must send a letter (called an 'instrument') of its ratification (variously called ratification, acceptance, approval or accession) to the Depository (the Secretary General, United Nations). A minimum number of such ratifications are required for entry into force. The legal agreements enter into force on the ninetieth day after the receipt of the minimum number of ratifications. A government cannot approve an agreement unless it has also approved all the previous agreements. For instance, to ratify the Copenhagen Amendment of 1992, a government also had to ratify the Vienna Convention, the Montreal Protocol and the London Amendment of 1990.

The entry into force of the Vienna Convention (and of the London, Copenhagen, Montreal and Beijing Amendments) was to be on the ninetieth day after the date of deposit of the twentieth instrument of ratification. The Montreal Protocol required 11 instruments of ratification, representing at least two-thirds of 1986 estimated global consumption of the controlled substances. Each government conducted its own internal procedure before sending such a letter.

There was considerable delay before the agreements entered into force. While the Vienna Convention has 184 ratifications so far and the Montreal Protocol 183, the Amendments to the Protocol have not yet been approved by some governments, even though the same governments joined the consensus for these agreements during the Meetings of the Parties. For example, the non-ratification of the Copenhagen Amendment by China legally implies that China has not bound itself to phase out HCFCs and methyl bromide included in the Protocol by this Amendment. This causes some concern. China informed the Secretariat in March 2002 that it intends to ratify the Copenhagen Amendment before the end of 2002.

Table 4.1 *Dates of agreement and entry into force of the Vienna Convention, Montreal Protocol and Amendments**

Agreement by Meeting of the Parties	Month of agreement	Date of entry into force	Number ratified (by 5 Dec 2001)
Vienna Convention	March 1985	September 1988	184
Montreal Protocol	September 1987	January 1989	183
London Amendment	June 1990	August 1992	163
Copenhagen Amendment	November 1992	June 1994	141
Montreal Amendment	September 1997	November 1999	80
Beijing Amendment	December 1999	February 2002	33

* As at 30 May 2002.
Source: Data from the Ozone Secretariat, UNEP.

their own convertible currencies if fluctuations in its inflation rate were less than 10 per cent for the preceding three years. The implementation would be reviewed by the end of 2001.

• Removal of three uses from the list of global exemptions for laboratory and analytical uses.

BOX 4.8 MONTREAL PROTOCOL: THE ART OF COMPROMISE/PEOPLE MAKE A DIFFERENCE

Paul Horwitz, International Advisor,
Stratospheric Protection Division, US EPA

The Montreal Protocol would never have succeeded as well as it did were it not for the willingness of participants to work hard to bridge differences and find some common ground on which they could agree. For the environmental purist, this tendency to negotiate had what was often considered to be negative effects. That said, I believe that the willingness to negotiate within the range of the possible, a factor that has historically been called 'the spirit of Montreal', is one of the reasons that the Montreal Protocol is often referred to as 'the one that works'.

In the early years of the Protocol, it was often UNEP Executive Director Mostafa Tolba who both set the agenda of the Parties and helped to bridge the gap in the positions among Parties. In fact, in this day and age in which Parties to Multilateral Environmental Agreements closely guard their sovereign prerogatives, it is hard to comprehend the impact that Tolba had on the Protocol negotiations. An example can be seen in the fact that for the first several years of the negotiations, the Parties actually disregarded the meeting agenda, and instead used the Executive Director's meeting note and his proposals as a road map and point of departure for discussion. Such notes now hardly get the passing interest of delegates. Another innovation of Tolba's, largely missing in today's MEAs, was his use of a small group of key delegates to informally explore the limits of the possible. This tool, referred to at the time as the 'Friends of Tolba' group, helped the Executive Director to carefully craft his proposals for consideration of the Parties. Indeed, after watching him do this on several occasions, it appeared that he could have had a specific calculation that he used to craft his proposals – one that found the midpoint of the rival positions, and then added 30 per cent for the environment. This manner of support helped push the parties in a pragmatic way toward consensus. It of course was a help to the Montreal Protocol that the simple adjustment process imbedded in the Protocol (which allowed for increased stringency of control without formal ratification of parliaments) freed Parties to make concessions with the understanding that the issue under discussion could be revisited in a future year.

In the latter years of the Protocol, much has changed. For the past decade, there has not been either the bridging help of Tolba or the stimulating feeling that the whole world was watching what was happening in our meetings. Indeed, since 1992, it has been possible to get the feeling that not much of the world is watching – perhaps not even senior representatives of the Parties negotiating at the meetings. In such a context, the desire for progress has often resulted in the Parties having to accept the aphorism that the perfect is the enemy of the good. This can be seen in the many Protocol decisions that are sometimes seemingly incomprehensible, and only really understood by the Parties who were central to the negotiations. While this is clearly a loss for the Protocol Parties, and is surely seen by the purists as a failure, it has allowed the Montreal Protocol to keep pace in a timely manner with evolving implementation issues. This has clearly benefited both the Parties and the global community as a whole.

- Urging the non-Article 5 Parties to develop strategies for management of CFCs, including for recovery or recycling, disposal and destruction.
- The assessment panels were requested to submit another comprehensive assessment by 2002. The Scientific Assessment Panel was requested to

continue its collaboration with the Intergovernmental Panel on Climate Change on aviation and global atmosphere. The Scientific Assessment Panel and the TEAP were requested to develop criteria to assess the potential ODP of new chemicals, and mechanisms to facilitate public–private cooperation in the evaluation of potential ODPs of new chemicals. The report was requested for the Meeting of the Parties in 2003.

- The Scientific Assessment Panel and the TEAP were requested to develop ways to involve the private sector in assessing ODPs.

The fifth meeting of the Conference of the Parties to the Vienna Convention,[21] held back to back with the eleventh Meeting of the Parties to the Protocol, endorsed the recommendations of the fourth meeting of the ozone research managers. It appealed again for assistance to the WMO and the developing countries to implement the research programmes.

MOP 11 was concluded with a ringing declaration of renewed commitment to the protection of the ozone layer of all the Parties.

TWELFTH MEETING OF THE PARTIES, OUAGADOUGOU, 2000: FURTHER ATTEMPTS TO TIGHTEN CONTROLS ON HCFCs[22]

Science of short-lived ODSs

The meeting of the Open-Ended Working Group in Geneva in July 2000 heard from the Scientific Assessment Panel on the conditions for short-lived gases containing chlorine or bromine to be harmful to the ozone layer. To be harmful, the substance must have a vapour pressure sufficient to generate a significant gas-phase concentration in the atmosphere, low solubility in water, a lifetime in the lower atmosphere long enough for it or its halogen-containing degradation products to reach the stratosphere, and a potential to release its bromine or chlorine atoms into the stratosphere. The short-lived compounds presented special problems. The accurate determination of an ODP depended on quantifying the amount of halogen delivered to the stratosphere and how it affected the ozone in the stratosphere. Because the transport to the stratosphere mainly occurred through the tropical tropopause, a short-lived species might undergo significant loss through various removal processes before reaching the tropics. The amount reaching the stratosphere was strongly dependent on the region and season of emission. The TEAP therefore was encouraged to submit the seasonal and latitude data for emissions for n-propyl bromide, which had an atmospheric life of about ten days.

The European Community tries again, unsuccessfully, to tighten regulations on HCFCs

The European Community quoted the decision of the meeting in 1995 that the control measures for Article 5 Parties for HCFCs would be reviewed in the year 2000. It proposed that the control measures be changed for Article 5 Parties by prescribing intermediate stages of a freeze in 2007 and other steps, without changing the final phase-out year of 2040. Many countries strongly opposed this proposal. The European Union also proposed a new decision on metered-dose inhalers to require each non-Article 5 Party to develop a national strategy by 31 December 2001; to encourage Article 5 Parties to do so by the end of 2004; and to require every country to report its progress to change to non-CFC inhalers. There were differing views on these proposals.

Twelfth Meeting of the Parties (MOP 12), Ouagadougou, 2000

The twelfth Meeting of the Parties took place in Ouagadougou, Burkina Faso, on 11–14 December 2000. Blais Compaore, President of Burkina Faso, opened the meeting, and Fidele Hien, Minister for Environment of Burkina Faso, assisted in the meeting. This was the first meeting in West Africa and in a country that consumed a very low volume of ODSs, about 30 tonnes in 1999. The decisions of the meeting were as follows:

• Countries were reminded that any metered-dose inhaler containing CFCs newly approved after the end of 2000 would not meet the essential-use criteria and therefore not be authorized; a list of such non-essential inhalers should be communicated by the Parties to the Ozone Secretariat, which should maintain a list of such products.
• Every developed country must submit a strategy of transition to CFC-free metered-dose inhalers by 31 January 2002, report every year on the progress of the transition, and encourage every Article 5 Party to do the same. The Executive Committee of the Multilateral Fund was requested to consider providing assistance to Article 5 Parties for implementation of this decision.
• The essential-use authorizations were made transferable to avoid unnecessary production.
• The TEAP was requested to establish a task force on destruction technologies to assess the status of such technologies. The TEAP was also requested to evaluate the feasibility for the long-term management of contaminated and surplus ODSs in all the countries, and to consider possible linkages to the Basel Convention on the Transboundary Movement of Hazardous Wastes.
• The Global Environmental Facility, which had announced that its assistance to the countries with economies in transition was completed, was requested to clarify its future commitment until the goal of total phase-out in all the countries was achieved.

Illegal trade
MOP 12 included a discussion at the meeting on illegal trade, which resulted in a decision, triggered by a proposal by Poland. Everyone recognized that control of trade between Parties of ODSs and of products containing ODSs was very important to promote implementation of the Protocol. It was acknowledged that effective control at national borders was extremely difficult because of: problems identifying ODSs and mixtures containing ODSs, complexity of customs codes, absence of an international labelling system, lack of trained

customs officers, and the need to approach the problems by cooperative action by all countries. A representative of the European Community mentioned new use controls in its regulations which would help to control international trade. The Parties asked the Ozone Secretariat to examine the options for studying the various connected issues such as national legislation, costs and benefits of labelling, and providing guidance for customs authorities. UNEP, as an implementing agency of the Multilateral Fund, was encouraged to continue with the customs training programmes it had already started.

Ouagadougou Declaration
The twelfth Meeting of the Parties ended with the Ouagadougou Declaration, which appealed for the continuing cooperation of all the countries until the ozone layer was restored to its original level.

MEETINGS IN 2001: THIRTEENTH MEETING OF THE PARTIES IN COLOMBO, NON-COMPLIANCE, NEW ODS[23]

The year 2001 was dominated by the issues of new, uncontrolled ODSs appearing in the market, use of ODSs as process agents, and the first cases of non-compliance by the Article 5 Parties. The meeting of the Open-Ended Working Group in Montreal in July 2001 considered various proposals on these issues, but no consensus was recorded.

Thirteenth Meeting of the Parties (MOP 13), Colombo, 2001

The thirteenth Meeting of the Parties was opened by the Prime Minister of Sri Lanka, Ratnasiri Wickremanayake, and took place amidst great fears about terrorism among many delegates, following the attacks on New York and Washington, DC, and the bombing of parts of Afghanistan by the USA and the UK. The meetings were held as scheduled with attendance by representatives of 109 countries and with a consensus on the ozone layer even among nations which differed greatly in their views on other issues. The Environment Minister of Sri Lanka, Dinesh Gunawardena, opened the preparatory segment.

The meeting approved the terms of reference for a study by TEAP on the replenishment of the Multilateral Fund for the years 2003–2005 and also for an evaluation of the financial mechanism by the year 2004. The last evaluation was in 1995. The meeting requested the treasurer to evaluate the fixed exchange rate system of contributions to the Multilateral Fund through an independent expert in time for the 2002 meetings. The Meeting of the Parties requested the TEAP to prepare a handbook on nominations for critical uses of methyl bromide, which is due to be phased out by 2005 by all the non-Article 5 Parties.

New ozone-depleting substances

The issue of new ODSs became particularly urgent because four new substances had been reported by the Parties: hexachlorobutadiene (Canada), n-propyl bromide (Canada), 6-bromo 2-methoxy naphthalene (The Netherlands), and halon-1202 (Israel). At the Open-Ended Working Group meeting in July, the Scientific and Technology Assessment Panels proposed that industries which proposed new substances likely to have ozone-depletion potential, in accordance with criteria to be set by the Scientific Assessment Panel, should support independent research to obtain information on the substances' ozone-depletion potential and submit it to the Ozone Secretariat. The TEAP further proposed that Parties control all substances with a chemical structure likely to deplete the ozone layer significantly and allow their use by adjustment of the control schedule or by essential-use exemption, after review of scientific and technical information provided by the company proposing their production. Many Parties supported an expedited procedure to control new substances. The European Community introduced a specific proposal for such a procedure.

Some countries, however, felt that the proposals raised issues of national sovereignty and represented a burden on the national science establishment. They felt that the concept of a blanket ban on substances with ODP above a given limit could not be accepted and that substances had to be identified specifically before they were banned. There had been a similar debate in 1987 (see Chapter 3). The Parties finally decided to:

* request the Secretariat to keep an up-to-date list of new substances reported by the Parties on the UNEP website and to distribute the current version of the list to all Parties about six weeks in advance of the meeting of the Open-Ended Working Group and the Meeting of the Parties;
* ask the Secretariat to request a Party that has an enterprise producing a listed substance to request that enterprise to undertake a preliminary assessment of the substance's ODP following procedures to be developed by the Scientific Assessment Panel, submit toxicological data on the listed substance if available, and, further, to request the Party to report the outcome of the request to the Secretariat;
* call on Parties to encourage their enterprises to conduct the preliminary assessment of the ODP within one year of the request of the Secretariat and, in cases where the substance is produced in more than one territory, to request the Secretariat to notify the Parties concerned in order to promote the coordination of the assessment;
* request the Secretariat to notify the Scientific Assessment Panel of the outcome of the preliminary assessment of the ODP to enable the Panel to review the assessment for each new substance in its annual report to the Parties, and to recommend to the Parties when a more detailed assessment of the ODP of a listed substance may be warranted.

The Meeting of the Parties requested the Ozone Secretariat to compile precedents in other conventions regarding the procedures for adding new substances, and to provide a report at the meeting of the Open-Ended Working Group in 2002. The meeting noted the TEAP's report that n-propyl bromide was being marketed aggressively and that its use and emissions in 2010 were currently projected to be around 40,000 metric tonnes. It requested Parties to: inform industry and users about the concerns surrounding the use and emissions of n-propyl bromide and the potential threat that these might pose to the ozone layer; urge industry and users to consider limiting its use to applications where more economically feasible and environmentally friendly alternatives were not available; and to urge them also to take care to minimize exposure and emissions during use and disposal. It requested the TEAP to report annually on n-propyl bromide use and emissions.

The European Community proposed a study regarding the need for HCFCs and options for speeding up their phase-out. The Article 5 Parties strongly opposed this, and the proposal was eventually dropped by the meeting.

The issue of process agents again came up for debate, as the TEAP reported many more applications of ODSs as process agents, and many Parties said they were not clear whether some uses could be classified as process-agent use. The Parties requested the TEAP to come up with a further report for a decision in the year 2002.

Good progress in developing countries: But 20 countries do not comply

The report of the Secretariat on data for 1999 was received with general satisfaction. Only 16 of the 170 Parties due to report for 1999 did not do so. The Article 5 Parties on the whole had reduced their production of CFCs from about 109,000 tonnes baseline (the average of 1995–1997) to 97,000 tonnes in 1999, and consumption dropped from 160,000 tonnes baseline to 121,000 tonnes, much beyond the freeze obligation of the Protocol.

However, 20 small countries had exceeded their consumption baseline, and Argentina had exceeded its production baseline. While the excess tonnage was not significant, there was alarm among many Parties that such non-compliance happened despite all the assistance given by the Multilateral Fund. During the discussions, two factors were mentioned: cheap and plentiful CFCs in the market due to illegal trade; and imports of used CFC equipment into the small countries, which increased the demand for CFCs. In the decisions of the Parties on this non-compliance, the non-complying Parties were cautioned to provide the Implementation Committee with a plan and benchmarks for return to compliance, and to note that the Parties could take other measures against non-compliance if it continued. These decisions were along the same lines as the decisions made about the Russian Federation and other countries of the former Soviet Union in earlier meetings. In the case of Article 5 Parties, however, the Multilateral Fund had already extended assistance whenever approached.

The Parties said that all the Article 5 Parties must, in order to comply with the control measures, now ensure that necessary policies to reduce the consumption of ODSs, including a licensing system for trade in ODSs, were in place and strictly enforced. An informal meeting of the ministers of 11 Article 5 Parties, held on the margins of the meeting to discuss the issue of compliance, also revealed the helplessness of Article 5 Parties to prevent excess imports of ODSs or the used equipment given their porous borders. The meeting requested the Ozone Secretariat to study, in consultation with others, the various aspects of illegal trade, including methods for controlling it.

Box 4.9 Illegal trade in ozone-depleting substances

Julian Newman, Environmental Investigation Agency

The signing in 1987 of the Montreal Protocol on Substances that Deplete the Ozone Layer marked the most ambitious attempt by the global community to control and eventually phase out widely used but environmentally damaging chemicals. The architects of this landmark agreement adopted a phased approach to reining in the use of ozone-depleting substances, gave developing countries a grace period, and put the onus on controlling production rather than consumption.

Although unforeseen in 1987, the basic tenets of the Montreal Protocol created fertile conditions for an illegal trade in ozone-depleting substances (ODSs) to emerge and flourish. By concentrating on the production side in industrialized countries, consumption could continue unhindered; by permitting the growth of production in developing countries at the same time, a cheap and plentiful supply of ODSs was guaranteed. Loopholes such as the unrestricted trade in recycled ODSs, continued production of ODSs in industrialized countries for essential uses and basic domestic needs, and the allowance of imports for repackaging under inward processing relief schemes all contributed to the growth of illegal trade.

The first major milestone in the Montreal Protocol coincided with the emergence of the first wave of ODS smuggling. In 1995, the European Union prohibited the production of CFCs, and the USA followed suit in 1996. At the same time, the Russian Federation was unable to meet its commitment to halt production because of its economic turmoil.

During 1995, a number of cases of illegal CFCs were pursued in the USA, centred around the state of Florida, and the port of Miami in particular. The scale of the illicit profits being made from this lucrative black market prompted one Florida State Attorney to describe CFCs as second only to cocaine as contraband being smuggled through Miami.

Although cases of CFC smuggling detected at this time varied in their methods, there were some unifying characteristics. Much of the illicit material entering the USA originated in Russia, and passed through the hands of European companies on its westward journey. It was imported into the USA under transit regulations, with the final destinations stated as developing countries in Latin America and the Caribbean. In fact, large quantities of CFCs never left North America and entered the black market in the USA instead. In one case, a Florida shipping company imported over 3000 tonnes of CFC-12 into the USA and filed false shipping manifests claiming the material re-exported. The manager of the firm was jailed for five years.

The scale of ODS smuggling prompted the US authorities to establish a Task Force on CFC Smuggling under the aegis of the Justice Department and involving the Environmental Protection Agency, the Federal Bureau of Investigation, the Customs Service and the Internal Revenue Service. Such a multi-agency approach facilitated the flow of information to enforcement agencies and stemmed the flow of black-market CFCs into Florida by a combination of intelligence gathering and heavy penalties for those caught.

The Task Force was also involved in operations to counter ODS smuggling across the US border with Mexico. By mid-1996, substantial quantities of CFCs were crossing the border and being distributed through automobile service shops in the southern states, particularly Texas. Traditional smuggling methods were used in the border area, with a few canisters of CFCs being hidden in car trunks, and people even crossing the Rio Grande carrying CFC canisters in backpacks.

The smuggling of CFCs also emerged in Europe in the mid-1990s. Again, much of the illicit material originated in Russia, although considerable amounts of material manufactured in Europe for export to developing countries were also laundered into the domestic market. The existence of a thriving ODS black market in Europe was signified by the rapid spread of fly-by-night brokers entering the CFC market in the mid-1990s. Several firms even set up offices in Estonia to move CFCs from Russia into the black markets of Europe and the USA. The UK played host to several major ODS brokers; the refrigeration industry in Spain denounced 30 firms it claimed were trading in illegal ODSs.

Initial seizures of illegal ODSs occurred in 1995 in Greece and Italy, but the true scale of ODS smuggling was not revealed until July 1997, when a major ODS smuggling ring was exposed. Through a network centred in Germany, a group of brokers conspired to bring in over 800 tonnes of illegal ODSs from China over a two-year period. Much of the material was falsely labelled as replacement chemicals and circulated to Germany, France, Italy, the UK, Greece, Hungary and the USA.

Illegal trade was a major issue at the Meeting of the Parties to the Montreal Protocol in 1997. The European Union announced its intention to ban the sale of CFCs unilaterally, a move which finally became law in October 2000, and the Parties agreed to implement a licensing system to control shipments of ODSs.

By 1997, the flow of illegal ODSs from Russia had fallen and, instead, unscrupulous brokers were turning their attention to China, now the world's largest producer of ODSs. A series of suspicious imports into the USA of Chinese CFCs and halons took place in 1997 and 1998, with the loophole of illegal trade in recycled ODSs being exploited. Illegal Chinese CFCs were also smuggled into Taiwan using an ingenious method of false cylinders. Chinese ODSs also entered the European black market, often falsely labelled as replacement chemicals.

By 1998, the surge of illegal ODS movements had subsided as improved enforcement and legislative measures combined to stem the flow. From its peak of around 20,000 tonnes a year in the mid-1990s, the illegal trade had fallen to around 5000 tonnes a year.

Yet in 1999, the first cases of ODS smuggling into developing countries were recorded, as a consequence of the production freeze which came into force in July 1999. Between mid-1999 and March 2000, more than 800 tonnes of CFCs were smuggled into India alone, equivalent to 12 per cent of the country's annual consumption. Illegal trade has also been detected in Pakistan, the Philippines, Malaysia, Vietnam and Indonesia.

The Parties to the Montreal Protocol are presently exploring further ways to combat the illegal trade of ODSs, through enhanced information-sharing and better customs coding. A new concern is the illegal trade in equipment using ODSs, often dumped from the industrialized world onto developing-country markets. Illegal trade still remains a threat to the success of the Montreal Protocol, particularly for the phase-out schedule for developing countries. However, many valuable lessons have been learned over the last 15 years regarding the control of trade in restricted chemicals and the importance of enforcement for the success of multilateral environmental agreements.

Russian Federation closes its production facilities

The thirteenth Meeting of the Parties also noted non-compliance by Armenia, Kazakhstan and Tajikistan, and urged them to stick to the benchmarks provided by them to return to compliance. The meeting noted the non-compliance by the Russian Federation, but also the good news that the Russian Federation had closed all its ODS-manufacturing facilities on 20 December 2000, meaning that it would return to compliance in 2001.

Johannesburg 2002 and collaborative working on international agreements

The European Community and its members brought to the notice of the meeting the upcoming World Summit on Sustainable Development to be held in Johannesburg in September 2002 to review the implementation of Agenda 21 approved by the Earth Summit in 1992. A representative of the European Community put the Montreal Protocol in the context of sustainable development and mentioned that the latest challenge of the Protocol was to phase out HCFCs. He reported that the EEC had adopted a regulation mandating the phase-out of HCFCs by 2010, 20 years earlier than mandated at the present time. At the urging of the Community and its members, the meeting recognized the need to consider ways to improve the effectiveness of international environmental agreements and to support appropriate collaboration and synergies that may exist between the agreements. The decision looked forward to the outcome of the World Summit and other bodies in this regard.

Chapter 5

Technology and business policy[1]

INTRODUCTION

In about 1900, methyl bromide and carbon tetrachloride became the first two commercial ozone-depleting substances (ODSs) used in fire protection, metal and fabric cleaning and pest control. However, it was the invention and mass marketing of CFCs in the 1930s that increased global emissions to levels that posed a threat to the ozone layer that was unappreciated at the time. In 1974, Mario Molina and F Sherwood Roland published their discovery that emission of CFCs would deplete the stratospheric ozone layer. In the next 13 years, the theory was verified, the Antarctic ozone 'hole' was discovered, and the Vienna Convention for the Protection of the Ozone Layer (1985) and the Montreal Protocol on Substances that Deplete the Ozone Layer (1987) were agreed upon by the world community. The 1987 Montreal Protocol, which had only mild controls on some ODSs, was strengthened with periodic amendments and adjustments from 1990 onwards to phase out a long list of ozone-depleting substances in both developed and developing countries.

All ozone-depleting substances controlled by the Protocol have common chemical ingredients of chlorine and bromine, but vary widely in chemical lifetime and toxicity. The most toxic ODSs, methyl bromide and carbon tetrachloride, were restricted in developed countries long before the Protocol, while the marketing of the least toxic CFCs and HCFCs was uninhibited by regulations, since they had no apparent adverse effects but had many desirable qualities for use in many applications. These differences and the importance of applications explain why each chemical has had a different commercial history.

Today, production of halons is virtually halted worldwide and the production of CFCs and methyl bromide is rapidly declining. As late as 1987, ODS manufacturers and customers claimed that safe alternatives and substitutes would never be available, but today only the methyl bromide industry makes this claim. This chapter tracks the invention, commercialization, proliferation and phase-out of each of the ozone-depleting substances controlled by the Montreal Protocol, and also the invention and commercialization of alternatives to the ODSs.

BOX 5.1 DISCOVERY AND COMMERCIAL DEVELOPMENT OF OZONE-DEPLETING SUBSTANCES, 1900–2000

1900	Carbon tetrachloride, for fire protection, metal and fabric cleaning, and pest control
	Methyl bromide for fire protection
1930s	CFC-11 and CFC-12 for refrigeration and air conditioning
	Methyl bromide as a fumigant for pest control
1936	HCFC-22 introduced for refrigeration and air conditioning
1940s	CFCs for pesticide aerosol products
	Methyl bromide for aircraft fire protection
	First air conditioning in luxury cars
1950s	R-502 (a mixture of HCFC-22 and HFC-152a) refrigerant introduced
	Proliferation of aerosol cosmetic and convenience products and air conditioning
	First use of CFCs in foams
1960s	Methyl chloroform and CFC-113 solvents
	R-502 (a mixture of HCFC-22 and CFC-115) refrigerant introduced
	First use of CFCs for foam insulation in refrigerators
	Halons-1211 and -1301 achieve commercial acceptance
1970s	Widespread use of methyl bromide as a pesticide for soils, stored agricultural products and other applications
	CFC-113 in electronics and precision cleaning
1980s	Market penetration of vehicle air conditioning
1985	Vienna Convention for the Protection of the Ozone Layer
1987	Montreal Protocol on Substances that Deplete the Ozone Layer
1990s	Decline, obsolescence and beginning of phase-out of ODSs
1994	Halon production halted in developed countries
1996	Production and consumption of CFCs, carbon tetrachloride and methyl chloroform halted in developed countries (with feedstock, process-agent and essential-use exceptions)
2000	ODS production halted in countries with economies in transition in December

---

COMMERCIAL HISTORY OF OZONE-DEPLETING SUBSTANCES

Box 5.1 summarizes the history of ozone-depleting substances, including the first date of commercialization and periods of rapid increase in sales and emissions.

In about 1900, methyl bromide became the first ODS to be commercialized; it was the last substance to be added to the Montreal Protocol in 1992. Carbon tetrachloride was the second ODS to be commercialized and was among those added to the list of controlled substances in the 1990 London Amendment. Chlorofluorocarbons (CFCs) are the most familiar ozone-depleting substances controlled by the Montreal Protocol, but were not commercialized until the 1930s. All of these substances are scheduled for phase-out under amendments and adjustments to the Montreal Protocol. There are three classes of exceptions to the phase-out: feedstocks, in which ODSs are almost entirely consumed; process agents, in which emissions are tightly controlled; and applications authorized by Meetings of Parties for essential use (such as metered-dose inhalers for asthma and cardiopulmonary obstructive disease).

The most toxic controlled ODSs – methyl bromide and carbon tetrachloride – were initially marketed into uses where the high levels of exposure resulted in death and injury (both immediate poisoning and long-term health effects), and subsequently suffered market and regulatory rejection for consumer uses. Other safer controlled ODSs – CFCs, halons, and methyl chloroform – enjoyed growth mostly unconstrained by toxicity concerns with aggressive marketing into a vast number of utilitarian and luxury uses.

Table 5.1 presents an overview of the many uses of controlled ODSs. Most 'historical' uses had already declined to very low global levels by 1987 when the Montreal Protocol was signed.

Table 5.1 *Uses of ozone-depleting substances controlled by the Montreal Protocol*

Commercial designation(s)	Refrigerant	Fire extinguishant	Solvent	Propellant	Foam-blowing	Process agent or feedstock	Pesticide	Miscellaneous
Carbon tetrachloride (Halon-104)	H	H	S			S(S)	S	H²
Methyl chloroform	H	H	S			S/S		M³
Methyl bromide (Halon-1001)		H				S(S)⁴	S	H⁵
CFC-11	S		S⁶	S		S/S		S⁷
CFC-12	S	M⁸		S	S			S⁹
CFC-113			S			M, N		
CFC-114	S			M				
CFC-115	S			M				
HCFC-22	S	M		M		S		
HCFC-123	N							
HCFC-141b			N		N			
HCFC-142b					M, N	S		
HCFC-225			N					
Halon-1211		S						
Halon-1301	M	S						
Halon 2402		S						

Key:
S   Significant use at signing of the Protocol
M   Minor use at signing of the Protocol
H   Historic use mostly discontinued prior to the Protocol
N   New use as a consequence of the signing of the Protocol
Note: Superscript numbers refer to the notes at the end of the book.

ODSs marketed for critical, cosmetic and convenience products

By the late 1980s, more than 250 separate product categories were made with, or contained, ozone-depleting substances. Some products were for cosmetic and convenience products where use was optional and alternatives were readily available. However, many other ODS products had become vital to society, with the implication that new alternatives would be necessary. The more critical uses included:

- medical applications, such as metered-dose medicine inhalers, sterilization, cleaning of heart pace-makers and artificial limbs, and blood substitutes;
- refrigeration for meat and fish processing, vegetable storage, frozen food, blood and medicines;
- comfort air conditioning, including in buildings and vehicles;
- foam insulation, for refrigerated appliances, building insulation and industrial applications;
- cleaning of critical electronic and mechanical components, including for weapons detection and guidance, safety systems for nuclear and hazardous chemical facilities, and aircraft flight control;
- fumigation for quarantine and pest control;
- fire protection and explosion suppression in telecommunications, naval and commercial shipping, aircraft, oil and gas processing and transport;
- safety foams used in vehicles as padding and structure.

In addition, ODSs had other uses in products including wineglass chillers, tyre inflators, dust blowers, toys, noise-making horns, tobacco puffing to reduce tar, and convenience and cosmetic aerosol sprays. These uses were deemed non-essential or frivolous by some environmental authorities and consumers.

Fire-extinguishing agents[10]

Seven halons controlled by the Montreal Protocol have been used as fire-fighting agents.[11] Carbon tetrachloride was introduced as a fire-extinguishing agent in around 1900, and by 1910 had widespread use in extinguishing gasoline fires from automobiles and other equipment with internal combustion engines. By 1917, however, carbon tetrachloride was suspected to harm human health; in 1919, the first recorded deaths occurred when two people were overexposed to the fumes of a carbon tetrachloride portable fire-extinguisher. Methyl bromide (halon-1001) gained fire-fighting popularity in the late 1920s and was used in both British and German aircraft and ships during World War II, but was never popular for use in portable extinguishers because it is more toxic than carbon tetrachloride. Bromochloromethane (halon-1011) was developed by Germany in World War II to replace methyl bromide.

In 1947, the Purdue University Research Foundation evaluated the extinguishing performance of more than 60 new candidate agents while the US Army Corps of Engineers simultaneously conducted toxicological studies. Four halons identified by this collaboration were commercialized: halon-1202, halon-1211, halon-1301 and halon-2402. Halon-1202 is the most effective fire-extinguishant but also the most toxic. Halon-1301 ranked second in fire-extinguishing effectiveness and is the least toxic. The US Army developed a halon portable fire-extinguisher for use inside armoured vehicles; the US Air

Force selected halon-1202 for military aircraft engine protection; and the US Federal Aviation Administration approved halon-1301 for commercial aircraft engine fire protection.

By the 1960s, growing concerns about the toxicity of carbon tetrachloride (halon-104), methyl bromide (halon-1001) and bromochlorodifluoromethane (halon-2402) led to their 'official' commercial finish as fire-extinguishing agents in developed countries. Limited use of halon-2402 continued in what was then the Soviet Union and countries dependent on Soviet weapons systems. From the 1960s until the Montreal Protocol, halon-1301 was marketed as a total flooding extinguishant for protecting computer and communications equipment, repositories of cultural heritage, shipboard machinery spaces, and petroleum pipeline pumping stations. Halon-1211 was widely marketed in portable fire extinguishers.

Pesticides[12]

Carbon tetrachloride was probably the first ozone-depleting substance used as a pesticide and an ingredient in pesticide products from its introduction in 1907 until the 1980s.[13] It is toxic and has been used to reduce the fire hazard of flammable pesticide ingredients such as carbon disulphide. Methyl chloroform has also been used as a solvent in pesticide formulations.

Methyl bromide is a naturally occurring substance that was initially commercialized as a fire-extinguishing agent (as discussed above). Since 1960, it has been used as a fumigant to control pests in dry foodstuffs, buildings, furniture, archives, and other products made from plant material. The use of methyl bromide as a soil fumigant increased dramatically in the 1980s. It is a potent 'biocide', killing most life-forms in most life-stages.

In early 1992, the Scientific Assessment Panel reported new evidence that methyl bromide had a very high ozone-depletion potential (ODP) and that emissions were probably high enough to contribute significantly to ozone depletion. The April 1992 meeting of the Open-Ended Working Group urgently requested a scientific and technical assessment. In June, the Scientific Assessment Panel and a joint Scientific Assessment and Technology and Economic Assessment Panel (TEAP) completed their emergency assessment and published a synthesis report confirming that methyl bromide posed a significant threat to the ozone layer and that technically and economically feasible technology was available to reduce its use and emissions in most applications. In July, the Assessment Panels presented their results to the Open-Ended Working Group. Methyl bromide producers, large agricultural customers and some ministries of agriculture disputed the scientific and technical conclusions.

In response to the technical complaints of the methyl bromide industry, the governments of The Netherlands and the USA organized workshops in The Netherlands and Italy that featured successful alternatives to methyl bromide. Participants heard directly from growers and examined the technology in detail. A consensus report was distributed in time for the 1992 Meeting of the Parties in Copenhagen where methyl bromide was added to the list of controlled

BOX 5.2 THE UNIQUENESS OF METHYL BROMIDE UNDER THE PROTOCOL

Linda Dunn, Industry Canada

I participated in the Montreal Protocol representing the interests of agriculture. As a senior official in Agriculture & Agri-Food Canada, I co-chaired the Canadian multi-stakeholder working group looking at methyl bromide issues and addressing Canadian agricultural business interests. I was a member of the Methyl Bromide Technical Options Committee, and of the Canadian delegation to the Open-Ended Working Group and Meetings of Parties where methyl bromide was an issue.

My perspective is that, although some producers and users took every opportunity to defend continued use of methyl bromide, on the whole users in Canada and most other countries exhibited concern for the environment and were motivated to protect the ozone layer. At all the critical times, Canadian industry rallied behind a progressive Canadian position supporting a phase-out.

The solid scientific base of the Montreal Protocol was instrumental in ensuring a positive, proactive positioning by Canada. The clear links between ozone depletion and human health – provided by the international scientific community – established the need for quick and decisive action. In Canada, there was little, if any, discussion on the validity and verity of these results.

Canadian agricultural interests trusted the assessment panels, and their conclusion that agricultural emissions contributed significantly to ozone depletion, because they included globally respected experts operating without political interference from the Parties. My task was made easier by the availability of Montreal Protocol experts to speak to Canadian stakeholders. Questions were answered frankly, in an easily understood manner, and in an open and transparent way.

If I had a dime for every time I said, 'Methyl bromide is not like CFCs or HCFCs – it's different,' I would be much richer than I am today. Everything from the actions of the molecule in the atmosphere to the end users and all points in between cried out for a different approach. First, no other substance included in the Protocol has substantial natural as well as anthropogenic sources, and no other controlled substance is so reactive in the atmosphere, introducing additional uncertainty into the calculation of ODP. Second, methyl bromide controls drastically affected certain portions of the agricultural sector which had been mostly unaffected by the controls on other ODSs. Third, methyl bromide was a vital part of the defence against foreign pests and the potential devastation that these pests can cause to domestic agricultural and forestry industries.

All of these differences made for a collision course, yet the very balanced approach taken by the Parties to the Protocol eased the tensions and diffused a potentially volatile situation. Parties recognized that methyl bromide was 'different', and: allowed exemptions for quarantine and pre-shipment uses which addressed agricultural security issues; put into place mechanisms for critical and emergency uses, guarding against unanticipated infestation; and generally recognized that alternatives needed to be tested and proven for different climatic, geographic and pest scenarios. This balance allowed environmental gains without undue economic and social consequences. This was a major coup for the Protocol.

Following national multi-stakeholder consultation, a Canadian working group was formed in January 1994 to address issues related to demonstrating and gaining registration of alternatives, deciding on research priorities, and assisting with the formation of negotiating positions at international meetings. Under the auspices of the working group, many cooperative projects have been conducted. Projects have included: demonstrations of the effectiveness of heat, phosphine and carbon dioxide in flour mills; the use of diatomaceous earth, alone and in combination with other substances; corrosion studies related to the use of phosphine; and various alternatives for shiphold fumigation. In all cases, industry has freely and willingly given its time, resources, expertise and equipment to try out new integrated pest-management approaches to eliminate the use of methyl bromide. The work by the Canadian Pest Control Association, its member companies and some highly innovative farmers has been extraordinary.

In 1997, Parties to the Protocol were considering more stringent controls on methyl bromide. This was the tenth anniversary of the signing of the Protocol, and the Meeting of the Parties was back in Montreal, Canada. A very hard-working group of Canadian and American researchers, pest control companies and associations, and alternative product providers put on a technology showcase which surprised and astonished many of the delegates and, I believe, contributed significantly to the decision to accelerate the phase-out of methyl bromide. 'Discover 97 – Methyl Bromide Alternatives Expo' demonstrated alternative products, techniques and approaches. Industry provided a delicious lunch and traditional Canadian hospitality in an informal setting, with hands-on professionals to answer questions. In my view, this event epitomized the Canadian agri-food approach to methyl bromide phase-out.

Note: See also Box 6.4 by Joop van Haasteren (Chapter 6).

substances. In 1993, the TEAP created the TEAP Methyl Bromide Technical Options Committee, which began a more detailed assessment of alternatives and substitutes to methyl bromide. A phase-out was agreed by the Parties in 1995 in Vienna and accelerated in 1997 in Montreal.

Refrigerants[14]

The first controlled refrigeration of perishable foods was accomplished in caves, cellars, wells or artesian 'springhouses'. Simple air conditioning was first achieved with flowing cold water or by evaporating water on porous surfaces. Natural ice was originally stored locally for warm-weather use and, by the early 1800s, globally marketed to temperate regions. Carbon tetrachloride and methyl bromide were the first ozone-depleting substances used as refrigerants, but only in obscure instances.[15] In the middle of the 19th century, mechanical ice-making equipment – primarily using sulphur dioxide, ammonia and hydrocarbon refrigerants – became competitive at locations far from unpolluted natural ice sources.

In the early 1900s, iceboxes competed successfully with electric refrigerators using flammable and toxic refrigerants. Iceboxes were highly reliable, but poor temperature control could lead to food spoilage, and ice could not keep food frozen. Refrigerators were more convenient, but had new risks of fire, explosion, toxic exposure, and food contamination and spoilage from refrigerant leaks. Refrigerant leaks were common from vibration, machine failure, defrosting accidents and servicing procedures. A leak of the most common refrigerants – sulphur dioxide and ammonia – typically required rapid evacuation, with people suffering from vomiting, burning eyes and painful breathing, but these accidents rarely resulted in death. In contrast, leaks of the less common refrigerant methyl chloride resulted in frequent deaths that were sensationalized by the home icebox industry.

Table 5.2 *Flammable and toxic refrigerants in use before CFCs*

Refrigerant	Flammability	Toxicity	Comment
Ammonia	Slightly	High, but noxious properties promote safety	Predominant in industrial and heavy commercial refrigeration; competitive in air conditioning; some light commercial and household refrigerators (particularly models using the absorption cycle)
Carbon dioxide	Extinguishes fire	Low	Competitive in air conditioning; high pressure requires sturdy construction
Dimethyl Ether	High	Moderate	One of the first refrigerants used
Ethyl chloride	None	High	Some light commercial and household refrigerators
Methyl chloride	None	High	Competitive in air conditioning
Methylene chloride	None	High	Distant second preference for light commercial and household; competitive in air conditioning; competitive in industrial and commercial refrigeration
Isobutane	High	Low	Some light commercial and household refrigerators
Sulphur dioxide	None	High, but noxious properties promote safety	Preferred for household refrigerators; strong odour warns of leaks but even small leaks spoil food; competitive in industrial and commercial refrigeration and air conditioning

Discovery of the CFCs, 1928

In late 1928, executives from General Motors (GM) and its refrigerator manufacturing division Frigidaire assigned Thomas Midgley of the GM Research Laboratory the task of inventing a non-toxic, non-flammable and non-corrosive refrigerant. Midgley determined that elements with boiling points appropriate for refrigeration were clustered on the Langmuir periodic table, which is arranged according to the number of vacancies in the outer shell of electrons. Working with Albert Henne and Robert McNary, Midgley ruled out unstable and inert elements, leaving carbon, nitrogen, oxygen, sulphur, hydrogen, and the halogens fluorine, chlorine and bromine. Others had dismissed fluorine because chemical substances containing fluorine are often toxic and/or corrosive.

Midgley and Henne, however, were familiar with Frederic Swarts' theory that the toxicity of fluorine could be negated if strongly bonded with chemicals that had complementary valences. Within two or three days of receiving their research assignment, they had identified chlorofluorocarbons (CFCs) as prime candidates and synthesized dichloromonofluoromethane (CFC-21) from carbon tetrafluoride. Within months, it was confirmed that chlorofluorocarbons would be non-flammable, non-explosive, non-corrosive, very low in toxicity, and odourless, and that their vapour pressures and heats of vaporization made them very suitable for refrigeration applications.[16] Within a year, GM patented the family of CFCs and perfected the manufacturing process for the first commercial substances (CFC-11 and CFC-12). On 27 August 1930, General Motors and DuPont formed a joint stock company – the Kinetic Chemical Company – to manufacture and market CFCs.[17]

In 1930, CFCs were considered perfect in every known way because stratospheric ozone depletion was neither understood nor anticipated. CFCs are chemically stable, non-corrosive, non-flammable, odourless, colourless, energy-efficient and inexpensive, and have very low toxicity. Refrigeration and air-conditioning equipment manufacturers and their customers came to think of CFC as 'wonder gas'.

CFCs rapidly captured much of the global refrigeration and air conditioning markets. Ammonia and hydrocarbons continued to be used as refrigerants in applications with substantial economic advantage and where the risk of human exposure was small and carefully managed. Energy-efficient applications for ammonia included refrigerated cold storage, ice- and ice-cream-making and breweries. Energy-efficient applications for hydrocarbons included niche applications in industrial processes such as petroleum refining. CFC sales increased with expanding population, wealth and consumerism.

The first motor vehicle air conditioning using an ODS refrigerant was on the 1939 Packard. Global application of air conditioning for motor vehicles grew slowly from the 1950s to the 1980s; by the time the Montreal Protocol was signed, more than 90 per cent of North American and up to 50 per cent of new Asian and European vehicles had air conditioning. The transition from CFC-12 to HFC-134a refrigerants was so successful that growth in vehicles with air conditioning has been unabated. Today, vehicle air conditioning is a standard feature on a majority of vehicles worldwide, providing air conditioning for comfort, and window defogging and demisting for safety.

Blowing agents in flexible and rigid foam

During World War II, Dow Chemical Company introduced the first foam product manufactured with CFC-12 under the brand name Styrofoam™. CFC-12 helps the polymerizing mixture to form closed cells containing CFC gas. The CFC-12 also improves the thermal insulating properties because it is less conductive than most other foam-blowing agents. Polystyrene foam insulation boardstock was invented in Sweden in the early 1940s but was further developed to an extrusion process in the USA. The original blowing agent was methyl chloride, not CFCs. Extruded polystyrene foam insulation made with CFC-12 was introduced in the early 1960s. CFC-12-based foam was extensively used in food service, packaging and insulation.

In 1959, Imperial Chemical Industries (ICI) implemented the first use of CFC-11 foam for insulation of the holds of ships used to transport meat from

Australia and New Zealand to the UK. In 1961, ICI implemented the first use of CFC-11 insulating foam to replace glass fibre in refrigerators manufactured at a Danish plant. By 1985, CFC-11-based foams represented a high market share in refrigerators, building insulation and other applications. In the late 1950s, flexible foams made from polyurethane quickly became the predominant cushioning material for furniture, bedding, carpet underlay, automobile seats and dashboards.

Solvents[18]

Carbon tetrachloride (CCl_4) was introduced in the early 1900s as a low-cost, colourless, odourless, and non-flammable liquid solvent to dissolve fats, oils and greases in metal and fabric cleaning. It is toxic when absorbed through the skin or when inhaled. At high temperatures, it reacts to form highly poisonous phosgene gas.

By the 1930s, carbon tetrachloride replaced gasoline as an agent for dry cleaning delicate fabrics and leather. However, by the 1960s, ODS 1,1,1-trichloroethane (methyl chloroform) had replaced carbon tetrachloride in most dry cleaning, only to be replaced later by non-ODS perchloroethene. In the 1970s, CFC-113 was also used to clean certain delicate fabrics, leather and garments with decorations that are adversely affected by other solvents. After the 1960s, carbon tetrachloride was prohibited in most developed countries for all consumer products due to its toxic effects on livers and kidneys.

Methyl bromide was used at one time as a degreasing solvent for wool. Methyl chloroform (1,1,1-trichloroethane) is a solvent that is relatively very low in toxicity. It grew in popularity in the 1970s and 1980s as a metal degreaser and adhesive ingredient when less expensive but more toxic chlorinated solvents were controlled by environmental authorities.

Until the 1970s, electronics and precision products were cleaned with alcohol, aqueous and chlorinated solvents. Once commercialized in the late 1970s, however, CFC-113 quickly dominated because it is a gentle solvent ideally suited to cleaning products with delicate components: it is compatible with a wide range of metals and plastics, dries quickly and has low toxicity.

Feedstocks, process agents and product ingredients[19]

Carbon tetrachloride is used for a large number of feedstock, process agent and manufacturing uses. It is a feedstock for the production of CFCs, hydro-chlorofluorocarbons (HCFCs), pesticides, hexachloroethane, tetrabromo-methane, pyrosulfuryl chloride and other products. It is a process agent in the production of chlorine, anti-corrosive coatings (chlorinated rubber), adhesives, pharmaceutical products, insecticides, synthetic fibre and ballistic armour. It is an 'involuntary' by-product of industrial processes, notably the production of chlorinated solvents and vinyl chloride. Carbon tetrachloride is also used to recover tin in tin-plating waste, in the manufacture of semiconductors and integrated circuits, and as a product ingredient in petrol additives and soap perfumery. Process-agent and feedstock applications of methyl bromide include use as a reagent and as a methylating agent.

INDUSTRY OPPOSITION AND THEN SUPPORT FOR
REGULATION OF OZONE-DEPLETING SUBSTANCES[20]

Stratospheric ozone depletion emerged as a public issue in the 1970s global debate on the environmental acceptability of supersonic transport aircraft (SST) that would fly in the stratosphere faster than the speed of sound. Environmental activists initially opposed the SST because 'sonic booms' would disturb peace and quiet, damage cultural and natural objects, and disrupt wildlife and ecosystems. However, opposition widened after atmospheric scientists warned that the global climate could be affected by high-altitude emissions of hydrogen oxides or water vapour, and hypothesized that SST nitrogen oxide emissions could deplete the ozone layer. Stratospheric ozone depletion was a contributing factor in the decisions of the USA and Soviet Union to abandon their SST programmes, though the British–French consortium commercialized the Concorde SST.[21]

Industry's initial resistance to regulation

Chemical companies first to suspect environmental impacts
Chemical companies were alert to the possible atmospheric and environmental consequences of CFC emissions as early as 1972. A letter from organizers at DuPont inviting scientists to the 1972 International Conference on the Ecology and Toxicology of Fluorocarbons outlined the concern: 'Fluorocarbons are intentionally or accidentally vented to the atmosphere world-wide at a rate approaching one billion pounds per year. These compounds may be either accumulating in the atmosphere or returning to the surface, land or sea, in the pure form or as decomposition products. Under any of these alternatives, it is prudent that we investigate any effects that the compounds may produce on plants or animals now or in the future.' After the 1972 conference, DuPont and other CFC producers formed a consortium coordinated by the Chemical Manufacturers Association (CMA) to investigate environmental properties of fluorocarbons.[22]

The Molina–Rowland hypothesis captures attention
When scientists F Sherwood Rowland and Mario Molina started to research the impact of CFCs on stratospheric ozone, their approach began with the questions: 'What is the ultimate fate of CFCs?' and 'Are there any consequences of their release to the environment?' (see Chapter 1). After concluding that CFCs would reach the stratosphere and affect the ozone layer, they familiarized themselves with the work of the US National Academy of Science (NAS) Climatic Impact Committee founded in 1971 and the preliminary findings of the joint NAS/US Department of Transportation Climatic Impact Assessment Project.[23]

When the Molina–Rowland hypothesis that CFC emissions would deplete stratospheric ozone began to gain attention in late 1974, CFC producers and industry customers quickly organized to defend CFCs, refocused industry research on key elements of the CFC/ozone-depletion theory, and established coordination with government research programmes. DuPont often took the industry lead in the USA.

Some countries take early action despite company arguments for delay
Even by the mid-1970s, it was known that CFCs were accumulating in the atmosphere.[24] However, CFC industry stakeholders and some scientists contended there was no need for an urgent response because ozone depletion was only a hypothesis based on chemical reactions observed in the laboratory and through computer models; there had been no atmospheric observations of the proposed ozone destruction cycle, and there were questions about the relative magnitude of CFCs as a source of stratospheric chlorine. CFC manufacturers and their customers argued for delay in a regulatory response for a few years until scientific research answered these outstanding questions (see Chapter 1).[25] Early on, industry estimated that hundreds of thousands of jobs and billions of dollars of economic wealth depended on CFCs.

Despite industry arguments to wait for more scientific evidence, public concern for the ozone layer was very high, particularly in the USA, Canada, Denmark, Finland, Norway and Sweden. US aerosol product sales slumped; the S C Johnson Company earned the historic distinction in June 1975 of being the first company to abandon CFCs. Within weeks, Sherwin-Williams, Bristol Meyers, and Mennen joined with S C Johnson in advertising for their CFC-free products and against the use of competitors' CFC products. In 1976, the US government announced plans to ban 'non-essential' aerosol products. Canada, Sweden, and then Norway, also announced bans, and many nations discussed further action.

Pressure builds for regulation

UNEP and environmental NGOs persevere
After the initial flurry of activity from 1975 to 1980, national governments took little additional action and public interest in ozone depletion faded. However, the United Nations Environment Programme (UNEP) and a few environmental non-governmental organizations (NGOs) persevered while industry built political coalitions and continued opposition arguments including the contention that controls were not scientifically justified. Government and academic scientists increased their investigations, and the Chemical Manufacturers Association sponsored scientific research that complemented government research efforts.

From 1975 to 1985, industry positions evolved with each new scientific study or scientific assessment. Industry carefully followed the scientific debate, highlighting disagreements and uncertainties and arguing that no action was needed until ozone depletion was verified and conclusively linked to CFCs and other ODSs. From an industry perspective, CFCs were innocent until proven guilty, and the strategy was to insist on a high standard of scientific proof. At that time, industry did not subscribe to the precautionary principle.

As scientific evidence mounted, industry also argued that alternatives were not currently available, would take years to develop and deploy, and were technically inferior and more expensive than CFCs. It was argued that public health and wealth would suffer without CFC products such as refrigeration, air conditioning, electronics, sterilization and medicine. Industry contended that the science did not justify commercialization of substitutes or alternatives.[26]

Despite continuing scientific uncertainty and debate, and little public interest, UNEP was proceeding methodically with its own efforts (see Chapters 2–4 and Appendix 1). In June 1973, at the first meeting of the UNEP Governing Council, UNEP Executive Director Maurice Strong cited 'damage to the ozone layer as a possible "outer limit" that, if breached, may endanger the continuance of human life on the planet'. From April 1975 on, UNEP organized meetings and workshops on ozone-depleting substances that included detailed discussions of technical and economic feasibility of alternatives, as well as regulatory incentives and prohibitions. In most cases, industry participation at these meetings consisted of prediction of dire economic and safety consequences, while government and academic experts presented estimates of affordable costs and administrative feasibility and simplicity.

Companies declare chemical substitutes too expensive
In early 1986, representatives of DuPont, Allied and ICI separately reported that between 1975 and 1980, they had identified compounds meeting environmental, safety and performance criteria for some CFC applications, but had terminated research and development when they concluded that none were as inexpensive as CFCs.[27] The reasoning for these decisions was that CFCs 'work so well, in fact, that chemical makers have done little to find alternatives'. DuPont, the world's largest producer of CFCs, ended a five-year US$15 million research programme in 1980, largely because preliminary results indicated substitutes and alternatives would be uneconomical to make, require difficult changes in manufacturing plants, take a minimum of ten years to develop and market, and involve controversial toxicity-testing for adverse health effects.[28]

At a UNEP workshop in Rome on 26–30 May 1986, DuPont stated that the costs of chemical substitutes for CFCs, mainly HFCs, would be three to six times the price of CFCs, while ICI estimated costs at eight to ten times the CFC price. DuPont justified its policy of not commercializing substitutes by explaining that:

> *'Neither the marketplace nor regulatory policy, however, has provided the needed incentives to make these equipment changes or to support commercialization of the other potential substitutes. If the necessary incentives were provided, we believe alternatives could be introduced in volume in a time frame of roughly five years.'*[29]

Thus, after a decade of industry opposition to regulation, industry claimed that it was the lack of regulation that prevented it from introducing products to protect the ozone layer.

Costs of ODSs could be unimportant or highly important, depending on the cost of the ODS as a percentage of the product price. Each sector faced changes in both capital investment and operating costs. Often, the capital investment cost per unit of ultimate output was minor, while the operating costs could remain significant. A typical automobile might cost US$10,000–30,000 and offer an air conditioner for US$1000 more, containing 0.5–1.5kg of CFC-12 at a wholesale price of less than US$1–2 per vehicle. Aerosol products, which comprised more than one-third of CFC use in 1985, cost less to produce without CFCs. However, in the insulation industry, with CFCs representing 20 to 30 per cent of the production costs of rigid insulating foam, manufacturers would lose markets to fibreglass, mineral wool, and other insulating choices.

Many workshops develop regulatory and technical options

Particularly after the Vienna Convention was signed in 1985, numerous workshops and informal meetings were held on regulatory options and technical issues (Table 5.2). By 1986, consumption of CFCs was increasing, and credible experts had forged a global scientific consensus that projected emissions would drastically deplete stratospheric ozone. In 1984 Japanese scientists published the first indication of Antarctic ozone depletion, but it was not until the 1985 measurement of ozone depletion over Antarctica by British scientists that the 'Antarctic ozone hole' captured the attention of scientists, politicians and the public.

Table 5.3 *UNEP technology and economic workshops and conferences 1975–1988*

Date and location	Meeting	Technical and economic topics
1–9 March 1977 Washington, DC, USA	Governing Council Conference	Selective product bans, containment, recovery and recycling
6–8 December 1978 Munich, Germany	Workshop on Regulatory and Voluntary Actions	Economic impact of product bans, voluntary production limits (capacity and output)
20–28 January 1982 Stockholm, Sweden	Ad Hoc Legal and Technical Experts	Chemical substitutes, bans, emissions limits, quotas or production caps with marketable or transferable permits, taxes on production and use, deposit-refund incentives to recycle, consumer education and labelling
26–30 May 1986 Rome, Italy	Workshop on Economics of Controls	Economics of regulatory controls
8–12 September 1986 Leesburg, VA, USA	Workshop on Demand, Technical Controls and Strategy	Technology, economics, environmental acceptability (toxicity, climate, energy efficiency)
13–15 January 1988 Washington, DC, USA	International Conference on Alternatives to CFCs and Halons	Technology, technical cooperation, and government policy for market transformation
19–21 October 1988 The Hague, The Netherlands	Conference on Science and Development, CFC Data, Legal Matters, and Alternative Substances and Technologies	Technology and economics with detailed presentations on solvents, foam insulation, small and large refrigeration and air conditioning, food packaging, fire protection and vehicle air conditioning

By 1985, some national governments were considering unilateral regulations for CFCs; some environmental NGOs were advocating CFC boycotts and other drastic actions; some environmental leadership companies were moving toward alternatives and substitutes; and some executives and investment counsellors were concerned about businesses' reputations and liability for health damages. The technology was commercially available to replace at least one-third to one-half of CFC uses, primarily in aerosol products, rigid and flexible foam, and solvents.[30]

Anticipating regulation, industry coalitions favour international solutions
In the face of increased scientific evidence, global public concern, and impending national or global controls, the industry coalitions fighting for CFCs began to change their strategies and, in some cases, to become adversaries of CFCs. In the USA, some industry groups representing ODS users changed their strategy and advocated an international solution. A panel of chemical experts sponsored by the US Environmental Protection Agency (EPA) announced that substitute chemicals were technically and economically feasible. In the autumn of 1986, simultaneously announcing support for international limits on CFC consumption, DuPont announced that substitute chemicals could be commercialized and marketed at prices that would be affordable to many applications. When the Montreal Protocol was signed in September 1987, industry began a transformation of attitude and technology.

INDUSTRY RESPONSE TO THE MONTREAL PROTOCOL: WHAT A DIFFERENCE A TREATY MAKES!

Prior to 1987, the majority of companies depending on CFCs had done little to investigate alternatives. In January 1988, the US EPA, Environment Canada and the Conservation Foundation organized the first annual technology conference on alternatives to ODSs, with more than 1000 participants. The scientific evidence for depletion of the ozone layer was no longer questioned, governments offered to collaborate on new technology, ODS-using companies sought alternatives, and potential suppliers were motivated by the market opportunity. Technical optimism was contagious:

- AT&T held a press conference to announce that a semi-aqueous solvent made from orange peels could clean electronics as well as CFC-113; the new technology was from an innovative small company, Petroferm.
- Solvent equipment suppliers announced methods to cost-effectively reduce CFC emissions by half or a third.
- Executives of the Mobile Air Conditioning Society, General Motors, and the EPA announced plans to commercialize CFC-12 recycling for motor vehicles. Dozens of experts stayed after the conference to begin work on the plan.

Within the year, US industry, environmental NGOs, and the US EPA announced the world's first voluntary national CFC phase-out in food packaging. Canada's Nortel and Japan's Seiko Epson announced corporate goals of a complete CFC-113 phase-out on accelerated schedules (see Boxes 5.3–5.5).[31]

The 15 March 1988 release of the findings of the International Ozone Trends Panel was a major turning point for the ODS manufacturing industry (see Chapter 1). For the first time, the presence of chlorine monoxide inside the Antarctic 'ozone hole' was clearly measured – a 'smoking gun' that proved to the satisfaction of industry that CFCs caused catastrophic ozone depletion. Within ten days of the release of the Ozone Trends Report, DuPont committed to an orderly global transition to a total phase-out of fully halogenated CFCs, signalling drastic changes in markets and technology. The Pennwalt Corporation almost immediately seconded DuPont's position and urged that CFC production be halted as soon as practical. By October 1988, ICI also announced support for a CFC phase-out. CFC prices were uncertain, and DuPont began to notify customers that supplies might be rationed.[32] (See Appendix 1 for other corporate and military milestones.)

The announcements by CFC producers supporting international agreements to limit global CFCs and the rapid rejection of those chemicals by their industrial customers had financial implications that further stimulated the quest for substitutes.[33]

International assessments provide information to decision-makers

Before the release of the three-volume report *Atmospheric Ozone: 1985*, organized by Robert Watson of the USA, there had been no comprehensive international assessment of ozone science, and various national assessments often sent conflicting messages to policy-makers. Watson took the initiative to recruit and involve scientists from many countries, and to consolidate various national and international science projects into a single assessment team for that project.

Prior to the Montreal Protocol, industrial customers of CFCs and halons often relied on chemical suppliers for information on the science of ozone depletion. Under the provisions of the Montreal Protocol, UNEP sponsored four global assessment panels composed of teams of experts on: atmospheric science; the environmental effects of ozone depletion; technical options to reduce and eliminate emissions of ozone-depleting substances; and economic implications of dealing with the issue. The initial leadership of these panels provided by co-chairs was essential to their success.[34] The panels have consistently provided independent policy-relevant but policy-neutral information on all aspects of the ozone-depletion issue.

Communication of information to decision-makers was also a key to success. As an example, Scientific Assessment Panel Co-Chair Daniel Albritton translated complicated scientific information into drawings of the atmospheric interactions that could be quickly understood by people trained in a wide variety of disciplines and with limited English-language skills. Industry representatives could master the fundamental scientific arguments and present the slides to their management. Thus, scientists successfully motivated and guided business in the search for alternatives.

BOX 5.3 ELIMINATING CFCS: MANAGEMENT PHILOSOPHY INTO PRACTICE

Tsuneya Nakamura, Past President, Seiko Epson Corporation

Tending to world environmental problems such as protecting the ozone layer, preventing global warming, and developing counter-measures for acid rain is the common responsibility of all of us on Earth. It is a task that should be confronted by concentrating the wisdom of mankind. These environmental problems now transcend national borders and countries; businesses and even individuals must behave as though the problem is their own.

The movement to eliminate ozone-depleting substances, which is being carried out within the worldwide environmental network featuring UNEP at the hub, is said to be a model illustrating the effectiveness of a coordinated effort in protecting the Earth's environment. Seiko Epson began to use CFC-113 solvents in the 1970s because we believed the chemical supplier's claim that CFCs were environmentally harmless and safe for workers. When the Montreal Protocol was adopted in September 1987, that claim crumbled around us. We, therefore, reached the conclusion quite naturally that we must stop the use of CFCs in order to protect the environment and to avoid the business consequences of tighter restrictions.

In December 1988, when Seiko Epson announced our phase-out plan to employees, suppliers, and the press, we received an unexpected positive response from Japanese mass media. I honestly was puzzled by why this decision was seen as unique and why it generated so much interest from others. As the President of Seiko Epson at that time, the fact that I so easily made my decision is deeply related to the condition of the environment and our traditional business policy.

In order to wrestle with the elimination of CFCs through the cooperation of the whole company, we established 'The CFC Phase-out Centre'. We decided on the goal of totally eliminating the use of CFCs within five years, by the end of March 1994. The employees responded dramatically, and we were able to eliminate CFCs in October 1992, one and a half years before our goal. Our success has had the happy result of strengthening the bonds between top management and employees. It brings me great satisfaction that, understanding my desires, the employees took up the challenge and succeeded in the technological research necessary to achieve a CFC-free company.

As Seiko Epson accomplished our own phase-out, we realized that we could speed the recovery of the ozone layer by sharing our experience and technology as widely as possible. We therefore chose to open to the public all of the alternative technologies we had painstakingly developed. As we began sharing our ozone-protection technology, grateful companies reciprocated by sharing their own proprietary technology and know-how. In the spirit of cooperation and a sense of commonality in technological exchange, we began focusing on the wish to eliminate CFCs from the Earth even one day sooner.

There are many, many problems which humankind must overcome, but I think that the kind of global technological cooperation experienced during the ozone-layer protection movement can serve as a valuable model for similar activities in the future. It is the responsibility of those of us living in the present to pass on to future generations a healthy environment. We intend to do what we can to continue this effort into the future.

Box 5.4 Eliminating ODSs: Lessons for the Future from Seiko Epson

Hideaki Yasukawa, President, *Seiko Epson Corporation*

In December 1988, when then-President Tsuneya Nakamura launched Seiko Epson's ODS-elimination programme, we faced a number of difficult technical challenges, none of which we knew how to solve. Nevertheless, a concerted effort brought steady progress: we successfully phased CFCs out of our domestic operations by October 1992, and by December 1993 we completely eliminated all controlled ozone-depleting substances – including 1,1,1-trichloroethane – from our manufacturing facilities around the world. Moreover, our major suppliers in Japan followed suit.

It was my decision some 20 years earlier as head of production engineering development at Seiko Epson to adopt the use of CFCs. At the time, CFCs were more expensive than the cleaning agents we had been using, mainly chlorine-base solvents such as trichloroethylene and tetrachloroethylene, and alcohol-base solvents, but CFCs also offered superior cleaning, drying and permeability characteristics. Just as important, the non-flammable, non-toxic chlorofluorocarbons offered safety, and I felt that worker safety should come before manufacturing cost. In the years following the decision to use CFCs, these solvents – with their superior characteristics – became an integral part of our manufacturing processes, a part we believed to be indispensable.

Manufacturers often become dependent on 'miracle' substances like CFCs, which satisfy the many requirements of cleaning and drying processes, with the unfortunate result that technical progress stagnates. This was the case at Seiko Epson. The active introduction of CFCs at Seiko Epson meant, in effect, that cleaning and drying process innovation came to a halt. Since the superb characteristics of CFCs almost guaranteed product quality, we could use them without ever having to fully research cleaning and drying process technologies that would just fit product quality requirements. Seiko Epson, as a rule, does not make radical changes in technologies and processes unless there is very good reason.

I think that the key to the success of Seiko Epson's ODS-elimination programme is that we made sure our people understood the importance of protecting the stratospheric ozone layer and that we then set a high target – 'total elimination' – and got right to work. Our people began working towards establishing CFC-free manufacturing processes to eliminate ODSs by the end of March 1994. The pledge of total elimination was a challenging target because we had just begun activities to reduce CFC use, and we had no more than a vague idea that we would use 'water' to achieve our goal; we had identified no particular alternative solvents or technologies.

Yasuo Mitsugi, a senior managing director who had overall responsibility for the work on the front lines, demonstrated strong leadership in devising alternatives to CFC-dependent processes. He was instrumental in fostering a sense of solidarity between management and labour and in promoting activities that culminated in the elimination of CFCs years ahead of schedule. This programme showed me just how important it is to identify environmental objectives clearly, to set the highest possible targets, and to charge ahead as a team towards the realization of those targets.

* Hideaki Yasukawa was head of manufacturing when then-President Tsuneya Nakamura announced the plan to phase out CFCs, and was president when the phase-out was achieved.

INDUSTRY AND MILITARY MOTIVATIONS FOR LEADERSHIP ON OZONE PROTECTION

Industry motivation for leadership

Industry leadership was an important ingredient in the accelerated and cost-effective phase-out of ODSs (see Appendix 1). Experts from government and NGOs who worked directly with companies on phase-out projects have documented that in many cases, the overwhelming motivation was to protect the ozone layer.[35] In many other cases, however, companies undertook voluntary efforts and leadership projects in response to a wide range of factors.[36]

Table 5.4 *Industry motivation to speed protection of the ozone layer*

Motivation	Underlying forces
Social	Some industry sectors or leadership companies had become increasingly powerful members of the global diplomatic and governance community, choosing to play the role of champion, of exemplar, in helping to forge environmental policy.[37]
Reputation and goodwill	The reputation of a company today is more than product performance and reliability. Customers care about human rights, environment and other aspects of sustainability. Because reputation affects sales, it also affects financing.
Regulatory	Early action initiated by industry avoids 'command and control' actions initiated by government that rigidly prescribes particular technical solutions, and allows industry to search and choose the solution that satisfies the environmental criteria at the least cost while maintaining or enhancing product performance and appeal. Head-starts avoid desperate technology choices.
Economic and strategic	'Pollution prevention pays' – less waste, more product and less disposal.
	Prompt response to new science can avoid liability for health and environmental damage.
	Phase-out can impress customers and stimulate invention of new technology (for use, sharing or licensing).
	Sector leadership allows experts to collaborate and share the expense of development.
	Supplier competition and economies of scale reduce costs.
Public relations	As pressure from environmental NGOs increases, protests can embarrass companies and damage reputations with consequences to market share.
	ODS emissions data for companies and facilities were publicly available in many countries.[38]
	Recognition from government and NGOs is free advertising, and environmental awards build reputation.
Human resources	Environmental leadership attracts the best applicants seeking careers; a sense of public benefit can be a powerful team-building and motivating force for innovation and performance.

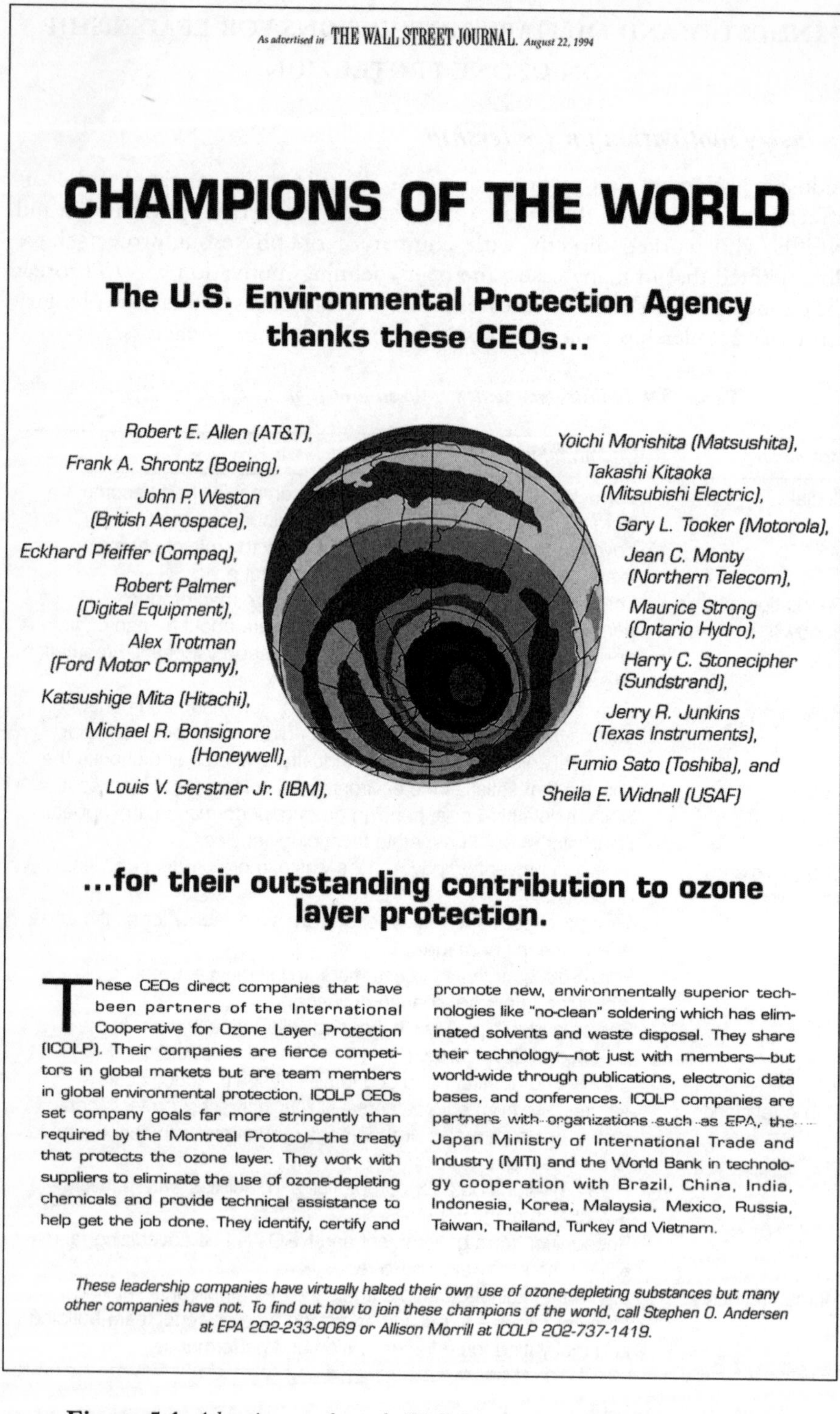

Figure 5.1 *Advertisement from the* Wall Street Journal, *22 August 1994*

Military motivation for leadership

Military organizations had been aware of ozone issues since the early 1970s, when it was hypothesized that supersonic transports would deplete stratospheric ozone, and scientists from the US National Aeronautics and Space Administration (NASA) had investigated whether rocket exhaust would be a contributing factor. As early as 1975, the North Atlantic Treaty Organization (NATO) was briefed on ozone-depletion issues by the EPA Administrator and was periodically updated.

By 1987, the US military had determined that every weapon system in its inventory depended on ozone-depleting substances (particularly CFC solvents and refrigerants and halon fire-extinguishing agents), and realized that military facilities and their contractors were the source of some of the highest ODS emissions.[39] Furthermore, stratospheric ozone depletion was an 'environmental security concern' because loss of agricultural productivity and adverse human health affects from ozone depletion would cause poverty, migration and other societal disruptions. Environmental crises destabilize democracies and create conflict. These economic and geopolitical instabilities increase the need for direct intervention by peacekeeping and humanitarian missions.

Early military technical leadership

The Office of the USA Secretary of Defense and its military services were generally ahead of even the most aggressive private corporations in supporting ozone protection. Senior military executives from Australia, Canada, Norway, the UK and the USA frequently used their considerable influence and financial and intellectual resources to support stratospheric ozone protection from 1986 to the present. Military leadership included policy directives, financing, staffing and creation of new special-purpose offices such as the US Navy CFC & Halon Clearinghouse. Phase-out efforts included research and development, changing military specifications to allow the use of ODS alternatives, and developing strategies to comply with the Montreal Protocol.[40]

This leadership first became conspicuous at the September 1987 meeting in Montreal where the Protocol was signed. US Air Force experts staffed a technical display showing how to reduce halon emissions by halting testing, training and accidental discharge, and by switching to chemical and not-in-kind substitutes.

After the Protocol was signed, military organizations faced some of the most technically demanding uses of ozone-depleting substances, including fire and explosion protection, weapons-systems electronics, and rocket and aircraft manufacture. These challenges were further complicated by harsh operating conditions, difficulty of repair of fielded equipment, and catastrophic consequences of technical failure. Moreover, the necessity to protect soldiers against enemy action makes military aircraft, ships, armoured vehicles and combat centres difficult to evacuate, requiring fire-extinguishing agents, refrigerants and solvents that are non-toxic and non-flammable.

The successful military strategy used both top-down and bottom-up authority, and relied on centres of excellence for innovation. Commitments, accountability and timeframes were directed from the top. Research, operations and management were inserted from the bottom. Centres of excellence (both technical and managerial) introduced stringent definitions and rigid policies on ODS use by military services. For example, by defining critical halon uses as only those on armoured combat vehicles, the US Army increased the pace of conversion to alternatives in other applications.

Civilian and military collaboration in the early phase-out of halon

Halons are remarkable for their effectiveness in extinguishing fires and suppressing explosions without significantly endangering people from chemical exposure. They were aggressively marketed into applications where they had clear economic and safety advantages, and into other markets where high price did not deter sales. Military organizations used significant quantities of halons to protect their personnel, infrastructure, aircraft, ships and tanks from fire and explosion. Civilian organizations used halon to protect a wide variety of equipment including manufacturing facilities, oil and gas production facilities, ships, aircraft and electronics equipment. Halon was strongly preferred to protect property in occupied spaces. Before it was recognized that halon was destroying the ozone layer, it was also widely used in testing and training.

Mostly for economic reasons, military organizations had already begun efforts in the mid-1980s to: identify, adopt and publicize halon alternatives; reduce halon use; and find even better agents than halons. Various military organizations contributed to technical and educational innovations through the use of simulators, computer modelling and surrogate agents.

In autumn 1986, the US EPA contacted the National Fire Protection Association (NFPA) and Gary Taylor, NFPA Chair of the Halogenated Fire Extinguishing Systems Committee. It was agreed that NFPA would act as a gathering point and clearinghouse for information with an immediate goal of reducing halon emissions. Dozens of meetings ensued. By 1987, the Committee reversed its recommendation for full-discharge testing of halon-1301 systems after installation and began engineering and demonstration programmes to continue to reduce halon use and emissions.[41]

Halon strategy from Victoria, Australia rapidly globalized

In May 1989, the Environment Protection Authority of the State of Victoria, Australia, organized the first international halon/ozone conference, 'The Future of Halons'. Australian citizens were demanding action on the ozone layer, and labour unions announced plans to refuse to install or repair halon systems. During the conference, the fire-protection industry abandoned its strong opposition to controls on halon, and issued a consensus conference agreement that 'halon consumption and emissions need to be reduced to protect the stratospheric ozone layer'. Victoria EPA published the detailed conference recommendations covering halon use, product labelling, fire-fighting training, recovery/recycling, decommissioning, storage, disposal and phase-out dates. These recommendations were the basis of the Australian approach of defining essential halon use through a panel of experts.

Professional organizations halt testing, training, and accidental discharges

There was early leadership to phase out ODSs from the US National Fire Protection Association (NFPA), insurance underwriters, industry trade associations, Underwriters Laboratories, and the enterprises most dependent on halon, including AT&T, GTE, British Petroleum Exploration, NASA, Nortel and US military services.

Once the scientific consensus was accepted and the Montreal Protocol signed, industry and military organizations moved quickly to minimize halon emissions. Fire-protection standards and codes were changed to discourage or prohibit the use of halons for testing and training; substitute tests such as the door fan test were developed to test fire-protection systems without releasing halons; and a greater focus was placed on improving fire-detection equipment in order to decrease the number of false halon discharges.

The search for halon alternatives quickly became a global effort among military organizations and the private sector, including Navy researchers in Norway, the UK and the USA. The Halon Alternatives Research Corporation was initially organized by the US Department of Defense, US EPA, the National Institute of Standards and Testing (NIST), and the fire-protection industry as a publicly funded research programme to develop substitutes for halon. The first project to develop and screen a large list of chemicals as potential substitutes was funded by a combination of the US military, EPA and halon users, and was performed at the National Institute of Standards and Testing and the University of New Mexico's Engineering Research Institute (NMERI). Once it became clear that industry was developing halon alternatives, the Halon Alternatives Research Corporation transformed itself into an international information clearinghouse, a facilitating organization focused on removing barriers to the development and approval of environmentally acceptable alternatives, and a coordinator of halon-bank planning and management programmes.

Halon recovery and recycling equipment globally developed and implemented

Technology to recycle halon-1211 was developed and commercialized by the US Naval Air Warfare Center Aircraft Division in Lakehurst, New Jersey. Members of the US Marine Corps were among the key leaders who not only worked with private industry to ensure commercialization of the equipment, but also travelled to many developing countries to demonstrate its use, often as part of a military-to-military exchange programme.

At the same time, Walter Kidde Aerospace, Aer Lingus, NASA, and others developed and commercialized halon-1301 recycling equipment that could separate halon from the nitrogen used for cylinder pressurization without emitting halon to the atmosphere. The US Army, with petroleum and other critical industry users, fast-tracked the development of a new standard for recycled halon-1301, encouraging the American Society For Testing and Materials to publish a standard of purity for recycled halon-1301 and rapidly citing that standard for military operations. The Halon Alternatives Research Corporation also completed a study that showed that halon banking within the USA was a viable concept, and detailed the blueprint for a halon-banking clearinghouse scheme that has been adopted by several countries. These efforts cleared the way for global halon banking as a viable technique for managing existing halons and ensuring an available supply of halon for critical uses without the need for new production. Under this concept, new halon production could be phased out, and recycled halon could be used to satisfy remaining critical uses. The US Defense Logistics Agency established the world's most extensive physical reserve of ODSs to supply critical military uses.

One significant critical halon use is in passenger aircraft. In 1992, the military, the Federal Aviation Administration (FAA), the International Civil Aviation Organization (ICAO), NASA and the private sector together began a programme to enable aircraft to obtain airworthiness certification without the use of halon. The Wright Laboratory Aircraft Halon Replacement Team used instrumented aircraft engines, nacelle mock-ups and other test equipment to develop new agents and delivery systems under the Joint Military Service Aircraft Halon Replacement Programme. In 1993, at the urging of the US Air Force, the Aerospace Industries Association and Halon Alternatives Research Corporation co-sponsored a two-day conference that led to the formation of the International Aircraft Systems Fire Protection Working Group under the leadership of the FAA. This group has produced minimum performance standards for replacing halon in cargo compartments, engines, passenger compartments and lavatories.

Development of alternatives to halon

The development by industry of hydrofluorocarbon and inert-gas alternatives to halon and the development of National Fire Protection Association standards for use of these agents gave the industry the confidence it needed to accomplish the accelerated halon phase-out. The two types of substance first developed as halon alternatives – HFCs and inert gases – are the most popular special-hazard fire-protection systems in use today. None of the alternative agents currently available is a drop-in replacement. Therefore, fire protection authorities and insurance underwriting organizations must be satisfied with the performance and safety of new alternatives in each system or application.

Fire-protection community promotes alternatives
By the mid-1990s, the fire-protection community had developed a plan of action to eliminate halon emissions except when used against fire, to identify and promote substitutes and alternatives, and to re-deploy existing halon to the most critical uses. In fact, it was the halon community itself that proposed the 1994 phase-out of halon production, two years before any other ODS. There was no industry opposition from manufacturers or users to the accelerated halon phase-out, but many diplomats and national environmental authorities were reluctant to take such drastic action, and some were confused by the very concept of military and industry environmental leadership. Participation by fire-protection

experts from developed and developing countries at crucial Meetings of the Parties ensured acceptance of the early phase-out date.

The fire-protection community, with financial assistance from the Multilateral Fund for Article 5 countries and from the Global Environment Facility for countries with economies in transition, will soon complete the global phase-out of halon production – the first of any controlled substance. However, the production phase-out of halon is only possible through halon banking. Today, critical halon uses are supplied from existing sources while the search for halon-1301 substitutes and alternatives continues. Critical civilian uses include commercial aircraft, oil and gas production facilities and pipeline pumping stations, and medical and communication facilities. Critical military uses include armoured vehicles, ships and aircraft.

PHASING OUT OZONE-DEPLETING SUBSTANCES FROM US MILITARY APPLICATIONS

Electronics and aerospace solvents

After the Montreal Protocol was signed, the magnitude of military influence on the global military and commercial use of ODSs as solvents became evident. A survey of industrial solvent uses in high-technology applications such as electronics discovered not only that the US military itself used significant quantities of ODS solvents, but also that military specifications and standards prescribing ODSs had been adopted as industry standards around the world. A team of global soldering experts estimated that up to half of the CFC-113 used globally could be replaced with available technically feasible alternatives if the military standards could be changed. Military organizations responsible for prescribing solvent use joined the effort to help define standards for cleanliness and materials compatibility, which the new alternatives would have to meet. A unique partnership was formed between the military, private industry and the US EPA to identify and verify the acceptability of non-ODS solvents for military uses. Private companies including AT&T, the Digital Equipment Corporation, Ford, IBM, Motorola, Nortel and Texas Instruments assigned their brightest engineers and environmental performance managers. The Defense Logistics Agency, the Defense Electronics Supply Center, and the Naval Air Warfare Center were leaders in changing the military specifications and standards on electronic components and products to eliminate ODS requirements.

Some ODS solvent uses proved to be particularly difficult to replace, such as cleaning oxygen life-support systems on board aircraft, submarines and in diving applications. These systems consisted of long runs of thin tubes, assorted valves and complex geometry. Any contamination posed possible flammability problems because of the oxygen-enriched atmosphere. The US Naval Sea Systems Command and its private partners developed and commercialized a non-flammable, ozone-safe cleaner that is easily recycled and disposed. Military experts made breakthroughs in non-ODS cleaning of sophisticated electrical,

BOX 5.5 COMPETITIVE ADVANTAGE THROUGH CORPORATE ENVIRONMENTAL LEADERSHIP

Margaret G Kerr, Senior Vice-President, Nortel (Northern Telecom), 1986–2001

I feel strongly about the business benefits of environmental leadership. As Senior Vice-President of Nortel from 1986 to 2001, I helped Nortel to perfect its belief that minimizing the impact of our products and operations on the environment is part of ethical responsibility as a global corporate citizen. Environmental protection is not just the 'right thing to do', it actually translates into competitive advantage for the company – lower operating costs, product and service differentiation, and improved corporate image. Most importantly, environmental protection can create customer value.

It's no secret that for a good part of the 1980s, most of those in industry who thought about environmental issues at all tended to have an image of them as costly problems that would have to be 'fixed' when they became too pressing. 'Environmental management' was really pollution control – putting scrubbers on smokestacks, treating polluted wastewater, and ensuring that hazardous waste was managed as safely as possible. These activities were a bottom-line cost to the company – things that industry chose to do in response to public pressure or was forced to do by increasingly stringent government regulations, but very definitely detrimental to the company's profitability.

When the Montreal Protocol was signed, Nortel started out thinking that the company's challenge was to find a less harmful alternative to the CFC-113 solvent used to remove flux residue from printed circuit boards. But instead of switching to a different solvent, Nortel ended up redesigning technology and processes to eliminate the need for cleaning altogether – and showed that environmental protection could actually save money.

In 1988 Nortel and Seiko Epson became the first multinational electronics and precision companies to make public our goals to phase out CFC solvents. This bold leadership made our intentions clear to employees, suppliers and our government partners. Nortel was a founding member of the Industry Cooperative for Ozone Layer Protection (ICOLP – now the Industry Cooperative for Environmental Leadership), the Corporate Sponsor of the Mexican Solvent Partnership, organizer of the Vietnam Pledge and other technology cooperations, and a participant in many of the other activities that earned the Montreal Protocol its reputation for efficiency and collaboration.

In addition to eliminating the use of this ODS years ahead of the deadline set by the Montreal Protocol, we managed to realize a pretty good return on investment that considerably heightened senior management's interest in pursuing environmental initiatives. Nortel spent US$1 million on research and development, but over the three-year duration of our 'Free in Three' project alone, we saved about US$4 million. The savings came from decreased solvent purchases, the elimination of cleaners and their associated operating and maintenance costs, and reductions in solvent waste for disposal. By maximizing the environmental and economic efficiency of our products at each stage in their life cycle, we added value for our customers and reduced costs for Nortel.

The CFC story didn't really end when Nortel became, in January 1992, the first multinational company in the electronics industry to eliminate CFC-113 from operations worldwide. In subsequent months, Nortel received several prestigious environmental awards, and earned a lot of positive media attention which, when you get right down to it, is free advertising for the company. For several years after its CFC phase-out, Nortel was still getting requests to write articles, give speeches or to be the subject of television spots on CFC elimination.

> More importantly, Nortel became deeply engaged in a process of sharing what we had learned with other countries, especially developing countries. Between 1992 and 1995, Nortel played a lead role in technology cooperation projects in Mexico, India, China, Turkey and Vietnam. These projects were supported by World Bank funding and involved close collaboration with local government and our partners in the Industry Cooperative for Ozone Layer Protection.
>
> Many companies have come to believe strongly that international cooperation between governments and industry is a highly practical way of resolving shared environmental problems. In part, companies devote substantial amounts of time and energy to this because it's part of the responsibility as a global corporate citizen. But the reasons are not just altruistic. Technology cooperation is a marketing tool: it builds goodwill and strong relationships with customers in emerging markets. The willingness to bring first-tier technology and our experience with environmental management to developing countries – helping them avoid some of the costly mistakes we've made – is a key driver in many of our joint ventures. The commitment to the environment is a part of the value proposition that is offered to potential customers.

optical and precision components. With private-sector partners, they conceived the plan to require the US Department of Defense to report phase-out progress to Congress and to identify barriers to prompt action. This self-imposed reporting mechanism motivated military managers to prioritize solving most ODS phase-out problems.

NASA, military and industry collaboration in aerospace solvent uses
After the US Clean Air Act was amended to include provision for ozone-layer protection in November 1990, a team of military and civilian experts was formed to coordinate the ODS-reduction efforts of the space shuttle, Titan, Delta and Atlas programmes. By August 1995, the use of ODSs in these programmes had been reduced to 1 per cent of 1989 levels, and the Parties to the Protocol granted an essential-use exemption for remaining minor uses. The Titan IV Programme ODS Reduction Team, with the Department of Defense, US EPA, Industry Cooperative for Ozone Layer Protection and NASA subsequently published a handbook, *Eliminating Use of Ozone Depleting Substances in Solid Rocket Manufacturing* (1996), available worldwide as a guide to reducing the use of ODSs in rocket motors.

Defence contractors also took the lead to eliminate ODS solvents from their production lines, often taking risks to convince their military customers to accept the changes they proposed. General Dynamics and Lockheed Martin rapidly eliminated more than 90 per cent of their ODS solvent use by painstakingly selecting alternatives that met all their performance requirements. They developed and implemented the non-ODS technologies for cleaning of gaseous oxygen and hydraulic tubing for aircraft and space launch systems. They later eliminated all ODS use. Technologies developed by military contractors have been implemented at major civilian and military manufacturing facilities around the world.

Refrigeration and air conditioning

The military also played a significant role in the phase-out of ozone-depleting chemicals as refrigerants. The US Navy began a fleet-wide conversion programme to change shipboard air conditioning and refrigeration systems from CFC-12 to HFC-134a. The military often faced greater challenges than civilian users due to the need for equipment to perform and maintain reliability in the harshest of wartime conditions aboard mobile weapons platforms such as ships, aircraft and armoured vehicles. In addition to the normal facility air conditioning and refrigeration systems, cooling is required on board aircraft, ships and armoured personnel carriers to keep critical weapons control and communications systems functioning.

The US Navy faced a particular challenge with many of its systems, which were specifically designed to use CFC-114. This substance was chosen for navy vessels because it provided quiet operation to escape sonar detection, reduced equipment volume; and offered compatibility with existing submarine atmosphere control equipment – all important features for warship systems. Since CFC-114 was not used extensively by the private sector for refrigeration, there was little incentive for commercial companies to develop alternative refrigerants to replace it. Also, since the primary use of CFC-114 by the private sector was for foam-blowing applications, once the foam industry converted to other alternatives, the availability of CFC-114 to support existing systems became a major concern. After work with several alternatives proved unsuccessful, HFC-236fa was identified as a viable alternative; a fleet-wide programme was undertaken to convert nearly 1100 existing air-conditioning chillers and refrigeration units on Navy ships to its use.

In parallel to the work to identify alternative refrigerants for existing equipment, non-CFC systems were developed for new ships. In 1995 and 1996, the US Navy provided support to Taiwan and Spain for conversion of their equipment from CFC-12 to HFC-134a.

ALTERNATIVES: CRITERIA AND EVOLUTION AFTER THE MONTREAL PROTOCOL[42]

The world had changed between 1900, when the first manufactured ODSs were commercialized, and 1987, when the global search for alternatives and substitutes began in earnest.[43] In 1900, there were very few laws for worker safety, public health and environmental protection. Highly toxic chemicals were commercialized without much testing, and there was little appreciation for long-term health or environmental consequences.

When the Montreal Protocol was signed in 1987, the choice of technologies was far more informed and far more complicated. Scientists, diplomats, environmental NGOs, and the public urged fast action to phase out ODSs, but were reluctant to compromise health and safety. Very-low-toxicity ODSs were used in emissive applications with high human exposure; environmental and safety authorities were unsure whether more toxic chemicals or more flammable

---

BOX 5.6 MEASURES OF ENVIRONMENTAL AND SAFETY ACCEPTABILITY AND TRADE-OFF

- Some ODS applications required alternatives with very similar properties; in many cases, these alternatives were not commercially available, and it would require several years to complete toxicity and environmental testing and to build new HCFC and hydrofluorocarbon (HFC) facilities.
- Suppliers of commercially available substitutes and alternatives wanted a rapid transition so they could capture markets before HCFCs and HFCs were available.
- Manufacturing companies depending on substantial quantities of ODSs had a strong preference for 'drop-in' chemical substitutes, particularly when the use of alternatives required substantial redesign and/or capital investment.
- Manufacturing customers using small amounts of ODSs for critical uses wanted market forces to allocate scarce supplies, while price-sensitive industries wanted allocations to help preserve markets and jobs.

---

substitutes could be safely used as alternatives. CFC and halon systems frequently discharged into occupied areas during use and service. Workers were exposed to particularly high concentrations of propellants, solvents and foam-blowing agents.

One serious complication was that both government regulations and private standards had become barriers to change. Companies were uncertain about the criteria governments and customers would apply for acceptable alternatives; in many countries, it was unclear how rapidly governments could act. There was also concern that new chemicals would need to satisfy multiple authorities, including governments, insurers and technical standards organizations.

Some markets for ODS substitutes and alternatives were quickly decided and rapidly transformed. For example, hydrocarbons were already used as aerosol propellants in several countries and were the clear economic and environmental choice for aerosol products and foam packaging. Flexible-cushioning foam production was quickly converted to already available substitutes, including water and methylene chloride. These changes rapidly decreased demand for CFCs in Europe by over 50 per cent by 1990–1992 with smaller reductions in demand in Japan, North America and other regions.

Some other markets for ODS substitutes and alternatives briefly competed, but well-organized partnerships of industry, government and NGOs then made their preferred choices and effectively ended debate about substitutes and alternatives in many industries. For example, most electronics, aerospace, and precision-cleaning companies rejected chlorinated and HCFC solvents in favour of alternatives such as no-clean soldering and aqueous solvents, while the global motor-vehicle air-conditioning industry unanimously selected HFC-134a systems.

In contrast, technical uncertainty, public demands by environmental NGOs, cost concerns and market competition resulted in fierce disagreement about the choice of substitutes for refrigerants and insulating foam, particularly for domestic refrigerators.

Companies collaborate in testing alternatives

Companies interested in new chemical substitutes faced the old challenge of satisfying the most stringent national environmental authorities, and the new challenge of the Montreal Protocol and the 1992 Framework Convention on Climate Change (with its 1997 Kyoto Protocol). New partnerships were formed as the ODS-using industry learned to work with government and academic experts to evaluate alternatives under the technical panels of the Montreal Protocol.

The chemical production industry responded to these challenges in various ways. In order to accelerate the pace of toxicity testing and reduce costs, 14 global ODS manufacturers with an interest in commercializing new HCFCs and HFCs formed in 1988 the Programme for Alternative Fluorocarbon Toxicity Testing (PAFT). Another significant response was the creation of the Alternative Fluorocarbon Environmental Acceptability Study (AFEAS) consortium formed in 1989 to determine the environmental fate, and investigate any potential impacts, of alternatives in cooperation with government agencies and academic scientists. This unprecedented cooperation shortened the time to commercialization of new HCFC and HFC chemical substitutes by three to five years.

At least six industry associations were started with the express goals of speeding the elimination of ODSs: the Alternative Fluorocarbons Environmental Acceptability Study (AFEAS), the Halon Alternatives Research Corporation (HARC), Halon Users National Consortium (HUNC), the Industry Cooperative for Ozone Layer Protection (ICOLP), the Japan Industrial Conference for Ozone Layer Protection (JICOP) and the Programme for Alternative Fluorocarbon Toxicity Testing (PAFT).

At least two other industry organizations transformed themselves from questioning to supporting regulations to protect the ozone layer: the Alliance for Responsible Atmosphere Policy (ARAP) and the Association of Fluorocarbon Consumers and Manufacturers (AFCAM). Several dozen other existing organizations created substantial internal subcommittees on ozone layer protection.[44]

Market and environmental uncertainty

Some alternatives have climate and ozone impacts too

HCFCs, the chemicals with low ozone-depletion potentials, were relied on as alternatives to CFCs, when a complete phase-out of CFCs was agreed upon in 1990. However, realizing that, in the long run, even the use of HCFCs was detrimental to the ozone layer, they were declared as 'transitional', pending adoption of ozone-safe technologies. In 1992, the fourth MOP imposed controls on HCFCs, mandating a gradual phase-out by 2030. The long period was allowed to enable industry to recover their investments in HCFCs. In 1990, HFCs were seen as the new 'wonder chemicals', completely ozone-safe, to be used as substitutes to CFCs in many applications. HFCs, however, have a high global warming potential. The Kyoto Protocol of 1997 under the UN Framework Convention on Climate Change lists fluorine-containing substances

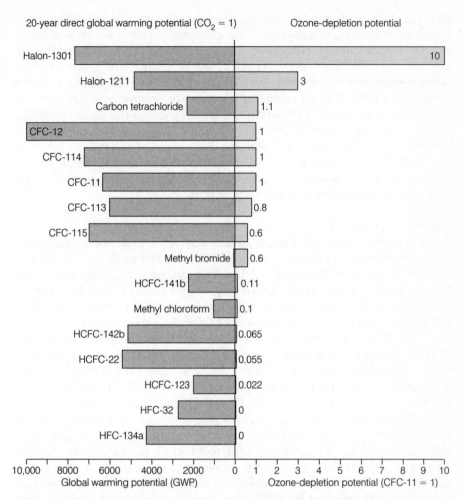

20-year direct global warming potential ($CO_2 = 1$)   Ozone-depletion potential

Figure 5.2 *Global warming potential of some ozone-depleting substances and alternatives*

Source: GWP data: Ozone Secretariat (1998) *Scientific Assessment of Ozone Depletion*, UNEP, Nairobi, Table 10.8. ODP data: The Montreal Protocol on Substances that Deplete the Ozone Layer.

hydrofluorocarbons (HFCs), sulphur hexafluoride (SF_6), and perfluorocarbons (PFCs) in the basket of greenhouse gases to be controlled and emissions reduced. Some current uses of HFCs are critical replacements for some ozone-depleting substances, such as health and safety aerosol products, specialized fire protection, military applications, and solvents. In applications where there is an energy-efficiency advantage, such as foams, refrigeration and air conditioning, HFCs can also contribute to efforts to reduce overall greenhouse gas emissions.

Choice of HFCs for refrigeration and foam-blowing

The two most controversial technical debates about substitutes and alternatives have concerned the choice of HFCs in specific applications. The first controversial application is hydrocarbons as refrigerants and foam-blowing agents in the manufacture of small refrigerators. The quantity of refrigerant is very small, suggesting that the risk of using flammable hydrocarbons is minimal; millions of hydrocarbon refrigerators have been sold to date in Asia and Europe without major incident. However, HFC manufacturers and some of their customers continue to argue that only HFCs are safe. The second controversial application of HFCs is for cosmetic, convenience or novelty products where HFCs offer a higher level of safety, but critics question the necessity of the product itself – for example, aerosol noise makers, single-use tyre inflators, dust blowers, wine chillers, bait guns, perfume dispensers and self-chilling beverage cans. A few aerosol products use HFC propellants because flammability is a concern and no alternatives are available – specifically, insect sprays and cleaners used around energized high-voltage power lines.

Uncertainty on HCFCs

Political, environmental, and media attention to ozone depletion helped speed the transformation in most sectors. However, some sectors that would depend on HCFCs for the CFC phase-out had slow progress as a consequence of regulatory uncertainty, negative advertising by promoters of non-fluorocarbon alternatives, and the campaigns by environmental NGOs to avoid both HFCs and HCFCs. Solvent users phased out CFCs with only minor use of HCFC-141b and HCFC-225 substitutes. Vehicle air conditioning made a global choice of HFC-134a as a CFC substitute. Some refrigerator manufacturers began using HFCs while others chose hydrocarbons. Some building-air-conditioning chiller manufacturers chose HCFC-123, while others chose HFC-134a. In 2002, many owners of existing CFC-11, CFC-12 and HCFC-22 air conditioners and reversible heat pumps, especially in developing countries, are still waiting to see which technology is environmentally and economically preferred. A geographic split developed in polyurethane rigid foam used for insulation. Europe developed hydrocarbon (pentane) technology as early as 1992; refrigerator manufacturers accepted this technology so quickly that conversion to pentane was nearly 100 per cent by 1997. This trend has been followed by many developing countries, notably China and India. US industry has used HCFC-141b and will follow with HFCs.

The HFCs and HCFCs: Current status

HCFC use currently falls into three categories:

1 Historic uses, including foam-blowing agents in packaging foam and some commercial refrigeration applications, where developed countries have eliminated use.
2 Transition uses that eliminated higher-ODP substances, including solvents and insulating foam, where developed countries are now shifting to ozone-safe alternatives.
3 Remaining historic uses, including building air conditioning, where developed countries are either fully containing low-ODP HFC-123 re-frigerants or shifting to new HFC refrigerants and blends.

Over-supply of HFC-134a has resulted in low prices. Some original HFC patents have expired after 17 years; companies are manufacturing HFC-134a in at least six countries: France, Germany, Japan, Italy, the UK and the USA.[45] China and India have developed new production processes for HFC-134a that will be financed by the Multilateral Fund. The concern over monopoly pricing of HFC-134a has abated as a result of multiple suppliers with different production processes, adequate supplies with competitive prices, and emerging new sources.

However, there are three or more application patents that give exclusive rights for the application of new HCFCs and HFCs. Honeywell has an application patent for the use of HFC-245fa in the manufacture of insulating foam, and Solvay, for HFC-365mfc. DuPont/Great Lakes has exclusive rights to use HFC-227ea as a fire-extinguishing agent. Honeywell/Asahi has exclusive rights to use HCFC-225 for electronics and precision cleaning.

HFC use in metered-dose inhalers

Patent protections for several CFC metered-dose inhaler formulations have expired, allowing the introduction of 'generic' versions of these medical inhalers that can be sold for the cost of production. Each year, more CFC-free metered-dose inhalers using HFC propellants are marketed, but prices still reflect recovery of the considerable research investment and profits from exclusive property rights. New forms of therapy such as injected, subdermal or oral drugs, as well as dry powder systems, will further compete with HFC metered-dose inhalers.

Other sectors

With few exceptions, companies have been able to phase out ODSs without unusual payment for intellectual property. A conspicuous exception is the payment by the Multilateral Fund of licensing fees to the tobacco company that invented the CFC-free technology to 'puff' tobacco to reduce the tar per cigarette. Cooperation in the food packaging, vehicle air conditioning, halon and solvent sectors has been particularly strong with:

*   public domain technology for HFC- and HCFC-foam food packaging, recovery and recycling of halons and CFCs, and no-clean soldering;
*   publicly available technical standards and training manuals; and
*   workshops, internships and company-to-company collaboration.

Chemical and not-in-kind alternatives

In non-Article 5 countries, approximately 80 per cent of ozone-depleting compounds that would be used today without the Montreal Protocol have been successfully phased out without the use of other fluorocarbons. ODS use was eliminated with a combination of 'not-in-kind' chemical substitutes, product alternatives, manufacturing-process changes, conservation and doing without. So far, 'in kind' HFCs have been used to replace about 8 per cent of the former ODS amounts, and 'in-kind' HCFCs have replaced approximately 12 per cent of former ODS amounts.

BOX 5.6 THE MINEBEA THAILAND PHASE-OUT AND DONATION OF TECHNOLOGY

Ryusuke Mizukami, Senior Managing Director, and Morio Higashino, ODS Phase-Out Manager, Minebea Thailand

The Minebea Group of Companies produces ultra-high-precision miniature ball-bearings used in applications such as videotape cartridges and hard disk drives for personal computers. Large quantities of CFC-113 and methyl chloroform were used during the washing process in order to meet customers' demands for rotational accuracy and cleanliness. Our largest manufacturing base in Thailand once consumed 40 per cent of the total CFC-113 and methyl chloroform used in Thailand.

During the early 1990s, Minebea became concerned for the ozone layer and decided to develop a washing method that would satisfy demanding quality criteria and also be more environmentally friendly. Among the many technical selections, it was decided to use water, which was most familiar to us. Thus we started the development of a water-based washing system.

Because the material used in a ball-bearing is extremely easy to corrode, we invented a washing machine using deoxidized water and vacuum degassing after water immersion. Experts from the Technical and Economic Assessment Panel Solvents Technical Options Committee who evaluated the technology at Minebea's Thailand factory during the March 1992 Japan–US–Thailand Trilateral Stratospheric Ozone Protection Conference were very helpful in recommending further design improvements to reduce operating costs and energy use. Our engineers were very excited to be part of a global endeavour.

Minebea installed this water- and hydrocarbon-based washing system at all of its plants worldwide, and by April 1993 completely eliminated ozone-depleting substances from all its production processes. Approximately US$40 million was invested. Thereafter, we decided to release this technology to the public as a contribution to global environmental preservation. The environmental, financial and performance benefits of the Minebea water-washing technology were promoted in publications and at technical conferences in Indonesia, Japan, Singapore, Thailand and the Untied States.

Because of our contributions, Minebea President Goro Ogino earned a Stratospheric Ozone Protection Award for individual leadership, and the entire Minebea Group earned the Stratospheric Ozone Protection Award and the prestigious Best-of-the-Best Stratospheric Ozone Protection Award. It was a very proud day for Minebea when the *Bangkok Post* reported:

> *'The Governments of Thailand, the United States and Japan are applauding the Minebea Company for completely halting the use in Thailand of solvents that deplete the ozone layer. Minebea had been the largest user in Thailand of the chemicals... Said the EPA's Dr. Stephen Andersen, "Minebea pioneered the use of new technology to clean precision bearings, and they are showing other companies how to use these technologies to protect the ozone layer."'*[*]

[*] *Bangkok Post*, 8 February 1994, 'Minebea Clears the Air', p1.

The Multilateral Fund has approved HFCs to replace 7 per cent of ODP tonnes, in which an ODP tonne represents a tonne multiplied by the ozone-depletion potential of the ozone-depleting substance. The Fund has also approved

HCFCs to replace 17 per cent of ODP tonnes in investment projects in Article 5 countries. In the refrigeration sector, 89 per cent of ODP tonnes of refrigerants were replaced with HFCs, while 66 per cent of ODP tonnes of foam-blowing agents were replaced with hydrocarbons.

TECHNICAL STRATEGIES TO REDUCE AND ELIMINATE OZONE-DEPLETING SUBSTANCES

With a few exceptions, the world initially pursued a common strategy of 'easy first, hard last' in the phase-out of ODSs. This approach translated into five ordered actions, as follows:

1   **Reduce emissions:** Simple housekeeping – such as closed chambers, careful fluid transfer, replacing threaded fittings with soldered joints, and checking for leaks with electronic meters – made substantial reductions in refrigerant, halon and solvent emissions. Greater reductions were made by eliminating accidental alarms and false discharges, and by halting testing and training with halons. Almost complete elimination of emissions was possible with design for containment, higher quality components and automatic leak monitoring.

2   **Recover and recycle:** Refrigerants and halons were typically vented during service and at the end of equipment life, and solvents were often disposed of when contaminated. There was little incentive to recover and recycle because the cost of recovery and recycling was perceived to be higher than the cost of new material; there was a reluctance to use recycled refrigerants in valuable equipment. Many regulations either discouraged recycling or specified the use of only newly manufactured chemicals. In response to concern for the ozone layer, industry quickly developed recovery and purification equipment that allowed ODS re-use. Taxes on ODSs in Korea, Singapore, Thailand, Taiwan and the USA provide a strong market incentive for recycled material.[46]

3   **Shift to commercially available alternatives and reconsider historic options:** Some sectors had commercially available alternatives to ODSs at comparable or even lower cost, but capital investment and market acceptance were required for transition to these alternatives. Hydrocarbons were available to replace CFC propellants in cosmetic and convenience aerosol products; not-in-kind pumps, sticks and creams were substitutes for aerosol dispensers. Hydrocarbons, ammonia, and HCFCs were commercial refrigerants. Aqueous and hydrocarbon solvents were available to replace CFC solvents. Dry-powder inhalers can replace CFC metered-dose inhalers for most patients. Hydrocarbons, carbon dioxide, and methylene chloride were alternatives for blowing some closed-cell foams; hydrocarbons and water were alternatives for most flexible-cushioning foams. However, hydrocarbons are flammable and require significant capital investments to maintain safety in plants and in servicing of refrigeration and air conditioning equipment; methylene chloride is toxic. Better engineering and higher quality components and control technologies enabled many

industries to revisit and reapply previously abandoned alternatives, with much better performance and greater safety than before.

4   **Eliminate processes dependent on ODSs:** The search for alternatives sometimes involved 'out-of-the-box' thinking that eliminated large, well-developed ODS applications. For example, flame-proofing and very early fire detection eliminated the need to extinguish fires; no-clean soldering eliminated solvent use; and ecosystem management using resistant plants, natural pesticides, crop rotation and predators eliminated methyl bromide use. Cutting and stamping oils were reformulated to provide rust protection or to evaporate; manufacturing was simplified to avoid multiple cleaning of parts and assemblies. Danish engineers invented inert-gas fire suppression that decreases oxygen to a concentration that extinguishes fire, but allows humans to survive by increasing concentrations of carbon dioxide to boost respiration rates.

5   **Commercialize new HFCs, HCFCs and PFCs:**[47] New halocarbon substances, including new low-ODP HCFC-123 and HCFC-225, intermediate-ODP HCFC-141b, and new applications of the previously used HCFC-142b and HCFC-22 as transition compounds and zero-ODP HFC-134a, continue to be important to the planning and execution of the ODS phase-out. Newly commercialized HFEs (hydrofluoro ethers) are seeing increased application after 2001, while PFCs have been less important but viable substitutes for niche applications. Other HFCs (HFC-245fa and HFC-365mfc) will be commercialized in the near future, primarily as substitutes for HCFCs used to expand insulating plastic foams. The advantages of HCFC and HFC alternatives are that they are chemically, physically and toxicologically very similar to the CFCs they replace. This simplifies re-design and minimizes potential problems with lubrication, materials compatibility and safety.

Later, companies 'leap-frogged' steps to select the best and most permanent solutions.[48] Better containment, recovery and recycling reduced CFC solvent and refrigerant emissions by up to 50 per cent or more; halting testing and training with halon, better containment, and elimination of false discharges eliminated up to 95 per cent of halon emissions. Existing substitutes and alternatives eliminated more than half of global use; cosmetic and convenience aerosol products alone were one third of global use. New technology was available for most remaining uses within five years. Each element of the phase-out strategy was successful.

Refrigerants: Alternatives and phase-out strategies

Before the Montreal Protocol was signed, CFCs were the substance-of-choice for refrigeration and air-conditioning applications. The most common ODS refrigerants were CFC-11, CFC-12, CFC-115 and HCFC-22. However, small amounts of CFC-114, halon-1301 and other refrigerants were used in speciality applications for air conditioning and refrigeration.

In 1987, there were three refrigerant strategies:

1   **Reduce emissions and recycle:** In 1987, the largest source of refrigerant emissions was intentional venting during service and disposal. Within one year, new practices were available to detect and minimize equipment leaks, and equipment was commercialized to recover and recycle ODS emissions. Implementation occurred because of a notable partnership between global vehicle manufacturers, vehicle service technicians, environmental NGOs and the US EPA to authorize the use of recycled CFCs under new-car warranty. Many vehicle companies subsequently required CFC recycling in their dealerships worldwide.

2   **Switch to existing refrigerants:** From 1987 until 1992, a handful of leadership companies – such as Sainsbury UK – switched from high-ODP CFCs to lower-ODP HCFC-22. Ammonia, already used in cold storage, ice- and frozen-food manufacturing, captured a larger market share. In this early stage, there were no conspicuous switches from CFCs to hydrocarbons. Following the commercialization and competitive supply of HFC-134a in 1990 and 1991 and in response to deliberations and amendments adding HCFCs to the list of controlled substances in 1992, HCFCs were generally selected only in applications with substantial technical and economic advantage. HCFC-123 continues to be preferred for air-conditioning chillers when energy efficiency is a priority. HCFC-141 use is declining even where allowed by environmental authorities as a drop-in replacement for vapour degreasing and cold cleaning. In 1992, Greenpeace and its private-sector partners in Germany introduced the non-ODS Greenfreeze™ hydrocarbon refrigerator technology, which by 1998 had captured impressive shares of small-appliance markets in most European countries, China and India. In 2000, several more multinational food service companies pledged to investigate non-fluorocarbon refrigerant alternatives and convert where economically feasible and where safety could be maintained. In 2001, the first Japanese companies introduced hydrocarbon refrigerators. However, hydrocarbons have not penetrated larger equipment where safety would be more expensive to achieve, although there are some examples of successful implementation.

3   **Commercialize new HCFCs and HFCs:** HFC-134a currently dominates as a replacement for CFCs in refrigeration and air conditioning, with the exception of large chillers where low-ODP HCFC-123 has significant energy-efficiency advantages. The HFC blends R-410A (a blend of 50 per cent HFC-32 and 50 per cent HFC-125) and R-407c (a blend of 23 per cent HFC-32, 25 per cent HFC-125 and 52 per cent HFC-134a) are expected to be the dominant replacements for HCFC-22 in residential and commercial air conditioning.

Now, after 15 years of intense global research, only six refrigerant options not controlled by the Montreal Protocol exist for the vapour-compression cycle: ammonia, carbon dioxide, hydrocarbons and blends, HFCs and blends, air, and water. Each one is a good refrigerant for certain applications, and each has its own environmental and/or safety advantages and disadvantages. Ammonia is toxic and flammable, and hydrocarbons are flammable, but both are energy-efficient in optimized systems and are insignificant contributors to climate change. HFCs have inherent safety advantages of non-flammability, but emissions significantly contribute to climate change and are to be controlled under the Kyoto Protocol. Air-cycle refrigeration has very poor energy efficiency but no direct environmental effects. Carbon dioxide is emerging as a highly energy-efficient, non-flammable refrigerant in cascade systems. However, carbon dioxide has low energy-efficiency in simple cycles like those of CFCs (transcritical applications at higher temperatures). When HFC and HCFC systems are managed for low emissions, the dominant factor in determining contributions to climate change is the energy-efficiency of the refrigeration or air-conditioning system.

Foam-blowing agents

The replacement of CFCs in insulating foams posed particular challenges. CFC-11 had been used for polyurethane (PUR) and phenolic (PF) foams, and CFC-12 for extruded polystyrene foams (XPS). In these foams, the blowing agent also acts as the insulating gas and it remains, substantially, in the foam cells. Such foams have a service life of 10–50 years.

The European Isocyanate Producers Association (ISOPA) was established in 1987 to speed the development and commercialization of new polyurethanes not dependent on ODSs.[49] It has been instrumental in accelerating the development and implementation of legislation to phase out CFCs and HCFCs. ISOPA signed voluntary agreements with the European Commission in 1992 to reduce consumption of CFCs in the foam industry, and negotiated and supported the European regulations that established the 1 January 1995 phase-out for CFCs and the 1 January 2004 phase-out schedule for HCFCs in foams.[50]

The first replacement option for polyurethane foams was the commercialization of 'reduced CFC' formulations that used half the CFC-11 level. This was adopted by the European refrigerator industry in January 1989. The next option, for all types of rigid foams, was to use HCFCs. In the case of extruded polystyrene foams, the choice was between HCFC-142b, HCFC-22 or a blend of the two. For polyurethane and phenolic foams, the main initial strategy was to use HCFC-141b, which was commercialized in 1991.

Environmental pressure in Europe resulted in the development and use of hydrocarbon blowing agents for both building and refrigerator insulating foams. For the former, isopentane and normal pentane have been used since 1992; this use has now spread to US industry. The Liebherr Company in Germany first used cyclopentane in refrigerator foams in March 1993; this use quickly spread as the standard throughout the European industry. A cyclopentane/isopentane mixture, giving better economics and technical performance, has now replaced cyclopentane technology. Other regions, with the exception of the USA, are following this European trend toward hydrocarbon. For example, by 2001, pentane-insulating foam captured about 70 per cent of the Chinese refrigerator industry.

ENVIRONMENTAL PERSPECTIVE ON SUBSTITUTES AND ALTERNATIVES

In 1987, most experts expected that compounds with properties very similar to those of CFCs would be required for most applications. Thus, it was anticipated that new and existing HCFCs and HFCs would replace CFCs in most applications. However, the phase-out of ODSs occurred faster, cheaper and better than expected. Many applications adopted process changes and product substitutes that eliminated ODSs without chemical substitutes; a large portion used non-fluorocarbon chemical substitutes, especially hydrocarbons. Excellent engineering, quality components, good housekeeping, and recovery and recycling reduced use and emissions in applications where HFCs and HCFCs were used.

The environmental acceptability of chemical substitutes depends on the direct and indirect environmental effects, human health and safety. The direct atmospheric effects are analysed in terms of ozone-depletion potential, global warming potential, atmospheric fate, and biological and ecosystem effects. The indirect atmospheric effects are the emissions resulting from generating the energy to power the application. Additional environmental concerns include resource consumption, aquatic and terrestrial toxicity, and sustainability. Human health and safety concerns include human toxicity and flammability. Further, technical performance includes measures of materials compatibility, physical properties such as boiling point, chemical properties such as solvency, and other measures depending on the application.

Environmental NGOs, government negotiators and regulators, UNEP and its assessment panels, chemical manufacturers and their customers, engineers, and academics were fully aware of these criteria. The US EPA conducts the most structured regulatory review of ODS alternatives under a programme called the Significant New Alternatives Policy (SNAP). EPA evaluates each alternative and publishes lists of acceptable options, often with use quali-fications, and unacceptable options prohibited from use.

The UNEP scientific assessment has always included detailed consideration of the impacts of substitutes on climate change, including presentation of the global warming potentials of proposed alternatives and substitutes. TEAP and its Technical Options Committees featured comprehensive sections in their reports on toxicity, flammability, global warming potential, energy efficiency and other factors. They repeated this information prominently in executive summaries. Assessment Synthesis Reports highlighted climate change, global warming potential, energy efficiency and other measures of environmental acceptability of substitutes and alternatives.

Responsible use of HFCs

In 1999, TEAP cooperated with the Intergovernmental Panel on Climate Change and published the cross-cutting report, *The Implications to the Montreal Protocol of the Inclusion of HFCs and PFCs in the Kyoto Protocol*.[51] TEAP defined and documented a set of 'Responsible Use Principles' that can lead to significant additional reductions in the emissions of HFCs used as substitutes for ODSs. These principles for responsible use include:

- Use HFCs only in applications where they provide safety, energy efficiency, environmental or critical economic or public health advantage.
- Limit emissions of HFCs to the lowest practical level during manufacture, use and disposal of equipment and products.
- If HFCs are to be used, select the compound with the smallest climate impact that satisfies the application requirements.

The TEAP report also found that the inclusion of HFCs, PFCs, and SF6 in the basket of gases to be controlled in the Kyoto Protocol need not have adverse implications for the implementation of the Montreal Protocol, provided that implementation by each country allowed HFC use where viable alternatives to ODSs were not available. The TEAP report also concluded that use of HFCs as substitutes for ODSs was not likely to impede Parties to the Kyoto Protocol from meeting emission-reduction obligations. Because HFCs are a very small part of national emission targets, HFCs can be more energy-efficient than ODSs, and emissions can be minimized.

Chemical manufacturers planning to market HFC and HCFC substitutes, in coordination with government programmes, financed the global studies of the atmospheric fate of fluorocarbon substitutes under the Alternative Fluorocarbon Environmental Acceptability Study and toxicity testing under the Programme on Alternative Fluorocarbon Toxicity Testing. Manufacturers of HFCs and new HCFCs emphasized in product and professional publications the advantages in terms of safety (very low toxicity and non-flammability) and performance (efficiency and materials compatibility) as well as global warming potentials of these chemicals. Distributors of HFCs and HCFCs sometimes failed to disclose product disadvantages, but there was no evidence that customers were unaware of this information. Professional association meetings and technical publications featured in-depth discussions of design and selection criteria for substitutes and alternatives.

Selective use of alternatives

When the Montreal Protocol was signed, ozone-depleting substances were used primarily in highly emissive applications. These applications could be classified according to when the emissions occurred. The emissions were immediate during solvent use and open-cell foam manufacturing, and when aerosol products were dispensed. The emissions were continuous, but at a low level, during the operation, service and disposal of refrigeration, air-conditioning and fire-extinguishing equipment. The emissions were delayed for insulating foam products with some emissions at the time of manufacture, small emissions during use of the foam, and virtually total emissions of the remaining CFCs when the foam was disposed of.

Most of the applications where ODSs were emitted immediately at use now rely on non-fluorocarbon alternatives or new substitute technologies. The

Ozone art by *Laila Nuri* (aged 8), Indonesia – UNEP DTIE OzonAction
Programme Children's Painting Competition, 1998

Ozone art by *Manuel Arcia* (aged 15), Panama – UNEP DTIE OzonAction
Programme Children's Painting Competition, 1998

Ozone art by *Meleko Mokgosi* (aged 16), Botswana – UNEP DTIE OzonAction
Programme Children's Painting Competition, 1998

'I fear there is a hole in the ozone layer sire'   *Brant Parker and Johnny Hart*

By permission of Johnny Hart and Creators Syndicate, Inc

Little Ozone Annie

'Little Ozone Annie'   *Scott Willis*

By permission of Scott Willis and *San Jose Mercury News*

'Naturally, the question arises as to *who fixed* the hole in the ozone layer?'
Don Wright

By permission of Don Wright and Tribune Media Services

'Detour: Hole in ozone ahead'   *Dana Fradon*

'Ignore 'em. What can happen?' *Mike Luckovich*

'I miss the ozone layer' *Bill Schorr*

'Wonder gas' CFCs invented, 1928 *Joseph Kariuki*

© 1998 UNEP. Ozone Secretariat, *The Ozone Story*, p6

UNEP Assessment Panel of Experts reports need for tougher controls, 1990 *Joseph Kariuki*

© 1998 UNEP. Ozone Secretariat, *The Ozone Story*, p20

Industrialized countries agree to phase out methyl bromide by 2005. Developing countries to do likewise by 2015 *Joseph Kariuki*

© 1998 UNEP. Ozone Secretariat, *The Ozone Story*, p29

Front cover of *Time* magazine, 17 February 1992

'Go with the floe'. Poster promoting Epson's commitment to the phase-out of ozone-depleting substances

By permission of Seiko Epson Corporation

'The sky's the limit'. Poster promoting Epson's commitment to the phase-out of ozone-depleting substances

By permission of Seiko Epson Corporation

Friends of the Earth supporters protest against ozone depletion significant exceptions are HFC-based aerosol cosmetic and convenience products, including dust blowers, tyre inflators, wine-bottle openers, and horns used for sporting events. Emissions have been greatly reduced in other applications such as refrigeration and air conditioning through better containment and recovery and recycling. Improved containment can also mitigate risks associated with flammable and/or toxic compounds such as hydrocarbons and ammonia. However, in applications with significant consumer exposure and where larger amounts of refrigerant are involved, non-flammable chemicals with low toxicity (eg fluorocarbons) mitigate safety risks.

ODS phase-out without environmental compromise

In the majority of cases, the replacement of ODSs has been accomplished with no compromise in any measure of environmental risk, and manufacture of equipment and safety performance is maintained or improved. However, in some cases the new technology has higher or new risk, primarily in manufacturing and servicing, compared to the technology dependent on ODSs; these risks have been managed. In still other cases, the new technology may be environmentally superior to the one it replaced, yet arguably still not acceptable under increasingly stringent environmental standards. It is not yet clear whether risks will increase with alternatives and substitutes for the critical military applications now dependent on banked ODSs.

Consider the following measures of environmental success with distinguishable environmental and safety acceptability:

- Clear environmental success is when a new technology is superior in some ways, worse in no way, and environmentally acceptable overall. Examples include replacing fire extinguishing with fire prevention, and replacing solvent cleaning of resin-based soldering with no-clean soldering.
- Imperfect environmental success is when a new technology is a step in the right direction – better in every respect, but not an entirely acceptable solution. For example, HFC-134a, as a substitute in vehicle air conditioning, is ozone-safe, energy-efficient, has one-fifth the direct global warming potential of the ODS it replaced, has a very low toxicity and is non-flammable. However, HFC-134a is still a potent greenhouse gas included in the substances to be controlled by the Kyoto Protocol. Emissions have been decreased further through system and component design, and by recovery and recycling.
- Ecologically compromised environmental success is when a new technology is better for ozone-layer protection, but ecologically worse in some other respect. For example, chemical pesticides that replace methyl bromide may be equally effective against target pests, but may have residues that are more toxic than the bromine products of methyl bromide decomposition or may be more hazardous to people applying the chemical or near the area being treated.
- Safety-compromised environmental success is when a new technology is environmentally superior, but worse for worker or consumer safety. For

example, substituting flammable pure ethylene oxide for non-flammable blends of ethylene oxide and CFC-12 in hospital sterilization is ozone-safe, but can compromise worker safety. The methylene chloride that replaced CFCs in the manufacture of flexible foam is ozone-safe but toxic if worker exposure is allowed during manufacture of the foam. Hydrocarbons used as refrigerants, aerosol propellants or foam-blowing agents are superior for ozone-layer and climate protection, but are flammable and can contribute to formation of photochemical smog in some locations. If emissions can be limited and safety risks can be fully mitigated, such technology is considered a clear environmental success – but success that technology can improve upon over time. Mitigation of flammability in refrigerators has been satisfied with small charges, quality construction, explosion-proof electronic components and worker training.

- Uncertain environmental success is when new technology has significant environmental, health, or safety concerns that offset the advantages to ozone-layer protection. For example, PFCs and SF6 are ozone-safe alternatives to ODSs, but are very potent greenhouse gases with atmospheric lifetimes of thousands of years.
- Unachieved environmental success is when no new technology to replace critical uses of ODSs has satisfied environmental, health or safety concerns. The Parties to the Montreal Protocol have allowed methyl bromide uses for quarantine and pre-shipment and have granted essential-use exemptions for specific medical, laboratory and analytical uses, rocket manufacture, torpedo maintenance and some fire-protection applications. One issue still under debate involves refrigerant applications of HCFC-123, which has relatively short atmospheric lifetime, a low global warming potential and a low ODP. There is no global consensus on whether the climate-protection advantages from the superior energy-efficiency of HCFC-123 used in building air conditioning are more important than the effect of very small emissions of this low-ODP substance on the ozone layer.

ECONOMICS OF PHASING OUT OZONE-DEPLETING SUBSTANCES

When the Montreal Protocol was signed in 1987, few alternatives to ozone-depleting substances were available, and some substitutes had not been even envisioned. Industry had predicted that technical progress would be slow, that costs would be very high, that valuable equipment would be abandoned and that safety might be compromised. As it turned out, progress was faster and costs were less than industry predicted. The cooperative working relationships – developed among industry, regulators, scientists and environmental NGOs – provided an environment that stimulated innovation. For some applications, particularly solvents, systems and processes were completely redesigned to maintain quality and safety, and, in some cases, reduce costs. In other applications, particularly refrigeration and air conditioning, refrigerant leaks were reduced, energy efficiency and reliability were maintained or improved, and safety was not compromised. In fact, new technologies are often more energy-efficient and cheaper to maintain, and produce systems that are more reliable than those based on CFCs and other ozone-depleting substances.

Today, industry, government, and non-government organizations all agree that the costs of eliminating ozone-depleting substances are far less than the consequences of ozone depletion. Prior to 1986, however, there was no agreement on the economics of taking action. In some cases, the costs were overstated because some of the investments would have occurred without the Montreal Protocol and because the benefits of lower operating costs, less maintenance, and higher product quality and reliability were not accounted for in economic models.[52] For example, the cost of fuel savings and lower maintenance paid for new refrigeration and air-conditioning equipment, and the increased reliability and performance of electronic equipment, paid returns beyond their transition costs. Aerosol manufacturers recovered their cost of investment, and customers received better value for money with environmental protection as a no-cost bonus.

Early industry estimates of the cost of alternatives and substitutes

From the early 1970s until 1986, most ODS-industry stakeholders considered ozone-depletion science to be inconclusive, and opposed any regulations on ODSs. By the late 1970s, more than 250 separate product categories were made with, or contained, ozone-depleting substances. Some of these products were vital to society, but many uses were for convenience or comfort. The CFC industry emphasized the most socially important uses of CFCs, while calculating the economic and employment contribution of all uses, including those that were socially less important or had competitive alternatives.

Critical uses with high social value and uncertain alternatives represented less than 50 per cent of CFC use in 1985, but improved containment practices could have reduced the amount used in these applications to less than 20 per cent. These critical uses included medical products, refrigeration, cleaning of electronic and mechanical computers and components, and fire protection. Medical applications included metered-dose inhalers for dispensing asthma and heart medicines, skin chillers for burn victims, sterilization, pacemakers, artificial limbs and blood substitutes. Electronic products included business and defence computers, and telecommunication systems. Fire protection with halon was used for nuclear and hazardous chemical facilities, air traffic control, aircraft and ship fire protection, and oil and gas processing and transport.

Desirable uses – primarily by wealthy customers – represented 20 to 30 per cent of CFC use in 1985 and included air conditioning of office buildings, vehicles and sports facilities, and thermal and sound insulation. In these applications, improved containment could have reduced emissions. Less important uses with competitive alternatives represented 25 to 30 per cent of CFC use in 1985. These uses included cosmetic and convenience aerosol products and novelty products such as wineglass chillers, noise-making horns, and foam party stringers.

In 1979, DuPont estimated US sales of CFCs at approximately US$375 million, with more than 650,000 US jobs related to CFC use and 240,000 domestic business locations using CFCs.[53] Estimates of the CFC industry's production, sales and employment levels in 1985 included one from the *Wall Street Journal*[54] of US$750 million in US sales.

The Alliance for Responsible CFC Policy,[55] an industry organization, estimated that in 1985:

- Approximately 10,000 US companies used CFCs to produce US$28 billion/year in products.
- The CFC industry provided 715,000 direct and indirect jobs.
- Two-thirds of those direct and indirect jobs were in the service sector.
- Halon protected human life and safety, and billions of dollars of property.
- Installed equipment worth approximately US$135 billion depended on CFCs.

In December 1987, an Alliance-sponsored study concluded that under the 1987 Montreal Protocol's terms to freeze halon production at base-year levels and reduce CFC production by 50 per cent, ODS prices would rise significantly even with rapid implementation of alternatives; US social costs of investment and depreciated capital would be US$5.5 billion, and US$9 billion would be transferred from consumers and user industries primarily to ODS and substitute chemical suppliers from 1990 to 2000.[56]

Retrospective industries' cost evaluations were presented at various occasions between 1993 and the ten-year anniversary of the Protocol in 1997. The electronics and aerospace sector reported impressive cost savings and improvements in product performance and reliability.[57] The motor vehicle and building air conditioning sectors reported higher reliability, lower operating costs, and satisfied customers. Chemical manufacturers reported high invest-ment, over-capacity and disappointing profits.[58]

Government estimates of the phase-out costs

Government estimates of the cost of protecting the ozone layer were typically part of a benefit/cost analysis that also estimated the damage avoided. The calculation of avoiding death and injury inherently required assumptions about the value of time and human life. Industry expressed many concerns with the parameters used in the US EPA studies, in particular the data linking the expected levels of ozone depletion with potential deaths from skin cancer.

In 1987, the US EPA estimated that the total cost of phasing out CFCs to the US economy would be US$28 billion, and the economic benefits of damage avoided would be US$6.5 trillion – a benefit cost ratio of 240 to 1.[59]

On the tenth anniversary of the Montreal Protocol in 1997, Environment Canada estimated that the global costs of phasing out ODSs would be US$235 billion, and the global benefits would be US$459 billion for the period between 1987 to 2060, excluding reductions in health care costs. This results in a net benefit of US$224 billion, plus health benefits.[60]

Concerns of windfall CFC profits not borne out

When the Montreal Protocol was signed, it was speculated that rising CFC prices would result in substantial windfall profits to producers. Preliminary calculations in 1988 by the Natural Resources Defense Council (NRDC) based on EPA price projections estimated 'wealth transfer' to CFC producers to be US$2.0–5.7 billion to 2000 for the scheduled 50 per cent reduction in CFC production and freeze in halon production, depending on how high CFC prices climbed.[61] Consultants to the Alliance for Responsible CFC Policy estimated that CFC producers would earn US$9 billion for the scheduled reduction to 2010, assuming no further reduction was scheduled.[62]

No retrospective study has recalculated these estimates based on the actual price increases and accelerated reduction schedule. Complete analysis would require knowledge of the value of abandoned property, and the increasing cost of producing decreasing quantities of CFCs as plant capacity became under-utilized. In the USA, CFC taxes captured some ODS sales revenue; the rapid elimination of the use of CFCs as aerosol propellants and open-cell blowing agents led to significant over-supply in Europe. However, in European and some other non-Article 5 countries, illegally imported CFCs resulted in the prices not rising significantly above the pre-Protocol prices for much of the phase-out period.

Other monopoly profit concerns

Many countries with economies in transition and developing countries were concerned that intellectual property laws would allow property owners to extract monopoly profits from the sale of alternatives and substitutes to ODSs (see Chapter 3). There was particularly high concern that patents for producing new HCFCs and HFCs would force countries to become dependent on foreign sources from a small number of multinational companies. In 1987, ODSs were produced in at least 19 countries: Argentina, Australia, Brazil, Canada, China, France, Greece, India, Italy, Japan, Mexico, The Netherlands, South Africa, South Korea, the UK, the USA, the USSR, Venezuela and West Germany. However, patents to produce the most promising halocarbon substitute, HFC-134a, using two chemical routes were initially controlled by companies in only three countries: Japan, the UK and the USA. There was great concern that one patented process would ultimately dominate.

Although the transition is not yet complete, it is now possible to make some early observations. Foremost, new HFCs and HCFCs have been far less important than originally expected, and non-halocarbon 'not-in-kind' alternatives that are not patented have been more important than originally expected. However, it is too early to know whether all developing-country markets will continue to have access to competitive supplies of these new chemicals.

Categories of industry leadership

Industry leadership is speeding protection of the ozone layer and helping to make alternatives affordable and environmentally acceptable.[63] This leadership can be described in several categories:

- Reversal of the previous position that regulation is not justified by scientific evidence, and calling for international limits such as the announcements by DuPont and the Alliance for Responsible Atmospheric Policy in 1986.
- Commercializing CFC-free products, such as the announcement as early as 1975 by S C Johnson that it was phasing out CFCs as aerosol propellants and switching to hydrocarbons.
- Publicly endorsing alternatives developed by others, such as AT&T's announcement that citrus-based aqueous processes cleaned electronics as well as CFC solvents.
- Breaking ranks with associations and competitors, such as the 1988 public pledges by Nortel and Seiko Epson to phase out CFC-113 solvents.
- Committing to a unilateral phase-out of CFC production, as DuPont did in 1988.
- Demanding and developing ODS management plans and recovery/re-use infrastructure and then sharing technology and training materials worldwide, such as Sainsbury's programme for supermarkets and the Coca-Cola Company's programme for beverage vending and dispensing equipment.
- Organizing environmental leadership companies into new proactive associations, such as the Industry Cooperative for Ozone Layer Protection and the Japan Industrial Conference on Ozone Layer Protection.
- Promoting technology cooperation, such as the ICI/Huntsman implementation of hydrocarbon foam in China, and the Japan/Thailand/USA partnership to phase out CFC refrigerators in Thailand.
- Developing new technology and allowing unrestricted global use, such as the Foodservice and Packaging Institute promotion of CFC-free foam packaging; CFC recycling equipment developed by the automobile industry and its suppliers; no-clean soldering developed by AT&T, Ford, IBM, Minebea, Motorola, Nortel, and others; and the donation of specific patents and know-how to the public domain by the Digital Equipment Corporation, Nortel and Seiko Epson.
- Pledging to investigate and adopt, where economically feasible and safe, non-fluorocarbon refrigerant alternatives, such as the announcements by McDonald's, the Coca-Cola Company and Unilever Foods in 2000, and the 'Vietnam Leadership Pledge' by more than 40 multinational companies to invest only in modern, environmentally acceptable technology in their Vietnam projects.

Measures of industry success

Most products previously dependent on ODSs are still on the market with indistinguishable or superior performance and comparable cost of ownership. Some ODS-free products have had a higher first cost, but greater energy efficiency, durability and/or reliability. No important product left the market because of the phase-out of ODSs.

Chemical substitutes are three to five times more expensive than ODSs but, with important exceptions such as halon fire-protection systems, chemicals are typically a small part of product cost. Higher chemical costs are offset by better manufacturing yields (as with solvents and foam-blowing), higher performance (in refrigeration, air conditioning, electronics, metered-dose inhalers and sterilization), and higher reliability and less maintenance (in refrigeration, air conditioning and electronics).[64]

After 1986, industry demonstrated its confidence in the scientific consensus and the Montreal Protocol by openly criticizing the few remaining sceptics on ozone science.[65]

Chapter 6

Implementation of the Montreal Protocol

INTRODUCTION

The implementation of the Montreal Protocol has benefited from the public and political reaction to the discovery of the Antarctic ozone hole, the health and ecological consequences of ozone depletion, and the scientific consensus that chlorofluorocarbons (CFCs) and other ozone-depleting substances (ODSs) were the cause of that depletion. The Protocol has elicited the cooperation of a large number of highly diverse organizations.

The UNEP Secretariat for the Vienna Convention and the Montreal Protocol, called the Ozone Secretariat, provides the requisite support to the Meetings of the Parties in strengthening the Protocol, monitoring implementation, and reporting problems to Parties when necessary. The Protocol's Implementation Committee views its mandate widely and takes a pro-active stand in anticipating problems and recommending solutions to the Meetings of the Parties. The Meetings of the Parties make decisions on non-compliance by the Parties in a spirit of sympathy, assistance and, where necessary, sternness, and help countries achieve compliance.

The Executive Committee of the Multilateral Fund for the Implementation of the Montreal Protocol, assisted by the Fund Secretariat, has given speedy approval to projects that lead to the phase-out of ODSs by the industries in developing countries that operate under Article 5, Paragraph 1. Article 5 specifies the concessions to the countries that qualify as 'developing', as agreed by a Meeting of the Parties, and whose consumption of ODSs per capita is below the limits specified in Article 5. The Global Environment Facility (GEF) has ensured that the Russian Federation and other countries with economies in transition return to the path of compliance despite their grave social and economic problems.

Many United Nations agencies are actively involved in the implementation of the Protocol. The UNDP, World Bank and UNEP are the implementing agencies of the Multilateral Fund and GEF. The United Nations Industrial Development Organization (UNIDO) is an implementing agency of the Multilateral Fund. The WMO is the coordinating agency for assessment of the state of the ozone layer. The World Customs Organization was the authority that harmonized the commodity-numbering system for monitoring trade in

ODSs. The WHO is the source of information on the adverse effects of methyl bromide. The FAO is the source of information about methyl bromide and its alternatives. Organizations involved in promoting the Protocol's implementation include: regional development banks such as the Asian Development Bank; regional United Nations economic and social organizations such as the Economic and Social Commission for Asia and Pacific (ESCAP) and the Economic Commission for Europe (ECE); and regional political organizations such as the African Ministers Council on Environment, the Council of the Arab Ministers Responsible for the Environment, and the Forum of Ministers of the Environment for Latin American and the Caribbean Countries.

The implementation of the Protocol is the sole responsibility of the national governments, but it is industries that have to make the technical changes. National governments have adopted many innovative policies and regulatory and financial measures that encouraged cooperation and built global momentum. The industries that used or manufactured ODSs and their organizations at the national and global level have helped each other to phase out ODSs, a sharp contrast from business-as-usual. Many technical professional organizations throughout the world have helped to set or alter standards and facilitate the introduction of new ozone-friendly technologies. Environmental non-governmental organizations (NGOs) have gone beyond their usual role of identifying and blowing the whistle on activities which harm the environment, and have taken the unusual path of identifying and promoting alternative technologies. The implementation of the Montreal Protocol has been, indeed, a worldwide movement.

STRUCTURE OF THE OBLIGATIONS OF THE MONTREAL PROTOCOL

The crux of the Montreal Protocol is the series of control measures for developed countries (specified in Article 2) and developing countries (specified in Article 5), for the ozone-depleting substances specified in Annexes A, B, C and E. The measures specify binding limits on the production and consumption of ODSs separately for the developing-country Parties operating under Article 5 and for other Parties who do not operate under Article 5. These binding limits were reduced over time until, by a specified year, the Parties had to phase out the production and consumption of controlled substances. The control measures of the 1987 Montreal Protocol were made stricter with the adjustments and amendments to the Protocol approved by the Parties in 1990, 1992, 1995, 1997 and 1999.

- Article 4 obliges all Parties to ban trade in ODSs with non-Parties. The Article provides for identification of products containing ODSs and products made with, but not containing, ODSs. Article 4A specifies the obligations for control of trade with the Parties.
- Article 7 mandates baseline and annual reporting by Parties of both production and consumption for all ozone-depleting substances.

- Article 8 forms the basis for action in case of any non-compliance with any of the obligations.
- Article 9 requires Parties to conduct research and development, to exchange information on alternatives to ODSs and control strategies, and to cooperate in promoting public awareness of the environmental effects of the emissions of controlled substances. It also requires each Party to submit within two years of entry into force, and every two years thereafter, to the Secretariat a summary of the activities it has conducted pursuant to this article.
- Article 10 prescribes the obligations of the Parties with respect to the financial mechanism, and Article 10A with respect to the transfer of technologies.

The control measures for the phase-out of production and consumption are specified on a basis of 'ODP-weighted' quantities for 'groups' of chemically similar gases. The quantities are calculated by multiplying the quantity of the controlled substance by the ozone-depletion potential (ODP) of that substance, and summing the total within the group. This flexibility allows each Party to choose to produce and consume the combination of controlled chemicals that best suits its needs. The Protocol encourages the practices of recovery and recycling by not counting those quantities recovered and recycled, and allows Parties to credit quantities destroyed against total production and consumption.

Obligations under the Protocol

The primary obligation on all Parties is the timely phase-out of ozone-depleting substances in accordance with the specified control schedules. This is achieved through adoption of substitutes and alternatives and by phasing out the production of ODSs. The data submitted annually by each Party are verified. This phase-out will ensure that the chlorine and bromine abundance in the stratosphere will be eventually reduced to naturally occurring levels. By the year 2050, it is anticipated that the chlorine and bromine abundance will fall from peak levels of 5ppb to the 1980 level of 2ppb – the level when the Antarctic ozone hole first appeared. If actions are taken promptly also to stabilize the concentrations of greenhouse gases, the ozone layer will be restored to its natural level.

THE ROLE AND ACTIVITIES OF THE MULTILATERAL FUND FOR THE IMPLEMENTATION OF THE MONTREAL PROTOCOL

The environmental benefits of the Multilateral Fund investments

By the end of December 2001, the Executive Committee had approved more than US$1.34 billion for the implementation of investment and non-investment projects. This level of funding is expected to result in the permanent, annual phase-out of about 200,000 ODP tonnes of ODS consumption and production. Projects approved by the Executive Committee have thus far resulted in the permanent annual phase-out of 141,531 ODP tonnes of ODS consumption and production, of which 138,418 ODP tonnes is from completed projects and 2995 tonnes from ongoing projects. About 57 per cent of the 3885 projects sanctioned have been completed.

Structure of the Fund and its executive elements

The Executive Committee

The Multilateral Fund meets the agreed incremental costs of the Article 5 Parties for complying with the control measures, which are the additional costs incurred when a company switches from an ODS technology to an alternative. The Fund also finances clearinghouse functions and the Fund Secretariat. The responsibility for overseeing the operations of the Multilateral Fund rests with an Executive Committee, supported by the Fund Secretariat. The Executive Committee: develops and monitors the implementation of specific operational policies, guidelines, and administrative arrangements, including the disbursement of resources; develops the three-year plan and budget for the Multilateral Fund, including allocation of resources among the implementing agencies; develops the criteria for project eligibility and guidelines or the implementation of activities supported by the MLF, and reviews the performance reports and expenditures of those activities; and reports annually to the Meeting of the Parties on the activities exercised, and makes recommendations as appropriate.[1]

The Executive Committee consists of fourteen members, seven from Article 5 Parties and seven from non-Article 5 Parties. The terms of reference require that decisions shall be taken by consensus wherever possible. Otherwise, decisions shall be taken by two-thirds majority of Parties present and voting, representing a majority of Article 5 Parties and a majority of non-Article 5 Parties present and voting – giving both donor developed countries and recipient developing countries effective veto power. All decisions, so far, have been made by consensus; there has never been a resort to voting on any decision by the Committee. The members are selected by each group of Parties and formally endorsed by the Meeting of the Parties. The Chair and the Vice-Chair of the Executive Committee are subject to annual rotation between the Article 5 and non-Article 5 Parties, while the group not providing the Chair selects the Vice-Chair.

Subcommittees of the Executive Committee

The Executive Committee has two subcommittees: the Monitoring, Evaluation, and Finance Subcommittee, and the Project Review Subcommittee. The Project Review Subcommittee allows three NGO representatives to attend meetings, but not to participate in the deliberations – one nominated by environmental NGOs, one by industrial NGOs and one by academic NGOs.

Implementing agencies

UNDP, UNEP, UNIDO and the World Bank implement the projects funded by the Multilateral Fund. In addition, a number of developed countries provide bilateral assistance to the Article 5 Parties.[2] Up to 20 per cent of the contribution due to the Fund can be spent by a contributor on regional and bilateral cooperation if it is related to compliance with the provisions of the Protocol and meets agreed incremental costs. In practice, the Executive Committee has to approve each of the bilateral activities before the expenditure on such activities is counted as a contribution to the Fund. This ensures that the activities are within the norms set by the Executive Committee and that bilateral or regional activities do not duplicate projects of the implementing agencies.

Article 5 countries can choose to collaborate with any of the four implementing agencies or with Parties offering bilateral assistance in preparing their country programmes or project proposals. UNDP, UNIDO and the World Bank prepare and implement investment projects, while UNEP undertakes information dissemination, capacity-building, institutional strengthening, networking and other assistance, particularly to those consuming low levels of ODSs. The Executive Committee has signed formal agreements with the implementing agencies specifying their obligations and privileges. The Executive Committee invited UNEP to be the treasurer of the Fund. Thus, UNEP is responsible for receiving and administering contributions, and disbursing funds to the Fund Secretariat and the implementing agencies based on the directives of the Executive Committee.

The Fund Secretariat

The Fund Secretariat is based in Montreal, Canada, and consists of a small number of internationally recruited professional staff and local support staff. The Secretariat is headed by the Chief Officer, nominated by the Executive Committee for appointment by the Executive Director of UNEP. The Chief Officer reports directly to the Committee. The Secretariat develops and manages plans and budgets, disburses funds, monitors the implementing agencies, prepares policy papers, and undertakes other activities as directed by the Executive Committee.

Preparation of country programmes

As an essential first step for obtaining financial assistance from the Multilateral Fund, an Article 5 Party is required to submit a government-approved country programme to the Executive Committee. The country programme is expected to include: a review of recent production, imports, applications and use of controlled substances by the main producers and consumers in the country; links to transnational producers or users; the policy and institutional framework governing controlled substances; and a statement of strategy for implementation of the Protocol. Each country is expected to provide an action plan, a timetable, a budget and a financing programme.

The Fund finances preparation of a country programme, technical assistance, institutional strengthening, project preparation, demonstration, training and agreed categories of project incremental costs. (Appendix 5 contains an indicative list of categories of incremental costs agreed to by the second Meeting of the Parties in London in 1990 and reconfirmed by the fourth Meeting of the Parties in 1992.) A general principle for the incremental cost approach is that the most cost-effective and efficient option should be chosen, taking into account the national industrial strategy of the Party. The infrastructure used for production of the controlled substances can be put to alternative uses, thus reducing capital abandonment. Savings or benefits that would be gained at both the strategic and project levels during the transition process are taken into account on a case-by-case basis.

Application of incremental costs

While the Parties approved the indicative list of categories of incremental costs (Appendix 5), the Executive Committee and the Fund Secretariat had to apply this list to specific projects in more than 100 widely differing developing countries. With no precedent to rely on, they had to: interpret the new concept of incremental costs to promote the fastest phase-out, sanction projects with broad geographic balance and, at the same time, use the available funds prudently and ensure that the new technology satisfied globally accepted environment, health and safety criteria.

In light of technological advances making new investment in ODS-free technology as affordable as the obsolete ODS technology, the Executive Committee decided not to consider projects to convert any ODS-based capacity installed after July 1995.

The vexing issue of whether or not to pay for technological upgrades, which are implemented in adopting new technologies, complicated the calculations of incremental costs. After much deliberation, the Executive Committee decided that technological upgrades that were unavoidable in the technical conversion were to be taken into account in determining categories of eligible incremental costs. It also decided that costs associated with avoidable technological upgrades should be considered ineligible and therefore not financed by the Multilateral Fund. It established a methodology for the quantification of technological upgrades as guidance in calculating incremental costs.

Economic and technical issues addressed by the Multilateral Fund

Technology transfer paid for only once in a country
The Executive Committee requested the implementing agencies to ensure that adequate guarantees were obtained from technology vendors when technology transfer was to be replicated within the country. Noting that several project proposals included technology transfer fees from the same supplier for projects in the same country, it requested implementing agencies and countries to negotiate fees jointly to cover groups of projects.

At its eighth meeting in October 1992,[3] the Executive Committee decided that proposals for research and development of substitutes and recycling and destruction equipment could be considered on a case-by-case basis, provided that the costs incurred were of an incremental nature.

No funds for transnational corporations or to taxes
The Fund decided not to finance phase-out of ODSs at enterprises that were wholly owned subsidiaries of transnational corporations or enterprises

permitted to operate in 'free zones' with output for export only. Funding proportional to the percentage of local ownership was provided for local enterprises partly owned by a transnational corporation.

The Multilateral Fund would not finance: taxes, duties or other such transfer payments; the loss of economic subsidies; or rates of return in excess of cost of capital which might incorporate non-economic financial effects such as administered prices or interest rates.

The Executive Committee endorsed guidelines for enterprises that exported part of their production to non-Article 5 countries to ensure that consumers in developed countries were not subsidized and that companies in developed countries were not subject to unfair competition.

Concessional loans

At its thirteenth meeting in July 1994,[4] the Executive Committee recommended that projects that realized net incremental savings would be good candidates for concessional loans from the Multilateral Fund or other funding sources. The Committee encouraged implementing agencies, regional development banks and other lending institutions to provide loans for ODS phase-out projects that were not eligible for grants from the Multilateral Fund. Many non-Article 5 Parties unsuccessfully tried to include part of the replenishment for the periods 1997–1999 and 2000–2002 as concessional loans. The developing countries strongly opposed this. They argued that prior to 1996 larger enterprises had been financed by the replenishments which had no loan component. They considered it unfair that smaller enterprises to be financed after 1996 would receive loans and not grants. So far, only one project in Turkey has been approved for a concessional loan.

CFC equipment to be destroyed after project completion

Because of concerns that the ODS-using equipment replaced by the projects could be reinstalled elsewhere and CFC use continued, the twenty-second meeting in 1997[5] and many subsequent meetings of the Executive Committee directed the implementing agencies to ensure that equipment replaced by the projects was destroyed or rendered unusable.

Small countries

Smaller countries repeatedly complained at the Meetings of the Parties and to the Executive Committee that they were neglected in the allocation of funds. The seventh Meeting of the Parties in December 1995 requested the Executive Committee to provide specific funding to low-volume ODS-consuming countries and regions for awareness-raising and training programmes, workshops, institutional strengthening and retrofitting in sectors vital to these economies.

The Executive Committee declared that annual consumption of 360 tonnes or less of ODSs qualified a country to be a low-volume ODS-consuming country and made several concessions for such countries, including separate allocations, higher cost-effectiveness thresholds and special projects such as refrigerant management plans to suit their needs.

Refrigerant management plans

For many of the Article 5 Parties with low volumes of CFC consumption, the main concern was about maintaining their refrigeration equipment when CFCs were no longer available. The Executive Committee approved preparation of refrigerant management plans for many such countries. The steps for formulation of a refrigerant management plan included review and analysis of refrigeration and air-conditioning sectors and subsectors, consumption estimates of CFC and HCFC refrigerants and their availability, identification of sources of refrigerant supply and distribution channels, and assessment of technical options to minimize ODS refrigerant needs such as good practices, recovery and recycling, conversion, retrofitting and replacements. The refrigerant management plans also looked at policy options such as voluntary programmes and agreements, legislation and regulations, economic instruments, formulation of a refrigerant management policy that included a training programme for refrigeration technicians and customs officials, and recovery and recycling systems.

Cost-effectiveness thresholds

The sixteenth meeting of the Executive Committee in March 1995[6] adopted sector and subsector cost-effectiveness threshold values capping the maximum amount that could be financed per ODP tonne reduced. An ODP tonne represents a tonne multiplied by the ODP of an ozone-depleting substance. For example, the threshold was US$4400 per ODP tonne phased out in the aerosol sector, while it was fixed at US$13,600 per ODP tonne for the domestic refrigeration sector.

In developing countries, a large number of enterprises using ODSs were small- or medium-sized, and lacked the awareness, information and capacity to utilize the assistance of the Multilateral Fund. The twenty-fifth meeting of the Executive Committee in July 1998 decided to include an allocation of US$10 million from the resource allocation for 1999 to facilitate pilot conversions of significant groups of small firms in the aerosol and foam sector in non-low-volume ODS-consuming countries. The cost-effectiveness threshold was relaxed up to 150 per cent of the level of the current cost-effectiveness thresholds for the relevant eligible subsectors.

Sector-based approach instead of individual projects

While in the initial years the Multilateral Fund supported only individual projects, it switched to a sector-based approach in the later years. At its twenty-eighth meeting in July 1999,[7] the Executive Committee decided to encourage cooperation between implementing agencies and the national governments in Article 5 countries to develop umbrella projects and sector approaches. The responsibilities of the international implementing agencies would focus on consulting, monitoring and auditing project implementation by domestic implementing agencies. China implemented the sectoral approach extensively. It agreed that in exchange for specified annual funding from 1998 to 2007 totalling US$62 million, subject to many conditions, it would reduce its production and consumption of halons to levels specified for every year, leading to a phase-out in the year 2010. China also committed to phase out completely the consumption of CFC-11 in the tobacco industry by the year 2007 for a total cost of US$11 million. The Executive Committee approved US$52 million for the phased reduction and complete phase-out of consumption of tri-chlorotrifluoroethane (CFC-113) and 1,1,1-trichloroethane, as well as the consumption of carbon tetrachloride, all used as cleaning solvents in China.

Technology choices: The debate on HCFCs and HFCs

For a long time, it had been well known that hydrocarbons such as cyclopentane and isobutane work well as foam-blowing agents and refrigerants for domestic refrigerators, but were not adopted because of concerns about their safety. Fluorocarbon manufacturers and many of their customers preferred the ozone-depleting HCFCs, or HFCs which are ozone-safe but have a high global warming potential, to replace CFCs. A leading environmental NGO, Greenpeace, brought together the Dortmund Institute of Germany and refrigerator manufacturer DKK Scharfenstein to develop and popularize the hydrocarbon technology as the environmentally ideal substitute for CFCs as refrigerants and insulating foam-blowing agents. Greenpeace called the technology 'Greenfreeze' which has now spread to most of the world, including Japan, but not North America (see Chapter 5 and Appendix 1). When developing-country projects planned with Greenfreeze came before the Executive Committee of the Multilateral Fund, questions were raised about the additional costs of safety precautions necessary to adopt this technology. The sixteenth meeting of the Executive Committee in March 1995 recognized that there were significant costs related to the provision of safety equipment, and agreed that those costs would not be included in calculating the cost-effectiveness of such projects, but would be considered in determining the funding level.

The fifteenth meeting of the Executive Committee in December 1994[8] decided to finance the use of HCFCs only if an enterprise certified in writing that it understood that HCFCs would have to be phased out ultimately, and that the Fund would not pay for a second transition from HCFCs to ozone-safe alternatives. The seventeenth Meeting of the Executive Committee in July 1995[9] decided that qualifying projects utilizing hydrocarbon technologies could be up to 35 per cent more expensive than projects using HCFCs or HFCs. From the sixth Meeting of the Parties in October 1994 onwards, some countries and environmental NGOs continuously urged the Executive Committee not to sanction ODS phase-out projects that used HCFCs or HFCs because, according to them, ozone- and climate-safe technologies were available for all applications. Some countries, however, held that HCFCs and HFCs were essential in some sectors for implementation of the scheduled CFC phase-out. The Executive Committee directed the implementing agencies to point out all the technological options and merits to the enterprises and to provide justification to the Executive Committee if HCFCs are chosen as alternatives to CFCs in any project.

Assistance for CFC production phase-out

Of the Article 5 Parties, only China, India, the Democratic People's Republic of Korea, the Republic of Korea, Argentina, Brazil, Mexico, Romania and

Venezuela produced CFCs. Only China and India produced halons. During the base period for Article 5 Parties from 1995 to 1997, the total production of CFCs was about 98,800 tonnes, and that of halons was about 41,000 tonnes. China and Romania combined produced about 800 tonnes of methyl bromide. Of these countries, the Republic of Korea agreed that it would not draw any assistance from the Multilateral Fund for implementing its ODS phase-out schedule. Brazil had only production plants owned by multinational companies and was therefore ineligible for assistance from the Multilateral Fund. The Executive Committee decided in 1998 that technical audits of the plants should be conducted to verify the reported data and to calculate the compensation for these countries to phase out their production of CFCs. Based on these audits, US$150 million was allocated in 1999 to phase out CFC production in China by the year 2010, with a set annual allocation and phase-out schedule. In 1999, a similar plan was agreed upon for India at a cost of US$82 million. Negotiations are currently taking place with Argentina along similar lines. These agreements account for 70,000 tonnes of CFC production, or 70 per cent of the total production of Article 5 Parties. Similar agreements were reached to phase out all production of halons.

Methyl bromide projects
The inclusion of methyl bromide as a controlled substance for Article 5 countries in 1995 led to an acceleration of work by the Executive Committee, which had already approved three regional data-collection and demonstration projects concerning methyl bromide in 1995. Methyl bromide projects were perhaps the most challenging area in which the Multilateral Fund had operated, because there was no 'one-size-fits-all' alternative for every application or crop. In addition, while industrial conversion through the replacement and destruction of existing equipment was a one-time event, there were few technical impediments to a farm reverting to use of methyl bromide in the years after a completed phase-out. Interim guidelines on methyl bromide demonstration projects were adopted at the twentieth meeting of the Executive Committee in October 1996,[10] and US$10 million for such projects was allocated for 1997–1999. A year later, in 1997, this commitment was increased by US$50 million. A strategy and set of guidelines for the whole sector, including a priority list of crops for soil fumigation, were approved in 1998 and modified and adopted in 2000; the first investment projects aimed at phase-out began with a project in the tobacco sector in Cuba in November 1998. Non-investment projects implemented by UNEP, such as those which create awareness and enhance local capacities, play a significant role. As of December 2001, a total of US$43.5 million had been allocated to demonstration, investment and training projects on methyl bromide for an estimated reduction of 1740 ODP tonnes.

Monitoring and evaluation of Multilateral Fund projects

The follow-up of project implementation was a general cause for concern in the first few years of the Multilateral Fund's operation. The period 1995–1999 saw the establishment and refining of a monitoring and evaluation system, as called for in the study on the effectiveness of the financial mechanism in 1995.[11] In 1997, a Monitoring and Evaluation Subcommittee was established, and a monitoring and evaluation post created within the Multilateral Fund Secretariat. The Subcommittee, like its counterpart for project review, met immediately prior to meetings of the Executive Committee and had seven members, drawn from the membership of the Committee.

An evaluation guide was prepared, intended to be a dynamic document subject to revision in light of experience. Henceforth, all project proposals were to include a standardized component on monitoring and evaluation, including milestones for the completion of the various stages of the project, which should provide sufficient information to enable a comparison between the approved proposal and the project-completion report. The purpose of the process was to ascertain whether the project objectives had been met in terms of the disbursement of funds, the amount of controlled substances phased out, and various other performance indicators. Agencies were to submit progress reports by May each year for all activities approved by the end of the previous year.

MULTILATERAL FUND REPLENISHMENT AND CONTRIBUTIONS

The Interim Multilateral Fund was set up at the second Meeting of the Parties in June 1990, with an allocation of US$240 million for 1991–1993. The Multilateral Fund has been replenished four times by the Parties for three-year cycles (Table 6.1).

Table 6.1 *Interim funding and replenishments of the Multilateral Fund*

Finance stage	Finance period	Amount (US$ million)
Interim	1991–1993	240
Replenishment 1	1994–1996	455
Replenishment 2	1997–1999	466
Replenishment 3	2000–2002	440

Sources: Reports of the Meetings of the Parties to the Montreal Protocol in 1990, 1993, 1996 and 1999.

The Multilateral Fund is financed by contributions from Parties not operating under Paragraph 1 of Article 5 on the basis of the United Nations scale of assessments. As of December 2001, the total income to the Multilateral Fund, including interest and miscellaneous income, stood at US$1.38 billion, while disbursements amounted to US$1.34 billion. Of the income, the contributions of the countries to the Fund stood at US$1.28 billion with US$52.50 million of that in bilateral contributions with Article 5 countries, representing about 4 per cent of the total; about US$99.6 million was interest earned since 1991. OECD countries contributed almost 100 per cent of their assessments. The arrears from the countries with economies in transition were US$96 million.

BOX 6.1 WORKING TOGETHER FOR THE MULTILATERAL FUND

Omar E El-Arini, Chief Officer, Multilateral Fund Secretariat

It is almost 12 years since the Parties to the Montreal Protocol established the Multilateral Fund in June 1990. The first 30 months of that period representing the transition from the interim fund to the establishment of the Multilateral Fund from 1 January 1993 by a decision of the fourth Meeting of the Parties (Copenhagen, November 1992) after the London Amendment had entered into force in August 1992, was very crucial. The Multilateral Fund's Executive Committee had to discharge its responsibilities under its terms of reference and those of the Fund in a comprehensive and effective manner. The Fund Secretariat started functioning in February 1991, the Committee concluded its agreements with the Fund's implementing agencies and with the Fund's Treasurer in the same year, it has laid down criteria and guidelines for project eligibility, country programme and work programme preparations and has developed other operational policies and guidelines.

During the eight meetings it had held during that short period, it also disbursed resources to the implementing agencies for the preparation and implementation of investment projects in Article 5 countries. In his report to the fourth Meeting of the Parties, the Chairman of the Executive Committee, at the time Ambassador Juan Antonio Mateos of Mexico, informed the Parties of the approval by the Committee of projects for the phase-out of more than 30,000 tonnes of ozone-depleting substances representing some 20 per cent of the total consumption of developing-country parties to the Protocol in 1992.

During and since that period, the Fund Secretariat had/has to examine each and every submission to determine its eligibility for funding and the amount of incremental costs to be recommended for the consideration and approval of the Executive Committee. In so doing, the burgeoning Secretariat had to work with the Fund's implementing agencies, who are very well established international organizations with definitive institutional culture and working procedures, to operationalize the Committee's decision on eligibility and incrementality of costs. The agencies viewed these operational issues from the experience of their organizations which contrasted with the Secretariat's understanding of the concept of incremental costs and their eligibility. These seemingly conflicting views enriched the process since they gave rise to a number of policy papers prepared by the Secretariat and sometimes by the Secretariat and the implementing agencies to address these views, which were considered by the Executive Committee and resulted in a set of rules and guidelines governing details such as cost-effectiveness thresholds for projects from different industrial sectors, technology upgrade arising from the acquisition of new and more sophisticated capital equipment for conversion projects, development of guidelines for the preparation of all kinds of projects, including the phase-out of ODSs in the production sector.

Over the years, a more entrenched cooperative attitude evolved which led the agencies and the Secretariat to agree on thousands of projects which facilitated the approval process of the Executive Committee and saved millions of dollars on costs that were agreed not to be incremental. Moreover, and through this evolutionary process, the Executive Committee was enabled to consider and approve multimillion-dollar sectoral phase-out agreements that will result in the phase-out of thousands of tonnes of ODSs in Article 5 countries, thus enabling these countries to meet their compliance targets.

Unlike the Secretariat and the implementing agencies, the Executive Committee's membership is subject to periodic change as new members replace those who have served their terms. This has not diminished the continuity in the Executive Committee nor its effectiveness. The Executive Committee has just concluded its thirty-sixth meeting which was held in March 2002 and approved some US$54 million for new projects and activities in Article 5 countries bringing the total approval since the establishment of the Fund to US$1.3 billion.

The most recent 30 months in the life of the Multilateral Fund has witnessed nearly total compliance by Article 5 countries with their first Montreal Protocol control measure. During the same period, the Executive Committee has had rich debates on a new strategy for the Multilateral Fund during the compliance period. The strategy draws on all the lessons learned during the grace period by all the Fund's stakeholders, Article 5 and non-Article 5 countries, members of the Executive Committee, the Fund's implementing agencies and the Fund Secretariat.

The challenge to all is to maintain the momentum that was painstakingly developed over the past 12 years in order to achieve the formidable task of bringing all Article 5 countries in timely compliance with all their obligations under the Montreal Protocol. The experience of the past 12 years should leave no doubt in anybody's mind that the international system can work very effectively and very transparently and with full accountability to find solutions for global environmental problems.

The terms of reference of the Multilateral Fund state that contributions should be in convertible currency or, in certain circumstances, in kind and/or in national currency. Payment may be made either in cash or through the use of promissory notes. It was required that promissory notes be issued with a fixed schedule of encashment and with the option of an accelerated schedule in light of needs. Until 1999, the pledges and contributions were in US dollars. In November 1999, the eleventh Meeting of the Parties agreed to use a fixed exchange rate mechanism on a trial basis for the 2000–2002 replenishment, in order to ease some of the administrative difficulties caused to Parties by having to make commitments in foreign currencies. The implementation of the new mechanism was to be reviewed at the end of 2001 to determine its impact on the operations of the Multilateral Fund and on the funding of efforts to phase out ODSs in Article 5 countries. The review by the thirteenth Meeting of the Parties resulted in a request to the treasurer, UNEP, to study this problem further using the services of a consultant.

IMPLEMENTING AGENCIES OF THE MULTILATERAL FUND

United Nations Environment Programme (UNEP)

UNEP's activities as an implementing agency of the Multilateral Fund, within its OzonAction Programme, fall into the following main categories:

- Acting as a global information clearinghouse providing information services to enable decision-makers to make informed decisions. Since 1991, the Programme has developed and disseminated over 100 individual publications, videos and databases that include public-awareness materials, the OzonAction newsletter, two websites, sector-specific technical publications for identifying and selecting alternative technologies, and guidelines to help governments establish policies and regulations.

- Training, at the regional and national levels, to build the capacity of policy-makers, customs officials and local industry to implement national ODS phase-out activities.
- Regional networks providing regular forums for Ozone Officers from Article 5 countries to meet to exchange experiences with counterparts from both developing and developed countries. UNEP currently operates eight regional/subregional networks involving 114 developing and 9 developed countries.
- Refrigerant management plans (RMPs), providing countries (presently 60) with an integrated, cost-effective strategy for ODS phase-out in the refrigeration and air-conditioning sectors.
- Country programmes and institutional strengthening to support the development and implementation of national ODS phase-out strategies, especially for low-volume ODS-consuming countries. The Programme has assisted over 90 countries to develop their national programmes and 76 countries to implement institutional-strengthening projects.

As of December 2001, the Executive Committee has approved 617 projects valued at US$55.8 million for UNEP to implement (ie about 4.9 per cent of all projects approved by the Multilateral Fund to date). UNEP has also leveraged resources for additional activities from many developed countries, including Australia, Canada, the European Commission, Finland, Italy, The Netherlands, New Zealand, Sweden, Switzerland and the USA. The Governments of Finland and Sweden have provided additional financial assistance to the OzonAction Programme to help encourage non-Parties to ratify the Protocol, and provide networking among countries, respectively. Since 1994, UNEP and UNDP have assisted countries with economies in transition through the Global Environment Facility with non-investment activities.

To respond to the changing needs of Article 5 countries, UNEP in 2002 has reorganized itself into a 'Compliance Assistance Programme' (CAP) by creating a core team with appropriate skills to help countries to achieve compliance.

United Nations Industrial Development Organization (UNIDO)

UNIDO became an implementing agency of the Multilateral Fund in 1992. The expertise available in UNIDO enabled it to respond quickly to the challenge. For example, UNIDO could act rapidly to the urgent need to phase out ODSs in the foam sector because of its previous experience in the plastics development and transformation industry.

UNIDO paid special attention to Eastern European countries that were slowly showing interest in the phase-out programme. With UNIDO's help, countries such as Macedonia and Croatia achieved an impressive ODS phase-out, ahead of their respective phase-out schedules. Currently, many efforts are being made to assist Bosnia Herzegovina and Yugoslavia, countries recovering from a devastating war.

UNIDO and UNDP complemented the work of the other implementing agencies working in the flexible-foam sector by introducing liquid carbon dioxide technology in order to avoid, wherever possible, use of methylene chloride, considered to be harmful to workers and the environment. UNIDO also managed projects to test methyl bromide alternatives available in industrialized countries under the different climate and geographical conditions of developing countries and, more important, to test their acceptability to local farmers. By 2001, UNIDO had implemented 23 methyl bromide demonstration projects.

The World Bank

The World Bank commenced its work as an implementing agency for the Multilateral Fund in 1991. This gave rise to a new challenge to the Bank: it had always dealt specifically with loans and was now faced with new policies and procedures to integrate into Bank operations. In addition, projects being approved by the Fund's Executive Committee were small in comparison to the multi-million dollar loans of the Bank. During the early years, client countries struggled with disbursement; problems with delays through financial intermediaries, and a general lack of knowledge and understanding of the Bank's system significantly slowed disbursement. The Bank worked to resolve these challenges by establishing financial intermediaries to give countries project ownership and help them build local capacity in line with its policy of 'national execution'. The introduction of umbrella grant agreements as the standard Bank mechanism for transferring Montreal Protocol resources to clients proved to be significantly more cost-effective and allowed the faster processing of investment projects.

During the first two or three years of implementation, all implementing agencies focused primarily on establishing country-level strategies through the preparation of country programmes. The first country programmes developed with the assistance of the World Bank were for Argentina, Chile, Egypt, Jordan, Tunisia and Turkey. Subsequently, the World Bank developed, approved and implemented individual investment sub-projects in more than 20 countries in all of the Bank's geographic regions. The project-by-project approach helped countries to get started in ODS phase-out and assisted them in phasing out consumption in large enterprises, primarily in the foam, aerosol and refrigeration sectors.

The sector approach, initiated by the Multilateral Fund Executive Committee, was the first Bank effort that sought government commitment to institute policies alongside projects financed by the Multilateral Fund to achieve complete ODS phase-out in a sector. This approach had its origins in a series of projects for China's commercial refrigeration sector in late 1994. The strategy included plant closures and conversion projects to improve the match of production capacity for compressors with market demand. More importantly, China committed itself to following through with these closures and other policies, in exchange for a financial commitment by the Multilateral Fund to assist in the phase-out of all CFC compressor manufacturing. After this experimental strategy, China and the Bank developed a more sophisticated sector plan for phasing out halons, which introduced performance-based implementation. This ensured that the government honoured its commitments before annual funding was provided.

The Bank also developed an auction system to channel grant funding to industries in Chile for the implementation of the Protocol. Through this mechanism, companies are invited to bid on cost-effective conversion projects. The bidding process usually lasts two months, during which time companies submit bids based on a cost-effectiveness threshold announced before each auction by the National Commission on the Environment, which administers the programme. Six auctions with the participation of more than 20 enterprises have taken place, resulting in the reduction of more than 350 ODP tonnes of CFCs at an average cost-effectiveness of nearly US$7.14 for a kilogram of ODS phased out. The intention behind competition is that for each round of bidding, a better cost-effectiveness value than the thresholds established by the Multilateral Fund will be achieved. In addition, lower administrative costs for preparation and supervision are assumed. After the closure of each auction, project proposals are evaluated for funding eligibility from the Multilateral Fund, technical feasibility, and the financial solvency of the companies involved. Subsequently, grants are allocated to the selected enterprises, starting with the most cost-effective bid, until funds are depleted or all selected projects have been served.

In 1992, the World Bank established a technical advisory team called the Ozone Operations Resource Group (OORG) to provide advice on the most cost-effective, environmentally acceptable alternatives and substitutes. This group of internationally recognized technical experts, primarily drawn from TEAP and its technical options committees, provides updates and information about aerosols, mobile air conditioning, refrigeration, foams, solvents, halons, production, methyl bromide and process agents.

United Nations Development Programme (UNDP)

UNDP implements investment, technical assistance and demonstration projects. It has received about US$350 million from the Multilateral Fund and another US$4.3 million from bilateral agencies. It has implemented 1440 projects in 78 developing countries to eliminate over 41,500 tonnes of ozone-depleting substances.

THE GLOBAL ENVIRONMENT FACILITY (GEF)

The countries of Eastern Europe and republics of the former USSR (Soviet Union) have faced great economic and social turmoil in their transition from communism to a market economy system in 1991. They were not classified as developing countries under the Montreal Protocol and had to fulfil the same phase-out schedule as industrialized countries, but could not meet those phase-out schedules. Most of these countries, not being classified as developing countries, were not able to use the resources of the Multilateral Fund. This was a continuous concern in the Meetings of the Parties and the Implementation Committee with regard to this non-compliance.

As a part of its mandate at the time of its establishment, the Global Environment Facility assisted these countries with economies in transition. It has cooperated closely with the Implementation Committee by making its support dependent on approval by the Committee. As a consequence, the Facility's operational strategy adopted in October 1995 made ratification of the London Amendment and Adjustments to the Protocol, which demands the phase-out of all major ODSs, a precondition for receiving the Facility's assistance for the implementation of phase-out programmes. The Implementation Committee has also asked all non-compliant countries with economies in transition to commit to clear phase-out schedules to comply with the Protocol and to provide benchmarks for measuring progress in the phase-out process. On that basis, the Meetings of the Parties, upon recommendation by the Implementation Committee, determined a number of related benchmarks for the non-compliant Parties.

Receiving Global Environment Facility assistance has involved going through a project cycle, including the following steps: preparation of a national country programme for the phase-out of ODSs; preparation of the 'GEF project'; project appraisal; and project implementation.

The Global Environment Facility has provided resources with which UNDP, UNEP and the World Bank prepare Facility projects based on the national country programmes, and implement the approved sub-projects contained therein. UNDP and UNEP have jointly provided assistance with Facility funding to eight countries with economies in transition, of which five have already implemented the projects. UNDP has been responsible for investment sub-projects, and UNEP has had the lead on country-programme preparation, institutional strengthening and capacity building, and information and training activities. After approval, the projects go through the respective approval procedure of each implementing agency. The approach of the implementing agencies during implementation also varies; while the UNDP country offices assist UNDP and UNEP, the World Bank acts through financial intermediaries, such as the national banks.

Assistance to countries in transition

Most of the countries with economies in transition receiving support from the Global Environment Facility have faced considerable difficulties in fulfilling their obligations under the Protocol to phase out the ODSs in Annexes A and B. Except for Hungary, Slovakia and Slovenia, which was reclassified as an Article 5 country after 1995, all other countries with economies in transition have at times been in non-compliance with the control measures of the Protocol. By 1997 and 1998, Bulgaria and Poland achieved compliance, and the Czech Republic nearly achieved compliance. Non-compliance continued in the other countries with economies in transition that have implemented Global Environment Facility projects, as well as Estonia, which is likely to implement its programme in future.

BOX 6.2 REFLECTIONS ON THE GEF ROLE IN PROTECTION OF THE OZONE LAYER

Mohamed El-Ashry, CEO and Chairman, Global Environment Facility

The contribution of the Global Environment Facility (GEF) to the protection of the ozone layer is one of its most important successes. As the source of financial assistance for phasing out ozone-depleting substances in Russia and 18 other countries with economies in transition (CEITs), the GEF has enabled these countries in most cases to meet the same stringent timetable required of the industrialized countries – more than a 90 per cent reduction so far. These projects have typically been carried out with strong local support and participation as countries realized that it was in their interest to phase out the use of chemicals banned in the industrialized countries. When, inevitably, technical problems arose, they have been addressed in a spirit of cooperation with an important role for technology transfer.

The history of the GEF is in many ways closely interwoven with that of the protection of the ozone layer, even beyond its participation in funding ozone-protection projects. The international negotiations that led to the GEF pilot phase were concurrent with those related to the creation of the Multilateral Fund, and there are common principles reflected in both agreements. For example, the principle that funds should be provided for the added or 'incremental costs' incurred to protect the global environment was included in both, reflecting the understanding that development should not be penalized by expenses that could not be justified by domestic benefits. A further shared principle was the utilization of existing institutions, primarily the World Bank, UNDP and UNEP, to prepare and implement projects. The importance of scientific and technical advisers is also an important feature in both systems. In both systems, the provision of resources is negotiated by donors on roughly a four-year cycle.

The evolution of the GEF was also shaped in some ways by reactions to the Multilateral Fund. One difference was the governance structure; the Multilateral Fund works on the United Nations formula of one country, one vote, while the GEF has a double majority system that reflects the interests of both donors and recipients. Another distinction that influenced GEF negotiations was the desire to avoid a proliferation of single-purpose funds. Global environmental problems are, we have learned, often difficult to treat in isolation. For example, we have been asked to help replace HFCs, chemicals introduced as replacements for ozone-depleting substances in refrigeration that unfortunately proved to contribute to climate change. The ability to look across global environmental issues is proving to be one of the most valuable features of the GEF.

The GEF role in financing ozone protection in the countries with economies in transition is also indicative of its flexibility to go beyond its role as a financial mechanism for global environmental conventions. This has also become the basis for GEF support of climate-change projects in the countries with economies in transition, which are similarly carried out consistent with but outside Convention mandates.

Looking to the future, I note that the Global Environment Facility has recently been asked to consider supporting financing projects for phasing out HCFCs and methyl bromide. One consequence of our past successes has been to make us more aware of remaining challenges. We hope to learn from the ozone story in meeting these growing needs.

Good results from GEF assistance

According to data reports under Article 7 of the Protocol, total consumption of CFCs in the countries funded by the Global Environment Facility decreased from about 190,000 tonnes in 1986 to 17,600 tonnes by 1999, a drop of more than 90 per cent. Production has been reduced accordingly. Of the four original producers of ODSs, only Russia sustained a considerable production capacity, which was phased out at the end of 2000.

The countries with economies in transition have implemented various supportive and innovative policies and measures, accompanied by institutional strengthening projects and UNEP's regional projects, with assistance from the Global Environment Facility. All the Parties assisted by the Facility have import/export licensing systems, most for ozone-depleting substances only; a few such as Belarus, Russia, Ukraine and Uzbekistan have licensing systems for products containing ODSs. Almost all these Parties have a system of import quotas; many have imposed import taxes. The Czech Republic, Hungary and Poland have banned the use of ODSs in aerosols.

Since 1992, the Global Environment Facility has provided US$140 million to 19 countries with economies in transition. In 1999, it completed its assistance programmes to these countries; it is now up to the countries to fulfil their commitments under the projects which will continue until 2003.

THE ROLE OF NATIONAL GOVERNMENTS

National governments of Parties to the Montreal Protocol took many kinds of measures to promote implementation. Summaries of national regulations are available from UNEP and industry sources.[12] The range of strategies undertaken include:

1 Production and trade
    – customs codes
    – permits for production, import, and export
2 Command and control
    – aerosol and non-essential product bans and use controls
    – sector phase-out schedules, typically with exceptions
    – refrigerant service procedures: venting prohibition, recovery/recycling, service training and certification, sale of ODSs only to certified technicians working in shops
    – recycling
    – manufacturing containment
3 Economic incentives and disincentives in the market
    – ODS taxes and fees, including trading and auctioning schemes
    – government procurement preference for ODS-free products
4 Public awareness and consumer empowerment
    – labelling requirements: official, third party, or self-certification
5 Industry education
    – workshops

- networking
- working groups
6 Alternatives approval
7 Infrastructure and logistics
- equipment standards
- halon banks
8 Research and development
9 Quasi-government activities
- standard associations
- health and medical registration
10 Voluntary government/industry programmes
- company pledges
- industry/government/NGO partnerships and associations

Regulatory innovation

Examples of regulatory innovation in ozone-layer protection include institutional structures, cooperation and advisory committees. Regulations have covered: controls; trade of ODSs; products made with or containing end-use restrictions (aerosol and non-essential); refrigerant service procedures, disposal, and recycling; alternatives approval; and halon banks.

In 1978, the USA became the first country to ban the manufacture and import of most cosmetic and convenience CFC aerosol products. Several other countries enacted similar bans by the end of 1990 (Sweden in 1979, Canada in 1980, Norway in 1981, Taiwan in 1983, Australia in 1989, Austria in 1989, Brazil in 1989, Indonesia in 1990, and Thailand in 1990).

In Thailand, Japanese and Thai manufacturers and the US EPA negotiated the first national phase-out of the production and import of CFCs in refrigerators. Thailand became the first developing country to enact an environmental trade restriction.

Taxes and fees

Many countries[13] have taxes or fees on ODSs that are intended to discourage use and raise revenue. In some cases, the revenue is spent on programmes to encourage ozone-layer protection, while in other cases the revenue is not targeted to environmental activities. The USA has one of the most substantial and therefore influential taxes on ODSs, imported products containing ODSs, and on 'floor stocks' of ODSs held for future use. When the US tax was first imposed, customers experienced dramatic increases in ODS prices, particularly in uses where ODS costs were a substantial portion of a product or service. Price signals encouraged early transition and improved the cost-effectiveness of the transition, since the avoided costs of using ODSs increased when they were taxed.

The Republic of Korea charged an ODS tax and dedicated it to research, education and technical assistance to companies making the transition away from ODSs.

The People's Republic of China State Environmental Protection Administration (SEPA) has used production quotas and bidding mechanisms to meet reduction goals and to distribute Multilateral Fund financing to enterprises in the halon and CFC production sectors. Production quotas are allocated to individual companies annually, with overall quotas reduced each year in accordance with the phase-out plan. Enterprises with a quota are allowed to trade their quotas with other producers or with the government. They can either trade their quotas for a particular year, or make permanent trades. Producers have an opportunity once each year to sell back their production quotas to the government in exchange for funding from the Multilateral Fund.

SEPA uses a competitive bidding process to allocate monies from the Multilateral Fund for a particular year. The bidding process takes place each year between SEPA and enterprises that produce halons or halon-based extinguishers and extinguishing systems. Participation in the bidding process is voluntary. Each participating firm offers a proposal to close or partly close its production line(s) with an 'offer price' in *yuan*/kg of ozone-depletion potential. SEPA then ranks the offer prices in order and selects enterprises with the lowest price offers based on the Multilateral Fund financing available to SEPA for that year. SEPA pays the bid-winning factories their offer prices; in return, those factories close or convert their halon production lines. Because the bidding is competitive, enterprises have incentives to bid close to their true estimated costs. The competitive bidding process may guarantee savings compared to the project-by-project approach.

Singapore enforced a reduction of ODS supplies through a bidding process designed to capture the 'monopoly rent' that chemical suppliers would have charged customers as ODS supplies became more scarce under the national phase-out. ODS customers were granted an annually decreasing portion of their previous use and were required to bid for permits to purchase additional amounts. The bid price was set at a level that kept the total ODS use within the Protocol limits. Revenue from the permits financed research and technology assistance to companies seeking alternatives and substitutes.

Singapore also developed a coordinated taxation strategy with the Ministry of the Environment, and a strategy implementing market-based and voluntary programmes to control supply and demand with the Trade Development Board, the Productivity and Standards Board, and the Economic Development Board. A key element of the two-fold strategy was the CFC tender and allocation system, which set an amount of CFCs that would be available for public tender and an amount that would be distributed based on historical use. In the tender exercise, CFC end-users and ODS distributors submitted bids specifying the quantities of ODS desired and the price they were willing to pay. The Tender Authority evaluated the bids and determined the 'market-clearing price' that would result in all the ODSs being purchased. Quotas were given to the highest bidders, but all winning bidders paid the same price. This strategy resulted in the maximum quantities being allocated to those companies that would pay the highest price, but at a cost that avoided unnecessary economic burden and the unfairness of customers more dependent on ODSs paying far more than other bidders.

The tender and quota allocation system led to a significant rise in CFC pricing, which in turn created strong incentives for recycling, conservation and transition to alternatives. The government used profits from the CFC tender to

BOX 6.3 THE IMPORTANCE OF AUSTRALIAN LEADERSHIP IN THE MONTREAL PROTOCOL

Stephen O Andersen, US EPA, and Helen K Tope, EPA Victoria (Australia)

Protection of the stratospheric ozone layer has been a particular imperative for Australia because it has a sunny climate and outdoor lifestyle with a population vulnerable to skin cancer and cataracts. Ozone-layer depletion has an immediate, direct effect on the health of Australians, and an indirect economic effect from damage to agricultural and natural ecosystems and from loss of fisheries dependent on Antarctic waters. Australians felt themselves more threatened and more vulnerable to ozone-layer depletion and consequently more linked to the environmental impact of global activities than ever before. As one of the world's largest per capita users of chlorofluorocarbons (CFCs) and halons in the late 1980s, Australia also recognized that it had an important role to play in responding to the global challenge of phasing out ozone-depleting substances.

Through the active participation and leadership of the Commonwealth Government, Australia has been a constructive and influential supporter of the Montreal Protocol from its beginning, and often served as a bridge between the concerns of developed and developing countries. Australia hosted critical informal meetings supporting the consensus agreements at Meetings of the Parties to the Montreal Protocol.

In 1988, the Environment Protection Authority – Victoria* (EPA Victoria) started developing an innovative model of strategic policy that addressed each industry sector in detail, specifying dates and milestones for a 95 per cent phase-out of CFCs and halons by 1996. The Victorian policy, released as a draft in 1989, prompted the development of a national policy. The Commonwealth and State Governments formed a task force to develop a comprehensive national policy with each State taking responsibility for work with a particular industry sector to refine phase-out milestones. The focus was to develop 'end-use' controls as a means of phasing out ozone-depleting substances in Australia. The government–industry consultations led to the national, sector-by-sector approach. This 'bottom-up' policy development was well tailored to the practical problems of managing the transition from ozone-depleting substances.

The Australian consultations involved a wide range of stakeholders. For example, the Australian Plumbers and Gasfitters Employees Union inspired a debate on environmental ethics when its members refused to install new halon systems in applications where alternatives were adequate, and agreed to maintain existing halon systems only when emission-reduction steps were taken. The union was particularly influential in gaining recognition that halon systems were not always essential and were add-on fire protection, and was successful in swaying industry views. In Australia, the union action discouraged halon use and stimulated the creation of a process to identify critical halon applications. In the rest of the world, the union action reminded the fire protection industry of the advantages of conceiving and implementing ozone-layer protection itself rather than waiting for others to suggest solutions.

The Australian Conservation Foundation accomplished one of the most highly successful and influential environmental NGO interventions with a delegation of Australian school students at the London Meeting of the Parties to the Montreal Protocol in 1992. The participation of the young people at a diplomatic negotiation had at least three important impacts: it reminded delegates that the Montreal Protocol is largely for the benefit of future generations; it captured the imagination of the press who found the Australian students technically prepared, highly concerned and optimistic; and it helped break the deadlock as meeting time was running out and disagreements of the proposed accelerated phase-out were highlighted in brackets. An impassioned Australian student was allowed to intervene, stressing that 'our fate lies in your square brackets' and pleading for the ultimately agreed 1996 phase-out.

The Australian phase-out strategy was remarkably successful: the ozone layer is protected; industry accomplished the transformation with minimal cost and disruption; customers of products made with or containing ozone-depleting substances were supplied with substitutes and alternatives at similar price and equal or higher performance; and government accomplished its objectives with minimal taxpayer expense or bureaucracy. The phase-out was celebrated in December 1995 at the seaside resort of Queenscliff, Victoria, at a hotel appropriately called the Ozone Hotel!

* EPA Victoria is a statutory authority, established under an Act of the Victorian State Parliament in response to community concern about pollution, which reports to the Victorian Parliament through the Minister for Environment and Conservation.

develop a new CFC-113 recycling system and to provide consulting services to industry for reducing and eliminating CFC use. The Singapore Products and Standards Board audited manufacturing processes, and certified products and processes as CFC-free. This certification provided a competitive advantage to companies that phased out their ODS use, particularly to avoid the US law requiring labelling of products made with or containing ODSs (see below). The Singapore Economic Development Board also offered tax and financial incentives to help small and medium-sized enterprises. The Singapore system achieved two highly desirable outcomes: equitable distribution of the limited quantity of available CFCs, and a strong market signal to induce ODS users to look into substitutes, conservation measures and recycling.

Product labelling

Some countries require products containing ODSs to be specially labelled; some also require products made with ODSs to be labelled.[14] Product environmental labelling educates consumers about the extent of ODSs in products and the environmental consequences of ozone depletion, empowers consumers to avoid and/or boycott ODS products, and encourages product manufacturers to halt ODS use to satisfy customers and avoid administrative expenses and penalties.

The US label says: 'WARNING! Contains (Manufactured with)..., a substance which harms public health and environment by destroying ozone in the upper atmosphere.' Marketing experts in the USA predicted that consumers would reject labelled products, particularly electronics products where brand competition is fierce, and toys where environment and safety are a priority.

In response to the predicted market and administrative consequences of the US labelling law, the Industry Cooperative for Ozone Layer Protection (ICOLP) undertook a global campaign to end ODS solvent use by its suppliers. The dozen multinational companies in the Cooperative from Canada, Japan, the USA and the UK wrote a joint letter to their thousands of suppliers world-wide informing them that they would purchase only components that were certified as ODS-free and not manufactured with ODSs.

BOX 6.4 THE METHYL BROMIDE PHASE-OUT IN SOIL FUMIGATION IN THE NETHERLANDS, 1979–1992

Joop van Haasteren, Ministry of Housing, Spatial Planning and the Environment of The Netherlands (retired 2001)

This is the story of how The Netherlands accomplished its own early phase-out of methyl bromide and influenced the phase-out decision under the Montreal Protocol. The experience of The Netherlands was a prelude to the experience in other countries dependent on methyl bromide and the negotiations under the Montreal Protocol. Health and atmospheric scientists would develop new evidence that methyl bromide was environmentally unacceptable, governments would announce strong action to protect the public, the methyl bromide industry would intervene politically, government would temporarily relax the regulation, scientists would confirm unacceptable risks, and new regulations would be proposed – over and over again. Fortunately, the drive for regulation gained strength from each new study, and pro-active growers and their suppliers developed technology to replace methyl bromide.

In 1978, methyl bromide came under simultaneous environmental, public health and agricultural scrutiny in The Netherlands. A study commissioned to investigate the risks of storage, transport and use of methyl bromide, and to advise on possible risk reduction, came to the alarming conclusion that human exposure to methyl bromide exceeded the Population Exposure Limit in communities near treated fields. Furthermore, victims would have no warning because the neurotoxic methyl bromide is odourless even in lethal concentrations.

In August 1979, the Environment Ministry proposed that methyl bromide fumigation not be permitted after August 1980 due to the combined risks to human health, food safety and soil and water pollution. After loud complaints from the agricultural industry, it was decided to establish an inter-ministerial working group to draft a final recommendation about further policy concerning the substance. This process took more than a decade, ultimately leading to a complete phase-out of methyl bromide as a soil fumigant in 1992 – coincidentally, also the year when the Scientific Assessment Panel identified methyl bromide as a significant ozone-depleting substance.

Looking back, it is a tribute to the technical innovation of growers in The Netherlands that the phase-out was so successful. Non-governmental organizations like the Stichting Natuur en Milieu (Foundation for Nature and Environment) and, to a lesser extent, the agricultural labour union (Voedingsbond FNV) were instrumental in generating and sustaining societal pressure to reduce the use of methyl bromide despite the influences of methyl bromide producers and agricultural organizations. The new substrate techniques have proved to be a very good alternative for the farmers, both technically and economically. Crop yields per hectare have increased considerably and the quality of the products has improved. The increase in farm income was significant.

In June 1992, I presented the case study of The Netherlands phase-out at the meeting of the Joint Scientific Assessment and Technology and Economic Assessment Panel, contending that what was done in The Netherlands could be accomplished worldwide.

One month later, in July 1992, the Assessment Panels presented their results to the Open-Ended Working Group. The technical conclusions were loudly disputed from the floor by executives from the Dead Sea Bromine Company who had been certified as members of the delegation from Israel. Dead Sea Bromine publicly questioned whether The Netherlands had accomplished what it claimed. Stephen Andersen of the US EPA answered their questions about the technical findings and asked The Netherlands to respond regarding our experience. Imagine how angry we were to have the integrity of our agricultural industry and Environment Ministry questioned. We reiterated that the technology was available and affordable and announced that workshops would be organized immediately to disclose in full all technical and economic details.

In October 1992, well before the Meeting of the Parties in Copenhagen, a travelling workshop was held in Rotterdam, The Netherlands, and Rome, Italy, to introduce the Parties of the Montreal Protocol to the successful alternatives to methyl bromide. Participants heard directly from growers and examined the technology in detail. A consensus report was rapidly produced, published and distributed to Parties. The report completely confirmed that The Netherlands phase-out was accomplished to the satisfaction of growers and that the costs were affordable.

The Netherlands is very proud of the leadership by its Environment Ministry, and the technical and economic prowess of its agricultural growers, but we are most proud that we stood up and prevailed against methyl bromide defenders. Small countries sometimes have great influence.

The companies in the Cooperative also offered technical cooperation and even financing to help suppliers in developing countries to make the transition away from ODSs. For example, Nortel financed an ODS solvent phase-out expert to assist Mexican companies; Motorola managers coordinated Cooperative experts through the Motorola manufacturing centre in Kuala Lumpur, Malaysia; Nortel and Seiko Epson offered factory tours and training in China; and the Cooperative published handbooks with the US EPA.

The Cooperative campaign was so successful that the US EPA took the unprecedented step of exempting electronics from the labelling requirement. In effect, companies devoted time and money to the phase-out, rather than devoting time and money toward paperwork to comply with the labelling regulations. Thus, the US national labelling law accelerated the ODS solvent phase-out worldwide.

THE ROLE OF GOVERNMENT AGENCIES AS CUSTOMERS AND MARKET LEADERS

Government agencies and military organizations were large consumers of ODS products in electronics, refrigeration, air conditioning, vehicles, furniture, and fire protection – and often influenced product design and manufacturing techniques. When these government organizations implemented their own phase-out programmes, they transformed markets – making it less expensive and more reliable for other enterprises to follow. Once markets were transformed away from ODSs, the phase-out became almost automatic.

For example, it was determined that the US military endorsement of the technical performance of CFC-113 solvent resulted in 50 per cent of its global use. Some companies used CFC-113 to qualify products for military markets and used the same cleaning process on products for civilian markets. Other companies used CFCs because the military had determined they were technically superior. The US Department of Defense worked with the EPA to certify

BOX 6.5 JAPAN'S ACTION TO PHASE OUT OZONE-DEPLETING SUBSTANCES

*Tetsuo Nishide and Naoki Kojima, former heads of the Ozone Protection Office, Ministry of International Trade and Industry, Japan**

Implementation of the Montreal Protocol in Japan was a very difficult task because CFCs and other controlled substances were key materials in almost every industrial and consumer use. Japan established the Ozone Layer Protection Law in 1988, which was repeatedly adjusted and revised according to the adjustment and revision of the Protocol. Although this law set strict limits on production and import of controlled substances, the government recognized that no law could achieve the goal of the phase-out unless businesses and citizens found affordable ways to make real reductions in use and emissions. Users would not halt the use of CFCs and other ODSs until they felt confident about the quality of their products manufactured using alternative substances or technologies. It often took the users of controlled substances a long time to test and verify the quality of alternatives.

In order to achieve the phase-out, the Japanese government took three coordinated actions that complemented and strengthened the Ozone Layer Protection Law. The first was to help to establish a new organization for better communication and cooperation among the related parties and also between the government and the private sector. The Japan Industrial Conference for Ozone Layer Protection (JICOP) was established in 1989 as the central vehicle for this purpose. The number of JICOP members reached 65, including all of the major industrial associations: Japan Electrical Manufacturers Association (JEMA), Electronics Industry Association of Japan (EIAJ), Japan Electronic Industry Development Association (JEIDA), Japan Refrigeration and Air-conditioning Industry Association (JRAIA), Japan Automobile Manufacturers' Association (JAMA), Japan Clock and Watch Association (JCWA) and Japan Fluorocarbon Manufacturers Association (JFMA). These association members represented all the major industries that needed to take action for ozone-layer protection. The government also secured the support of government corporations, such as the Small and Medium Enterprise Corporation, to organize seminars and to provide individual consultation using a registry of Japanese technology experts. NGOs like JICOP and JEMA worked very hard and functioned excellently. The Japan Industrial Conference on Cleaning (JICC) was also established in 1994 by companies using ODSs as solvents and cleaning agents.

The second Japanese action was an annual conference at which the Minister of International Trade and Industry directly persuaded the presidents or chief representatives of various industries to accelerate the necessary actions to phase out ODSs. The government, with strong support from JICOP and other industrial associations, also designated each September as 'ozone layer protection month', with many seminars presented all over Japan. The industrial associations themselves made brochures and newsletters to circulate information to their members and their customers. The government asked the local governments to support the implementation of the policies in each locality.

The third action was policy measures to encourage the phase-out. The government supported the development of new technologies, and offered tax incentives and low-interest loans from the government financing organizations. JICOP organized a special working group and developed technology manuals with detailed information on technologies and alternative products. These manuals were distributed at seminars for the ODS users and were also used as textbooks for technology-transfer seminars and training courses in developing countries. Over 8000 copies of the Phase-out Manual of 1,1,1-trichloroethane were sold to small manufacturers seeking substitutes and alternatives.

*Naoki Kojima headed the Ozone Protection Office in the Ministry of International Trade and Industry from 1989 to 1991, Yoshihiko Sumi from 1991 to 1992, and Tetsuo Nishide from 1992 to 1994.

Box 6.6 China's experience: The mending of the sky by present-day Nu Wa

Qu Geping, former Minister of Environment of China, present Chairman of the Environmental Protection and Resources Conservation Committee of the National People's Congress of China*

According to ancient Chinese legend, a big hole once appeared in the sky, causing great calamities all over the land and depriving the people of their means of survival. A goddess of that time named Nu Wa stepped forward bravely, determined to save the masses from this grave danger. She toiled day and night with all her energy, and in the end succeeded in mending the hole in the sky with five coloured stones. The catastrophes were thus averted, and the people of the land returned to their peaceful life. This mythological story of 'Nu Wa mending the sky' has long been passed on from generation to generation in China.

As a large developing country, with the largest population in the world, China has attached great importance to the issue of ozone-layer protection. At the conference held in London in March 1989, the Chinese delegation stated clearly its position in support of the purposes and principles of the Vienna Convention and the Montreal Protocol. However, it also pointed out that the Protocol did not adequately reflect the principle of equality that 'those who release more should reduce more', and stressed that the developed countries should shoulder the major responsibility for the destruction of the ozone layer and provide the developing countries with financial resources to assist them in their efforts to convert to alternative products and technologies. Thereafter, at the Helsinki Conference held in April of the same year, the Chinese delegation put forward its proposal on the establishment of an international fund for the protection of the ozone layer, which subsequently received wide support.

At the invitation of Mostafa Tolba, I attended the Conference of the Environment Ministers convened by UNEP in Geneva in July 1989. That conference played a very important part in modifying the provisions of the Protocol and in promoting the participation of developing countries in the work under the Protocol.

The State Environmental Protection Agency (SEPA) in China has accomplished the following five tasks for the protection of the ozone layer:

1   A report was compiled detailing the findings of a study on the response to and control of the substances that deplete the ozone layer in China.
2   To facilitate the signing of the Protocol, SEPA proposed to the State Council that the country should join the Convention first and then seek to modify the provisions of the Protocol. This proposal was subsequently accepted by the State Council.
3   SEPA has sought to have the Protocol modified based on facts and reality. The original version of the Protocol had defects with respect to the following issues: differentiation of responsibilities for the destruction of the ozone layer; establishment of a financial mechanism for protection of the ozone layer; source and method of financing; facilitation of scientific and technological exchanges; technology transfer; and guarantee of equal status for all countries, especially developing countries, in international cooperation in these areas. At the London Conference held in June 1990, the participants, after incorporating various changes, finalized the Protocol, and thereby put in place the fund mechanism and the mechanism for transfer of technology. The Chinese delegation at the Nairobi Conference formally announced China's official decision to join the revised Protocol.

4   In July 1991, China's Leading Group for the Protection of the Ozone Layer was established, and assumed the tasks of formulating programmes and follow-up actions. SEPA was designated the Group's leader. The Group had 12 members, including such establishments as the Foreign Ministry, the State Planning Commission, the State Science and Technology Commission, and the Ministry of Finance. The State Council approved the country programme in August 1992.

5   Both the Convention and the Protocol are signed for implementation by sovereign states. Taiwan is an integral and indivisible part of Chinese territory. It has been agreed that mainland China is responsible for collecting and reporting various relevant data in Taiwan.

In November and December 1999, the Meeting of the Executive Committee of the Multilateral Fund, the fifth meeting of the Conference of the Parties to the Vienna Convention, and the eleventh Meeting of the Parties to the Montreal Protocol were held back to back in Beijing. The convening of these meetings marks the significant progress that has been achieved in the joint efforts of the international community to protect the ozone layer, and signifies that its work in this area is entering a new stage.

Developing countries, including China, committed themselves to freezing, by 1 July 1999, their production and consumption of CFCs at 1995–1997 average levels. In fact, China attained this target ahead of schedule and is now endeavouring to meet the final targets of 50 per cent reduction by 2005 and 100 per cent phase-out by 2010. With financial support from the Multilateral Fund, China has phased out more than 50,000 tonnes of ozone-depleting substances. The State Council approved, in November 1999, a revised country programme, thereby making the sectoral phasing-out mechanism the most important method of CFC phase-out in our country. Where economically and technologically feasible, some of the ozone-depleting substances used in sectors such as fire-fighting, automotive air conditioning and cleaning could be phased out earlier than scheduled.

As the old Chinese proverb goes, 'where there's a will, there's a way'. The initiative pioneered by UNEP to protect the ozone layer, once it was debated among all parties concerned on an equal footing and institutionalized through rational mechanisms, eventually succeeded in prompting countries to take meaningful action and achieved tremendous progress in the field of ozone-layer protection. Humankind collectively shares the sky above. The protection of the environment of our planet requires the joint efforts of all countries. In this regard, the successful implementation of the Vienna Convention for the Protection of the Ozone Layer and the Montreal Protocol on Substances that Deplete the Ozone Layer has provided us with an excellent example.

* The author of this article is the recipient of both the 1992 UNEP Sasakawa Prize and Japan's Blue Planet Prize for 1999.

alternatives and then prohibited the use of CFCs. This and other industry actions accelerated the commercialization and implementation of ozone-safe solvents.

The Norwegian Navy developed and demonstrated halon-free fire protection for ship engine rooms on combat vessels. This technology was rapidly accepted for civilian applications because combat vessels face more demanding fire threats.

BOX 6.7 A FRENCH PERSPECTIVE ON NEGOTIATIONS FROM 1991

Lawrence Musset, Delegate of France to the Ozone Negotiations

An initial regulation allowed the implementation of the Montreal Protocol within the European Communities from 14 October 1988. This regulation was then revised in 1991, 1992, 1994 and 2000. It often required stricter controls than the actual Protocol. As one of the main producing countries, France has always endeavoured to take strict measures for the protection of the ozone layer while allowing Elf Atochem, one of the main producers of chlorofluorocarbons (CFCs) and halons in the world, and the Solvay Company the time required to develop substitutes. It has been essential to ensure that producers do not fall behind their competitors and develop more promising substitutes.

Negotiations have, in fact, always been more difficult with the users of the substances, notably in the area of refrigeration and air conditioning. The interests of producers and users have been in conflict. It has been in the interest of the producers to eliminate CFCs and replace them with substitutes while the users of these substances have seen only additional costs resulting from their elimination. These conflicts of interest have resulted in numerous inter-ministerial meetings and necessitated the arbitration of the Prime Minister's office on several occasions.

Moreover, a number of industrialists have argued against enforcement of control measures on the basis of the heated scientific debate in France on the destruction of the ozone layer.

Two major national events are noteworthy by virtue of their early timing. The first of these events was without doubt the signing of several voluntary agreements on 7 February 1989. Five conventions were signed by public authorities, Elf Atochem and 26 professional organizations representing the users of CFCs and halons. The signatories of those conventions were the ministers of environment, industry, the interior and agriculture. The convention signed with the producers of aerosols allowed them to affix on their aerosol canisters not containing CFCs the logotype 'protects the ozone layer' representing a hand above the Earth.

The conventions of 1989 quickly became obsolete given the acceleration of the statutory schedules for reduction of the production and consumption of the substances. They were very useful, however, for sensitizing the public and professional organizations and preparing the latter for the subsequent hardening of the measures.

The second major event was the adoption in December 1992 of a decree making it obligatory for enterprises registered in the prefecture, to recover refrigeration fluids including CFCs, hydrochlorofluorocarbons (HCFCs) and hydrofluorocarbons (HFCs).

There were other events worthy of mention, such as the creation in 1988 of the French Halons-Environment Technical Committee, and the signature in 1993 of the Convention on Refrigeration Fluids, and in 1994, of the Recycling of Halons Charter.

Some initiatives failed, however, no doubt because they were premature. For instance, in 1990, the minister responsible for environmental matters proposed to gradually prohibit the use of CFCs in new products as substitutes became available, but was unable to obtain agreement on this proposal. At that time also, France was not able to convince the European Community to adopt a harmonized system of assistance in the recovery, recycling and destruction of CFCs based on their taxation.

Contrary to conventional thinking, the case of the protection of the ozone layer is not yet closed, even in the developed countries. It is extremely difficult to achieve total elimination of a substance with multiple uses, particularly when these uses affect sectors as vital as the refrigeration of food and the treatment of asthma. It is therefore necessary to process the many requests for special dispensation.

BOX 6.8 LESSONS FROM THE PHASE-OUT OF ODSs IN THE REFRIGERATION SECTOR IN TURKEY

Fatma Can, Environment Ministry

In 1997 and 1998, Turkey faced considerable difficulties in implementing its ozone-depleting substance (ODS) phase-out strategy in the refrigeration sector. These were a result of long delays in the approval of its submissions to the Multilateral Fund.

Turkey ratified the Montreal Protocol in 1991. Its industry is relatively advanced compared to other Article 5 countries. Because of this, the strategy adopted was to achieve virtual elimination of ODS use in manufacture by 1 January 2000, with the bulk of use to be eliminated by 2005.

Turkey prepared its country programme for this phase-out, and applied for financial support and technical assistance from the Multilateral Fund to achieve this challenging goal. Among its key components was a ban on the use of ODSs in the manufacturing of refrigeration equipment by 1 January 2000. Only use in the servicing of existing equipment was to be permitted from that date.

Following the successful phase-out of ODSs among the larger refrigeration producers in 1996 and 1997, Turkey turned its attention to small and medium-sized enterprises with a programme of training workers in the servicing of new non-ODS-using equipment (HFC-134a) and on recycling and recovery of ODSs in the refrigeration sector. To this end, preparation was started on a new set of projects in spring of 1997.

A survey undertaken in 1998–1999 identified around 325 small and medium-sized enterprises, which were grouped according to specific types of commercial refrigeration activities; three umbrella projects were prepared.

The subsectors of commercial refrigeration and stationary air conditioning are facing a general lack of appropriate training, the uncontrolled release of ODS emissions during servicing, and slow retrofits due to the lack of knowledge by service workers. The service companies that could potentially undertake such actions are not equipped either to recover the CFCs or to evacuate and charge HFC refrigerants. Attempts made have resulted in both venting of CFCs and improper functioning retrofitted installations, which also have adverse environmental effects, since energy efficiency is often compromised. In addition to the negative environmental impact, the lack of HFC servicing equipment has an adverse social effect, since less reliable retrofitted CFC installations are used to conserve food and provide other services.

A properly designed recovery/recycling/reclaim/disposal scheme would strongly support the Turkish efforts in successful implementation of the National Ozone Policy. When fully implemented, it could provide sufficient quantities of recycled/reclaimed CFCs to cover the service need. Further, the scheme would enable Turkey to reduce future emissions of CFC alternatives, which contribute to global warming.

THE ROLE OF INDUSTRY AND INDUSTRY NON-GOVERNMENTAL ORGANIZATIONS

Implementation of the Montreal Protocol required unprecedented efforts from industry both nationally and internationally. Each new technology required development; testing for performance, safety, and environmental acceptability; commercialization; implementation; and upgrading. Technologies sometimes evolved so quickly that companies eliminated CFCs and then promptly invested in the next generation of technology.

Private companies and industry NGOs were often faster to act on phasing out their ODS use than governments; they sometimes utilized strategies that were instrumental in accelerating protection of the ozone layer, but were inappropriate for governments.[15] One strategy was to advertise against products that depleted the ozone layer in order to persuade customers to choose ozone-safe products. For example, once they switched to CFC-free aerosol products or not-in-kind alternatives, S C Johnson and Gillette advertised aggressively against CFC aerosol products. Another strategy was to gain public support for ozone-safe products with safety hazards greater than ODS products. For example, Greenpeace and its business partners promoted Greenfreeze™ technology utilizing hydrocarbon refrigerants and hydrocarbon foam.

Some phase-out issues were inherently crosscutting. Because a wide variety of applications evolved to take advantage of the physical and chemical properties of CFCs, it was understandable that hydrochlorofluorocarbons (HCFCs) and hydrofluorocarbons (HFCs) designed to perform like CFCs would be technically suitable for many of the applications they were needed for. Furthermore, the toxicology of the replacement chemicals would be identical, regardless of the chemical process used in their manufacture. This situation led to global programmes for cooperative testing of the toxicology and environmental acceptability of new substances.

Cooperation instead of competition

Eleven chemical producers (Allied Signal, DuPont and LaRoche of the USA, Asahi Glass and Daikin of Japan, Ausimont of Italy, Elf Atochem of France, Hoechst of Germany, ICI and Rhone-Poulenc of the UK, and Solvay of Belgium) jointly funded a research and assessment programme to determine the potential environmental effects of alternative fluorocarbons and their degradation products. This Alternative Fluorocarbon Environmental Acceptability Study (AFEAS) was initiated in 1988, and its results were incorporated in the 1989 TEAP report and in subsequent assessments. The research was conducted at universities, private research organizations, and government laboratories, and results were rapidly disseminated to environmental authorities, public interest groups, chemical manufacturers, and chemical customers. The Programme for Alternative Fluorocarbon Toxicity Testing (PAFT) was a cooperative effort of 16 of the leading CFC producers formed in December 1987 to expedite the development of toxicology data for alternatives. The chemicals tested include HFC-32, HFC-125, HFC-134a, HCFC-123, HCFC-124, HCFC-141b, and HCFC-225(ca and cb). The results were publicly communicated as soon as available and incorporated in the TEAP's reports.

The selection of lubricants for HFC-134a was another example of cross-sector efforts. Fluorocarbon manufacturers had some metallurgy and materials compatibility expertise, but lacked experience in lubrication and friction science. In the 1980s, it had been decades since a new refrigerant had been commercialized. Automobile manufacturers seeking a global solution organized a project to jointly develop technology for vehicle air conditioning, while refrigerator manufacturers worked with universities, associations, and professional organizations. Frequent joint meetings, including an annual technology conference in Washington, DC, allowed vehicle and refrigerator experts to benefit from each other's work.

Likewise, when the electronics industry developed the technology to replace halon in the protection of electronics equipment, halon was banked for redeployment to critical uses such as aircraft fire protection.

REGULATIONS FORCE NEW TECHNOLOGIES

National implementation also often stimulated the commercialization of important new technology that was globally applied. Some regulators call this process 'regulatory forcing'. However, under the spirit of cooperation of the Montreal Protocol, industry was working much faster than regulatory requirements. Thus, the process might more accurately be called industry leadership or 'regulator forcing', in which industry advocated faster timetables and stricter controls than government. The following are a few examples of new technologies that came about because of regulations in these countries:

- An Australian government agency – the Department of Administrative Services Centre for Environmental Management (DASCEM) – developed halon recovery, purification, and destruction equipment, and later transferred it to private ownership. The Association of Fluorocarbon Customers and Manufacturers (AFCAM) created a national recycling system and established refrigeration training and certification.
- Brazilian engineers at the Arbor Works of the Ford Motor Company developed and implemented the first low-solids flux soldering that could be either cleaned without ODSs or not cleaned at all.
- Denmark conceived and commercialized fire protection using inert gases. In this process, argon and other non-flammable gases are injected into protected spaces, reducing the oxygen to concentrations until the fire is extinguished. Simultaneously, carbon dioxide is released to stimulate rapid breathing and allow humans to survive the reduced oxygen levels.
- In France, Atochem introduced solvent recovery and containment services that allowed customers to achieve emission reduction goals while seeking technology to eliminate CFC-113 and methyl chloroform solvent use.
- Lufthansa engineers in Germany developed ODS-free aircraft maintenance procedures and joined with other innovative airlines and aircraft builders to share and exchange technology. This cooperation avoided duplication and encouraged choice of the most environmentally and financially acceptable alternatives.
- India's chemical manufacturing association, the Refrigerant Gas Manufacturers Association (REGMA), allocated the chemical manufacturing quota among companies, avoiding time-consuming controversy and starting cooperation on commercialization of chemical substitutes.
- Asahi Glass in Japan invented and commercialized the very-low-ODP HCFC-225 solvents that served as a technically perfect substitute for uses

of CFC-113 when other alternatives were not available. Seiko Epson invented centralized ODS-free cleaning facilities to allow small- and medium-sized enterprises to phase out ODSs at an affordable cost. Minebea invented aqueous cleaning for miniature ball bearings. Nissan was the first global automobile manufacturer to pledge a complete CFC phase-out, motivating suppliers to invest in new CFC-free technologies.

- The national association of aerosol product manufacturers in Mexico invented open-air factories that allowed the safe use of hydrocarbon propellants without the expense of explosion proofing and mechanically ventilating enclosed buildings. The association gained the agreement of all companies to halt CFC use; it also joined government advertising campaigns to gain customer acceptance of new products that had more active ingredients but weighed less, because hydrocarbons were much lighter than the CFCs they replaced.
- Saab Scandia of Sweden developed ODS-free aircraft maintenance procedures and shared its technologies with Thai Airways.
- In the USA, EPA and its partners AT&T, Ford, IBM, Motorola, and Nortel developed no-clean soldering. With partners from global automobile manufacturers and their suppliers, the US EPA, the Mobile Air Conditioning Society, the Society of Automotive Engineers, and Underwriter Laboratories developed CFC recycling.
- The UK supermarket Sainsbury's implemented leak reduction and system monitoring for existing refrigeration systems, and accelerated the replacement of CFC systems with HCFC alternatives.
- Venezuela's largest aerosol product company, Spray Quimica, halted CFC use and pressured European cosmetic companies to halt sales of CFC aerosol products to developing countries. As a result of this industry leadership, Venezuela and Mexico halted sales of CFC aerosol products sooner than most companies operating in European countries.
- Vietnam worked with various companies, the Industry Cooperative for Ozone Layer Protection, and US EPA to promote a 'leadership pledge' by multinational companies not to increase the dependence of Vietnam on ozone-depleting substances. This action speeded up ozone layer protection and avoided expansion of ODS use that would have later required financing by the Multilateral Fund.

On the occasion of the tenth anniversary of the Montreal Protocol, the US EPA presented 71 awards to the individuals, companies, and organizations that had made the most significant contributions to stratospheric ozone layer protection. These winners, judged by previous winners of the Ozone Protection Awards, were called 'Best-of-the-Best'. The corporate and military award winners were: Asahi Glass, Coca-Cola, DuPont, Hitachi, IBM, ICI, Lockheed Martin, Lufthansa, Minebea, Mitsubishi Electric, Nissan, Nortel, Raytheon TI, Seiko Epson, Thiokol/NASA, 3M, and the US Department of Defense, including Air Force, Army, and Navy organizations.

Table 6.2 *Examples of bilateral cooperation on ozone protection*

Bilateral co-operating country	ODS sector/activity	Article 5 country/region
Australia	Halon Banking, Methyl Bromide, Recovery/Recycling, Refrigerant Management Plans, Technical Assistance, Training	Argentina, India, Indonesia, Kenya, Vietnam, African Region (1), Asian Region (3)
Austria	Training	Romania
Belgium	Foam	Bolivia
Canada	Halon Banking, Methyl Bromide, Recovery/Recycling, Refrigerant Management Plans, Solvents, Technical Assistance, Training	Antigua/Barbuda, Belize, Benin, Brazil, Burkina Faso, Chile, China, Cuba, Georgia, Guyana, India, Jamaica, Kenya, Moldova, Uruguay, Global (1), Latin American Region (1)
Denmark	Recovery/Recycling	China
Finland	Refrigerant Management Plans, Training	Namibia, Nicaragua, Panama, Global (1)
France	Chillers, Country Programmes, Foams, Halon Banking, Methyl Bromide, Mobile Air Conditioners, Ozone Unit, Refrigerant Management Plans, Recovery/Recycling, Refrigeration, Solvents, Technical Assistance, Training	Central African Republic, China, Costa Rica, Cote d'Ivoire, Ghana, Iran, Jordan, Lao PDR, Lebanon, Madagascar, Malaysia, Mali, Mauritania, Morocco, Senegal, Syria, Thailand, Vietnam, African Region (2), Asian Region (2)
Germany	Aerosol, Foams, Country Programmes, Halon Banking, Methyl Bromide, Recovery/Recycling, Refrigerant Management Plans, Refrigeration, Technical Assistance, Training	Algeria, Angola, Botswana, Brazil, China, Egypt, Ethiopia, Gambia, India, Iran, Jordan, Kenya, Lebanon, Lesotho, Malawi, Mauritius, Morocco, Mozambique, Namibia, Nigeria, Oman, Philippines, Seychelles, Swaziland, Syria, Tanzania, Uganda, Yemen, Zambia, Zimbabwe, Global (4), Asian Region (2), European Region (1)
Japan	Country Programme Update, Foams, Refrigeration, Solvents, Training	China, Nigeria, Sri Lanka, Thailand, Asian Region (2)
Italy	Foam, Methyl Bromide, Refrigeration	India, Romania, Yugoslavia
Poland	Training	Vietnam
Singapore	Training	Asian Region (2)
South Africa	Training	African Region (1)
Sweden	Halon Banking, Networks, Refrigerant Management Plans, Technical Assistance, Training	Philippines, Thailand, Asian Region (2)
Switzerland	Refrigerant Management Plans, Technical Assistance, Training	Argentina, India, Senegal, Global (1)
UK	Training	Mexico
USA	Country Programmes, Halon Recovery/Recycling, Mobile Air Conditioner Recovery/Recycling, Other Recovery/Recycling, Refrigeration, Technical Assistance, Training	Argentina, Chile, China, Colombia, Costa Rica, Dominican Republic, Ecuador, Guatemala, Jamaica, Malaysia, Mexico, Panama, Philippines, Thailand, Trinidad/Tobago, Turkey, Uruguay, Venezuela, Vietnam, Global (1)

REGIONAL AND BILATERAL COOPERATION

The Regional Conferences of Ministers took great interest in the implementation of the Protocol and discussed the issue regularly. The African Ministerial Conference on the Environment, at its meeting in Dakar in October 2000, heard presentations on the ozone issue from UNEP; ministers expressed strong appreciation for the financial and technical assistance being extended to them from the Multilateral Fund. They recommended that the experience of the Protocol be transferred to other environmental agreements. The inter-sessional Committee of the Forum of Ministers of the Environment for Latin America and the Caribbean Countries, at its meeting of October 2000 in Mexico, also heard a presentation and discussed ozone issues. The Executive Bureau of the Council of the Arab Ministers Responsible for the Environment in March 2001 urged all Arab countries to effectively implement the Protocol.

Many bilateral aid agencies developed programmes to assist implementation in developing countries; several industrialized countries have opted to use part of their contribution to the Multilateral Fund for bilateral assistance. These programmes were credited as a part of their contribution to the Multilateral Fund. Sweden and Finland gave funds over and above their contribution to the Fund.

THE ROLE OF CONFERENCES AND WORKSHOPS

The character and importance of conferences and workshops changed dramatically once the Montreal Protocol was signed. Businesses dependent on ODSs became more interested in technical options, and suppliers of new technology aggressively pursued the new customers. The premier conference, 'Substitutes and Alternatives to CFCs and Halons' was held on 13–15 January, 1988, organized by the US EPA and co-sponsored by Environment Canada and the Conservation Foundation, which was then headed by William K Reilly who was later appointed Administrator of the US EPA. This conference became an annual tradition, most recently expanding its focus to include climate protection, in addition to ozone-layer protection. It is the platform for political and technology announcements, and the organizing site for partnerships and other collaboration. It was informally integrated into the Montreal Protocol network through co-sponsorship and international agenda planning teams, with TEAP and technical options committee members often setting the agenda and presenting summaries of the most appealing new technology.

THE ROLE OF ENVIRONMENTAL NGOS

Environmental non-governmental organizations (NGOs) typically concentrate on public awareness and political action. However, in the case of ozone-layer protection, environmental NGOs also undertook innovative partnerships and technology promotion (see Chapter 9). Environmental NGOs helped implement the Montreal Protocol in a variety of ways that include these illustrations:

Box 6.9 Japanese leadership in technology cooperation

Yuichi Fujimoto, Japan Industrial Conference for Ozone Layer Protection, Naoki Kojima, Petroleum Energy Center, Tetsuo Nishide, Japan Ministry of the Environment, and Tsutomu Odagiri, Japan Industrial Conference on Cleaning

Japan created a public/private technology cooperation partnership that (1) joined with US EPA in technical seminars and leadership initiatives in Thailand and Vietnam; (2) sponsored extensive annual training programmes for experts from developing countries; and (3) is now funding projects in China under terms of bilateral assistance.

Japan, cooperating with US EPA, has held technical seminars in Southeast Asia annually since 1992[16] and organized technical seminars in Hong Kong, Nanjing, Shanghai, and Shenzhen, China.[17] Speakers participated voluntarily from big enterprises in Japan, North America, and Europe, including many members of the Solvents Technical Options Committee. The governments of the venue countries provided seminar halls, and recruited technical and corporate participants. Experts visited local manufacturers to give recommendations to promote ODS phase-out. Speakers from the Multilateral Fund and implementing agencies explained how companies could receive financing for the incremental costs of phase-out. Follow-up meetings were held to promote and complete phase-out of ODSs.

The Thailand leadership initiative was organized after a 1991 analysis concluded that Japanese companies were responsible for up to 50 per cent of ODSs consumed in Thailand, and that North American and European companies were responsible for at least 25 per cent. In response to this alarming situation, Japanese and US government and industry partners created the 'Thailand Pledge' by multinational companies to eliminate ODSs within one year of their corporate or national schedule (whichever was sooner).

At a 1992 meeting in Bangkok, Japan promised to phase out ODSs from household refrigerators manufactured in Thailand by the end of 1996. Japanese companies proceeding with the conversion to alternatives suffered a setback when new HFC compressors supplied from a local Thai company failed durability tests essential for total quality assurance of refrigerators. The reliability problem was caused by contamination of the lubricant and unacceptable wear of rolling parts. The Thai manufacturer of the faulty compressors and the US company licensing the compressor design asked the Government of Thailand to delay the CFC phase-out until they could solve the design problem. In a show of strong environmental resolve, the Government of Thailand rejected the request and authorized Japanese companies to import compressors if necessary to meet the December 1996 goal. In order to meet the schedule but at the same time protect the jobs of Thai workers, Japanese manufacturers, Hitachi, Matsushita, Mitsubishi, Toshiba and the Japan Electrical Manufacturers Association (JEMA), decided to provide the local compressor supplier with engineering and technology necessary to improve the quality of the compressors.

Japanese manufacturers in Thailand succeeded in the phase-out of ODSs from the household refrigerators by the end of 1996, one year later than that in Japan. It was the first sector phase-out in any developing country, and the Government of Thailand protected its CFC-free manufacturers with the first environmental trade barrier enacted by a developing country.

In August 1994, Stephen Andersen of the US EPA, Margaret Kerr of Northern Telecom, Yuichi Fujimoto of JEMA, Shinichi Ishida of Hitachi, and Viraj Vithoontien of

UNEP visited Vietnam to discuss and promote ozone-layer protection. Vietnam was now just starting to use ODSs such as CFCs. The delegation visited a factory where new German, Japanese and Korean automobiles were being fabricated with CFC-12 air conditioning systems. The team instantly realized that such uses could become substantial as Vietnam rapidly industrialized. It was very important for the global environment and economic efficiency to stop ODSs from the very beginning. The delegation immediately scheduled meetings with relevant Vietnamese ministries and came to agreement on a strategy.

The Vietnam Pledge was announced by 43 leading companies from Canada, Germany, Japan, Korea, Sweden, Taiwan, the UK and the USA[18] in September 1995:

> 'Our company pledges to invest only in modern, environmentally acceptable technology to avoid the use of chlorofluorocarbons (CFCs), halons, carbon tetrachloride, and 1,1,1-trichloroethane. Our company also pledges to limit the use of transitional substances such as hydrochlorofluorocarbons (HCFCs) when suitable replacements become available. We also encourage our joint ventures and suppliers to make this pledge.'

There were two major agreements at the Meeting of Parties in London in 1989: one was to phase out CFCs in 2000, and the other was to finance technology cooperation with developing countries through an Interim Multilateral Fund. However, much time would pass before the London Amendment would be ratified and more time before the Multilateral Fund would begin operation. On the other hand, the Antarctic ozone hole was increasing every year, CFC consumption in developing countries such as in Southeast Asia was expanding rapidly, and only a few countries had ratified the Montreal Protocol.

Again, Japan took strong and immediate action by organizing intensive training courses for ozone protection authorities from developing countries. In October 1990, the first training course started and continued annually every year until 2001. Government officers of national ozone layer protection offices are trained in Tokyo and several other locations in Japan. These courses have been funded by the Ministry of Economy, Trade and Industry (METI) and Ministry of the Environment (ME), and implemented by the Japan International Cooperation Agency (JICA) under official development assistance by the Japanese Government. Japan Industrial Conference for Ozone Layer Protection (JICOP) and Japan Environmental Sanitation Center (JESC) have cooperated to make the curriculum, and select lecturers and sites to visit. Trainees appreciate the curriculum and acquire comprehensive knowledge to implement the Montreal Protocol. From 1990 until 2001, 267 people from 41 countries have attended these courses.[19] Half of these countries were not Parties when the courses started, but now all are Parties.

Japan is proud of its early and strong support for the Montreal Protocol, for its leadership with small- and medium-sized Japanese companies, and especially for the technology cooperation with developing countries.

- Friends of the Earth (FoE) organized boycotts of CFC foam food packaging, helped persuade McDonald's to support alternative foam packaging, and joined an agreement of the American food packaging industry to replace CFC foam packaging with similar packaging made with HCFCs and hydrocarbons. American food packagers shared this technology; McDonald's insisted that its owned and franchized restaurants worldwide make the change to a CFC-free alternative. Suppliers to

McDonald's provided CFC-free packaging to all their customers, with the result that CFC food packaging was phased out. Friends of the Earth also promoted CFC recycling from motor vehicles, and persuaded state and national lawmakers to prohibit CFC venting.

- Greenpeace formed business alliances to produce domestic refrigerators without CFCs as refrigerants or foam-blowing agents, persuaded German customers to accept the added risk of flammability in these refrigerators, worked with safety authorities to approve practical engineering mitigation, and pressured governments and companies to embrace the new technology.
- The Environmental Investigation Agency exposed widespread illegal markets for ODSs in Europe and North America, motivated national and international law enforcement agencies, and helped undertake 'sting' operations that trapped people who responded to offers to buy or sell ODSs under the pretence of a fictitious company.
- The Center for International Environmental Law co-sponsored leadership conferences that explained to companies how fast action to protect the ozone layer could avoid citizen lawsuits that might seek compensation for damages from ozone layer depletion.
- The Pesticide Action Network, Californians for Alternatives to Toxics, and Friends of the Earth organized coalitions of environmental, labour, family, and food safety organizations to oppose the use of methyl bromide.

THE ROLE OF PROFESSIONAL MEMBERSHIP ORGANIZATIONS

Dramatic market transformations, such as ODS replacement, require changes in environmental, safety, and compatibility standards, as well as in education and training. Government authorities manage some standards; however, special-purpose standards organizations, industry associations, or professional associations privately manage many standards. For example, the Society of Automotive Engineers (SAE) manages standards that prescribe custom fittings for each new vehicle air conditioning refrigerant, and the National Fire Protection Association prescribes procedures for fire protection without halon. Many other organizations – including the American Society of Heating, Refrigeration, and Air Conditioning Engineers (ASHRAE); the Institute for Interconnecting and Packaging of Electronic Circuits (IPC); International Institute of Refrigeration (IIR); and Underwriters Laboratory – all undertook substantial efforts to allow the use of various alternatives to ODSs.

THE IMPORTANCE OF AWARENESS CAMPAIGNS

During the 1970s, only environmental NGOs, United Nations agencies, and governments undertook ozone protection awareness campaigns. With the exception of a few companies with ozone-safe alternatives to ODS aerosol products aggressively advertised, urging consumers to do their part for ozone protection. After 1987, individual companies, industries, and artists undertook awareness campaigns involving thousands of messages presented in a dozen or more languages. The awareness campaigns ultimately included publications, art, posters and banners, advertisements, badges, songs, calendars, videos, computer sites, and crossword puzzles. Beginning in 1998, Seiko Epson published large art-quality posters with messages such as, 'Help pitch in! Together, we can clean the earth — Epson to ban CFCs in 1993' and 'The Sky's the Limit – The Cost Of Ozone-Depleting Chemicals Is Stratospheric.' Motorola distributed thousands of badges using international symbols reading 'No CFCs'. Dozens of countries conducted national children's art contests; the Japan Save the Ozone Network organized a children's choir and distributed its CD worldwide; and UNEP and several other organizations created children's colouring books and puzzles. The US EPA ran a banner advertisement in the *Wall Street Journal* thanking leadership companies for their accelerated phase-out of ODSs (see Figure 5.1).

UNEP *public awareness-raising activities*

UNEP has been instrumental in the campaign to raise public awareness about the hole in the ozone layer. Apart from its many technical reports, UNEP has published a myriad of booklets, brochures and posters aimed at the public, *The Ozone Layer*, *Action on Ozone*, *The Ozone Story*, to name a few. In 1994, the General Assembly of the United Nations designated, on the initiative of Venezuela, 16 September as the International Day for the Preservation of the Ozone Layer (Ozone Day). From 1995, the Ozone Secretariat, with the help of the Communications and Public Information Division of UNEP, orchestrated the observance of the day by all governments and international organizations as an occasion to spread awareness. Each year a theme was chosen, such as the adverse effects of ozone depletion or availability of ozone-safe products, and wide publicity given through posters, radio and TV programmes, special editions of the UNEP's flagship magazine *Our Planet*, workshops, seminars, product exhibitions, essay competitions, painting competitions, and rallies by school children. In some countries such as China, high officials, including ministers of the government, joined public rallies urging the public to prefer ozone-safe products. The OzonAction Programme of UNEP helped organize such events throughout the world. It prepared and distributed many publications on how every human being could contribute to the protection of the ozone layer and on how every office can be made ozone-safe.

In recent years UNEP and the Television Trust for the Environment have worked together to produce a series of films on ozone for BBC World's Earth Report, which are broadcast weekly to 178 countries worldwide. Examples of such films include: *The Greenhouse Effect*, 1988, *Saving the Ozone Layer: Every Action Counts*, *The Montreal Protocol*, 1997, and a 30 second public service announcement: *Healthy Harvest: Alternatives to Methyl Bromide*. UNEP has also honoured key players in the field of science who have contributed to the protection of the ozone layer and whose work has led to appropriate responses. The recognition has taken the form of presentations of the Global 500 Award to worthy recipients and also the most prestigious United Nations environment prize of all, the UNEP Sasakawa Environment Prize, worth US$200,000, which was given to Molina in 1999.

Awards can be considered part of an awareness campaign, but they also reward and inspire actions to protect the ozone layer. Appendix 6 lists the winners of the most prestigious awards:

Table 6.3 *Key ozone and environment awards*

Award name	Organization presenting
1985 Ozone Award	UNEP (Vienna Convention Tenth Anniversary)
1987 Ozone Award	UNEP (Montreal Protocol Tenth Anniversary)
Global 500	UNEP (Individual Achievement Award)
Stratospheric Protection Award	US EPA (globally awarded each year since 1990)
Best-of-the-Best Stratospheric Protection Award, 1987	US EPA (presented on the occasion of the Tenth Anniversary of the Montreal Protocol for the most significant contributions to date)
Stratospheric Protection Awards	Brazil, Vietnam, Egypt, China
Twentieth Century Award for Environmental Leadership	Mobile Air Conditioning Society
Stratospheric Protection Award	Industry Cooperative for Ozone Layer Protection
Nobel Prize in Chemistry	Nobel Prize Committee, Stockholm
Japan Ozone Layer Protection Award	Nikkan Kogyo Shimbun/Japan Ministry of International Trade and Industry
Sasakawa Environment Prize	Sasakawa Foundation, Japan and UNEP
Global Video Competition	UNEP OzonAction Programme
Children's Painting Competition for Protecting the Ozone Layer	UNEP OzonAction and the Egyptian Environmental Affairs Agency

The importance of education and training

In most countries, improved servicing was the source of early and substantial emissions reductions. Improved servicing resulted from better training; new leak detection, recovery and recycling equipment; quality parts; and government and industry standards prescribing best practices. Trade and professional associations, labour unions, and private companies primarily undertook training. Many countries created certification programmes that rewarded advanced training and prohibited servicing of ODS equipment by untrained technicians.

Chapter 7

Compliance with the Montreal Protocol

INTRODUCTION

Compliance with the Protocol consists of compliance with all the mandatory provisions of the Protocol, including the control measures, control of trade with non-Parties, and reporting under Articles 7 and 9. The Parties to the Protocol have to demonstrate their compliance with these provisions of the Protocol, and also have the right to verify compliance by other Parties. Generally international treaties themselves set out the procedures for dealing with disputes between Parties regarding compliance with those treaties. Article 11 of the Vienna Convention and the arbitration procedure under the Article approved by the first Meeting of the Conference of the Parties to the Convention are such procedures. They are applicable to disputes regarding the Montreal Protocol also.

The Parties to the Montreal Protocol, have, in addition, set up another more helpful and quicker process for dealing with disputes regarding compliance through an Implementation Committee of the Parties. The Implementation Committee considers any complaints by one Party regarding compliance by another Party and also any non-compliance brought to its notice by the Secretariat or by the non-complying Party itself. The committee makes recommendations to the Meetings of the Parties to the Protocol, which will make the decisions on such non-compliance.

The findings on compliance rely on timely reporting of data on production and consumption of ozone-depleting substances by all the Parties. The Parties found that the reporting requirements under Article 7 require establishment of complicated reporting systems within their countries. While some succeeded in installing these systems quickly, many, including most developing countries, found this difficult. The record on reporting was dismal at first but improved gradually to a satisfactory status now, due to the efforts of many.

The compliance of the Parties so far has been remarkable on the whole. The production and consumption of CFCs, for example, has come down by more than 85 per cent. Scientists have measured the decline of rate of growth of these substances in the atmosphere. The Montreal Protocol has been declared an unprecedented success. However, there have been some exceptions such as the Russian Federation and a dozen other countries which did not comply. The Implementation Committee had the task of bringing them back to compliance.

The Implementation Committee began work in wholly uncharted territory, dealing with reports from over a hundred countries on 95 chemicals. It had to deal with the widely varying conditions of different countries. It had to be firm with the non-complying countries such as the Russian Federation by pointing out their deficiencies, but at the same time not being so negative that assistance to these countries was abandoned as hopeless. This delicate exercise was managed well. The committee had the satisfaction of noting a steady downward trend in ODS consumption in these countries, and the closure of all the CFC and halon production facilities in the Russian Federation by the end of 2001. The same issues have also arisen in some developing countries in 2001. This chapter covers the story of the progress in data reporting and compliance by the Parties to the end of 2001.

REPORTING ON COMPLIANCE MEASURES

Reliable and timely reports by the Parties on production and consumption of the ozone-depleting substances (ODS) in accordance with Article 7 of the Montreal Protocol are the basis for monitoring the implementation of the control measures and compliance with the provisions of the Protocol. Countries are obliged to report to the Ozone Secretariat the production and consumption of the nine categories of controlled substances contained in annexes to the Montreal Protocol.

The Ozone Secretariat receives these data from Parties. Parties report bulk ODS imports and exports with sub-categories for new substances, recovered or reclaimed substances, and exempted categories, including feed stocks and process agents, permitted essential uses, and methyl bromide for quarantine and pre-shipment applications. Total production is reported, irrespective of the purpose to which the ODS is finally used; the amounts destroyed are reported separately, since a quantity of ODSs destroyed using approved technology can be deducted from production and consumption totals. The data on imports from, and export to, non-Parties are checked to ensure that the trade controls under Article 4 are fully implemented.

In addition to the ODS reporting requirement under Article 7, the Meetings of the Parties have prescribed, from time to time, reports on other aspects of implementation. The sixth Meeting of the Parties in October 1994 asked the Parties to list and report capacity of reclamation facilities in their countries. The seventh Meeting of the Parties in December 1995 requested the Parties: to report on measures taken to regulate imports and exports of products and equipment containing substances listed in Annexes A and B; to develop halon management plans and ODS licensing schemes; and to report on process-agent use and emissions. The tenth Meeting of the Parties in November 1998 requested all non-Article 5 Parties to report their strategies for management of halons and regulations that mandated the use of methyl bromide for the exempted quarantine and pre-shipment applications.

Parties that have obtained exemptions for essential uses must report annually on a prescribed form; all non-Article 5 Parties must report steps to eliminate the use of ODSs for laboratory and analytic use, and the quantity of use and emissions of ODSs as process agents and feed stocks. The transfer of hydrochlorofluorocarbon (HCFC) consumption or production between Parties in the process of industrial rationalization must also be reported to the Secretariat. Non-Article 5 Parties are to report a summary of requests from developing countries for supplies of ODSs.

Countries that have received the assistance of the Multilateral Fund periodically report to the Fund Secretariat data on total consumption divided into specific sectors such as aerosols, foams, fire-fighting, refrigeration, solvents, fumigation and other uses.

Difficulties in reporting data

Since the beginning of the data-reporting system in 1989, Parties have experienced difficulties in collecting and reporting their data. While data on production could be collected easily in view of the small number of facilities producing ODSs, the data on imports and exports proved difficult for countries, particularly those with many border points, importers and exporters. An ad hoc group of the Parties discussed the issue of the difficulties in 1989 and made many suggestions to the countries to improve the reporting of data (see Chapter 3).

The accuracy of data collected directly from importers and exporters depends on the willingness of traders to cooperate. Many Parties have implemented a system of licensing whereby any trader in ODSs is compelled to obtain approval of the government for that trade. This has proved helpful in collecting data. The Montreal Amendment of 1997 made such licensing systems compulsory for the Parties.

The collection of import and export data from customs departments proved to be particularly difficult in developing countries because international customs codes combine ozone-depleting and non-ozone-depleting substances into a single category. The lack of customs-code numbers under the harmonized system of the World Customs Organization for mixtures proved to be a particular source of difficulty. These problems have now been solved to a certain extent. The customs departments of most developing countries keep the data on imports and exports of ODSs separately and provide them to the ozone programme agencies, but only after a considerable time lag. Furthermore, many of the ODSs are marketed under company brand names and sold in mixtures with only part of the product being an ODS, making collecting data on such mixtures particularly difficult. For example, a mixture called R-406A is marketed under a dozen different trade names.

THE ROLE OF THE IMPLEMENTATION COMMITTEE

The mandate of the Montreal Protocol's Implementation Committee is to consider submissions by Parties or the Secretariat regarding non-compliance with the provisions of the Protocol. However, since its inception in 1990, it has proactively considered all aspects of implementation, instead of merely

Box 7.1 Harmonized international customs codes for ozone-depleting substances

Michael Graber, Deputy Executive Secretary, Ozone Secretariat

The Harmonized System (HS) is a complete product classification system, which intends to cover all imported or exported goods in international trade. The HS was designed as a 'core' system of headings and subheadings that are to be used internationally by all the countries that have adopted the system. However, countries can make further subdivisions at the national level according to their needs. The HS went into effect on 1 January 1988, with the entry into force of the International Convention on the Harmonized Commodity Description and Coding System, or simply the Harmonized System Convention (HSC).

The HS is periodically updated by the World Customs Organization Council, which is the governing body of the HSC, generally once every four years. Amendments to the HS nomenclature that relate to substances controlled by the Montreal Protocol were introduced in Chapters 29 and 38 of the HS on 26 June 1990 and became effective from 1 January 1996. The relevant HS heading for all the pure substances controlled under the Montreal Protocol is: 29.03 'Halogenated derivatives of hydrocarbons'.

In 1989, the Parties to the Montreal Protocol requested the World Customs Organization Council to create separate code numbers under the HS for each of the ozone-depleting substances (ODSs) controlled by the Montreal Protocol at that time under Annex A. In response, the Council provided four separate international codes for the five Annex A, Group I substances (CFCs) and one additional code for the three halons together. In 1990, the Parties informed the Council that it would be useful to have separate codes for the other substances now controlled by the Montreal Protocol (ie, the Annex B substances: ten more CFCs, carbon tetrachloride, and methyl chloroform, and the Annex C substances: 34 HCFCs). The Council found that the importance to trade of most of the substances on the new list was too limited to warrant separate specification, in particular with regard to the HCFCs. As a compromise, in 1992, the Council concluded a recommendation on separate national codes for each individual substance in Annex B, Group I (other fully halogenated CFCs), methyl chloroform, and carbon tetrachloride. Methyl bromide, which was added to the list of substances controlled by the Montreal Protocol in 1992, was also already included in the international HS.

On 20 June 1995, the World Customs Organization Secretariat rephrased its 1992 recommendation to introduce at the national level two separate codes, one for all the HCFCs and one for all the HBFCs. On 15 July 1999, the Council approved this rephrased recommendation. In 1997, the ninth Meeting of the Parties to the Montreal Protocol requested the World Customs Organization Council to revise its decision of 20 June 1995 by, instead, recommending national codes for the most commonly used HCFCs. In response, the Council approved, on 15 July 1999, customs codes under the Harmonized System to all the pure substances controlled by the Montreal Protocol.

The ninth Meeting of the Parties to the Montreal Protocol in 1997 further requested the World Customs Organization to work with major ODS suppliers to develop customs codes for ODSs that are commonly marketed as mixtures, for use by national customs authorities. The Ozone Secretariat sent a list to the World Customs Organization Secretariat of six categories of mixtures containing ODSs of importance in trade that should warrant specification in the HS, as well as their uses; the World Customs Organization is further investigating this issue.

responding to submissions. The Ozone Secretariat annually collects production and consumption data from the Parties, as specified by Article 7, and places an analytical report based on these data before the meetings of the Implementation Committee and the annual Meetings of the Parties to the Protocol. This report: contains the non-confidential data on ODSs from all the countries for the base year and the most recent year; comments on the timelines of reports; and points out any deviation by any Party from the obligations of the Protocol. The Implementation Committee reviews the report and makes its recommendations to the Meeting of the Parties on the data reported. The Implementation Committee meets twice each year. Countries that persistently fail to submit reports are frequently invited to meetings to explain their non-reporting. Such a summons generally results in the defaulting Party reporting its data.

Status of reporting

The meetings of the Implementation Committee and the Meetings of the Parties in the initial years noted the great shortfall in data reporting and went into each case at great length.[1] For example, the seventh meeting of the Implementation Committee in Bangkok in November 1993[2] had invited Belarus, Burkina Faso, Costa Rica, Iran, Italy, the Maldives, the Syrian Arab Republic, Togo, Trinidad and Tobago, and Ukraine to appear before the committee to explain their non-reporting of data. Two of the Parties reported the required data before the meeting; the developing countries promised to report data as part of their country programmes on implementing the Protocol. The developed countries of Belarus and Italy explained their difficulties, proposed remedies, promised to report their data by February 1994, and promised to be punctual in the future. The Russian Federation, a member of the Implementation Committee, explained that the legal and administrative machinery for collecting information had yet to be established and was complicated by the break-up of the former Soviet Union.

By 1994,[3] data reporting had improved, and only Algeria, Antigua and Barbuda, the Central African Republic, Iran, and the Syrian Arab Republic had to appear before the Implementation Committee to explain their non-reporting. The twenty-first meeting of the committee in November 1998[4] noted that 140 Parties reported their data for 1996, a considerable improvement over the previous years. The twenty-seventh meeting in October 2001[5] noted that all but 16 of the 170 Parties had reported their 1999 data.

RESULTS OF IMPLEMENTATION, 1989–1999

Phase-out by industrialized countries

The 24 countries in the Organisation for Economic Co-operation and Development (OECD) produced nearly 90 per cent of the CFCs in the world in 1986: 908,000 tonnes of the total world production of 1.07 million tonnes. The Montreal Protocol mandated a freeze of production and consumption of CFCs in 1989, a 75 per cent reduction by 1994, and complete phase-out by 1996. The

Table 7.1 *Summary of essential-use exemptions granted by the Meetings of the Parties (metric tonnes authorized)*

Annex	Group	1996	1997	1998	1999	2000	2001	2002
A	I	13,869.61	13,530.58	11,990.35	9882.33	8366.7	6792.8	5996.0
A	II	352.00	300.00	255.00	160.00	90.00		
B	III	59.70	60.50	60.10	59.60	58.40		

Source: Data from Ozone Secretariat, UNEP.

actual production in 1994 was about 184,000 tonnes, a reduction of 80 per cent and therefore exceeding the mandate of 75 per cent. The 1996 production was about 34,000 tonnes, to satisfy the exempted essential uses approved by the Meetings of Parties and to export to meet the basic domestic needs of developing countries, as permitted by the Protocol. The essential-use applications included medical aerosol products, the manufacture of solid rocket motors, and laboratory and analytical uses. Since then, authorized essential use was gradually reduced with the introduction of CFC-free medical aerosol products and progress on eliminating other ODS essential applications. Thus, the phase-out of CFCs by industrialized countries was achieved as mandated by the Protocol.

Similar dramatic reductions have occurred for halons used for fire-fighting applications. Original production of about 158,000 tonnes was reduced to 53,000 tonnes in 1993 and completely discontinued in 1994 – two years prior to the CFC phase-out. The production phase-out in 1994 was accomplished by recovering halons from applications where alternatives existed, placing the halons in 'halon banks', and re-deploying the banked halon to applications of halons where alternatives are still not available, such as fire-fighting on aircraft and at petroleum facilities. There are minor essential-use exemptions for halons used only in the Russian Federation and not available from the halon banks.

The phase-out of Annex B substances – carbon tetrachloride, methyl chloroform and other CFCs – was also accomplished by the industrialized countries by 1996. Hydrobromofluorocarbons (HBFCs) had very few uses and were phased out by 1996 without any Party requesting an essential-use exemption. The implementation of control measures on HCFCs and methyl bromide is now in progress. A freeze on HCFC consumption by 1996 has been implemented; the next step is a 35 per cent reduction by 2004. A 25 per cent reduction of production and consumption of methyl bromide by 1999 has been successfully achieved.

Compliance by countries in transition

The countries of Eastern and Central Europe and the former Soviet Union not classified as developing countries are required to observe the control measures of other industrialized countries under Article 2 of the Protocol.[6] However, the political and economic turmoil in 1990 from the end of communism and introduction of market-based economies resulted in initial non-compliance by all these countries, with the exception of Hungary and Slovakia. However, by

1998, Bulgaria, Poland and the Czech Republic achieved compliance; with the assistance of the Global Environment Facility (GEF), all countries are expected to return to compliance by about 2003–2005.

Although short of complete compliance, by 1997 these countries had recorded a 90 per cent reduction from their reported 1986 ODS consumption of 190,000 tonnes – partly due to their economic decline and partly due to their efforts to phase out the substances. Their 1999 consumption was further reduced to about 17,600 tonnes.

Compliance by developing countries

The developing countries with annual per capita consumption of Annex A substances less than 0.3 kilograms (and less than 0.2kg of Annex B substances) were allowed to delay the beginning of their implementation of control measures by ten years. Thus their freeze of CFCs started on 1 July 1999. The freeze in halon production and consumption was scheduled to start on 1 January 2002. The first control measure for carbon tetrachloride will be an 85 per cent reduction on 1 January 2005. The methyl chloroform freeze will start on 1 January 2003; the freeze of HCFCs will be in the year 2016, and for methyl bromide, 2002. The base level for Annex A substances was the average of 1995–1997 figures; for Annex B, it was the average of 1998–2000; for HCFCs, it is the average of 2015; for methyl bromide, it is the average of 1995–1998. The developing countries are allowed to increase their production or consumption to meet their basic domestic needs. Since the term 'basic domestic needs' has never been clearly defined, this, in effect, meant that the developing countries could increase their production and consumption within the limits set by Article 5 until their year of freeze.

The production of CFCs by developing countries, which was about 45,000 tonnes in 1986, reached an average of 108,000 tonnes for 1995–1997 and was reduced to about 97,000 tonnes in 1999. Consumption reached a peak of about 158,000 tonnes in 1995–1997 compared to about 138,000 tonnes in 1986, a growth of about 16 per cent in ten years, or an annual growth rate of less than 1 per cent. During the same period, the countries which accounted for most of the consumption of developing countries, such as India and China, accounted for annual economic growth rates of more than 6 per cent and even more for the sectors where ODSs are used, such as refrigerators. The relatively small increase in consumption of ODSs is credited to the phase-out projects launched in these countries by the Multilateral Fund and to the awareness about the ozone problem created by the UNEP programmes financed by the Fund. The increase of more than 100 per cent in production of ODSs by the developing countries' producers partially replaced exports from the non-Article 5 Parties.

The data submitted to the Ozone Secretariat for 1999 showed the total production of CFCs by the Parties operating under Article 5 to be 97,250 tonnes, a reduction of about 15 per cent, compared to the mandate of a freeze. Consumption, to be frozen at the 1995–1997 level of 158,000 tonnes, was reduced to about 120,000 tonnes. Halon production of about 45,000 tonnes in 1995–1997 was reduced to about 25,000 tonnes in 1999.

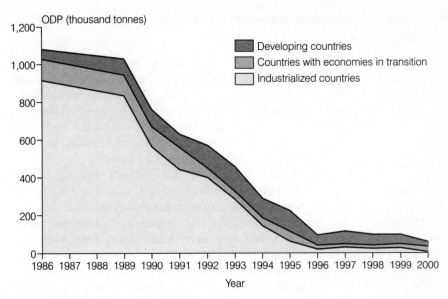

Source: Data from Ozone Secretariat, UNEP.

Figure 7.1 *World production of CFCs by type of economy, 1986–2000*

NON-COMPLIANCE BY PARTIES WITH ECONOMIES IN TRANSITION

Implementation Committee keeps a watch

Beginning in 1993, the Implementation Committee kept a particularly close watch on non-compliance because the first phase-out was scheduled to be achieved for halons by 1994. The Russian Federation confirmed to the seventh meeting of the committee in November 1993 that it would still be producing halons in 1994, and thus would be in non-compliance. A representative of the Russian Federation argued before the committee that the year in which the Soviet Union had split up, 1990, should be treated as the baseline year and that the Federation could not supply data for 1986 or 1989. The committee explained that there was no provision in the Protocol for changing the base year. The representatives of the Russian Federation, Belarus and Ukraine at the meeting explained the difficulties of the republics of the former Soviet Union in complying with the Protocol and fulfilling their obligations to phase out ODSs.

In 1994, the Ozone Secretariat and UNEP held a meeting of countries with economies in transition in Minsk, Belarus, with the participation of the United Nations Development Programme (UNDP), the World Bank, the GEF Secretariat, and technical experts from many sectors. Each country reported on the status of its ratification and its perceived difficulties in implementing the Protocol. All participating countries declared to the meeting their intention to implement the Protocol, but also their inability to do so in view of their unsettled social and economic conditions.

Russian Federation and others of Eastern Europe give notice of non-compliance

At the eleventh meeting of the Open-Ended Working Group in Nairobi in June 1995, a statement was made by the Russian Federation on behalf of itself, Belarus, Bulgaria, Poland and Ukraine explaining why those countries could not comply with the phase-out schedules for the controlled substances of Annexes A and B. A letter from the Russian Federation addressed to the Executive Director of UNEP in the same year pointed out the difficulties in those countries in implementing the control measures of the Protocol.

The Secretariat placed the issue before the eleventh meeting of the Implementation Committee in Geneva in August 1995,[7] pointing out that the statement and letter constituted a submission under Paragraph 4 of the non-compliance procedure, which permitted Parties to approach the Implementation Committee if they concluded that they were unable to comply with their obligations under the Protocol. The Implementation Committee would then consider this submission and make recommendations for a decision by a Meeting of the Parties. The Implementation Committee consulted with other states included in the statement of the Russian Federation.

Poland reported that it had fully complied with the Protocol, but had found it necessary to submit an essential-use nomination for 100 tonnes of CFC-12 to be used to service existing refrigeration equipment. Belarus, Bulgaria and Ukraine expected that they would be able to comply, with assistance from the GEF. The committee also heard from representatives of three non-Party republics of the former Soviet Union – Armenia, Georgia and Kyrgyzstan. The committee advised these republics to ratify the Protocol and place their problems before a Meeting of the Parties as the other republics of the former Soviet Union had done. The committee requested the Russian Federation to give specific details about how it proposed to implement the control measures.

Implementation Committee investigates

The twelfth meeting of the Implementation Committee, in December 1995,[8] considered the timetable of implementation by the year 2000 given by the Russian Federation. The Russian Federation did not specify the quantity of controlled substances it would be exporting and the quantity it would be consuming itself. The issue of the Russian Federation's illegal exports to the non-Parties among the republics of the former Soviet Union was also raised. The President of the committee clarified to the Russian Federation that neither the Implementation Committee nor the Meeting of the Parties could formally grant an extension of the phase-out since the Protocol did not provide for such extensions. The President also sought details regarding recycling and reclamation facilities in Russia, and asked why recycled ODSs could not be used to close production in Russia.

The representative of the Russian Federation explained that because it was a large country with many financial and administrative problems, it was difficult for the government to regulate the production and consumption of controlled substances or to assist with the transition to alternatives. Even though the country had sufficient technical expertise, it needed money to produce the alternatives, the representative said. The enterprises in Russia were in a free-market mode and found it more profitable to export recycled substances than to sell them within the Russian Federation. Administrative problems made regulation of industry difficult; the regulatory apparatus, which had operated before 1991, no longer existed. According to the representative, the transition to alternatives had been slowed by concerns about reliability, toxicity and effects on the labour markets; the government was afraid of creating social disruption from inadequate access to essential uses of ODSs.

GEF to the rescue

The representatives of the World Bank and GEF reported on efforts to prepare ODS-elimination projects for the Russian Federation and cautioned that penalties for non-compliance would hinder the Russian phase-out of ODSs. The GEF further said that it would wait for Implementation Committee findings regarding the quality of submissions by the Russian Federation before proceeding with projects for that country. The committee recommended to the seventh Meeting of the Parties that the Russian Federation be extended international assistance, and decided to keep a constant watch on the progress of implementation by that country.

Implementation Committee monitors every year

The thirteenth meeting of the Implementation Committee in March 1996[9] considered the report of the ad hoc working group of the Technology and Economic Assessment Panel (TEAP) on issues for the countries with economies in transition. The TEAP noted that some countries of the former Soviet Union were reluctant to ratify the Montreal Protocol and the London Amendment, because they were concerned that they would not be able to fulfil their control obligations, nor would they be able to pay their contributions to the Multilateral Fund. The committee considered the information provided by Belarus, the Russian Federation, and Ukraine, and recommended to the Meeting of the Parties that the GEF fund projects in these countries. The committee also considered the inability of Lithuania, Estonia and Latvia to comply with the Protocol; Estonia was a non-Party, but Latvia and Lithuania had ratified the Protocol. The committee recommended that assistance be provided to these countries and that they keep the committee informed of further progress in implementing the Protocol. Thereafter, the committee reviewed the compliance status of each of the countries in transition every year and made recommendations to the Meetings of the Parties.

Success in the Russian Federation

The twenty-third meeting of the Implementation Committee in 1999[10] heard from the GEF regarding the special support given for closing down the ODS-production facilities of the Russian Federation. The Russian Federation reported that three of the seven main production facilities had already been eliminated. The committee, at its twenty-sixth meeting in July 2001, heard from the Russian Federation that all production of CFCs and halons had closed in December 2000.

A framework for dealing with non-compliance

The committee at its twentieth meeting in July 1998[11] decided on a framework for dealing with non-compliance. The first step was identification of a Party in non-compliance. The second step was a review by the committee of the country's plan to achieve compliance. The third step was to specify benchmarks, which would be incorporated into a recommendation by the committee for a decision on the country by a Meeting of the Parties. These benchmarks could include policy measures or reduction and phase-out steps, which the country in question would undertake by a certain date in order to receive the committee's endorsement. Countries that implemented the agreed commitments of their country plans would be treated as being in compliance and recommended for assistance from the GEF or the Multilateral Fund. If the country deviated from those commitments, the Implementation Committee would recommend additional measures such as further assistance or restrictions on exports to the Parties. The Implementation Committee felt that this framework would create clear performance standards, allowing objective evaluation of a Party's efforts.

COMPLIANCE BY DEVELOPING COUNTRIES OPERATING UNDER ARTICLE 5

Proactive stance of the Implementation Committee

Even though implementation of ODS control measures by developing countries was scheduled to start in 1999, the Implementation Committee took an active earlier interest in the performance and problems of these countries to help them report data and comply with the Protocol. At its third meeting in April 1992,[12] the committee noted that the country programmes sponsored by the Multilateral Fund would be of great use to the developing countries to organize collection and submission of data; it asked to be kept informed of the progress of such country programmes.

The Multilateral Fund Secretariat and its implementing agencies have attended meetings of the Implementation Committee since its fourth meeting in September 1992,[13] and have presented reports on country programmes and other progress to the Implementation Committee. At every subsequent meeting, this interaction between the implementing agencies continued. The eighteenth meeting of the Committee in June 1997[14] considered the discrepancies between the data submitted by countries to the Ozone Secretariat and to the Fund Secretariat, and requested the two secretariats to take action to resolve the differences.

Non-compliance by some developing countries

The issue of compliance by developing countries operating under Article 5 surfaced directly in 2001 with the review of data for the years 1999 and 2000. The first control measure for these countries was to freeze the production and consumption of CFCs for the period 1 July 1999 to 30 June 2000 at the level of the average for 1995–1997. The reporting period so far had been calendar years, and many of the developing countries did not report for the first control period of July to June. The 1987 Protocol prescribed the July–June control period, but the 1990 amendment to the Protocol changed the third control period to the calendar year by making the second control period one-and-a-half years. The same rules applied to Article 5 Parties. Of the 120 Article 5 Parties that submitted their data for 1999, 23 Parties reported 1999 consumption data which exceeded the baseline data for 1995–1997. The Secretariat demanded from these Parties the figures for the control period of 1 July 1999 to 30 June 2000 to verify compliance. Some did not submit these figures. Four of the Parties – Belize, Cameroon, Ethiopia and Peru – which submitted the figures for the control period, exceeded their base-period consumption; Argentina exceeded its production quota. Fifteen countries[15] did not report their control-period data, creating potential non-compliance by them.

Caution administered by the Implementation Committee

The Implementation Committee took a stern view of non-compliance cases and recommended decisions[16] to the thirteenth Meeting of the Parties similar to those taken in the case of Parties with economies in transition. While administering caution to the non-complying Parties, continued assistance by the Multilateral Fund was also recommended.

THE RESPONSE OF THE MEETINGS OF THE PARTIES TO NON-COMPLIANCE

Based on the recommendations of the Implementation Committee, each Meeting of the Parties took decisions on non-compliance by Parties:

- Armenia ratified the Protocol in 1999 and was in non-compliance. The thirteenth Meeting of the Parties noted that Armenia did not report its data and recommended international assistance, once Armenia ratified the London Amendment.
- The tenth Meeting of the Parties in November 1998 noted the commitments of Azerbaijan to phase out CFCs by 1 January 2001 and recommended international assistance if it stayed on course. A similar decision was taken with regard to Belarus by the seventh and tenth Meetings of the Parties in December 1995 and November 1998.
- The seventh Meeting of the Parties in December 1995 noted that Bulgaria, Poland, the Russian Federation and Ukraine declared the possibility of their non-compliance. The eleventh Meeting of the Parties in November 1999

BOX 7.2 KEEPING THE TREATY FLEXIBLE

Peter Sand, Former Chief, Environmental Law Unit, UNEP

Successful models have a way of developing their own dynamics, irrespective of parental guidance. For example, Patrick Szell (Head of the Legal Directorate of the UK Department of Environment, member of the delegation of the UK to the ozone meetings for many years and Chairman of the Legal Drafting Group at the Montreal protocol meetings from 1989) cautiously speculates that there really is nothing to commend the Montreal Protocol as a model for the 1994 Convention on Desertification; little does he seem to realize that the Desertification Secretariat has already embarked on the preparation of 'procedures to resolve questions of implementation' under Article 27 of the Convention, drawing heavily on Patrick's pioneering work with the Montreal and Geneva procedures. With his usual modesty, he also fails to mention that the Lucerne Conference of European Environment Ministers in 1993 called for the development of 'non-confrontational' compliance procedures (à la Montreal) for all multilateral environmental agreements – which is why such procedures are now popping up in the Economic Commission for Europe (ECE) draft convention on environmental information and public participation. So, the 'Montreal model' keeps multiplying and cloning itself; and I sometimes picture Patrick as the sorcerer's apprentice who started it all and who would now wish to reach for his magic wand to get those genies back into the bottle.

There is another problem with our Montreal package deal, however, which goes to the heart of the non-compliance procedure as it is now evolving. The Implementation Committee, whose accomplishments both Patrick Szell and Duncan Brack (Head of Sustainable Development Programme, Royal Institute of International Affairs, and a scholar of international environmental affairs who has written extensively on global environmental agreements, including the Montreal Protocol) acknowledge, typically evaluates compliance by 'problem countries', and has repeatedly had occasion to do so with regard to Eastern Europe. Obviously, a harsh finding of non-compliance would deprive those countries of the 'carrot' of GEF funding; and so the Implementation Committee tends to lean over backwards to find them more or less in compliance – possibly in order to keep their goodwill, and not to lose them as treaty Parties altogether.

It is true that such pragmatic interpretation keeps the treaty flexible. In affirming their power to make their own evaluation of what constitutes compliance, the contracting Parties may (arguably) even grant exceptions from the strict application of treaty rules. Yet, consensual redefinition of treaty standards – however well meaning, albeit for the sake of avoiding conflict – also tends to 'soften' the entire treaty regime, and thereby risks weakening its effectiveness in the long run. What that reminds me of is a subtle Italian way of describing justice in Sicily, the land of the Godfather:

> *La legge e applicata al nemico – ma interpretata all'amico*
> (Law will apply to the enemy – but be interpreted to a friend)

Source: cited in Le Prestre, P G, Reid, J D and Morehouse, E T Jr (eds) (1998) *Protecting the Ozone Layer: Lessons, Models and Prospects*, Kluwer Academic Publishers, Boston.

noted with appreciation the work done by Bulgaria in cooperation with the GEF and recommended international assistance if Bulgaria stayed on course. Poland did not come to further notice since it then complied with the provisions of the Protocol.

- The Czech Republic came to the attention of the eighth, ninth and tenth Meetings of the Parties in 1996–1998 for small infractions. The Meetings decided that no particular action was necessary, and that the Implementation Committee would continue to review the country's status.
- The tenth Meeting of the Parties in November 1998 noted the reduction targets Estonia set for itself to phase out Annex A substances by the year 2002 and recommended international assistance if it stayed on course.
- Kazakhstan, which ratified the Protocol in 1998, was in non-compliance but reported a plan for return to compliance by 2004. The thirteenth Meeting of the Parties in 2001 decided to monitor the plan and recommended international assistance.
- Latvia and Lithuania came to the attention of the Meeting of the Parties three times: at the eighth Meeting in November 1996, to be recommended assistance by international agencies; at the ninth Meeting in September 1997, to note their timetables for ratification of the London Amendment; and at the tenth Meeting in November 1998, to note their commitments made to phase out ODSs by the year 2000 and to recommend international assistance if they stayed on course.
- The Russian Federation came to the notice of the Meetings of the Parties in almost every year from 1995 on because of its large size and production capacity. The Meetings continuously encouraged international assistance to Russia after noting the specific promise of the Russian Federation to comply by the end of 2000. The thirteenth Meeting of the Parties in 2001 noted its non-compliance in 1999 and 2000, but was pleased to note that Russia's ODS production facilities closed in December 2000.
- Tajikistan, which ratified in 1998 was in non-compliance but reported a plan for return to compliance by 2004. The thirteenth Meeting of the Parties in 2001 decided to monitor the plan and recommended international assistance.
- The eleventh Meeting of the Parties in November 1999 noted the work done by Turkmenistan to develop a country programme and its promise to phase out Annex A and Annex B substances by the year 2003. It recommended financial assistance if Turkmenistan stayed on course.
- Ukraine came to the attention of the seventh and tenth Meetings of the Parties in December 1995 and November 1998. Ukraine was encouraged to stick to its promise to phase out the consumption of Annex A and Annex B substances by 1 January 2002 with the assistance of the GEF Secretariat.
- The tenth Meeting of the Parties in November 1998 noted the programme of Uzbekistan to reach compliance by the year 2002 and encouraged international assistance if it stayed on course.

The twelfth Meeting of the Parties in December 2000 was concerned about the statement of the GEF Secretariat at the eleventh Meeting that its assistance to the countries with economies in transition was complete. It requested the Facility to clarify its future commitment to providing continued assistance to these countries for phasing out all ODSs.

The thirteenth Meeting of the Parties in October 2001 noted non-compliance by Article 5 Parties Argentina, Belize, Cameroon, Ethiopia and Peru in the control period 1 July 1999 to 30 June 2000. Argentina failed to freeze its production of CFCs at the baseline level and, instead, increased production in 1999 by nearly 10 per cent. The other four countries increased their consumption in 1999 compared to the baseline. The thirteenth Meeting of the Parties urged these Parties to submit a plan of action to return to compliance and to adopt necessary policy measures to reduce their production and consumption of ODSs. Such a plan of action could include import quotas to freeze at the baseline level, and a ban on imports of ODS equipment. It also cautioned them that other measures, such as ensuring that excessive supplies of CFCs to the Party are ceased, could be taken in future if non-compliance continues.

The meeting also noted, on the basis of their data for 1999 or 2000, potential non-compliance on the part of 15 more developing countries which had not reported their data for the control period of 1 July 1999 to 30 June 2000. The meeting urged these countries to report the data for the control period and cautioned them of other measures in the future. A number of the Parties explained that their plans for controlling consumption of refrigerants to service existing equipment through recycling had not yet been completed and, where completed, were not functioning successfully since cheap CFCs were plentiful.

CONCLUSION

The compliance of the Parties with the Articles of the Montreal Protocol has been very good in general. The Implementation Committee and the meetings of the Parties have continuously reviewed the state of compliance and taken timely decisions to assist the Parties towards better compliance. The reporting of the production and consumption data for ODSs by the Parties has steadily improved over the years. There have been no instances of violation of the trade provisions of the Protocol. The industrialized countries have contributed fully the resources approved by Meetings of the Parties for the Multilateral Fund. Industrialized countries and larger developing countries have fully complied with the Protocol, including the first stage of the control measures applicable to developing countries – freeze of production and consumption of CFCs.

However, a number of the smaller developing countries with low volume consumption of CFCs have been unable to observe the freeze. These countries have provided a number of explanations for their inability to comply – easy availability of CFCs at low prices, inability to effectively control inflow of CFCs, large imports of used equipment that relies on CFCs, and increases in the demand of CFCs as a consequence of increasing economic prosperity. UNEP had urged the Parties to initiate measures to further reduce the production of CFCs to match the consumption needs. This step would raise the prices of CFCs, making alternatives to CFCs attractive and would reduce the inflow of CFCs to the developing countries. The Executive Committee of the Multilateral

Fund is continuously reviewing the problems of developing countries in complying with the Protocol and has approved projects to assist countries in introducing appropriate policies and regulations to promote compliance. The Multilateral Fund is also funding capacity-building in these countries such as training of the relevant officials and technicians.

Most of the countries of Eastern and Central Europe had not been classified as developing countries and were thus expected to follow the same control schedule as industrialized countries. The political and economic transition in the early 1990s from communism to a free-market system threw these countries into chaos for some years and they were unable to implement the phase-out of many of the ODSs by 1996 as scheduled. However, the patient guidance by the Implementation Committee and financial assistance by the GEF have enabled all countries, excepting the countries of the former USSR, to return to compliance quickly. This assistance is now helping the countries of the former USSR to return to compliance with the control measures. Their production of CFCs and halons has stopped and their consumption is reduced by more than 90 per cent. It is expected that they will be in complete compliance by 2003.

Chapter 8

Media coverage of the ozone-layer issue*

INTRODUCTION

Prior to the 1960s, industry and business interests dominated the debate about the impact of economic development and the environment. However, the visibility of environmental issues was raised dramatically by the growth of television as a national and international medium and by news-making environmental activists such as Rachel Carson, author of *Silent Spring*. Since the late 1960s, when scientists first speculated that supersonic aircraft would deplete the ozone layer, the mass media have played a major role in shaping awareness and perceptions of the scientific, diplomatic, commercial and technical challenges of protecting the ozone layer.[2]

This chapter quantifies how the media have covered the ozone layer, presents case studies of media coverage of nine seminal events in the history of protecting the ozone layer from 1974 to 1999, and links media coverage to public support and policy analysis.

ANALYSIS OF MEDIA COVERAGE

The influence of news coverage[3]

News, as reported by the media, is a socially constructed product influenced by a host of political, economic and ideological factors. News is often more than just reporting what happened and why; it is also a force in 'public agenda-setting' that both mirrors public priorities and actually influences them. The more coverage the media give a topic, the more likely it is that people will think about that topic, and the more the media may also influence what people think. By their selection of newsworthy events, journalists define critical issues and force policy-makers to justify themselves to a larger public. Different interest groups – including political institutions, industries, environmentalists and professional associations – use media coverage as an arena in which to convey their political and economic views to the public. It is in this context that the ozone-layer issue has played out in the media for more than 30 years.

---

* This chapter was written by Don Smith and Penelope Canan.[1]

The media coverage of environmental issues is particularly important because most people learn about risks to public health and the environment through the media rather than by personal experience. Furthermore, public concern for the ozone layer is reflected in the frequency and global distribution of media stories and by poll results.[4]

Quantitative analysis of media coverage for the ozone layer

Analysis of more than 43,500 electronically accessible articles about the ozone layer published from 1970 to 2000 reveals the key role of media coverage of the stratospheric ozone issue.[5]

Number of published articles as a measure of media and public concern
Media coverage:

- grew steadily in the 1980s, was at its highest levels in 1989 and 1992, ebbed toward the end of the 1990s, but remains at a steady level between those two extremes today;
- was concentrated in North America and Europe, although stories from other regions – Asia in particular – grew through the 1980s and 1990s;
- had a wide variety of story focus, with 25 per cent of the 43,550 stories focused on the reporting of atmospheric science;
- shifted from the science debate in the 1970s and 1980s to coverage of legal/regulatory and business/industry issues in the 1990s;
- cited government organizations as the most frequent primary source, with business/industry sources second; and
- focused most often on chlorofluorocarbons (CFCs) used as refrigerants and aerosol propellants followed, to a much lesser extent, by halon fire-fighting chemicals and methyl bromide pesticides.

A breakthrough in public understanding of science
The combined influence of mass media coverage placed the ozone-layer issue directly in the public debate and on the policy-making agenda. '[T]he evidence suggests that the ozone hole coverage has achieved a breakthrough in the public understanding of the basic science and policy issues.'[6]

The first story in the database of 43,550 ozone-layer stories published between 1970 and 2000 was published in the *New York Times* on 28 May 1970. Three hundred and twenty-four other stories were published in the 1970s; 7399 in the 1980s; 33,164 in the 1990s; and 2662 stories published in 2000 alone. The first record of a UNEP ozone press release was 22 October 1981, reporting findings from the conference in Copenhagen organized by the Coordinating Committee for the Ozone Layer (CCOL).

There was consistent quarter-by-quarter growth in the number of stories from 1986 to 1989. The number of stories was greater than 1000 in six quarters of the 1990s, including the quarters of the second Meeting of the Parties in London, the fourth Meeting of the Parties in Copenhagen, and the ninth Meeting of the Parties in Montreal.

Detailed analyses of coverage in ten leading publications

More detailed research examined how ten of the world's leading publications or news services covered the ozone layer issue: the *Financial Times* (UK); the *Guardian* (UK); the *Independent* and the *Independent on Sunday* (UK); Inter Press Service (Italy); Japan Economic Newswire (Japan); the *Los Angeles Times* (USA); the *New York Times* (USA); TASS (USSR/Russia); the *Toronto Star* (Canada); and the Xinhua General News Service (China). Not all of these publications were on the database for the earliest years of the ozone-layer controversy but all were on the database from 1982. Those on the database before 1982 showed only very limited publication of ozone stories until 1985, strong coverage from 1986 to 1989, and declining coverage in 1990 and 1991, and the single year of the most significant coverage in 1992.

Since 1993, there has been a gradual decline in coverage, with the significant exception that Xinhua published 93 stories in 1999 compared to 87 in 1990, probably because the eleventh Meeting of the Parties to the Montreal Protocol was held in Beijing that year. Figure 8.1 illustrates how the ten publications or news sources covered the ozone issue from the point at which each source was available in full-text on the database.

Where and how the stories appeared

Detailed analysis of 282 stories from the general database of 43,550 stories indicated that: 81 per cent of stories were published in 'popular' news sources – daily newspapers, news magazines and broadcasts; 9 per cent in 'technical'/ 'trade' publications; 5 per cent in 'business' news sources; and 4 per cent in 'professional' sources. The popular press maintained ozone coverage during the entire time period – the 1970s to 2000 – while the technical and trade press had declining coverage from the 1980s to the 1990s.

Of these 282 stories, 72 per cent were in newspapers, 10 per cent in magazines, and 4 per cent each in broadcasts and newsletters, 1 per cent each on the Internet and in journals; and 7 per cent in publication types that could not be identified. Of the stories, 69 per cent were news stories; 16 per cent feature articles; 10 per cent opinion; and 2 per cent editorials and letters to the editor.

The primary sources of the 282 stories were authorities from: government (15 per cent); business/industry (7 per cent); the United Nations (5 per cent);[7] academia/research communities (4 per cent); NGOs (3 per cent); and the Technology and Economic Assessment Panel (TEAP) (1 per cent). Figure 8.2 presents the relative importance of various types of primary sources.

Twenty-four different NGOs were mentioned in the sample of the 282 stories. The three most frequently mentioned NGOs were Greenpeace, Friends of the Earth and the World Resources Institute. Twenty-one other NGOs were mentioned in at least one story.[8]

The ozone layer issue was the main focus in 76 (27 per cent) of the 282 stories in the selected sample. In that sub-sample of 76 stories, science was either a primary or secondary theme in 42 per cent of the stories, legal/regulatory in 41 per cent, business/industry in 34 per cent, health/social in 33 per cent, technology in 23 per cent, policy debate in 22 per cent, and the Protocol/United Nations Environment Programme (UNEP) in 5 per cent.

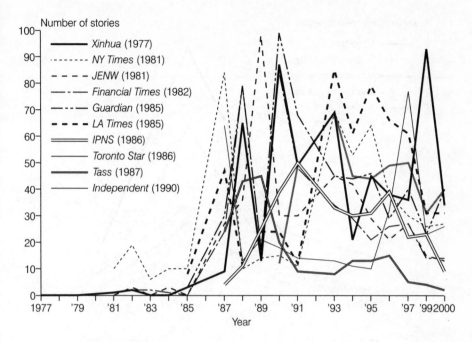

Figure 8.1 *News coverage of ozone depletion, 1997–2000*

MEDIA COVERAGE OF SEMINAL OZONE-LAYER EVENTS

A sample of stories was read and analysed to determine how the media covered nine seminal events in ozone-layer history (Table 8.1), and how that coverage changed over time.

Table 8.1 *Nine seminal events in the story of the ozone layer*

Event	Timeframe
Molina–Rowland hypothesis	1974–1975
US ban of CFC aerosol products	1977–1978
Discovery of Antarctic ozone hole	May–December 1985
1987 Montreal Meeting and Agreement	August–November 1987
1989 London Conference	February/March 1989
1990 London Meeting of the Parties	1990
1992 Copenhagen Meeting of the Parties	1992
1997 Montreal Meeting of the Parties	1997
1999 Beijing Meeting of the Parties	1999

That analysis of the nine seminal ozone-layer events found that:

- Media coverage played an important role in a sequence of events that led to the US ban of CFC aerosols in the late 1970s. However, in the aftermath of the ban, the media also played a role in leading people to believe, mistakenly, that the ozone layer had been protected.

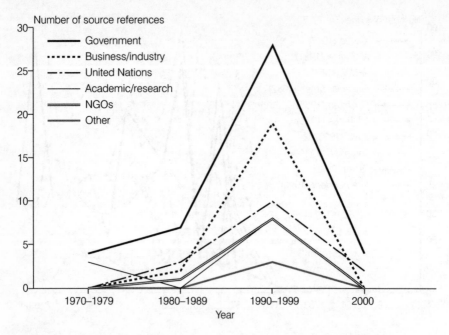

Figure 8.2 *Sources of news coverage by sector, 1970–2000*

- In the run up to the signing of the Montreal Protocol, the media played a vital role in building public concern, which stimulated political support.
- There was a relatively small number of individuals who achieved widespread media attention during the nine seminal events. Among these individuals were Maneka Gandhi, Indian Environment Minister; F Sherwood Rowland, pioneering ozone scientist; Margaret Thatcher, UK Prime Minister; and Mostafa Tolba, Executive Director of UNEP.
- Prior to the London Conference on Saving the Ozone Layer in 1989, environmental NGO activities were relatively insignificant in media reports, but escalated rapidly thereafter, often becoming the 'news angle' for stories about the ozone layer. Some environmental organizations, such as Greenpeace and Friends of the Earth, achieved media status as both 'eco-alarmists' and technology advocates.
- Editorial coverage of the amendments to the Montreal Protocol was especially high in the early 1990s, but ebbed considerably by the time of the Beijing Meeting of the Parties in 1999.

The rest of this chapter details the media coverage of each of the nine seminal events.

THE MOLINA–ROWLAND HYPOTHESIS, 1974–1975

The June 1974 publication of the hypothesis by F Sherwood Rowland and Mario Molina of ozone depletion from CFCs in *Nature* received scant coverage, but the September 1974 press conference at the American Chemical Society Annual Meeting received extensive media coverage, and captured the attention of American consumers.[9] Media coverage from 1974 to 1975 was careful to point out that Molina and Rowland were presenting a hypothesis that had yet to be confirmed, referring to it in some coverage as 'controversial'.[10] As a result, various approaches were taken to present the underlying issues. For example, *Business Week* (17 February 1975) reported, '[S]ome scientists claim, the ozone so vital to the earth's ecosystem is threatened by one of the most mundane artifacts of western civilization: the aerosol spray can'. According to *Chemical Week* (11 June 1975), Molina and Rowland had 'conjured up [their] now-famous theory that the ozone layer in the upper atmosphere may be shrinking under the action of free chlorine atoms originating from chlorofluorocarbon propellants.' However, even in this early period, some media organizations were willing to be unequivocal in their characterization of the environmental risk. For example, *Science Digest* reported (January 1975), 'The combined effect of billions of aerosol cans releasing their chemical propellant into the atmosphere is gradually eroding the ozone layer.'

The industry arguments

The media also presented industry's side of the argument. For example, *Chemical Week* (16 July 1975) quoted British meteorologist Richard S Scorer as referring to the ozone-depletion theory as 'a science-fiction tale...a load of rubbish...utter nonsense,' and characterizing the Molina–Rowland computer model as a simplistic representation of 'exceedingly complex chemical and meteorological processes'. S Robert Orfeo, Director of Applied Research in Allied Chemical's Specialty Chemicals Division, was quoted in *Aerosol Age* (June 1975), arguing that there was simply insufficient information on which to base new policies: 'The validity of the Rowland–Molina hypothesis has not been established... As a matter of fact, detailed analysis of the available ozone data indicates that the ozone level in the stratosphere actually increased in the 1960s, a period of high production of chlorine and chlorine-containing products.' Some stories noted the economic disruption that could be caused by a ban on the sale of aerosols using fluorocarbon propellants. *Chemical Week* (1 October 1975) reported a spokesman for DuPont as saying in a Wisconsin legislative committee public hearing that gross sales of fluorocarbon aerosols were about US$45 million per year in Wisconsin alone.

News stories often framed the Molina–Rowland hypothesis as a debate among scientists, environmentalists and industrialists. *Chemical Week* (16 April 1975) reported that at the 169th national meeting of the American Chemical Society, 'The continuing debate over the possibility of danger that chlorinated fluorocarbons pose to the protective ozone in the stratosphere was punctuated by sharp disagreement between... F S Rowland and the technical director of DuPont's Freon Products Division, Raymond L McCarthy... Rowland asserted the danger is so great that use of fluorocarbons should be banned in aerosol spray cans immediately. But McCarthy claimed the hazard is being overstated.' Another journalistic approach was to report the story in the context of a 'complex fight... between environmentalists, who want the fluorocarbons

---

BOX 8.1 CRITICISM OF MOLINA AND ROWLAND

A 1986 news story looking back in time reported that Molina and Rowland had been roundly criticized by industry chemists for their 'exaggerated claims'. Moreover, the story noted, 'The two were viewed as something akin to renegades, traitors to their profession. Scientists and officials at DuPont, the world's largest producer of the chemicals involved, were particularly caustic, suggesting there was something inherently wrong with their colleagues' computer models.'

Source: Froelich, W (1986) 'Ozone hole may be omen; South Pole discovery worries some scientists', 9 February, *The San Diego Union-Tribune.*

---

banned immediately, and the aerosol industry, which believes the ozone depletion theory is mostly speculation that lacks experimental verification.'[11]

Some 1974–1975 news coverage emphasized environmentalists' claims of serious health impacts associated with a reduced ozone layer. For example, *Newsweek* (23 June 1975) reported that, 'Continued use of [CFCs] at the present rate... environmentalists argue, could cause a 13 per cent depletion in the ozone layer and lead to perhaps 80,000 extra cases of skin cancer each year, and possibly other effects as well.' *Business Week* (17 February 1975) reported on a Harvard study that suggested that fluorocarbons could result in a 16 per cent depletion of the ozone layer within 25 years, thus resulting in more than 100,000 additional cases of skin cancer per year.

Industry reaction

The media also reported that the aerosol industry feared a difficult time ahead with continued use of fluorocarbon propellants because of increased media coverage of the ozone issue. *Chemical Week* (11 June 1975) reported that, '[A]nxiety about the ozone question is believed to be one factor in the recent decline in the US aerosol sales.' The impact of increased reporting on the issue was noted in a later *Chemical Week* story (15 October 1975) in which Montfort Johnson, vice-president of Peterson/Puritan, one of the largest aerosol fillers in the USA, was quoted as saying that the main cause of the market slump in aerosols was 'the adverse public relations' which aerosols were receiving. Industry leadership in abandoning use of CFC propellants also received media coverage. For example, *US News & World Report* (29 September 1975) noted that, 'Johnson Wax Company announced [in June 1975] that it had replaced the few aerosol products in its line that used fluorocarbons. Its spray cans now carry the label: "Use with confidence, contains no Freon® or other fluorocarbons claimed to harm the ozone layer".' *Chemical Week* (16 July 1975) also reported that Sherwin-Williams had discontinued its use of fluorocarbon propellants in its aerosol products.

Media reaction in Japan

Media interest in the Molina–Rowland hypothesis was most pronounced in North America, but there was also strong international concern for the ozone layer. *Chemical Week* (11 June 1975) reported that a study had been initiated in West Germany to evaluate the impact of fluorocarbon gases on the upper atmosphere. The controversy as it had played out in the USA 'caused a panic in the industry' in Japan, Sadaji Takada of the Aerosol Industry Association of Japan said later in *Chemical Week* (15 October 1975), which reported that, 'Nervous European aerosol manufacturers...are starting to feel the heat from the fluorocarbon controversy in the US'.[12]

US BAN ON CFC AEROSOL PRODUCTS, 1977–1978

The effort to ban CFC aerosol products in the USA reached its zenith in 1977 and 1978. The extensive media coverage of the underlying controversy played an important role in the events leading up to the eventual ban. 'Media coverage helped trigger a sequence of events that made it easier for both aerosol manufacturers and policy-makers to envision – and economically justify – a rapid shift away from CFC propellants,' according to the *Spray Can Ban*.[13]

Media coverage fostered changes in consumer demand by pointing out that aerosol sprays that used CFC propellants were quickly losing market share. For instance, *Business Week* (28 February 1977) reported that the Gillette Company had been 'hurt badly...by the controversy over fluorocarbon propellants used in aerosol sprays and their depletion of the ozone layer of the earth's atmosphere... The company's Right Guard deodorant, the market leader with 30% of the business two years ago, has slipped to about 20%. And its Soft & Dri deodorant and anti-perspirant has also lost a share of the market, as concerned customers switched to sticks and roll-ons.'

Media coverage of the CFC aerosol ban reflected public pressure on the government to act. For instance, US Food and Drug Administration Commissioner Donald Kennedy said in the spring of 1977 that depletion of the ozone layer 'could increase the incidence of skin cancer worldwide, cause changes in the climate and have other undesirable effects'.[14]

Industry reaction

The media reported that despite the fact that manufacturers had complained about the cost of searching for substitutes, the end result might actually benefit consumers. 'Mechanical pumps and hydrocarbon sprays, the two most common

---

BOX 8.2 LEADERSHIP COMPANIES LISTENED TO CUSTOMERS

'We were a little like the railroads who didn't realize they were in the transportation business. We thought we were in the aerosol business because 80% of all users preferred aerosols and we were the leader in that segment. But when the ozone controversy broke, we found out we were really in the underarm business.'

Source: Derwyn F Phillips, President of Gillette Toiletries Division, cited in 'After the diversification that failed', 28 February 1977, *Business Week*, p58.

replacements, cost less than chlorofluorocarbons, and could reduce the price of aerosol sprays by nearly 15 per cent,' *Newsweek* reported (9 May 1977). The media also reported that some companies were seeking to use the ozone-layer controversy to their commercial advantage.

The media reported that despite the action in the USA to ban CFCs in aerosols, other parts of the world were taking a 'wait and see' approach. For example, *Chemical Week* (4 May 1977) reported that European countries were not going along with the US view of fluorocarbon propellants and 'will wait until more data from studies under way are available.' Similarly, *Business Week* (30 May 1977) reported that the industry complacency in the USA to the ban 'contrasts sharply with the negative reaction the ban has received overseas. For example, Richard Scorer, Technical Committee Chairman of Britain's Clean Air Council, says Britain will continue to sell aerosols. He claims US officials "had not even listened" to scientific arguments.'

Chemical Week (2 November 1977) noted that there were CFC-free aerosol products, but that, 'Strategies to control emissions from refrigerators, freezers and air conditioners and production of rigid and flexible foams pose costly technical problems.' A story in *Business Week* (5 December 1977) put the issue in this context: 'Worried makers of refrigerators and air conditioners are struggling to find ways – even stopgap ways – to stem leaks of chlorofluorocarbon refrigerants into the air. Other manufacturers need to trap similar chemicals – so-called blowing agents – evaporating from their production lines. If they cannot change their ways soon, these companies fear, the Environmental Protection Agency [in late 1978] will order them to control emissions.' *Business Week* (13 March 1978) reported that, 'pathetically' insufficient data on the amount of chlorofluorocarbon gases that could escape in the atmosphere from refrigerators, air conditioners, and production lines for foaming plastics have led the Environmental Protection Agency to delay indefinitely its proposed regulations on those uses of the chemicals.'

There were also reports that industry and government were beginning to work together on the underlying issues. For example, in the context of EPA consideration of a regulation to govern the use of CFCs in refrigerators and air conditioners in the spring of 1978, Herbert T Gilkey, executive director of public affairs of the Air-Conditioning & Refrigeration Institute, was quoted in *Business Week* (13 March 1978) as saying, 'At least we're talking to each other now, rather than at each other.'

Over-emphasis on aerosols

Once the US ban on aerosol products was in place, the news media may have inadvertently played a role in leading the public to believe that the entire ozone-depletion issue had been solved. This was further exacerbated by the diminishing interest in the story by the media. '[A]s the novelty of the story wore off, the press lost interest and failed to describe the growing complexity of the issue as it unfolded over the next few years,' according to F Sherwood Rowland.[15]

As the effort to ban CFC aerosol products proceeded, the media continued to characterize the threat at hand in various ways. *Chemical Week* (23 February

---

BOX 8.3 ADVERTISING NEW-FORMAT PRODUCTS, WITHOUT CFCS: ULTRA BAN II

'Bristol-Myers has launched a $17-million national advertising campaign to promote Ultra Ban II, hydrocarbon-propelled aerosol anti-perspirant. The company was one of the leaders in introducing non-aerosol anti-perspirants...when the controversy over the possible effects of fluorocarbon propellants on the ozone began affecting sales. But it says that 'the present 60 million aerosol deodorant and anti-perspirant users will stay with aerosols, and consumers who have drifted away will come back' to reformulated products. John B Carles, group product manager, says hydrocarbon-propelled aerosols should level out at 50% of the $750 million deodorant and anti-perspirant market [in 1977].'

Source: 'Ultra Ban II debuts', 9 November 1977, *Chemical Week*.

1977) wrote that, 'fluorocarbons from aerosols...are believed to damage the ozone layer in the stratosphere'; called it a 'complex issue' in another story (7 September 1977), and a 'theory' in another (2 November 1977). The *National Journal* (16 December 1978) reported that, 'Man's use of fluorocarbons as a propellant in spray cans and as a refrigerant has been implicated as the chief threat to the ozone layer.' However, some media organizations were less uncertain in their characterization of the matter. For instance, *Facts on File World News Digest* (21 May 1977) reported that 'scientists had concluded that fluorocarbons [damage] the ozone layer of the earth's atmosphere'.

The ban on fluorocarbon propellants also served as the basis for editorials. For example, *Business Week* (30 May 1977) opined, 'The ban on fluorocarbon propellants for aerosol sprays will be disruptive and expensive for the industry...but it can serve as a model for sensible regulation that minimizes both the economic and environmental damages.'

THE DISCOVERY OF THE ANTARCTIC OZONE HOLE, 1985

There was almost no media reaction to the 1984 findings of significant ozone depletion over Antarctica published by Shigeru Chubachi from data of the Japanese Scientific Stations in Antarctica, and initially little media reaction one year later to the publication of research that confirmed the depletion by the British Antarctic Survey in *Nature*.[16] But on 30 May 1985, a story in the *Guardian* began:

> *'The vital and protective layer of ozone – a reactive form of oxygen – in the Earth's atmosphere has diminished by a third in the decade 1972/1982. Or at least that is what has happened to the ozone over the Antarctic, in the Antarctic springtime, according to measurements made by a group of three scientists from the British Antarctic Survey... The decline in ozone...matches a detectable increase in the amounts of chlorofluorocarbons, or CFCs (aerosol propellants) in the polar atmosphere in the same period, the British group claims.'*

The story went on to suggest that while most environmentalists had long been warning that CFCs would prove a danger to the ozone layer, 'until these Antarctic results most scientists had thought measurable effects were unlikely for decades'. The story quoted one of the scientists, Jonathan Shanklin, as saying, 'There's no obvious source for the increased CFCs we've detected other than manmade sources.' Two months later, *The Economist* (13 July 1985) characterized the survey's results as 'disturbing' and went on to report, 'Those who worry that pollutants could damage the ozone layer now have more cause to worry than before.'

The findings of the British Antarctic Survey were also cited in a story in the *New York Times* (7 November 1985) which reported, 'Satellite observations have confirmed a progressive deterioration in the earth's protective ozone layer above Antarctica, according to scientists who analyzed data recently sent back from space.' The consequence of the findings, as interpreted by that story, was that the results, taken together, 'have persuaded some researchers that the ozone loss is proceeding much faster than expected.' Moreover, there was speculation that the findings would increase pressure from environmental groups to reduce the use of or ban CFCs. In this regard, the *Guardian* (30 May 1985) reported that pressure from environmentalists would result even though 'the world-wide impact of the ozone decline over the Antarctic is not yet clear – and may be only small.' However, the article noted that increased ultraviolet penetration had already been detected in the Antarctic, 'and there may be legitimate fears for its effect, say, on phytoplankton in southern oceans, the primary food source for the abundant fish, krill, birds, seals and whales of the area.'

NEGOTIATING AND SIGNING THE MONTREAL PROTOCOL, 1987

Strong media support

The media's role was important in the negotiating and signing of the Montreal Protocol on 16 September 1987. Beginning in the mid-1980s, the scientific consensus about the Antarctic ozone hole led to the Western mass media's almost 'apocalyptic mood'[17] characterizing the issue. This resulted in renewed and increased attention on environmental issues 'among the general public as well as among the politico-administrative elites'.[18] '[T]he power of knowledge and of public opinion was a formidable factor in the achievement at Montreal,' according to Richard E Benedick.[19] 'A well-informed public was the prerequisite to mobilizing the political will of governments and to weakening industry's resolve to defend the [ozone-depleting] chemicals... The media, particularly press and television, played a vital role in bringing the issue before the public and thereby stimulating political interest.'

Just prior to the Montreal meeting, coverage of scientific issues continued to attract considerable media attention. The *Guardian* (7 August 1987) reported that: 'British scientists are calling for an immediate 85 per cent cut in the use of chlorofluorocarbons (CFCs) used as aerosol propellants and foam-blowing agents after establishing that the chemicals are responsible for a big hole in the stratospheric ozone layer over Antarctica. The demand [by] the British Antarctic Survey will put extra pressure on the UK and [European Economic Community] governments at an international protocol in Montreal next month to control CFCs.' Some media understood the magnitude of what was about to happen. The *Toronto Star* (7 September 1987) reported, 'This is a rare spectacle: a decision by almost all the important countries to stop bickering and trying to get an edge on each other, to do something together to fight what they perceive as a common threat.'

Media coverage of the Montreal meeting stressed its importance. Tom McMillan, Canada's Environment Minister, in an opening address, was quoted by the Associated Press (AP) (16 September 1987) as saying, 'We must not fail, for nothing less than the future of the planet Earth is at stake... A planetary time bomb is ticking away, and the pace is accelerating. The implications transcend national boundaries.'

The day the Protocol was signed, the AP reported, 'Major industrialized nations and members of the developing world today approved an accord aimed at slashing production of chemicals damaging the Earth's ozone layer.' The Inter Press Service (16 September 1987) reported that, 'The intense negotiations...which led to the treaty are the culmination of five years of world-wide talks involving scientists, politicians and industrial experts.' The magnitude of the agreement was made clear by a statement from Winfried Lang, chairman of the Diplomatic Conference for the Protection of the Ozone Layer, who was reported in that story as saying, 'The world is signaling to itself that certain kinds of chemicals are no longer acceptable.' The *Washington Post* (17 September 1987) noted that China and India were not among the signatories: 'The Indians did not participate in the conference. The Chinese participated but did not sign the accord, and diplomats noted that Beijing's ambitious plans to provide refrigerators to its bulging society could include a role for CFCs.' The signing of the Protocol was also the subject of news coverage on television broadcasts.

Media praise for the protocol

Media coverage of the closing session emphasized the significance of what had been achieved. For example, the Inter Press Service (16 September 1987) quoted Mostafa Tolba, Executive Director of UNEP, as saying, 'We have shown that we care – that we want to give...young people a world worth living in... We now have a respectable legal document – the protocol. But the legal document, any legal document, is only as good as the parties are willing to make it...Protocols don't save environments. People save environments. This protocol is a point of departure. It is the beginning of the real work to come.'

The *Financial Times* (17 September 1987) emphasized the 'historic' nature of the agreement and pointed out that it was 'the first treaty to control a global air pollutant'. A front-page story in the *New York Times* (17 September 1987) said, 'Hailing a milestone in international cooperation to safeguard the environment, delegates from rich and poor nations approved an agreement...intended to protect the earth's fragile ozone shield.' The *Guardian* (17 September 1987) reported, 'A global treaty to protect the ozone layer was agreed [to] in Montreal...The first world-wide pact ever signed to reduce pollution, it was hailed by the environment group, Friends of the Earth, which also warned that the treaty's compromise terms were inadequate to save the ozone layer from depletion.'[20]

'Initial media reactions to the Montreal Protocol...were highly favorable and recognized its precedent-setting nature,' Richard E Benedick later wrote.[21] For example, a *Newsweek* headline (28 September 1987) stated that the Protocol was 'An Exemplary Ozone Agreement,' and a story reported, 'It was a milestone in the annals of international politics: diplomats saw the future, didn't like it and decided to change it.' *Time* (28 September 1987) noted that '[M]ost of those present at Montreal praised the agreement'; *Chemical Week* (30 September 1987) reported that observers had characterized the event as 'an unprecedented international consensus'; the *Nation* (10 October 1987) said the Protocol represented a 'pioneering international agreement... The potential importance of the agreement cannot be over-emphasized. Never before have the world's principal industrial and developing nations agreed to control an aspect of their economies in response to an intangible long-term threat to the environment.'

Editorials written in the run up to and following the signing of the Montreal Protocol recognized the importance of the event, and were also generally positive. For example, prior to the conference, the *Toronto Star* (12 September 1987) wrote:

> '[A]s delegates begin to arrive in Canada, there is still no agreement on just how far they are prepared to go. While there is widespread consensus on the need to reduce emissions, the major producers seem to want someone else to take the lead. The US, in particular, has been backing away from a specific timetable for production cutbacks, fearing a loss in market share. Amid rumours that their own multinationals would attempt to circumvent the agreement by setting up CFC plants in countries not bound by it, the Americans appear to be pushing for a toothless deal.'

In the period before the Protocol was signed, *The Times* (11 September 1987) opined, 'The speed with which the West's industrial powers have co-operated to protect the Earth's atmosphere against harmful solar radiation is impressive. The environmental summit next week in Montreal is to be welcomed. Indeed, in comparison to the response to similar issues in the past, the reaction to the charges about compounds in certain aerosol sprays is almost worth a celebration.' In the aftermath of the signing of the Protocol, the *Washington Post* (18 September 1987) suggested that the ozone treaty might become an example for other similar agreements: 'We don't know if it can, but it is an extraordinary achievement on its own terms, the more so because of how quietly it was brought about. A major environmental threat has apparently been deflected with very little of the shouting that usually accompanies such problems – maybe because there was so little shouting. Good for everyone involved.' The *New York Times* (19 September 1987) editorialized that, 'Under the threat of the ozone hole, the countries meeting in Montreal have been frightened into salutary action.'

Continuing coverage of ozone issue

Ozone depletion continued to attract widespread visibility in the period after the Protocol was signed. For example, a *Time* cover story (19 October 1987) was entitled: 'The Heat is On: Why the Ozone Hole is Growing'. In addition, several weeks after the conclusion of the Montreal Protocol, the media reported on a new study by the US National Aeronautics and Space Administration (NASA) about ozone depletion. The *Boston Globe* (1 October 1987) reported, 'The 'hole' over the Antarctic in the Earth's vital ozone layer has worsened significantly this year, and there is strong evidence that widely used man-made chemicals are the cause, scientists [have] said.' Reporting on the same study, the *Los Angeles Times* (1 October 1987) said:

> '*What has baffled scientists since 1985, when a British team first noted the Antarctic ozone hole, has been the fact that Antarctic ozone levels have been depleted so rapidly while changes elsewhere have been difficult to discern. The new NASA experiments suggest that the explanation lies in an "exquisite interplay between meteorology and chemistry," said Dan Albritton of NOAA [National Oceanic and Atmospheric Administration]. The severe cold and particular wind patterns in the atmosphere above Antarctica help to transform the chlorofluorocarbons into active chlorine compounds that destroy ozone, said Albritton and his colleague from NASA, Dr Robert Watson.*'

Despite the signing of the Protocol, some media organizations in 1987 continued to be cautious in their treatment of the underlying science related to the issue. For example, *Chemical Week* (30 September 1987) reported that CFCs 'are suspected of damaging the stratospheric ozone layer that helps shield the earth from dangerous ultraviolet rays'. The *Los Angeles Times* (1 October 1987) said, 'Scientists have long recognized that chlorofluorocarbons…might destroy ozone.' However, others, such as *Time* (19 October 1987), were bolder in stating the causal link: 'Mounting evidence has demonstrated that under certain circumstances [CFCs], rising from the earth high into the stratosphere, set off chemical reactions that rapidly destroy ozone. The precise chemical process is still uncertain, but the central role of CFCs is undeniable.'

LONDON CONFERENCE ON SAVING THE OZONE LAYER, 1989

On 5–7 March 1989, the UK government and UNEP co-sponsored the London Conference on Saving the Ozone Layer that attracted representatives from more than 120 nations, 'an important political milestone'[22] in the effort to protect the ozone layer. In the run up to the conference, several topics attracted special attention from the media, including: the importance of the impending conference; a significant ozone-related agreement reached by the European Economic Community just days before the convening of the conference; a report about a six-week survey of ozone levels over the Arctic; and a change in the US position about the CFC phase-out.

Media stress need to enrol developing countries

Several of the UK's leading publications published editorials drawing attention to the importance of the conference. For example, the *Financial Times* (2 March 1989) wrote, '[N]ext week's Ozone Conference in London must be regarded as a possibly useful initiative that could have a positive effect…[T]he single most important diplomatic action that can be taken to protect the ozone layer is to convince as many countries as possible that the danger exists. Only then will there be any chance that some of them do something about it.' The editorial went on to suggest that the support of developing countries was essential to the success of the ozone-layer effort and, as such, the selection of President Daniel Arap Moi of Kenya for the opening speech was 'an inspired choice'. An editorial in *The Times* (4 March 1989) commented, 'It would be wrong, as some environmental groups have done, to dismiss the conference as a publicity exercise by a Government seeking 'green' credentials. Publicity, the airing of the scientific and industrial evidence, is essential.'[23]

EEC's 2000 phase-out decision hailed

Another focus of media coverage was the agreement by the Environment Council of the European Economic Community to eliminate by the end of the century gases that deplete the ozone layer. 'The European Community yesterday took a big step in defence of the world's threatened ozone layer by agreeing to a complete ban on the production of a range of chlorofluorocarbon industrial gases (CFCs) by 2000. The decision is bound to increase international pressure for a world-wide agreement to outlaw the gases,' the *Guardian* reported (3 March 1989). In a front-page story, the *Los Angeles Times* (3 March 1989) characterized the action as 'a surprise pact to virtually eliminate household gases used in refrigerators and spray cans that are destroying the Earth's protective ozone layer.'

Good coverage for science and the London Conference

There was also media interest in a new scientific report about ozone depletion over the Arctic. Based on the findings, the Xinhua News Agency (18 February 1989) reported, 'The ozone layer over the Arctic has been seriously depleted but no ozone hole has been found,' while the *Independent* (18 February 1989) reported, 'The atmosphere over the North Pole contains concentrations of ozone-depleting chemicals 50 times greater than anyone had expected, according to an international team of scientists… The scientists' report… confirms fears that the same mechanisms are at work in the northern hemisphere as those that produced the ozone "hole" over Antarctica.'

Meanwhile, an announcement by US President George H W Bush that he would support a global phase-out of CFCs, if safe alternatives were available, attracted media attention. In a front-page story, the *Los Angeles Times* (4 March 1989) reported, 'Heeding disturbing new warnings that the Earth's ozone layer is being destroyed at a startling pace, President Bush called Friday for a ban on use and production of ozone-depleting chemicals by the turn of the century, if safe alternatives can be developed.'

Once the London conference was underway, media attention shifted to other key topics including: several significant speeches from, among others, Prime Minister Thatcher, Kenyan President Moi, Prince Charles, and a group of key scientists; the matters of sharing technology and providing financial aid to developing countries to enable them to phase out CFCs; and the Soviet Union's reluctance to sign the Protocol.

Prime Minister Thatcher's welcoming speech attracted much media attention. She was quoted by the Xinhua News Agency (5 March 1989) as stressing the need 'to pool our knowledge and experience, to learn from each other, to improve understanding of the serious implications of the damage done...and to pave the way for further, concerted action.' The *Toronto Star* (6 March 1989) reported that the Prime Minister told the delegates, 'We need to go further and act faster, to accept higher targets and shorter deadlines...please do not set your sights too low.' *The Times* (6 March 1989) reported the Prime Minister as saying, 'Those whose products, and whose use of those products, are doing the most damage...have a heavier responsibility than others. It falls to them to do most to remove the causes of the problem.' The *Daily Telegraph* (6 March 1989) noted that the Prime Minister stated, '[T]he problem is not one for governments alone. It will require co-operation with science and industry as well, and I am very glad both are well represented at this conference.'

Kenyan President Moi's opening remarks to the conference also received wide media coverage. For instance, President Moi was quoted by the Xinhua News Agency (5 March 1989) as saying, 'We have assembled here to discuss the important and urgent matter of how we can together redress the damage man has caused to the ozone layer... The message is clear... We tamper with the natural forces of climate at our own peril. There can be no winners from the damage that man continues to inflict upon his one and only planet.' According to the *Toronto Star* (6 March 1989), President Moi said, 'The world community, especially the industrialized nations...must be prepared to bear the burden of conserving the ozone layer equitably with the less industrialized nations. They, too, must make sacrifices.' The *Financial Times* (6 March 1989) reported President Moi as saying, 'It is not a matter for nicely worded conventions and protocols. It is not a matter of concern to only a few members of an exclusive club. All members of the international community have a duty to protect the ozone layer.'

The speech by Prince Charles attracted particular media attention. In a front-page story the *Daily Telegraph* (7 March 1989) reported, 'The Prince of Wales threw down a direct challenge to world governments...to introduce an immediate ban on chemicals that destroy the ozone layer' and went on to quote the Prince as saying, 'I am not entirely sure if I represent the man in the street in this matter, but part of the problem has been to convince the man in the street that unless the ozone layer is protected, he won't be able to stand in the street without wearing sunglasses and a thick coating of No. 16 sun blocker (and that will just be in winter).' In another front-page story, the *Financial Times* (7 March 1989) reported the Prince as saying, 'As far as the contribution governments can make is concerned, I do not believe it is enough to rely on voluntary action alone. There should be an obligation to intervene and when appropriate to accelerate or enforce environmental measures.'

There was also keen media interest in the remarks of several scientists who had played key roles in the ozone-layer issue. For example, a story in the *Toronto Star* (6 March 1989) reported that F Sherwood Rowland had told the conference that it would take about 200 years to repair the damaged ozone layer. Similarly, the remarks of Joseph Farman, one of the scientists from the British Antarctic Survey who in 1985 published measurements of the Antarctic ozone hole, also attracted significant coverage by the media. *The Times* (6 March 1989) reported that Farman 'told delegates in a voice quivering with emotion: "Production of all long-lived halocarbons which can carry chlorine or bromine to the stratosphere must be run down, worldwide, as quickly as possible. Only then will the atmosphere return to a safer state".' Robert Watson of NASA was quoted as saying, 'What does it take to protect the ozone layer? Today, there are three parts per billion of chlorine in the earth's atmosphere. Before the ozone hole over Antarctica developed, there were only two parts per billion. Even with a fully-ratified Montreal Protocol, the amount of chlorine will increase to at least six to nine parts per billion. In other words, the chlorine will triple over the next few decades.'[24]

In Thatcher's closing speech, which generated considerable media attention, it was reported that she warned that '[E]ven if all destructive chemicals were banned tomorrow, it would take 100 years before the ozone layer would be restored to the level of the 1960s... Such is the extent of the damage we have already done... For centuries, mankind has worked on the assumption that we could pursue the goal of steady progress, without disturbing the fundamental equilibrium of the world's atmosphere and its living systems. In a very short space of time, that comfortable assumption has been shattered.'[25]

Developing-country position highlighted

One of the substantive issues discussed during the conference that garnered major media attention was the developing world's position that it deserved help from the developed world in dealing with ozone depletion. On the matter of sharing technology, the media reported that on the first day of the conference, delegates from the developing world emphasized the need to share the technology of the developed countries. The Xinhua News Agency (5 March 1989) reported, 'Delegates from developing countries, while fully agreeing with the necessity of protecting the ozone layer from further damage, expressed the view that developed countries should share with them the technology for replacing CFCs.' Moreover, Xinhua reported the next day the remarks by Liu Mong Pu, head of the Chinese delegation, who said, 'We hope those who have developed alternatives would transfer gratis the technology to the signatories (of the Montreal Protocol), especially the developing countries, so that the use of CFCs could be substantively reduced throughout the world.' Mong Pu went on to say, 'This is the glorious mission the developed countries owe to the human history. They should not simply lecture the developing countries to do this and not to do that.' There was also coverage of a call by representatives of developing countries for an international fund financed by industrialized countries to provide developing countries with more expensive substitutes for CFCs at no cost. Mong Pu was quoted by the *Washington Post* (7 March 1989) as saying, 'The developed world's accumulation of a great amount of wealth was accompanied by the pollution and destruction of the environment... Now these countries can use past accumulated wealth to manage the environment... Such is not the case for the developing countries.'

That story also reported that Ziaur Rahman Ansari, Indian Environment Minister, was in favour of industrialized countries establishing a fund to make substitute chemicals available and more affordable for the developing world, saying, 'Lest someone in this conference think of this as charity, I would like to remind them of the excellent principle of "polluter pays" adopted in the developed world.' In a front-page story, the *Los Angeles Times* (7 March 1989) noted that the positions taken by India and China, in particular, were crucial to the effort to reduce CFCs and reported, 'While the two countries now account for about 10% of the world's production and use of CFCs, the potential for growth as China and India industrialize to meet the expectations of burgeoning populations is considered immense.'

Media attention was also attracted by the Soviet Union's reluctance at the conference to endorse the treaty. The *Washington Post* (7 March 1989) reported, 'Although Soviet President Mikhail Gorbachev has called for quickened international action to deal with worsening environmental conditions, [Vladimir] Zhakharov [a member of the Soviet delegation] indicated that Soviet officials are not yet convinced that man-made CFCs account for ozone depletion.'

In the aftermath of the conference, the media reported growing public awareness of ozone issues and the political consequences of this awareness. The *Financial Times* (8 March 1989) wrote, 'the world conference on protecting the ozone layer...called for a revision of the Montreal Protocol to speed up the programme for phasing out ozone-harmful chemicals. Delegates also called at the final session...for a new international aid programme to help developing countries find substitutes for chlorofluorocarbons.' The *Los Angeles Times* (8 March 1989) reported, 'A major international environmental conference ended here Tuesday after 13 more nations agreed to join a global accord to save the Earth's ozone layer from destruction by harmful chemicals. Their commitment to reduce the consumption of those chemicals by 50% by the turn of the century came in the wake of a drive by major industrialized countries...to go even further by totally banning the chemicals by the year 2000.'

The media's characterization of the results of the conference was generally favourable. For example, the *New York Times* (8 March 1989) reported that, 'Prime Minister Thatcher... who was host of the conference, succeeded in one of her major aims to get more countries to agree to sign [the Montreal Protocol].' The *Los Angeles Times* (8 March 1989) reported that UNEP Executive Director Tolba said in a speech closing the conference, 'Finally, the alarm bells are ringing loud enough for the global public and heads of state and government to hear... We have affirmed our intention to declare a truce in our assault on the atmosphere.' Similarly, a story about the conference in the *Financial Times* (8 March 1989) was headlined, 'Solutions Elude Delegates but Event is "Success".'

Writers of editorials were also favourable in assessing the results of the conference. *The Times* (8 March 1989) editorialized:

'This week's international conference...had all the potential to be one of those North/South confrontations which have given gatherings of the kind a bad name... It may, instead, have marked the critical turning point in the battle to save the ozone layer. Last Sunday morning, only 33 of the world's governments had signed the Montreal Protocol... Yesterday, 20 more, including Brazil, had committed themselves to do so and as many more again had promised to consider it seriously. Even China and India, which had hitherto held aloof, hinted that they might join the club, provided aid and technology were forthcoming from the West.'

The *Daily Telegraph* (8 March 1989) wrote, 'The volume of influential voices swearing allegiance to Mother Earth and the defence of her realm grows ever louder. This is welcome.'[26]

At the same time, the growing public awareness about ozone depletion, and the consequent impact on world governments, received media attention. For example, the *Christian Science Monitor* (8 March 1989) reported, 'Growing public awareness of the threat to the ozone layer has pushed many governments to affirm their support for...the Montreal Protocol... This week 20 more countries announced their endorsement of the agreement, bringing the total number to 53, and over a dozen others have indicated they probably will sign.' A leading article in the *Daily Telegraph* (8 March 1989) reported that Prime Minister Thatcher had said, 'The power of public opinion and of the consumer is already making itself felt.'

During this period, coverage of the ozone-layer issue was characterized by a growing confidence on the part of the media that there was a direct relationship between CFCs and ozone depletion. For example, the *Sunday Telegraph* (26 February 1989) reported that CFCs were 'responsible for depletion of the ozone in the upper atmosphere.' The *Washington Post* (3 March 1989) reported in a front-page story, '[CFCs] do not break down in the atmosphere like other chemicals, but float into the stratosphere 20 miles high and erode the gaseous veil of ozone that screens out dangerous solar rays.' However, a report from the UPI news service (3 March 1989) was not so definitive: 'Scientists believe CFCs deplete the ozone layer.'

SECOND MEETING OF THE PARTIES, LONDON, 1990

The US policy

The second Meeting of the Parties took place in London on 27–29 June 1990. In the run-up period and during the pre-conference working sessions to the meeting, media attention was dominated by stories involving: a sudden change in US policy on establishing a new fund to help developing nations phase out ozone-destroying chemicals; unhappiness among some developing countries about certain aspects of the changed US position; and new scientific reports about ozone depletion.

There was widespread media coverage of the events surrounding the decision taken by the Bush Administration less than two weeks before the beginning of the conference to reverse course and support new financial aid to developing countries as a part of an amendment to the Montreal Protocol. Some media reported that the change in policy was made all the more significant because just weeks before, in early May, the USA had made a decision to oppose participation in a US$100 million fund that had been under discussion at ongoing UN-sponsored talks during the previous year.[27] The importance of this issue was underscored by an article in the *Daily Telegraph* (15 June 1990) which noted, 'First world aid is viewed as essential to gain the support of countries such as China and India for the Montreal Protocol process at the London meeting which begins next week... It is understood that India and China are unlikely to be wooed by the programme's negotiators if America does not come up with its share of extra funding.'

The *Independent* (15 June 1990) reported that:

> *'Britain has been putting pressure on the United States government to save an international ozone layer conference in London next week from breaking down. [Among the] key objectives...[is] to set up international funding whereby developed nations will pay poorer countries compensation for investing in alternatives. In preliminary negotiations, all the developed countries apart from the US agreed this funding should be over and above the money spent on third world aid. If the US cannot be persuaded to change its mind, developing nations may refuse to agree to the much stronger curbs needed to protect the ozone layer from accelerating destruction.'*

Once the Bush Administration had changed its position, press attention switched to coverage of the reversal. In a front-page story, the *New York Times* (16 June 1990) reported, 'Reversing a policy decision, the White House said today that it would support a new international fund to help poorer countries phase out chemicals that are destroying the earth's ozone layer.' The *Boston Globe* (16 June 1999) reported in a front-page story, 'Bowing to pressure from allies, UN officials and members of Congress, the Bush Administration reversed itself yesterday and said it will support a new fund to help poor nations phase out ozone-destroying chemicals... Yesterday's announcement represented the second about-face in administration policy on the issue in the last six weeks.'

A story related to the change in the US position on establishing the international fund, to be called the Interim Multilateral Fund, was to arise just days before the full conference convened. The story involved unhappiness among developing countries with some aspects of the new American position. As reported by the Press Association Newsfile (22 June 1990):

> *'Talks in London over plans to save the ozone layer suffered a blow today when the Chinese threatened to block moves to ban the chlorine gases at the heart of the problem...[I]n a veiled warning issued today, China said it may not sign the Montreal Protocol...if the United States insists on tough conditions over aid to developing countries. The warning could signal a rift between the Third World and Washington which is insisting on conditions of how aid is spent... The row over US demands began on Thursday when it*

was found there were strings attached to its U-turn in agreeing to provide extra cash to all the Third World to develop safe substitutes for chlorine gases.'

Under the headline, 'US clashes with Third World over ozone fund', the *Financial Times* (22 June 1990) reported:

'Proposals to establish an aid fund to protect the ozone layer ran into trouble at an international conference in London yesterday when the US clashed with representatives of Third World countries. A group of developing countries protested that the conditions the US was attaching to the fund were unacceptable. The US argued that, as potentially the biggest donor to the fund, it should have the biggest share of the votes on the executive council that will administer the money. It also wants a permanent seat on the council.'

The status of the ozone layer

The release of several new scientific reports on ozone depletion was also the focus of media attention. In a front-page story about a report by the UK Stratospheric Ozone Review Group, the *Financial Times* (21 June 1990) reported, 'The ozone layer is being depleted faster than expected, according to a report which Mr Chris Patten, Environment Secretary, described yesterday as "stark reading".' *The Times* (21 June 1990) reported, 'The vast "hole" in the ozone layer over Antarctica will not return to normal for 60 years even if the world community acts now to phase out ozone-destroying chlorofluorocarbons (CFCs), British scientists said yesterday... Scientists of the UK Stratospheric Ozone Review Group said in their 1990 report that even if the protocol target to phase out CFCs by 2000 was achieved, it would be 2050 before the chlorine level in the atmosphere returned to its normal state.'

Once the ministerial conference was underway, the media's focus continued to be on the disagreement between the developing and developed countries about the ozone layer issue in general and the establishment of the international fund in particular. The agreement to phase out CFCs ahead of the schedule first set out in the Montreal Protocol also received wide media coverage, as did the highly visible role in the negotiations played by Maneka Gandhi, the Indian Environment Minister.

Developing-country arguments

Media coverage of the opening day of the conference focused on developing countries' claim that the developed countries were treating them unfairly. The Xinhua News Agency (27 June 1990) reported that the developing countries had accused the developed countries of subjecting them to 'environmental colonialism', a charge that was also levelled at the USA by Malaysia on the second day of the conference and reported by the *Los Angeles Times* (28 June 1990). The Inter Press Service (28 June 1990) reported that Datuk Amar Yong, Malaysia's environment minister, said, 'If we are really serious about the danger to human beings, plants and other life forms, we must regard the depletion of the ozone layer as a crisis situation... To save the world, it is a small price for rich countries to pay... My government has a duty to ensure that our continued support of the protocol should not lead to the demise of our industries.' On the second day of the conference, the Xinhua News Agency (28 June 1990) reported, 'China warned today that any obstacles in the way of technology transfers of ozone-friendly substances and financial assistance for that transfer would ruin the efforts for worldwide ozone protection.'

Multilateral Fund

The proposed Interim Multilateral Fund also generated much media interest. For example, the Inter Press Service (27 June 1990) reported:

> *'Britain will contribute nine million dollars to help the Third World phase out the use of chlorofluorocarbons...Prime Minister Margaret Thatcher announced here today. The amount is Britain's contribution for the first three years of a $240 million global fund for CFC elimination currently being discussed... The prime minister said it was the duty of industrialized countries to help the Third World obtain and adopt the substitute technologies which will enable them to avoid the environmental mistakes of the north. "An important part of that will be to help them financially, so they can meet the extra costs involved," she said.'*

The press also reported disagreement about the control and size of the fund. *The Times* (28 June 1990) reported that Maneka Gandhi of India had said that the US$40 million that India would receive in the initial three years of the fund:

> *'would not be enough. She referred to figures from British accountants Touche Ross, who estimated that it would cost India $350 million to phase out CFCs... "You have to give us more," Mrs Gandhi said. "You've seen that India needs $350 million, but instead you're giving us $40 million if we sign and then you expect us to buy technology to provide for our end-users".'*

The *Financial Times* (28 June 1990) quoted Gandhi as saying, 'The West has caused the problem and must help us clean it up... People are picking on us as the villains of the piece.' Gandhi was also reported as suggesting that the developed nations give the developing nations the technology to make ozone-friendly substances: 'If you continue to clutch your patents to your chest, you may not have a world which you need patents for... We do not have 200 years to catch up. Maybe you should give us some of your knowledge now.'[28]

London Agreement hailed

On the final day of the conference, there was considerable coverage of the agreements that had been reached to amend the Montreal Protocol. In characterizing the agreements, the *Washington Post* (30 June 1990) reported in a front-page story, 'Government ministers from 92 countries agreed today to establish the world's first global environmental fund to help save the ozone layer and to speed the elimination of chemicals that damage the Earth's protective shield.' In a front-page story, the *New York Times* (30 June 1990) reported, 'In a landmark agreement, most of the world's nations vowed today that by the end of the century they would halt the production of chemicals that destroy the atmosphere's protective ozone layer. The agreement followed a last-minute concession by Washington. The USA agreed to provide technological help to poor nations in phasing out the production and use of chlorofluorocarbons.' *The Times* (30 June 1990) described the agreement as 'a watershed in co-operation between the developed and developing countries over protection of the environment.' The *Los Angeles Times* (30 June 1990), in a front-page story, reported that the agreement marked 'a dramatic acceleration' in the pact's previous phase-out schedule, and that delegates had added new chemicals to the restricted list.

In addition, much media attention was focused on the respective positions of China and India. The *Independent* (30 June 1990) published a front-page story under the headline, 'Deal to save ozone layer agreed; India and China accept decision to phase out CFCs after tense talks at environment conference.' With respect to India's position, the *Washington Post* (30 June 1990) reported that while the establishment of the new global fund had largely been guaranteed when the USA withdrew its objections, India had held out, seeking language that would allow it to abrogate the pact unilaterally if it determined that funding was not sufficient; while the final agreement did not permit abrogation, it did allow a developing-world country to appeal if it determined that it had been short-changed.

The early phase-out objective also received much media attention. UPI (27 June 1990) reported that Prime Minister Thatcher said, 'We need to speed up action, and set ourselves higher targets and shorter deadlines for reducing and eventually eliminating CFCs and halons, and extend the Montreal Protocol to cover other substances. So far, we've done well, but, as many of us used to have written on our school reports, "could do better".' On the other hand, the *Washington Post* (28 June 1990) reported, 'The United States came under strong criticism [on the second day of the conference] from environmental activists and Third World states, who charged that Washington and some other industrialized nations were delaying agreement to accelerate a world-wide ban on chemicals that deplete the earth's protective ozone layer.'

Media coverage of Maneka Gandhi was particularly extensive. The *Los Angeles Times* (3 July 1990) said Gandhi 'became a veritable media star,' while UPI (30 June 1990) said she 'attracted a virtual rugby scrum of journalists around her during her frequent passes through the [conference] lobby' late on the final day of the conference.[29] Moreover, the *Independent* (30 June 1990) wrote, 'One of the most interesting figures at the Montreal Protocol conference…has been a small, freckle-faced woman in a sari. This is Maneka Gandhi.' Gandhi's negotiating prowess was highlighted in the story in the *Los Angeles Times*, which reported that she 'fielded questions from the press, exhibited knowledge of the subject, and was cagey enough not to tip a negotiating hand. At times, she would say something and flash a knowing smile at her inquisitors as if they had just shared a private joke.' In one of the more memorable quotes of the conference, UPI (29 June 1990) reported Gandhi as saying, 'We did not destroy the ozone layer. [The West has] already done that. Don't ask us to pay the price... You can't just give us a small sum of money and say that's it. We're not beggars.'

Media point out NGO role

One aspect of the media's coverage of the 1990 event that was different from all earlier ozone-layer events was the increased amount of attention received by non-governmental organizations. Richard Benedick has suggested that:

> '*For their part, environmental organizations were demonstrating more sophistication than had been the case during the process leading up to Montreal. Both before and during the London meeting, Friends of the Earth International, Greenpeace International, and the Natural Resources Defense Council...held press conferences and circulated brochures and briefing sheets to the public, the media, and officials to match the customary public relations output of industry.*'[30]

The *Guardian* (26 June 1990) reported that 'Environmental groups...are strongly represented at the conference.' The NGOs that attracted the most coverage at this meeting were Friends of the Earth, Greenpeace and the Natural Resources Defense Council. In most instances, the references to NGOs came in the form of sources for stories. For example, in a *New York Times* story (16 June 1990) about the Bush Administration's policy reversal on establishment of a new international fund to help developing countries phase out CFCs, David D Doniger of the Natural Resources Defense Council was quoted as saying, '[The White House has] closed the hole in ozone policy that they opened themselves.' In the aftermath of the agreement, some NGOs were openly critical of what had been achieved. A Friends of the Earth campaigner said the agreement did not go far enough, and was quoted by the Press Association Newsfile (29 June 1990) as saying, 'If governments only do this, the Antarctic ozone hole will never recover. This is carte blanche in legal terms to the chemical industry to do as it pleases.'

The media also covered the release of a report by Greenpeace that was critical of the international community for not phasing out CFCs faster, and another report released by Friends of the Earth which said that a fund to assist developing countries in addressing CFC problems should be established by the developed countries.[31] NGO-related street theatre also attracted media attention. UPI (27 June 1990) reported, for example, that a group of about 20 protestors dressed as penguins and toxic-waste handlers paraded outside the opening day of the ministerial conference, and quoted one Friends of the Earth campaigner as saying, 'I'm dressed as a penguin to show that the Antarctic has the most severe ozone depletion and ultraviolet light, which puts all animal and marine life at risk... We're trying to impress upon delegates that we want a much tougher protocol than it looks like they are going to provide us with.'

Media coverage in the wake of the conference was both widespread and favourable. For example, in a front-page story, the *Los Angeles Times* (30 June 1990) reported, 'The delegates' actions, expected to be formally ratified by their governments, were hailed as a major advance in the worldwide efforts to restore the ozone layer.' Two days later, another front-page *Los Angeles Times* story began, 'For 10 days last month, delegates from nearly 100 nations labored on the banks of the River Thames across from Parliament to write a new chapter in environmental diplomacy. They succeeded. Agreement was reached on unprecedented amendments to the Montreal Protocol binding the industrialized West and developing nations in a single global cause – protection of the Earth's eroding ozone layer.' Meanwhile, *Time* (9 July 1990) characterized the results of the conference as 'a historic improvement on the already tough Montreal Protocol of 1987'. There were also laudatory opinion pieces and editorials written about the outcome of the conference.[32] However, there was also media coverage of critical comments about the conference. For example, the *Christian Science Monitor* (2 July 1990) reported that two NGOs, Friends of the Earth and Greenpeace, had criticized the agreement for not going far enough. Moreover, the *Independent* (1 July 1990) reported that Joseph Farman, one of the British scientists who reported the Antarctic ozone hole, had described the new agreement as 'very disappointing'.

Ozone depletion as an issue continued to receive major media attention. For example, UPI (17 June 1990) reported that 'scientific studies [were] clearly showing ozone depletion is worse than previously believed'. The *Financial Times* (21 June 1990) reported, 'Unusually for a global environmental problem, a scientific consensus about the cause of ozone depletion quickly formed. The culprits were identified as chlorofluorocarbons (CFCs) and some other substances widely used in refrigeration, air conditioning, industrial cleaning, fire-fighting and aerosols. This consensus resulted in swift action to curb the use of CFCs.' In an editorial commending the decisions taken in London, the *Journal of Commerce* (10 July 1990) said, 'Do CFCs destroy the ozone layer and thereby expose humans to dangerous increases in solar radiation? The scientific evidence is still coming in. But the evidence of hazards posed by CFCs is strong – and the world's politicians deserve credit for refusing to ignore them.'

FOURTH MEETING OF THE PARTIES, COPENHAGEN, 1992

The fourth Meeting of the Parties took place in Copenhagen on 23–25 November 1992. In the run up to the meeting, media attention focused on several stories, including: the rationale for the upcoming ministerial meeting; a World Meteorological Organization (WMO) report on the worsening level of ozone depletion; and the debate on whether methyl bromide should be phased out and, if so, on what schedule.

Coverage of the science

The rationale for the Copenhagen meeting was noted in several media reports. The *Independent* (17 November 1992) reported, 'The production controls agreed [to] in London two years ago only allow a very slow recovery, with ozone holes continuing through most of the century. In fact, rising production of other chemicals not covered by the Montreal Protocol, whose ability to damage the ozone layer has only recently been appraised, may mean it will never recover.' Moreover, *The Times* (23 November 1992) wrote that in light of new scientific discoveries of the depletion of the ozone layer over the northern hemisphere, 'The news gives added urgency to the [Copenhagen] meeting.'

The WMO report, which had been released on 13 November 1992, received widespread media attention. For instance, the Press Association Newsfile (16 November 1992) reported that:

> *'The fourth round of the Montreal Protocol talks...comes amid growing alarm in the scientific community over the extent of ozone depletion. [The WMO report] said there had been dramatic ozone depletion early in the year over northern Europe, Russia and Canada and above inhabited areas of southern Argentina and Chile in the autumn...The UN's World Meteorological Organization...said that between January and March, average levels across a belt covering Moscow, major Scandinavian capitals and cities of Northern Germany, and above most of Britain were between 15% and 20% below normal.'*

A third topic receiving media attention before the meeting was methyl bromide. *Chemical Marketing Reporter* (16 November 1992) noted, 'A coalition of environmental and pesticide reform organizations issued a report last week calling for the rapid phase-out of methyl bromide, a widely used farm chemical that has been linked to depletion of the stratospheric ozone layer.' On the other hand, *Chemical Week* (18 November 1992) reported that, 'The Methyl Bromide Working Group..., representing US producers, says the world's oceans are the chief generator of atmospheric methyl bromide, and it is funding research into the chemical's role in ozone depletion.'

Robert Watson's speech and the Greenpeace siren

Once the meeting was underway, several particular events generated media coverage. The first involved a speech by Robert Watson of NASA. The *Daily Telegraph* (24 November 1992) began its story about Watson's speech, 'Ultraviolet rays passing through an Antarctic ozone hole are killing plankton in the Antarctic ocean, raising the possibility of damage to the food chain and enhanced global warming, Robert Watson, a leading authority on the ozone layer, warned yesterday.' A UPI story (24 November 1992) on Watson's speech said, 'The intertwined food chain that starts with microscopic plants is already suffering because of the thinning of the Earth's ozone layer, a NASA scientist warned at an international conference.' The story involving Greenpeace related to a 'deafening siren' that Greenpeace activists installed near the conference, as reported in the *Toronto Star* (24 November 1992): 'The siren...was intended as an "ozone alarm," said the group.' The media reported that as a result of the siren, 12 Greenpeace activists were arrested.

Results of the meeting

In the aftermath of the meeting, there was widespread international news coverage of the results of the conference. Stories noted that the assembled nations had agreed to speed up from 2000 to 1996 the elimination of CFCs, which was perhaps the most visible topic covered by the media. For example, the *New York Times* (26 November 1992) reported in the lead paragraph in a front-page story, 'Spurred by recent evidence that the earth's protective ozone layer is being depleted even more extensively than feared, the nations of the world agreed yesterday to bring forward yet again the deadline for halting the production of the most important ozone-destroying chemicals... The chemicals affected [are] mainly chlorofluorocarbons.' *The Times* (26 November 1992) reported, 'Environment ministers and officials from nearly 100 countries yesterday more than doubled the speed of phasing out chlorofluorocarbons.'

While the agreement to cover hydrochlorofluorocarbons (HCFCs) within the Protocol was mentioned in some news accounts, the failure to achieve an agreement about methyl bromide attracted considerable media attention. For example, *New Scientist* (28 November 1992) characterized the discussions about methyl bromide as 'bitter wrangling'. UPI (25 November 1992) quoted an unnamed UN source as saying that the inability to reach a phase-out agreement on methyl bromide 'had been a serious failure of the conference'.

NGO criticism of results

There was less media coverage of the agreement reached by Parties in Copenhagen to make the Multilateral Fund permanent than of the London agreement to create the Interim Multilateral Fund. Media coverage of the Copenhagen Meeting of Parties focused on criticism by the NGO community of the stringency and pace of the phase-out. For example, Agence France Presse (25 November 1992) quoted a Greenpeace spokesperson as saying that the conference had 'given a green light' to the ozone layer's destruction. A spokesperson for the Environmental Defense Fund was quoted on the front page of the *New York Times* (26 November 1992) as saying, '"[The final agreement is] basically half a loaf...I'm glad they did what they did with the major ozone depleters." But on the issue of methyl bromide, he said, "they left an important part of the problem unfinished (controls for developing countries) and they're going to have to revisit it soon".'

As with past Meetings of the Parties, there was considerable editorial interest in the outcome of the conference. Editorial reaction to the Copenhagen amendment was generally favourable. For example, the *Irish Times* (26 November 1992) wrote, 'The Copenhagen agreement...is a welcome recognition of the danger [ozone-depleting chemicals] pose to the Earth's environment and a useful reminder that international co-operation to tackle them can work to everyone's advantage.' The *Financial Times* (26 November 1992) wrote, 'International conventions on the environment are too prone to self-congratulation. However, after this week's United Nations conference in Copenhagen...congratulation is deserved. The 1987 Montreal Protocol...was tightened for a second time. It stands as one of the world's most successful responses to an environmental threat.' However, some editorials were more critical. For example, the *Guardian* (26 November 1992) said:

> *'Stand by for a bigger hole in the ozone layer. The chemical industry won its confrontation with the green lobby in Copenhagen yesterday. Delegates from the 95 nations who signed the Montreal Protocol protecting the ozone layer have turned a blind eye to the newest chemical threats by setting phase-out dates that ignore science and put the world at renewed risk. The green lobby has every right to feel depressed because, until yesterday, the ozone story was one of the most enlightened chapters in the environmental book.'*

Coverage of ozone science continued to be characterized in various ways during this period. For example, a story in the *Independent* (17 November 1992) said, 'Scientists agree that the damage to the ozone layer...will grow through the rest of the century. Nothing can be done about the long-lasting chemicals, especially chlorine and bromine that industry and commerce have already put into the atmosphere.' The *Financial Times* (17 November 1992) reported, '[The Montreal Protocol's] proposals have won strong international support, partly because of the scientific confidence of the link between ozone depletion and [chemicals that damage the ozone layer] and partly because of the availability of substitutes.'

NINTH MEETING OF THE PARTIES, MONTREAL, 1997

Coverage before the Meeting

In the run up to the ninth Meeting of the Parties, held in Montreal on 15–17 September 1997, the media focused on two topics: the widespread illegal trade of chlorofluorocarbons and the resultant increase in pressure on the upcoming Meeting of the Parties to address this problem; and the impending policy debate on methyl bromide.

BOX 8.4 CONSUMER RESEARCH HIGHLIGHTS BOTH CONCERN AND LACK OF UNDERSTANDING

As the countries assembled in Copenhagen considered what amendments to make to the Montreal Protocol, there was media coverage of a new Opinion Research Corporation poll, which measured responses to questions about the ozone layer. US Newswire reported that, 'survey shows that Americans still have a long way to go in understanding the sources and effects of CFCs. The survey, commissioned by the Consumer Aerosol Products Council...found that 67 per cent of Americans surveyed consider themselves "extremely concerned about the environment". But among those who are "extremely concerned": almost one-half do not know how CFCs affect the environment; almost one-half cannot name the most common uses of CFCs.'

Source: 'Concern for Ozone Depletion Clear, Knowledge of CFCs Mostly Cloudy, Survey Shows', 25 November 1992, US Newswire.

Illegal trade in CFCs

Much of the media coverage relating to the illegal trade of CFCs was based on a report issued by the NGO Environmental Investigation Agency (EIA) which found Europe-wide smuggling of CFCs. For instance, the *Financial Times* (4 September 1997) reported, 'New evidence of Europe-wide smuggling of CFC gases yesterday increased pressure on western governments to tighten a ban on substances that are destroying the earth's protective ozone layer... The EIA published its findings in a report timed to influence a meeting of governments in Montreal next week to review the effectiveness of an international treaty to phase out CFCs.' The *Guardian* (4 September 1997) reported in its lead paragraph on the story, 'Evidence of a large-scale illegal trade in CFC chemicals has been uncovered by a British environmental group. Between 6,000 and 20,000 tonnes of CFCs...worth up to 90 million [pounds], are smuggled into Europe each year, often through Britain, according to official estimates.' One of the most reported reactions to the report came from John Gummer, former UK environment secretary, who said, 'If you traffic in CFCs, you are, in a real sense, trafficking in the lives of our children.'[33]

Methyl bromide

The other major story prior to the meeting related to how the topic of methyl bromide would be handled at the conference. The *Financial Times* (8 September 1997) reported that, 'An important aim of the conference is to secure a deal for developing countries to ban methyl bromide. But even the [developed] countries are finding its use hard to kick.' Underscoring the importance of the methyl bromide issue, Elizabeth Dowdeswell, Executive Director of UNEP, was quoted by the Canadian Press Newswire (15 September 1997) as saying, 'We can undermine the 10 years of good work that we've done with CFCs if we do not tackle the issue of methyl bromide.'

Coverage after the meeting

The media reported decisions to reduce CFC smuggling, the phasing out of methyl bromide, and the failure to reach an agreement on phasing out HCFCs. There was widespread coverage of the agreement to introduce a licensing system to curb smuggling of CFCs, under which licenses would be needed either to import or to export CFCs from countries that had signed the Protocol. The *European Report* (20 September 1997) reported that the new system should allow for more effective monitoring of the illegal trafficking. The *Financial Times* (19 September 1997) quoted a regulatory affairs spokesperson for ICI, one of the chemical companies that manufactured substitutes for CFCs, as saying, 'This shows that governments are finally taking illegal trade very seriously.'

The media also gave significant coverage to the decision reached to phase out methyl bromide. For example, the *Financial Times* (19 September 1997) reported that the agreement to phase out methyl bromide by 2005 in the developed world and by 2015 in developing countries 'fills in the last gap in the 1987 Montreal Protocol'. However, an editorial in New Scientist (20 September 1997) suggested:

BOX 8.5 OZONE FATIGUE?

Ozone Depletion Network Online Today reported on 12 September 1997 that John Passacantando, executive director of the Washington-based NGO Ozone Action, had stated that the public was no longer pressing politicians about ozone-depleting chemicals. *The Economist* (13 September 1997) similarly wrote, 'Another problem [with the Protocol] is ozone fatigue. Governments tend to think of the ozone hole as yesterday's problem, and there may not be the resolve to finish the business.'

'*A dangerous void is opening up at the heart of global environmental policy. It was evident this week, when ministers from around the world seemed apathetic in the face of calls from scientific advisers to the Montreal Protocol for fast phase-out of...methyl bromide... The contrast with Montreal a decade ago, when the world decided to act on the thinning ozone layer, is instructive. At the time, there were scientific doubts about the phenomenon. But the world decided to act anyway – just in time, as it now appears.*'

On HCFCs, *Chemical Week* (24 September 1997) reported, 'Despite pressure from the European Commission and environmental groups to accelerate the phase-out of ozone-depleting...HCFCs under the Montreal Protocol, the schedule will not be changed.' In its lead paragraph in a story about the issue, Ozone Depletion Network Online Today (24 September 1997) reported, 'Air conditioning and refrigeration equipment owners breathed a sigh of relief last week when delegates to the 10th anniversary meeting of the parties to the Montreal Protocol defeated a European Union...proposal to accelerate the ban on all...HCFC production to 2010 instead of 2030 as stated in the treaty.'

The coverage of ozone science continued to evolve. For example, *Business Week* (29 September 1997) reported, 'Science clearly proves that CFCs...eat away at ozone.' The *Independent* (4 September 1997) reported that CFCs 'consume the Earth's protective ozone shield'. According to *The Economist* (13 September 1997), CFCs were 'ozone-eating compounds'. With respect to methyl bromide, it was reported by the Canadian Press Newswire (15 September 1997), 'Methyl bromide has been shown to destroy ozone much more rapidly and efficiently than...CFCs.'

ELEVENTH MEETING OF THE PARTIES, BEIJING, 1999

The eleventh Meeting of the Parties took place in Beijing from 27 November to 3 December 1999. Media coverage prior to the ministerial part of the meeting focused on the release of a new UNEP ozone study and actions taken by the host Chinese government in response to ozone depletion. For example, the Deutsche Presse-Agentur news service (17 November 1999) reported that according to the new UNEP report, the Antarctic ozone hole covered 22 million square kilometres which was 'more than twice the size of mainland China'. Klaus Töpfer, Executive Director of UNEP, was quoted by Ozone Depletion

Network Online Today (23 November 1999) as saying that the size of the hole offers proof that 'much more needs to be done' to save the ozone layer.

The other prominent story, the actions taken by the Chinese government to protect the ozone layer, found more attention in the Chinese media. For example, *China Daily* (25 November 1999) reported that Xie Zhenhua, minister of the State Environmental Protection Agency, had said, 'With the efforts of enterprises and the government, we have frozen the consumption and production of ozone-depleting substances this year... We have every reason to believe that we can reach the standards stipulated in the Montreal Protocol by 2010.' In a story published by the China Business Information Network (25 November 1999), it was reported that Zhenhua 'made positive remarks on the role the [Montreal Protocol Multilateral Fund] played in protecting the ozone layer'.

Replenishment of the Multilateral Fund

The future of the Multilateral Fund attracted significant media attention during the Meeting of Parties. Reuters (2 December 1999) reported that the developing countries were seeking US$700 million. When it was agreed to replenish the fund with an additional US$440 million over the next three years, the lead paragraph of a story in the *New York Times* (4 December 1999) reported, 'An environmental conference of 129 countries...has agreed to provide an additional US$440 million over the next three years to help poor countries stop using chemicals that harm the protective layer of ozone in the upper atmosphere.'

Other issues

There was some media attention to toughening the controls on HCFCs. A story in *Chemical News & Intelligence* (3 December 1999) about the final agreement reached on HCFCs said, 'Under the [Beijing] amendment, trade in HCFCs... will be banned in countries that have not yet ratified the Protocol's 1992 Copenhagen Amendment, which introduced the HCFC phase-out. UNEP said this will provide an incentive to those countries to ratify the Protocol as soon as possible.'

The news media also reported a shift in emphasis from the phase-out in developed countries to phase-out in developing countries. K Madhava Sarma, Executive Secretary of the Ozone Secretariat, was quoted by Agence France Presse (4 December 1999) as saying, 'Phasing out CFCs...in developing countries is by far the most important next step in protecting the ozone layer.' The *New York Times* (4 December 1999) reported that Shafqat Kakakhel, Deputy Executive Director of UNEP, said, 'The developed countries have almost completely phased out ozone-destroying chemicals. Now it's time for the developing countries.'

In the aftermath of the meeting, there was considerable media attention to NGO criticism of the meeting. For example, the Agence France Presse (4 December 1999) wrote:

Box 8.6 Novelty and paradox made ozone depletion a news story

Geoffrey Lean, Environment Editor, Independent on Sunday, UK

The threat of ozone depletion first received wide attention in the popular media as the by-product of what then seemed a much greater issue, the future of supersonic air travel. Controversy was already raging about the cost, noise, sonic booms and fuel consumption of the Concorde and the planned Tupolev 144 and Boeing equivalents as the 1960s gave way to the 1970s, when their potential effects on the ozone layer became publicized. Those effects became irrelevant to most of the media as it became clear that large fleets of the planes would never fly.

The issue took off a second time in the media only after the publication of Molina and Rowland's paper in *Nature* in June 1974 and the subsequent news conference at the November 1974 meetings of the American Chemical Society. Again the media were slow to spot the story in *Nature*, taking three months to cover it on the front page of the *New York Times*; European newspapers were even further behind. Media executives then became fascinated by the story, for it had the elements of novelty and paradox that the media find so attractive. The idea that something as insignificant as the chemicals used in hairsprays and hamburger cartons could be endangering so distant and so vital a planetary shield caught their imagination. But their attention faded almost as fast as it had arisen. The debate confused the media, overwhelmingly populated by arts graduates, who found it hard to understand the process of scientific debate – a process that was to be repeated over global warming 15 years later. The limited steps to control CFCs in the USA and European Community in the 1970s also appeared to them to solve any problem that might have existed. For years after, it was extremely difficult to get space for stories about ozone depletion. The media – always more interested in events than processes – ignored the painstaking series of discussions under UNEP's auspices; even the Vienna Convention went virtually unnoticed.

It took the discovery of the ozone hole to reawaken the media, though again, they were slow to react. Here was evidence of an effect, something people could almost visualize. Intense interest replaced apathy, environmental NGOs began campaigning intensively, and the media became engaged. Newspaper pressure in Britain played a part in getting the government to change its policy, facilitating agreement in Montreal. Yet coverage of the Montreal Protocol was still limited, especially compared to the attention given to recent negotiations on the Kyoto Protocol. Interest grew at the subsequent meetings, as the original agreement was strengthened. This was perhaps the most interesting phase of all, for it was the very success of the process that was attractive, giving the lie to the belief that the media are interested only in bad news. What they like is a dramatic, unusual story, and this, both the scientists and the negotiators provided.

'Non-governmental organizations Saturday slammed a just concluded UN-sponsored meeting on eliminating ozone-depleting chemicals…blaming developed nations for a lack of funding and international diplomats for caving in to corporate interests. "Industry took care of themselves, diplomats took care of their jobs, China took care of the meeting arrangements, and no one took care of the Earth," Larry Bohlen, director of health and environment programmes at Friends of the Earth, [said].'

According to a Reuters report (2 December 1999), 'Non-governmental organizations called the Beijing conference a failure,' while the US Newswire (2 December 1999) reported a spokesperson for Friends of the Earth US as saying, 'The most successful international environmental treaty has hit the Great Wall of industrial interests and government indifference.' An editorial in *Gulf News* (6 December 1999) stated:

> *'The urgency of the Protocol cannot be over-stressed. Developed countries have already phased out [ozone-depleting chemicals], while developing countries are committed to do so by 2010. Where [ozone-depleting chemicals] are concerned, there can be no "us" and "them". The developed countries will suffer just as much as the developing, regardless of who is at fault. It is in the interests of the planet as a whole to eliminate [ozone-depleting chemicals], ignoring corporate interests.'*

Characterization of the underlying causes of ozone-layer depletion continued to evolve during this period. Media coverage emphasized the causal connection between CFC emissions and the depletion of the ozone layer. For example, Deutsche Presse-Agentur (17 November 1999) reported, 'The ozone is destroyed by chemicals such as chlorofluorocarbons.' The Xinhua News Agency (22 November 1999) reported, 'The ozone layer…has had large holes eaten out of it by substances such as chlorofluorocarbons.' On the other hand, the AP (5 December 1999) was more circumspect in reporting, 'Widespread use of CFCs and other chemicals are suspected of thinning the ozone.'

Chapter 9

Environmental NGOs, the ozone layer and the Montreal Protocol*

INTRODUCTION: NGOS AS 'SHAPERS OF POLICY'

Non-governmental organizations (NGOs) operate outside the realm of government and are characterized by their non-profit status. NGOs involved in the sphere of the Montreal Protocol are typically drawn from the sectors of environment, industry or health. Environmental NGOs have, in some cases, a value-based orientation or a cadre of volunteers.[2] UNEP played a crucial role in providing access to both environmental NGOs and industry NGOs at international negotiations on stratospheric ozone as an 'honest broker' of knowledge and information.[3]

> *'NGOs are no longer seen only as disseminators of information, but as shapers of policy and indispensable bridges between the general public and the intergovernmental processes.'*
> Kofi Annan, UN Secretary-General (1998)[4]

Environmental NGOs have actively participated and played an important role in efforts to protect the stratospheric ozone layer.[5] They have been involved from the start of public concern about ozone depletion when, in the mid-1960s, scientists and citizens organized to oppose the noise and other environmental effects of supersonic transport.[6]

In connection with work on the Montreal Protocol, NGOs have: participated in the Open-Ended Working Group and Meetings of the Parties as observers and interveners; identified and promoted policies and measures to implement the Protocol faster than mandated; advocated ozone and climate-safe and environmentally sustainable technologies; forged alliances with industry and governments; served as watchdogs not only on illegal trade but also on many other issues; and generally striven to ensure that the goals of the Montreal Protocol are achieved.

This chapter describes how environmental and health NGOs participated in a wide spectrum of activities from education to advocacy and how their strategy and tactics evolved over the period of controversy.

---

* This chapter was written by Corinna Gilfillan.[1]

THE ROLE OF ENVIRONMENTAL NGOS IN THE
OZONE CAMPAIGN

Environmental NGOs popularized the Molina–Rowland hypothesis and demanded action

Environmental NGOs helped to organize the 1974 press conference at which the ozone-depletion hypothesis of scientists Mario Molina and F Sherwood Rowland was promoted, advocated early action by UNEP, and organized the legal interventions and public boycotts of CFC aerosol products which led to the first national prohibitions of CFC products. Environmental NGOs remained active in national and international efforts leading up to the Vienna Convention and the Montreal Protocol, but did not attend the meeting at which the Vienna Convention was signed. After 1987, the number of activists and organizations increased dramatically to strengthen and implement the Montreal Protocol.[7]

> *'Natural Resources Defense Council (NRDC) staff scientist Dr Karim Ahmed was a chemist acquainted with Dr Rowland and quickly appreciated the environmental significance of the CFC threat. When the* Nature *article initially failed to receive much attention, Dr Ahmed arranged a technical session and press conference at a meeting of the American Chemical Society that generated the first national and industry press coverage.'*
> Alan Miller, former NRDC attorney[8]

From the very beginning, environmental NGOs were involved in raising public awareness about the environmental and health impacts of ozone depletion, and pressuring industry and governments to take action to phase out ozone-depleting chemicals. However, over the years, their involvement has graduated to more than raising public awareness.

> *'Environmental NGOs are mandated by their supporters to protect the environment. They act as the watchdogs for the global commons – the oceans, lands and the atmosphere. They hold governments and corporations accountable.'*
> János (John) Maté, Greenpeace International[9]

Environmental NGOs lose interest after initial success, pick up again in 1989

Although environmental NGOs were often active, there was a conspicuous lack of participation in 1985 when UNEP energized the diplomatic process with the signing of the Vienna Convention, and in 1987 when the Montreal Protocol was signed:

> *'It is hard to believe that as recently as 1985, not a single environmental group attended the signing of the Vienna Convention; and only a handful*

were here in Montreal ten years ago. The milestone in this development can be dated from Prime Minister Margaret Thatcher's 1989 Conference on Saving the Ozone Layer, which took place several months before the Protocol's first Meeting of the Parties, and which was attended by more than ninety environmental NGOs.'

Richard Benedick, former US Ambassador[10]

Environmental NGOs were not always able to maintain public and government interest and are not always able to raise sufficient funds to continue necessary momentum. Many have wondered whether NGOs will stay actively involved as the phase-out of methyl bromide and essential uses is complete in developed countries, but the phase-out of all ODSs continues in countries with economies in transition and developing countries.

> *'The US ban on aerosol propellant uses of CFCs in 1978 may have ironically led to a rapid reduction in public interest because of the perception that the problem had been addressed. For several more years, NRDC was virtually the only environmental organization actively following the issue. When EPA in October 1980 proposed a cap on US production of CFCs, NRDC filed one of less than a dozen comments submitted in support, while chemical companies and industrial users of CFCs filed thousands of opposing comments.'*

Alan Miller, former NRDC attorney

When environmentalists did not attend the meeting at which the Vienna Convention to Protect the Ozone Layer was signed in 1985, the then head of the legal division in UNEP who helped to prepare the text, Peter Sand, wrote an article about the agreement, noting that environmentalists would have to become more involved if effective international action was ever to come about.[11]

A variety of actions by diverse NGOs

Environmental NGOs' strong influence on the Montreal Protocol is the result of a variety of actions taken by diverse groups of environmental NGOs from around the globe.[12] International environmental organizations have pressured governments to create and implement an international treaty for protecting the ozone layer. The work of smaller organizations focusing on the national and local levels has also been key to building political support for action on the international level. (A CD available from the Ozone Secretariat includes a list of NGOs involved in ozone protection.)

The nature of NGOs derives from three main concepts: single-issue focus; ability to enhance the transparency of dominant actors, states, intergovernmental organizations and multinational corporations; and the multinational character of NGOs.[13] Environmental NGOs have assumed an important place in issue identification, agenda-setting, policy formation, normative development, institution-building, monitoring and implementation.[14]

To promote the phase-out of ozone-depleting substances on the local, national and international levels, environmental NGOs employed a variety of strategies:

- raising awareness and generating media coverage;
- advocating policy measures;
- advocating development of alternative technologies;
- pressuring industry and government to take action;
- collaborating with industry and government to find solutions;
- boycotting ozone-depleting substance (ODS) products/creating consumer demand for ozone-friendly products; and
- monitoring implementation of the Montreal Protocol.

The diversity of NGOs and strategies included the following:

- Environmental law organizations intervening in legislative, administrative and judicial processes to control the use and marketing of ODS products (eg the Natural Resources Defense Council).
- Environmental policy and research organizations that concentrated on policy analysis and communication (eg Center for Science and Environment, Investor Responsibility Research Center, World Resources Institute).
- Large environmental organizations with international networks that launched national and international ozone-protection campaigns demanding action (eg Greenpeace, Friends of the Earth, Consumer Unity and Trust Society).
- Organizations that focused on environmental issues that could have beneficial or adverse impacts by actions under the Montreal Protocol and/or had expertise and experience applicable to the Protocol (eg Pesticide Action Network's work on alternatives to methyl bromide, the Environmental Investigation Agency's focus on illegal trade).
- Grassroots organizations that focused on protecting the ozone layer through public protests, product boycotts, workshops for small farmers, street theatre, reports for policy-makers, and public awareness (eg Red de Accion Sobre Plaguicidas y Alternativas in Mexico).

> *'The primary motivation behind 'direct actions' and 'public confrontations' is to bear witness to activity that is ethically and environmentally "wrong," and when possible, to physically interfere with such activity through non-violent means. Confrontations are symbolic of the inherent conflict between the short-term interests of the perpetrators of a damaging activity, and the long-term ecological interests of the planet.'*
> János (John) Maté, Greenpeace International

Environmental NGOs took different approaches in promoting the phase-out of ozone-depleting substances, depending on their strengths and financial capacity, and the political, economic and social situation in their country and region.[15] Some environmental NGOs were more strident in their advocacy efforts,

Photograph by Greenpeace

Figure 9.1 *Greenpeace banner at DuPont water tower*

Box 9.1 Japanese society and the Greenpeace campaign to protect the ozone layer

Yasuko Matsumoto, Fellow, National Institute for Environmental Studies, Tsukuba, Japan*

The people of Japan have great respect for nature but have not always been strong when it comes to fighting for global environmental protection. It has been my honour to be part of the Greenpeace campaign that had two sides in Japan. On one side, we used every non-violent tactic to pressure companies to change their destructive ways. On the other side, we did our best to give Japanese companies the information and ideas they needed to take actions that would help protect the ozone layer and arrest global warming, and we encouraged the Japanese leadership to promote new technologies.

Until January 1990, the issue of ozone-layer depletion was the province of an exclusive group composed mainly of companies using CFCs, a handful of consulting experts, and the government officials in charge of the issue. However, in 1990, Greenpeace simultaneously launched ozone and global warming campaigns in Japan, making the issues much more public.

At that time, only one Japanese environmental NGO, the Citizens Alliance for Saving the Atmosphere and the Earth, was campaigning to protect the ozone layer and sending questionnaires to some Japanese companies. There was surprisingly little public or media concern about the consequences of ozone-layer depletion; there was no tense expectation or heightened consciousness in anticipation of the June 1990 London Meeting of the Parties scheduled to negotiate whether and when CFCs and other ODSs would be phased out.

Greenpeace Japan's highest priority was to organize public opinion and public action to demand government leadership at the London meeting. The first tasks were to inform the public about the latest scientific findings showing that the ozone-layer problem had not been solved, and to translate into Japanese Greenpeace International materials providing the latest information on negotiation issues and progress surrounding the Montreal Protocol. Greenpeace campaigners personally visited the offices of the national newspapers and other primary media, providing accurate scientific and technical information and explaining the importance of the June conference in London and the need to phase out ODSs quickly. Although this approach required a great deal of time and effort, Greenpeace reasoned that media concern would be the fastest way to build public opinion in the short time available.

Greenpeace used this same strategy of facts and persuasion at the time leading up to the Copenhagen Meeting of the Parties in 1992. Greenpeace representatives visited Japanese media offices in Tokyo and their branch offices in London to have them cover the conference in Copenhagen. Once the ozone-layer issue drops off the radar screen, people lose their awareness of it.

Greenpeace people in Europe had a hard time understanding the business culture in Japan, because they had often been successful by persuading a few leadership companies to support environmental protection and then encouraging or pressuring more reluctant companies to follow the leaders.

Despite the pack mentality, a few companies exhibited rare courage in ozone-layer protection. In December 1988, at a time when nearly all global companies were opposed to a CFC phase-out, Seiko Epson (Suwa City, Nagano Prefecture) was the first company in the world to develop a policy to discontinue CFC use. In April 1994, Matsushita Electric and a few others broke away from the pack of chemical manufacturers and other refrigerator manufacturers and began using cyclopentane in place of HCFCs as the blowing agent for consumer refrigerator insulation.

Methods used to obtain citizen support differ from one country to the next. One characteristic that sets Japanese society apart from Western societies is a cultural climate in which consensus is usually sought, and confrontation and aggressiveness are disliked.

While in the West confrontation attracts strong public support and is therefore an important part of the Greenpeace methodology, Greenpeace Japan modified its tactics for the Japanese milieu. Another important element of a successful campaign in Japan is accurate information and data. For Greenpeace to make its position acceptable to Japanese society, it must continually back its assertions with accurate information and data.

The Greenfreeze campaign started by supplying appliance makers and the related government offices with information and data provided by Greenpeace Germany. In late April 1993, we invited a German engineer and Greenpeace Germany representatives to conduct a Greenfreeze exhibit in Tokyo, which in four days had about 700 visitors, many of them from makers of consumer appliances and other CFC users. There were of course some critical comments, but most of the questions from engineers were sincere and straightforward, and Greenfreeze received coverage in the *Asahi Shimbun*, *Mainichi Shimbun*, and other major national news media. The exhibit made it possible for some consumer appliance makers to talk among themselves as individual companies instead of as industry associations.

A few months after the exhibit, Greenpeace Japan launched a campaign in which consumers sent postcards to appliance makers, directly asking them to produce Greenfreeze refrigerators. In April 1994, Matsushita Electric announced that it would sell a refrigerator whose insulation used a hydrocarbon as the blowing agent, and Sharp Corporation followed this with a similar announcement of its own. The pack fell apart when these companies broke ranks.

The switch to Greenfreeze would not have happened without the Greenpeace campaign in Europe and Japan. Although Greenpeace Japan was less rapid than Greenpeace in Europe in popularizing Greenfreeze, today nearly all Japanese appliance makers produce refrigerators with hydrocarbon-blown insulation. In late November 2001, three manufacturers (Matsushita, Toshiba and Hitachi) announced that they would start selling refrigerators that use hydrocarbon refrigerants in early 2002.

Finally, the Greenfreeze campaign was of great significance because it showed the importance of interconnectivity among policies and measures on different environmental issues, and it demonstrated that solutions can be found for the connected issues of climate change and ozone-layer depletion.

* From 1990 to 1998 Yasuko Matsumoto was Atmosphere Campaigner for Greenpeace Japan.

building political support for action by publicly pressuring governments and industry to take action. These activities involved grassroots organizing and actions including public education campaigns, peaceful protests and civil disobedience, and media coverage to highlight the need for action by government and industry.

Table 9.1 *Examples of the role of administrative and civil law in ozone-layer protection*

Year	Plaintiff/ Complainant	Defendant/ respondant	Issue	Resolution
1974 and 1975	NRDC	US Consumer Product Safety Commission (CPSC), Food and Drug Administration (FDA) and EPA	NRDC petitions and re-petitions US CPSC, FDA, and EPA to ban CFC aerosol products	CPSC bans manufacture of CFCs for non-essential aerosol products by 21 October 1978, manufacture of such products by 15 December 1978, and interstate shipment and import of products by 15 April 1979
1983	NRDC	US EPA	NRDC begins administrative procedure to compel EPA to determine whether action is necessary to protect the ozone layer as required by the Clean Air Act	Court orders EPA to demonstrate diligence in evaluating the consequence of continued emissions of ODSs
1985	NRDC	US EPA	NRDC gives EPA official notice of its 'intent to sue' to compel protection of the ozone layer	Government agrees to 'Ozone Protection Plan' including a timetable for action
1988	NRDC	US EPA	NRDC demands EPA to increase stringency and chemical coverage in order to protect the ozone layer adequately, as required by the Clean Air Act	Complaints satisfied when EPA calls for faster and further reductions of CFCs and halons and the control of methyl chloroform and carbon tetrachloride
1991	DuPont	Friends of the Earth	DuPont refuses to allow stockholder resolution urging faster phase-out	US District Court sides with DuPont in 1991 but in 1992 the Security and Exchange Commission rules that DuPont was wrong, saying 'timing of the CFC phase-out has strategic, long-term implications for the company's business and raises other significant policy issues'

1991	Hoechst	Greenpeace	Hoechst sues Greenpeace for billboards blaming ozone depletion on the actions of company executives	After 8 years of litigation, the German courts rule that company executives are 'public figures' and therefore subject to such scrutiny
1992	Greenpeace	ICI	Greenpeace complains to the UK Advertising Standards Authority that advertising HCFCs as 'ozone friendly' and HFCs as 'environmentally acceptable' is misleading	The UK Advertising Standards Authority rules in favour of Greenpeace in four out of five complaints of misleading advertising
1992	Greenpeace, Ozone Action and other ENGOs	Manufacturers of ODSs	ENGOs prepared to sue chemical manufacturers for damages from skin cancer and other consequences of ozone depletion	Suit is abandoned when attorneys advise ENGOs that courts may require specific proof that each skin cancer was caused by ozone depletion
1992 to 1993	NRDC	US EPA	NRDC demands EPA phase out methyl bromide as required by the Clean Air Act for substances with ODP > 0.2	EPA agrees to phase out methyl bromide
1995	Ozone Action, Environmental Law Foundation and Earth Day 2000	Amana, General Electric, Whirlpool and several retailers	ENGOs sue claiming that 'CFC-free' labels on products containing HCFCs are misleading by implying that the product does not deplete the ozone layer	Manufacturers agree to explain HCFC content and to provide a $100,000 grant to research alternatives to ODSs. Retailers agree to display conspicuous labels
1999	Environmental Working Group, Friends of the Earth, Pesticide Action Network, and Pesticide Watch	State of California	ENGOs sue the State of California to reduce the exposure from methyl bromide used as a pesticide	A superior court judge rules that the State of California must adopt regulations to protect the public from exposure to methyl bromide

LITIGATION AND COLLABORATION:
COMPLEMENTARY APPROACHES

Litigation for government action

In Germany, the USA and the UK, citizen organizations used litigation to compel government and corporations to action. Table 9.1 presents a summary of significant litigation for ozone-layer protection.

> *'NRDC sued the Environmental Protection Agency in 1984 under the US Clean Air Act, a powerful environmental law that allows 'citizen suits' when the government does not carry out its duties. In this case, the government had published a finding that the ozone layer was still at risk even after the United States had banned CFC aerosol sprays, because the use of ozone-depleting chemicals was still growing in other industries and around the world. Under the American clean air law, that finding of risk obligated the EPA to impose further restrictions. NRDC and EPA resolved the litigation with a settlement agreement that set a schedule for performing new scientific risk assessments, holding domestic and international scientific meetings to build consensus on the risks, and making decisions on further domestic regulations. Aware that this was a global problem, NRDC timed domestic deadlines to coordinate with plans to resume international treaty negotiations in 1986.'*
>
> David Doniger, NRDC attorney[16]

Other environmental NGOs worked more collaboratively with government and industry in developing policies and identifying alternative technologies for protecting the ozone layer. These activities included organizing awareness-raising programmes, publicly supporting industry's efforts to protect the ozone layer and producing technical and policy reports proposing solutions.

> *'The government of Germany hosted the "Second Ad Hoc Meeting On Ozone Depletion" in Munich in December 1978, and hired NRDC to review domestic responses of industrialized countries as a background document for the meeting. Notably, the NRDC authors – who were likely to know of NGO activities – cite few instances where environmental groups had any ongoing advocacy role on the issue.'*
>
> Alan Miller, former NRDC attorney[17]

Regional approaches of environmental NGOs

There were regional differences in environmental NGOs' approaches to the ozone-depletion problem. For example, environmental NGOs in Africa, Japan and Latin America tended to take a more collaborative approach with their governments, while environmental NGOs in Europe and North America publicly pressured their governments to take action.

Combinations of strategies

Environmental NGOs selected a combination of strategies appropriate for specific sectors. In some sectors (eg halons, solvents, vehicle air conditioning), industry was proactive and needed little motivation in developing effective, economical alternatives. Therefore, it was most effective for environmental NGOs to work as partners with government and industry in speeding the transition. In other sectors (eg methyl bromide), strong industry opposition to a phase-out required environmental NGOs to be more confrontational and to focus more on pressuring industry and governments by generating media coverage, creating alliances and organizing grassroots activities.

RAISING AWARENESS AND GENERATING MEDIA COVERAGE

Environmental NGOs have demonstrated the ability to reach out and educate the public and key target groups about the environmental and health impacts of ozone depletion and the importance of taking action. Public concern built the political support for creating the Montreal Protocol, and establishing and meeting specific timetables for phasing out ODSs. Environmental NGOs translate scientific and technical information to explanations that can be easily understood by journalists, the general public and industry.

> *'Environmental NGOs synthesise viewpoints, assemble, digest and process information into easily understandable problem definition and policy options. They also try to ensure transparency of the negotiation processes and demand accountability from negotiators.'*
>
> Joyeetz Gupta, Frederic Gagnon-Lebrun[18]

> *'"Brand name" environmental NGOs – such as Amnesty International, Friends of the Earth, Greenpeace, Sierra Club, and World Wildlife Fund – have earned a far greater level of trust than some of the most well-respected global multinational companies such as Ford, Microsoft, G-7 governments and global media... NGOs are trusted nearly two to one to "do what is right" compared to government, media, or corporations. Nearly two-thirds of respondents say that corporations only care about profits, while well over half say that NGOs "represent values I believe in"... They speak directly to consumers, appealing to emotions through simple and concise themes.'*
>
> Edelman Public Relations Worldwide[19]

Awareness of environmental threats raises public concern that influences political agendas and persuades politicians to promote environmental cooperation on the international level.[20]

> *'One of the most striking and effective activities of environmental organizations has been the "ringing of the alarm bell", making people who are becoming simply passive TV-audiences and active consumers wake up to the fact that humankind and the planet are inescapably one, and "bearing witness" to environmental atrocities.'*
>
> M Wuori (1997)[21]

Photograph by Greenpeace

Figure 9.2 *Greenpeace activists wearing protective clothing at ICI facility*

Environmental NGO awareness campaigns include: producing and disseminating materials such as reports, brochures, posters, badges and songs; generating media coverage; organizing meetings, conferences and workshops; organizing grassroots activities; and building coalitions of diverse stakeholders.

- At the preparatory meeting of the fourth Meeting of the Parties to the Montreal Protocol in 1992, activists from Friends of the Earth (FOE) International demonstrated outside the conference centre dressed as penguins, and handed out bottles of sunscreen to delegates as one part of a Europe-wide day of action to protect the ozone layer. These actions generated media coverage and put pressure on policy-makers at the meeting.
- Japan's Save the Ozone Network (JASON) produced a song called *What's the Ozone Layer?*, and a picture book to educate children and parents about ozone depletion and encourage action to protect children from the sun. JASON performed its ozone song and dance during the eighth Meeting of the Parties to the Montreal Protocol. It has also organized international environmental NGO forums, and an ozone-protection fashion show.
- Friends of the Earth (FOE) Canada produced a 'Blue Skies' public service advertisement that was broadcast on national television stations through 1996.
- In 1998, FoE USA delivered ozone-friendly tomatoes from Florida to members of the US Congress to demonstrate that alternatives to methyl bromide exist. This resulted in articles in several prominent newspapers in Florida and was part of a broader campaign to maintain the 2001 ban on methyl bromide in the USA.
- From 1999 to 2000, the Centre for Environmental Education in India carried out an awareness-raising programme to educate schoolchildren about ozone depletion. An 'educator's kit' and 'school kit' were developed and widely disseminated, and four workshops were organized to educate teachers about how they could teach children about ozone depletion. These activities are still being implemented and are helping to build public support in India for phasing out ODSs.
- In 2000, the Consumer Unity and Trust Society–Centre for Sustainable Production and Consumption (CUTS-CSPAC) organized events in several cities in India to put pressure on multinational refrigerator companies manufacturing and selling 'Ecofrig' hydrocarbon refrigerators in Europe and certain parts of America, while selling hydrofluorocarbon-based refrigerators in developing countries such as India. In the event organized on the occasion of the International Ozone Day, CUTS held mock funeral processions of a refrigerator on a deathbed in Calcutta and Jaipur. This attracted huge media attention and created immense pressure on manufacturers.
- The American Lung Association took the lead in encouraging the use of CFC-free metered-dose inhalers. It created a stakeholder group composed of patients' organizations, and developed and disseminated recommendations on transition strategies and other policies necessary to ensure a smooth and timely transition to CFC-free inhalers. These activities were critical to building support for that transition.

336 Protecting the Ozone Layer

- Red de Accion Sobre Plaguicidas y Alternativas (RAPAM) in Mexico has worked to raise awareness among policy-makers and the public about the methyl bromide phase-out. In 1999, RAPAM released a report entitled 'Use and Alternatives to Methyl Bromide in Mexico' which monitored the accuracy of the information provided by the Mexico Ozone Protection Unit to the Montreal Protocol Secretariat, and advocated multiple stakeholder consultation in the design and implementation of a national methyl bromide phase-out plan. RAPAM also organized workshops about the effects of methyl bromide and alternatives for small flower growers' associations, and recommended experts to participate in UNEP training programmes.

In 2001, agricultural and environmental NGOs were working in partnership with the UNEP Division of Technology, Industry and Economics OzonAction Programme to raise awareness about the methyl bromide phase-out and available alternatives in ten developing countries (Chile, Costa Rica, the Dominican Republic, Ethiopia, Kenya, Malawi, the Philippines, Thailand, Zambia and Zimbabwe). The Methyl Bromide Communication Programme was the first project approved under the Multilateral Fund to utilize the expertise of environmental NGOs in helping developing countries to phase out ODSs. These environmental NGOs are carrying out a variety of activities, including developing and disseminating brochures, holding meetings and workshops with farmers, and conducting media campaigns.

ADVOCACY WORK ON POLICY AND ALTERNATIVE TECHNOLOGIES

Advocating policy measures

Environmental NGOs have advocated specific policy measures they believe are necessary to protect the ozone layer. Their positions were often based on scientific information on ozone depletion and the latest technological developments on alternatives. Many environmental NGOs have been strong proponents of setting specific phase-out dates for ozone-depleting chemicals, providing funds for developing countries to meet the phase-out requirements, reviewing and revising phase-out schedules based on evolving scientific and technological developments, and replacing ozone-depleting chemicals with environmentally sustainable alternative technologies.

> 'NGO advocacy of a fund for assisting developing countries with technology conversion, at a moment when government negotiators lacked creative solutions and leadership, enhanced NGO influence on the issue as well. It is impossible to assess how much the achievement of the 1990 London agreement can be attributed to NGOs, but there is certainly enough evidence to suggest that it happened sooner because of the NGO campaign.'[22]

In September 1987:

> '*NRDC hailed the Montreal Protocol as "a major half-step forward" and immediately renewed its campaign for a full phase-out, both domestically and globally. In 1990, NRDC joined with congressional leaders of both parties – most notably Senators John Chafee, Max Baucus and Al Gore, and Representatives Henry Waxman and Sherwood Boehlert – in crafting new Clean Air Act amendments to phase out CFCs in the United States.*'
>
> David Doniger, NRDC attorney

Environmental NGOs advocated policy measures by issuing policy reports, sending letters to governments, meeting with government officials, participating in Montreal Protocol negotiations to promote policy positions, building alliances, building public support for policy initiatives, and generating media coverage.

On 8 September 1987, 59 environmental NGOs from around the globe (Argentina, Australia, Brazil, Canada, Ecuador, Greece, India, Indonesia, Japan, Korea, Malaysia, Mauritius, Mexico, New Zealand, Nigeria, Sweden, Switzerland, Thailand, the USA, Uruguay and West Germany) issued a policy statement entitled 'Ozone depletion and climate change: A statement on CFCs and related compounds by the international environmental community'. This statement outlined the environmental and health risks posed by ozone depletion and climate change, and the need for governments to take rapid action to reduce emissions from CFCs and other ozone-depleting compounds.

At the 1989 ozone negotiations in Helsinki, 93 environmental NGOs from around the world released a statement entitled 'Safeguarding the ozone layer and the global climate from CFCs and related compounds.' This statement outlined the environmental and health impacts of ozone depletion and climate change and called for a 'crash programme' to: phase out CFCs and other ozone-depleting chemicals by 1995; ban the use of any new ozone-depleting chemicals; create an international fund to assist developing countries in protecting the ozone layer, financed partially by a fee on CFC producers; and enforce trade sanctions to ensure that countries became signatories to the Montreal Protocol.

Environmental NGOs from around the world have participated in Montreal Protocol negotiations and meetings of the Executive Committee of the Multilateral Fund to pressure governments to adopt policies to protect the ozone layer. The NGOs have carried out numerous activities at these meetings, including: meeting with government officials; disseminating reports and materials outlining NGO positions; holding press conferences; organizing demonstrations or civil disobedience; and producing *ECO*, a newsletter that presents the environmental NGOs' perspective on the negotiations.

In 1993, the Methyl Bromide Alternatives Network (MBAN), a coalition of 16 environmental, farmer, labour and consumer NGOs, was created to promote the phase-out of methyl bromide in the USA and globally. In 1995, MBAN contacted 1000 scientists, policy-makers and journalists about the dangers of methyl bromide use and the availability of environmentally sustainable alternatives.

In 1995, NGOs formed the International NGO Alliance to Protect the Ozone Layer, an informal network of hundreds of NGOs around the globe working on or interested in ozone-layer protection issues. Coordinated by Pesticide Action Network North America (PANNA), this Alliance jointly sent letters to key government leaders involved in the Montreal Protocol negotiations to urge them to agree to policy measures for a methyl bromide phase-out. In 1997, the Alliance sent letters to 17 countries urging them to ban methyl bromide.

JASON has lobbied and collected 140,000 signatures in support of legislation in Japan to prevent the release of ozone-depleting chemicals. Its advocacy and awareness-raising activities have led to 1000 local governments (of the total 3000) beginning programmes to recover CFCs from household refrigerators and air conditioners, and some of them enacting ordinances to require CFC recovery.

Advocating alternative technologies

Environmental NGOs have sometimes joined partnerships with government and industry to develop, commercialize and promote alternatives and substitutes to ozone-depleting substances.[23]

> 'We chose to see the demands of the environment protection groups as a challenge... We attacked the problems rather than setting up a defence. It gave us an excellent edge in the development and use of the Greenfreeze Technology... It is good for the environment – and it proves to be good business in a number of key markets that also want to protect the environment.'
> Erling Damkaer, Managing Director, Als Vestfrost

> 'Greenpeace's most notable accomplishment was the introduction of Greenfreeze hydrocarbon technology in refrigeration.'
> János (John) Maté, Greenpeace International

Greenpeace has aggressively promoted the adoption of 'Greenfreeze' refrigerators using hydrocarbon refrigerants and insulating foam.[24] Greenpeace has worked with companies in Europe and Japan to produce this technology and is also expanding its work to China and other developing countries. Greenpeace has produced case studies, videos and numerous technical reports highlighting how these technologies work, and promoted them at the Meetings of the Parties, and other ozone meetings and conferences. Two 1999 videos called 'COOL technologies: Working without HFCs' and 'Back to the future: Working safely with hydrocarbons' were jointly produced by Greenpeace, UNEP Division of Technology, Industry and Economics' OzonAction Programme, and Deutsche Gesellschaft fur Techische Zusammenarbeit (GTZ).

The Consumer Unity and Trusts Society (CUTS) launched a campaign in India to promote 'Ecofrig', a hydrocarbon-based energy-efficient, ODS-free refrigerator. It organized national consultations with top government officials, refrigerator and compressor manufacturers, and representatives from consumer and environmental groups to inform them about environmentally friendly refrigerators. CUTS also developed promotional material for informing and educating consumers on Ecofrig and held technical sessions by speakers of various research institutions to look at technical issues. CUTS held regional and national workshops to disseminate information, and developed a large network of like-minded grassroots NGOs to disseminate information throughout India. Indian consumers were able to buy their first Ecofrig refrigerators from one of the largest manufacturers in India in 2001.

Friends of the Earth (FOE) in the USA and Canada – with funding from the governments of the USA, Canada and Switzerland – published a study in 1996 entitled 'The technical and economic feasibility of replacing methyl bromide: Case studies in Zimbabwe, Thailand and Chile'. This was one of the first technical studies focused on the use of methyl bromide in developing countries. The report identified feasible alternatives, provided detailed analysis of the costs and performance of alternatives, and built support on the international level for a methyl bromide phase-out in developing countries.

In 1997, Pesticide Action Network/North America (PANNA) and Health and Environment Watch organized a technical workshop on 'Existing and potential alternatives to methyl bromide in cut flower production in Kenya'. This workshop educated cut-flower growers in Kenya about the methyl bromide phase-out, and presented detailed information on how growers in Colombia produced flowers without methyl bromide. Participants included technical experts and researchers from Colombia's cut-flower industry, farmers, NGOs, UNEP officials and government representatives.

OzonAction and FOE USA organized briefings by technical experts and scientists to inform Members of Congress and their staff about ozone-depletion issues. For example, OzonAction organized a briefing on Capitol Hill at which Mario Molina presented information about methyl bromide's impact on the ozone layer, and an expert from the Methyl Bromide Technical Options Committee (MBTOC) discussed alternatives to methyl bromide.

WORKING WITH INDUSTRY AND GOVERNMENT

Pressuring industry and government to take action

> '*It has been a great experience to work with international environmental organizations such as the Pesticide Action Network and Friends of the Earth on alternatives to methyl bromide. This cooperation allowed us to get information directly from governments and United Nations technical experts that we use for public education at national level. Access to technical information made it possible to demand public participation and accountability of the ozone protection policy in Mexico.*'
>
> Fernando Bejarano, Coordinator of Red de Accion sobre Plaguicidas y Alternativas en Mexico (RAPAM.PAN Mexico)

Environmental NGOs have pressured industry and governments to take steps to protect the ozone layer at the international, national and local levels. This has been accomplished by publishing reports on the issue and need for action, organizing grassroots activities to educate the public, exposing and publicizing companies and governments whose actions (or lack of actions) are damaging the ozone layer, generating media coverage about these issues, and lobbying government and industry officials.

In 1975 NRDC urged the public to shun CFC aerosol products, and successfully sued the US government to ban CFC aerosol products. Consumer and political campaigns in Canada, Norway and Sweden also resulted in bans of CFC aerosol products.

In 1986–1987, children and grassroots organizations in the USA successfully campaigned for McDonald's to stop using CFC packaging for its fast-food products. Lois Gibbs of the Citizens' Clearinghouse on Hazardous Wastes also sent a letter to McDonald's raising this issue, which generated significant publicity and concern among the public. As a result, McDonald's pledged to phase out its own CFC use and to require its suppliers to eliminate CFC use in foam packaging within 18 months. This corporate leadership contributed to the agreement by foam food-packaging manufacturers to phase out CFCs within the year.

Three weeks after the Montreal Protocol was signed, 33 Friends of the Earth International affiliates passed a resolution making ozone-layer protection their top priority, urging a worldwide ban on CFC aerosol products. In the UK, The Netherlands, West Germany, Belgium, Australia, France, Cyprus, Hong Kong, Malaysia and other nations where aerosols were still the major use of CFCs, these groups called on industry to phase out CFC aerosol products, and organized consumer boycotts. In The Netherlands, aerosol manufacturers agreed with the environment ministry to stop using CFCs by the end of 1989. In Britain, the Prince of Wales voiced his support for the upcoming boycott – and industry announced a phase-out just three days before the consumer boycott was to begin.

The achievement of a 50 per cent reduction in CFC emissions:

> *'has actually been made possible by the thousands of ordinary consumers and environmentalists whose concerted pressure persuaded the aerosol manufacturers to phase out their use of ozone-depleting CFCs by the end of this year (1989)…Human beings can be rightly proud of their inventiveness. We thought the world belonged to us. Now we begin to realize that we belong to the world. We are responsible to it, and to each other. Our creativity is a blessing, but unless we control it, it will be our destruction.'*
>
> Charles, Prince of Wales

While the International Conference on Saving the Ozone Layer was being held in London in 1989, Greenpeace activists scaled the sides of an ICI chemical plant in Denmark and hung a banner that read, '100% halt to CFCs now'. The action was covered in the British press and helped draw greater attention to industry's role in causing ozone depletion.[25]

In 1997, the Transnational Resource and Action Center and Political Ecology Group published a report entitled 'Bromide barons: Methyl bromide, corporate power and environmental justice', which accused methyl bromide

Photograph by Greenpeace

Figure 9.3 *Sign in German at Kali Chemie facility*

producers and distributors of obstructing global efforts to phase out methyl bromide. The report was widely distributed to policy-makers and grassroots activists.

In 2001, CUTS-CSPAC, along with South Asia Watch on Trade, Economics and Environment (SAWTEE) and the UNEP OzonAction Programme, organized the South Asian Consultation on Atmospheric Issues to increase awareness of parliamentarians and elected legislators in South Asia on ozone depletion and interconnected issues of climate change. These law-makers are expected to serve as pressure groups on government and industry by bringing up the issues in the parliament and state legislative assemblies. The conference led to formation of a caucus of elected legislators, civil societies, grassroots groups, and media to facilitate information exchange and develop a common stand on atmospheric issues at the regional level in the near future.

Collaborating with industry and government to find solutions

Another effective approach to environmental advocacy on ozone protection involved working collaboratively with industry and government to identify solutions to phase out ozone-depleting substances and find alternative substances and technologies. Such collaborative projects often led to win–win solutions: the ozone layer was protected, and the alternative technologies that were developed were often equally if not more effective than the ODS-dependent technologies.

In 1988, the US EPA, FOE USA, the Center for Global Change, NRDC and the Environmental Defense Fund (EDF – now Environmental Defense) formed a partnership with the food-packaging industry to phase out CFCs voluntarily in the manufacture of foam packaging. The foam manufacturers agreed to end the use of CFCs within one year (by December 1988) and agreed to use HCFC-22 only as an interim alternative. The companies committed themselves to developing a safer and more environmentally sustainable alternative to HCFCs, and a working group was formed to address this issue. The environmental NGOs publicly praised the industry's efforts to protect the ozone layer and urged other industries to follow suit. This agreement led to similar voluntary phase-out agreements in other countries, including Canada, The Netherlands and the UK, and promoted the development of environmentally sound foam-blowing agents.

FOE USA implemented the Halon Recovery Campaign, a grassroots initiative to promote the collection of halons in communities in the Mid-Atlantic region of the USA. This programme, which was supported by the US EPA and the Halon Recycling Corporation, helped to prevent further emissions of halons, and supported industry's efforts to meet critical-use needs and manage the existing stock of halons in an environmentally responsible manner.

A coalition of environmental organizations, electric utilities and governments developed a US$30 million 'Golden Carrot' prize to encourage companies to create CFC-free and energy-efficient appliances. This provided a powerful incentive for companies, leading to Whirlpool's introduction of a new model of refrigerator that exceeded energy-efficiency standards set by the US Department of Energy.

Comite Nacional Pro Defensa de la Fauna y Flora (CODEFF) worked closely with the Government of Chile to raise awareness about the methyl bromide phase-out and alternatives. It organized a seminar on methyl bromide and destruction of the ozone layer targeted to methyl bromide users in 1997, which was a critical step in Chile's efforts to replace methyl bromide. In 2000, CODEFF also worked jointly with the UNEP Division of Technology, Industry and Economics' OzonAction Programme and Government of Chile to organize the Regional Policy Development Workshop to Assist Methyl Bromide Phase-Out in Latin America, where countries identified policy measures and national action plans to phase out methyl bromide.

BOYCOTTING ODS PRODUCTS AND CREATING DEMAND FOR OZONE-FRIENDLY PRODUCTS

At the September 1974 American Chemical Society Press conference promoting their ozone-depletion theory, scientists Mario Molina and F Sherwood Rowland recommended regulatory controls on CFCs. Later, NRDC scientist Karim Ahmed and attorney Thomas Stoel met with Rowland and agreed that a ban on CFC aerosol products was the best initial strategy.[26] By November 1974, NRDC had announced its intention to petition the US government to ban cosmetic aerosol uses. These actions increased public awareness, including within the entertainment industry. In February 1974, characters in the American hit television show *All in the Family* debated whether the deteriorating world was a proper place to raise children, and accused CFC aerosol hairsprays of depleting the ozone layer. As product boycotts were organized, the S C Johnson Company broke ranks with competitors and pledged to halt the use of CFCs.

Environmental NGOs generated public demand for ozone-friendly products by targeting awareness-raising activities to consumers, and by advocating labelling requirements that informed consumers whether a product was produced with or without ozone-depleting substances. These actions pressured companies to stop using ozone-destroying chemicals.

By late 1989, FOE, in coordination with other organizations, persuaded more than 24 US cities to pass laws to reduce ozone-depleting gases more quickly than regulations outlined in the Montreal Protocol, to ban CFCs from packaging foam, and to recover and recycle CFCs from all products that contained them. These local regulations complicated national marketing and servicing, and persuaded industry to advocate more stringent national regulations that could satisfy local regulators.[27]

In 1994, several members of the MBAN sued the US EPA to require the labelling of foods grown with methyl bromide. While labelling was never required, the lawsuit helped to generate pressure on methyl bromide users to switch to alternatives.

In 1995, OzonAction and the Environmental Law Foundation sued Whirlpool and other companies for placing misleading 'CFC-free' labels on refrigerators produced with HCFCs. The environmental NGOs argued that the label was misleading advertising because it implied that the product didn't deplete the ozone layer. In 1995, a settlement required manufacturers to disclose the fact that HCFCs were used if claims were made that a product was CFC-free.

The Sustainable Tomatoes Campaign, a coalition of labour, environmental and consumer groups, has worked since 1998 to raise awareness among consumers in Florida about the environmental and health impacts of methyl bromide use on tomatoes. Its activities have included releasing a report on the hazards of methyl bromide use in Florida, disseminating brochures and stickers to consumers in Florida, and generating media coverage.

MONITORING IMPLEMENTATION OF THE MONTREAL PROTOCOL

Environmental NGOs have not only pressed for regulations and an international treaty for ozone-layer protection, but have also been active in ensuring that governments comply with their treaty commitments. Environmental NGOs have closely monitored the implementation of the Montreal Protocol, exposed loopholes in the treaty, and held countries and companies accountable to commitments they have made.

The Environmental Investigation Agency (EIA) has exposed illegal trade in ODSs and proposed specific recommendations for addressing illegal traffic. EIA issued a report called *Chilling Facts About a Burning Issue: CFC Smuggling in the EU* which documented an illicit market in CFCs throughout Europe. In 1997, EIA also gave a presentation on the illegal trade problem to the US Federal

Task Force on Illegal CFCs. EIA's efforts to expose this problem have played a key role in the Parties adopting measures to combat illegal trade in CFCs and other ODSs.

FOE International has monitored the performance of the Multilateral Fund and advocated ways to increase its efficiency and effectiveness. In 1992, FOE released a report entitled *At the Crossroads: The Multilateral Fund of the Montreal Protocol*, which evaluated the Fund's performance. FOE has also monitored whether industrialized countries have paid their contributions to the Multilateral Fund on time and has publicly exposed the countries that have been slow in making payments through press releases, newsletters and other publicity actions.

PANNA monitored the implementation of methyl bromide alternative projects under the Multilateral Fund. PANNA and FOE USA worked with 'Environmental NGO Contact Groups' in Mexico, Chile, Senegal and Malaysia to monitor the projects, provide input on project proposals, promote sustainable alternatives to methyl bromide, and participate in Executive Committee meetings. PANNA also published a report entitled *Funding a Better Ban: Smart Spending on Methyl Bromide Alternatives in Developing Countries*, which provides recommendations on how the Fund can effectively promote the adoption of methyl bromide alternatives.

CONCLUSION

Environmental NGOs are widely credited with effective advocacy of stratospheric ozone protection, including traditional roles in raising public awareness, and new roles in policy development and technology advocacy. This experience increased their political clout and fundraising ability, and has allowed greater participation in pending global environmental issues.

> *'Following this meeting [the 1989 London Conference], the NGOs extended their collaboration to preparations for the UN Conference on Environment and Development and to negotiations for the climate change, biodiversity, and other environmental conventions…NGOs became increasingly sophisticated in the negotiating process.'*

> Richard Benedick (1998)[28]

In the proposed Sixth EU Environmental Action Plan, the European Commission suggested that the European Community continue to provide financial support to environmental NGOs to facilitate their participation in the environmental policy-making process. It acknowledged that NGOs have played an important role in broadening the environmental dialogue in the EU, in channelling the views of the 'person in the street' to decision-makers, in participating in expert or technical groups, and in monitoring the implementation of legislation.[29]

Chapter 10

Conclusion: A perspective and a caution

'*Now there is one outstandingly important fact regarding Spaceship Earth,
and that is that no instruction book came with it.*'

R Buckminster Fuller, 1895–1983[1]

THE SUCCESSES OF THE OZONE REGIME

Mario Molina and Sherwood Rowland started a revolution when they published
their discovery of the link of chlorofluorocarbon 'wonder chemicals' to
depletion of the stratospheric ozone layer in 1974. The revolution has finally
involved not only many thousands of industries but also the international
cooperation between governments, scientists, technologists, NGOs, UN
organizations and the media. Over the last 15 years, a fascinated world has seen
governments, including those at the opposite ends of the political or economic
spectrum, agree totally on their resolve to protect the ozone layer.

Extraordinary and unprecedented cooperation has been the hallmark of
ozone-layer protection. An Executive Committee of the Multilateral Fund,
composed of equal numbers of representatives from developing and developed
countries, grants money collected from the developed countries to developing
countries – with no conditionality except effective utilization of funds and no
discrimination except on the basis of need.

The ozone treaties have been successful

- There is near-universal participation in the agreements: 184 countries
  ratified the Vienna Convention and 183 ratified the Montreal Protocol. Only
  11 small countries have not yet ratified the Protocol (see Figure 10.1).
- Fruitful collaboration has been established between scientists, governments,
  NGOs, media and UN organizations – science leading to awareness,
  awareness leading to policy, policy leading to implementation, and implementation leading to ozone-layer protection.
- The Multilateral Fund has functioned very effectively to support the phase-
  out projects in developing countries. The industrialized countries have paid
  their due contributions, US$1.3 billion so far, to the Multilateral Fund.

Almost all of the 140 developing countries implement ozone phase-out programmes in their countries with the assistance of the Fund and annually report data on the production and consumption of 95 controlled ozone-depleting substances.

- The industrialized countries have almost completely phased out the controlled substances. They have reduced their consumption of CFCs, for instance, from 1 million tonnes in 1986 to about 10,000 tonnes for essential uses now. The total global production and consumption has come down by more than 85 per cent.

- The Protocol's facilitative approach to enforcement is working. The Russian Federation and some countries of Eastern Europe have been unable to comply with their phase-out schedule of 1996, but are on their way to compliance by the year 2003, with assistance from the Global Environment Facility. The CFC and halon production facilities of the Russian Federation were closed in December 2000.

- Consumption of CFCs by developing countries increased from about 138,000 tonnes in 1986 to about 161,000 tonnes in their base period of 1995–97. It then decreased to about 123,000 tonnes in 1999, even though the present obligation under the Montreal Protocol is only for a freeze. This decrease is largely due to the phase-out projects funded by the Multilateral Fund.

- The abundance of ozone-depleting compounds in the lower atmosphere peaked in 1994 and is now slowly declining, proving that the Protocol is indeed working.

- The maximum ozone depletion will occur within the current or next two decades and, with full implementation of the Montreal Protocol, the Antarctic ozone hole will disappear by the year 2050.

- If there were no Montreal Protocol, the chlorine and bromine in the atmosphere would have been five times larger than that of today. Ozone depletion would have been about ten times greater than today, ultimately resulting in many more millions of cases of skin cancer and eye cataracts, and possibly irrecoverable damage to agriculture and the ecosystems.

WHY WAS THE OZONE REGIME SUCCESSFUL?

What accounts for this success? Is the success replicable for other global environmental issues? Was the ozone issue very 'simple' and capable of easy solution compared to, say, the issues of climate change or biodiversity? How did the decision-makers of the world come to understand the complicated atmospheric science? How did the usually irreconcilable national interests achieve congruence? Why did the industry abandon opposition and enthusiastically promote regulations on CFCs? Is it because the industry had alternatives on which it could make greater profits than on ozone-depleting substances? Could economic instruments have tackled the issue better than the regulation of the Montreal Protocol? Many scholars and analyses have addressed these questions.[2] Let us summarize the history so far, so that readers can derive the answers for themselves.

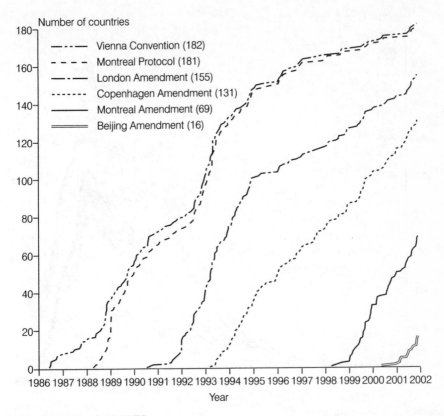

Source: Ozone Secretariat, UNEP.

Figure 10.1 *Ratification status of the Vienna Convention, Montreal Protocol and amendments, November 2001*

The seriousness of the issue

The depletion of the ozone layer is the most serious global challenge faced by humankind so far. If allowed to continue without any action, the depletion would have resulted in excessive ultraviolet radiation from the sun. This, the scientists are reasonably certain, would have resulted in millions more cases of skin cancer, eye cataracts leading to blindness, and loss of immunity among humans. It also would have reduced plant yields, damaged plastics and caused unforeseen adverse consequences for ecosystems. A longer inaction would have led to more serious damage. In short, it would have been a disaster of global proportions (see Chapter 1).

Was it an 'easy' issue?

The discovery of the link between CFCs and ozone depletion was made in 1974 and it took seven years to start international negotiations on the issue in 1981. It took four more years of negotiations to arrive at the framework Vienna Convention in 1985, then two more to arrive at a mild first step, the Montreal

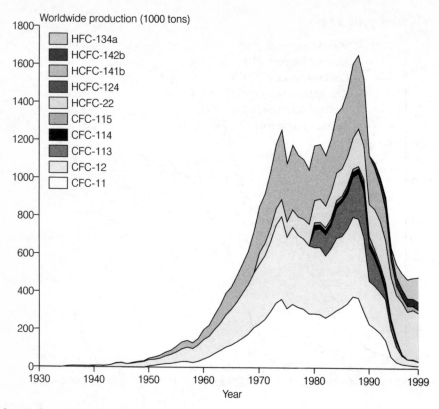

Worldwide production (1000 tons)

Source: Ozone Secretariat, UNEP.

Figure 10.2 *Worldwide production of ozone-depleting substances, 1930–2000*

Protocol, in 1987. After 1987, the Protocol became strengthened incrementally and the phase-out schedule for methyl bromide for developing countries could be agreed to only ten years later in 1997. The ozone-depleting substances were 'wonder chemicals' used in thousands of industries with billions of dollars worth of output. Almost every household in the world had some product which used ODSs and many more that were made with ODSs.

The use of these substances was diverse in different countries. In North America, refrigeration and air conditioning were considered the most important uses of ODSs, and cosmetic aerosol uses were considered 'frivolous'. However, in Europe air conditioning was not considered important and use in perfume aerosols was considered essential. Japan considered ODS use in making electronic products the most important. The awareness in the countries too was different. It was widespread in North America due to its strong media and NGOs but less so in Europe and Japan. The developing countries had a completely different perspective of equity in taking advantage of the cheaper ODSs for their economic development. The painstaking negotiations over the years brought out the differences, which were overcome only after much hard work on the part of the negotiators (see Chapter 2).

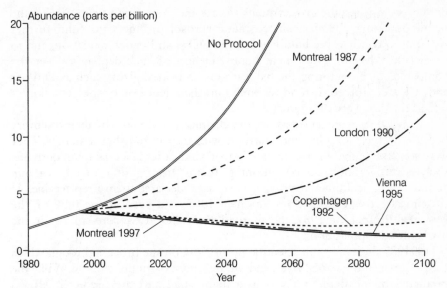

Source: Synthesis of Assessment Panel Reports, 1998, p155.

Figure 10.3 *Effect of the international agreements on stratospheric chlorine and bromine*

The reaction to scientific discoveries: Acceptance of precautionary measures

The reaction of the world community to the 1974 discovery of the connection between the human-made halocarbon chemicals and the depletion of the stratospheric ozone layer has been unprecedented. The scientists Mario Molina and Sherwood Rowland blew the whistle publicly at once and the media gave the discovery its due importance. The NGOs took up the cause, and public

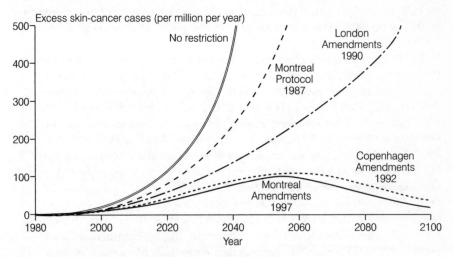

Source: Synthesis of Assessment Panel Reports, 1998, p22.

Figure 10.4 *Predicted excess cases of skin cancer, 1980–2100*

opinion compelled the governments to take the first steps to curb use of the ozone-depleting chemicals in cosmetic aerosol products, to fund further research to confirm the link between the ODSs and ozone depletion, and to predict the adverse environmental consequences of such depletion. Even the companies manufacturing the halocarbons, industrial giants such as DuPont and ICI, were forced to fund research into the science of ozone depletion as well as the discovery of alternatives.

Many ODSs were hailed as wonder chemicals at the time of their discovery because of their long life and inertness with respect to other materials. They were so versatile that they were used in many industries essential to modern life, such as refrigeration, air conditioning, foams, fire-fighting, metal cleaning, computers, agriculture and storage. Their users declared them irreplaceable. At the time of the Vienna Convention in 1985, and even at the time of the Montreal Protocol in 1987, the alternatives to the ODSs were still not in the market for all applications.

In 1987 there was still scientific uncertainty on the process through which CFCs damaged the ozone layer, and on the extent of future damage. While the predicted ozone depletion was in a high range and the predicted effects horrendous, the depletion was only beginning to be observed and the adverse effects had not yet begun (see Chapter 2).

Amazingly, the governments put their trust in mathematical models which predicted the ozone depletion and its adverse effects based on several assumptions. Scientists persuaded the governments that if the world waited for complete scientific certainty and for the adverse effects to manifest themselves, it would be too late to reverse the process. The long life of CFCs ensured that the damage would continue long after the stoppage of use of the chemicals, and the adverse effects would be severe before the ozone layer recovered. The world community accepted that actions had to be taken as a precautionary measure, as mentioned in the preamble to the Protocol. In 1992, this was incorporated in the declaration of the Earth Summit in Rio de Janeiro, Brazil, as Principle 15 that, in cases where serious harm is threatened, positive action to protect the environment should not be delayed until irrefutable scientific proof of harm is available.

*'Although scientific evidence that human activities were causing stratospheric ozone depletion was quite robust in the late 1980s, there were a number of sceptics who said "wait for perfect knowledge; there is uncertainty in the ozone models." Unfortunately the sceptics were absolutely right. The models were inaccurate. They **underestimated** the impact of human activities on stratospheric ozone. This means that even with the Montreal Protocol and its adjustments and amendments, society will have to live with stratospheric ozone depletion not only over Antarctica, but over all of the globe, except for tropics and subtropics, for at least another 50 years. Some of the same sceptics are now saying that not enough is known about climate change.'*

Robert Watson, Co-Chair, Scientific Assessment Panel,
Montreal Protocol, former Chairman, Intergovernmental Panel
on Climate Change, cited in *Partnerships for Global Ecosystem
Management*, World Bank, 1997

Could the ozone layer have been protected through economic instruments?

Modern thinking is in favour of economic instruments to achieve social objectives. Regulations are seen as blunt instruments, costly and inefficient. Economic instruments can be effective for dealing with environmental problems if the causes and effects of the problems are precisely known and costs and benefits can be calculated for each solution. These have been used effectively in many countries to tackle problems such as emissions from factories. Many countries have used some economic instruments to implement the Montreal Protocol, for example Singapore and the USA (see Chapter 6). However, on a global level, there has not been much progress on such instruments. Currently, the climate change treaty is attempting to implement some instruments for tradeable permits. There was no way the external costs of damage to humanity through ozone-layer depletion could be internalized in the costs of ODSs nor the products using them. Regulation was felt by the governments to be essential to phase out the production and consumption of ODSs. Economic instruments are very useful to implement the regulations.

LESSONS FROM THE DEVELOPMENT OF THE MONTREAL PROTOCOL

Acceptance of common but differentiated responsibility

From the beginning of the treaty-making process, everyone recognized that a small number of countries were responsible for a large portion of ozone depletion. Of the nearly 180 countries in the world, about 30 developed countries produced and consumed 90 per cent of the world's ODSs. The developing countries consumed only about 10 per cent. The developed countries have had the benefit of the cheap ODS technologies much longer than the developing countries. Developed countries also had the companies with technologies for the alternatives. When the developing countries argued that they needed the new technologies, transfer of financial resources to adopt the new technologies, and time to make the transition, the world community saw the justice in the arguments and provided for a grace period of ten years for the developing countries (Article 5 of the Protocol), assistance for technology transfer (Article 10A) and a financial mechanism (Article 10) to facilitate the adoption of the new technologies.

The world recognized the common responsibility of all countries to protect the ozone layer and also recognized that the richer countries, which caused more damage, had more responsibility. This has since been articulated as Principle 7 of the Rio Declaration. It requires states to cooperate in a spirit of global partnership to protect the environment. Yet, because states have contributed differently to global environmental problems, they should have common, but differentiated, responsibilities (see Chapter 3).

Richer countries lead

Through the Protocol, the richer countries assumed the responsibility of going through the transition to ozone-safe chemicals first, to transfer ozone-safe technologies and to contributing to the Multilateral Fund to meet the incremental costs of developing countries in complying with the control measures. The richer countries also contributed to the Global Environment Facility to assist the countries with economies in transition (such as the Russian Federation) to phase-out ODSs.

Importance of public opinion, media and NGOs

The desire to protect humanity from adverse effects of ozone depletion was the clear motivation for the near-universal participation. This desire was strengthened in most of the industrialized countries by the enormous publicity on the issue through the mass media (see Chapter 8). This influence of the media was particularly high in North America. From 1974, the ozone-depletion issue appeared in many plays, films and popular songs. The ozone depletion was seen by many as a paradigm of modern technology, coupled with greed, destroying nature and natural resources. The emergence of this issue gave great impetus to the environmental movement, whose dire predictions of environmental disasters were previously viewed sceptically by the influential decision-makers of the industrialized countries. The NGOs put enormous pressure on the world community through a variety of strategies (see Chapter 9). The horrifying predictions of galloping skin cancers due to ozone depletion, given wide publicity by the media, undoubtedly made the issue one of concern to every citizen of these countries and compelled the governments to take an active part in developing the ozone agreements and ratifying them quickly. In developing countries, the Multilateral Fund spread awareness and information through the programme of OzonAction, implemented by UNEP (see Chapter 6).

The NGOs adopted a number of methods – media campaigns, litigation, public protests and demonstrations – to spread awareness and to pressure governments to regulate the ODSs towards a phase-out. They adopted an adversarial position with the industry initially but their relationship began to change when Greenpeace collaborated with a small refrigerator company in Germany to develop the ozone-safe hydrocarbon technology, 'Greenfreeze'. This technology was adopted by most major companies in Europe and was transferred to many developing countries too. Other NGOs also concentrated on publicizing ozone-safe technologies. Some of the NGOs, such as the Environment Investigation Agency and Ozone Action, concentrated on exposing illegal trade in CFCs (see Chapter 9).

Flexibility in implementation

The control measures of the Protocol, while mandating a phase-out schedule for each basket of chemicals, left it to the governments to devise their own strategies for implementation. The governments resisted the temptation of imposing a sector-wise phase-out schedule or other restrictions on the use of ODSs, realizing that countries have different needs, and that cost-effective phase-outs could be achieved only if each country followed a strategy best suited to its conditions. Each country adopted its own mix of instruments to implement the control measures – regulations, policies, taxes, quotas, standards, and tradable permits best suitable for itself.

In some countries, like Germany, the industry came forward voluntarily to phase out earlier than mandated by the Protocol. The innovation of an 'adjustment' under Article 2, Paragraph 9 made it possible to reach a compromise on a reduction schedule that was considered to be weak by some since that schedule can be strengthened through an appropriate decision by a Meeting of the Parties to the Protocol at a later date. Such an adjustment can be decided with a two-thirds majority, and a majority of both Article 5 and non-Article 5 Parties, and will be binding on all the Parties without each Party having to ratify the adjustment. This resulted in faster phase-out, as alternatives matured and as science demanded a faster phase-out to heal the ozone layer earlier.

FEATURES OF THE PROTOCOL PROMOTING PARTICIPATION

The world community realized early that universal participation is necessary to phase out ODSs. Even if most countries phased out ODSs, the countries remaining outside the process, even if insignificant now in terms of their consumption of ODSs, could increase their consumption and could, in time, again damage the ozone layer. Hence the negotiators of the Montreal Protocol incorporated many features that promoted participation by everyone.

Trade measures

Many countries ratified the Protocol agreements in their own interest, to avoid the trade sanctions of Article 4. The equipment using CFCs – refrigerators, air conditioners and fire-fighting equipment – was in use in all the countries of the world and every country required these chemicals to maintain the equipment. In addition, many countries manufactured equipment using ODSs. The number of manufacturers of the ODSs, however, was only 15. Eight developing countries manufactured many, but not all, the ODSs through their own companies. All the Parties who manufactured were required to stop their exports of ODSs to non-Parties by 1 July 1990. Thus all the countries that did not manufacture ODSs but were dependent on imports were compelled to join the Protocol. The Parties, in their third meeting in 1991, made a list of products containing CFCs that now could not be imported by non-Parties to the Protocol. Thus even countries like South Korea, who made their own ODSs but who had a large export market of goods with CFCs like automobiles with air conditioners, were forced to ratify the Protocol.

A few countries, like some of those formed from the former USSR, depended on illegal imports for some years but the assistance available from the Multilateral Fund and the GEF to poor countries made it pointless for any country to remain out of the Protocol. While the control of trade under Article 4 is with non-Parties, the Montreal Amendment in 1997 introduced some control on trade with Parties too. There had been some academic discussion on whether the trade measures of the Montreal Protocol are consistent with the principles of the World Trade Organization and the General Agreement on Tariffs and Trade. There is a consensus that there is consistency and, in any case, no dispute was ever raised in the WTO regarding the trade measures of the Protocol.

The Multilateral Fund

The ten-year grace period given by the Protocol to the developing countries for implementation of the control measures and the establishment of the Multilateral Fund, to be contributed to by the developed countries, made it easier for all the developing countries to ratify the agreements. The Multilateral Fund is the most successful single-issue fund so far. The second Meeting of the Parties to the Montreal Protocol approved the setting up of the Fund through the London Amendment. Realizing that the amendment may take some time to enter into force (the Amendment approved at the end of June 1990 entered into force in August 1992 after the required 20 ratifications), the meeting approved an Interim Fund with an allocation of US$160 million for three years, to be increased to US$240 million if more Parties ratified the Protocol. This unprecedented step created confidence among the developing countries that the industrialized countries were serious about the Fund. The step also gave a great push for immediate implementation of the Protocol. Its management through an Executive Committee, which has an equal representation among the contributors and the receiving countries, is an innovative mechanism. The almost 100 per cent contribution of the contributions assessed for the industrialized countries by the Meetings of the Parties has rarely been seen in other similar Funds. The distribution of the Funds to developing countries has been only on the basis of assisting these countries to implement control measures. There has not even been a whisper of bias on political or ideological grounds. The Executive Committee has functioned effectively, approving funding for projects at its regular meetings and monitoring the progress regularly. The concept of 'incremental costs', which the Protocol had introduced, had no precedent and the Executive Committee, assisted by the Fund Secretariat, had done much original work to clarify the concept in an impartial and creative manner. The fund fixed cost-effectiveness thresholds and took up the more cost-effective projects first to maximize the benefit to the ozone layer. In addition to funding investment projects to phase out the ODSs, the Fund financed institutional strengthening in developing countries and promoted adoption of the right regulations and policies to phase out ODSs. Its initial project-wise approach has now given way to a sectoral phase-out approach (see Chapters 4 and 6).

A facilitative non-compliance procedure

The non-compliance procedure approved by the Parties to the Protocol recognized that non-compliance by a Party could be due to its inability to implement the Protocol for reasons beyond its control. The Implementation Committee consisting of ten Parties, established by the Parties to the Protocol, would inquire into non-compliance by any Party and recommend action by the Meetings of the Parties. The non-compliance procedure stipulated that assistance to the non-complying Party to comply and a caution to that Party should also be considered apart from punitive measures like applying trade measures by declaring the non-complying Party as non-Party to the Protocol. The procedure recognizes that a non-complying Party damages the ozone layer and it is more important to bring it back to compliance than to punish that Party.

When the countries of the former USSR and of Eastern Europe suffered the pains of economic, political and social transition from 1990 and were unable to meet their schedules of phase-out, the Parties to the Protocol and the Implementation Committee faced their first test of this procedure. They recognized that the non-compliance was because of the problems of sudden collapse of social, political and economic systems and transition to a market economy, and arranged for assistance by the Global Environment Facility (GEF) to those new countries which were not classified as developing countries and hence not eligible for assistance from the Multilateral Fund. (Georgia, Kyrgyzstan and Moldova of the former USSR, and Albania, Romania, Yugoslavia, Bosnia and Herzegovina, Croatia and Macedonia of Eastern Europe were recognized as developing countries.) At the same time, they insisted that each of these countries provide benchmarks for return to compliance and made financial assistance contingent on observance of these benchmarks.

When the Russian Federation exported its recycled ODSs to other countries, even while it produced new ODSs for its own use in violation of the Protocol, the Parties approved, as a part of the Montreal Amendment, a provision (Article 4A) to prevent such exports. All the new countries formed from the former USSR (except those classified as developing countries) ratified and implemented the Protocol willingly because of the assistance of GEF. They are all expected to return to compliance by the year 2003. Thanks to this assistance, the Russian Federation closed its HCFC and halon production facilities in December 2000. The Implementation Committee is following a similar facilitative procedure for developing countries (see Chapter 7).

Pragmatic consensus at every stage

At every stage of negotiation there was a conscious effort to attract countries to a consensus, even if the course to be adopted was not the best course to protect the ozone layer but only the next best. An international agreement on a global issue such as the protection of the ozone layer would not be effective if too many countries remained uncommitted. And many countries would not join if they feared that their competitors would not do the same.

Vienna, 1985

In 1985, when there was staunch opposition to controls on CFCs from the European Community, only a framework Vienna Convention of a general resolve to protect the ozone layer was adopted to keep all countries in agreement. However, a resolution was adopted to continue with negotiations to avoid losing momentum while the process of ratification of the Convention by the countries was in progress. Meanwhile, the scientific case was being strengthened.

Montreal, 1987

Studies and workshops in 1986 and 1987 demonstrated the need for action to protect the ozone layer. However, by the time of the final meeting of plenipotentiaries in September 1987, there were doubts in some countries about the link between CFCs and ozone depletion. Hence, only a mild first step of a 50 per cent reduction in CFCs and a freeze on halons was adopted in the Montreal Protocol to accommodate the opinions of sceptics. A provision was added to allow countries to add the production of CFCs and halons from factories under construction on, or contracted prior to, 16 September 1987 (the day the Montreal Protocol was signed) *and provided for in the national legislation*. This was done to suit the convenience of the then USSR, which had a planned economy and argued that its factories were a result of national legislation which a government cannot violate (Article 2, Paragraph 6). However, the innovative Article 2, Paragraph 9 of the Protocol allowed for adjustment of the control measures on substances already listed in the Protocol based on scientific and technological assessments carried out under Article 6.

London, 1990

Soon after the agreement on the Protocol in September 1987, new scientific discoveries, including the 'smoking gun' of the link between CFCs and ozone depletion brought all the industrialized countries to favour a full phase-out. However, the developing countries wanted more assurance of technology transfer and financial assistance. The London Amendment of 1990 provided both, to secure the agreement of the developing countries. The London Adjustment and Amendment provided a phase-out schedule for all substances based on the use of HCFCs, which are also ozone-depleting but less so than CFCs, in place of CFCs.

Copenhagen, 1992

There were already doubts on the wisdom of the use of HCFCs for the replacement of CFCs, and the fourth meeting in 1992 provided for the phase-out of HCFCs at a slow pace. There were differences between the European Union on one hand, which wanted faster phase-out of HCFCs, and North America and Japan, which favoured a slow phase-out in order to allow industries to recover their full investment in HCFCs as the replacement of CFCs.

The very definition of baseline of HCFC was distinctive. For all other substances, the baseline is the consumption of those substances in a particular year; for example, the baseline for CFCs is consumption in 1986, the baseline for methyl chloroform is consumption in 1989. However, the baseline for HCFCs is the consumption of HCFCs in 1989 *plus* 3.1 per cent of the consumption of CFCs in 1989. Why was the consumption of CFCs brought in? This was to settle the differences between those countries that had a high consumption of HCFCs in 1989 (the Europeans) and those countries that had a low consumption of HCFCs in 1989 (Japan for instance). Those who had a low consumption of HCFCs in 1989 argued that those with high consumption had an unfair advantage since it would be easier for the high consumers to reduce their consumption of HCFCs. They argued that the consumption of

CFCs in 1989 also should be given a weight in the baseline since the HCFCs will replace the CFCs. Complicated as this might sound, this paved the way for compromise.

Why 3.1 per cent and not, say, 3 per cent? This is to boost the baseline to suit the convenience of all those who consumed less HCFCs or had already phased out CFC aerosol products that required no HCFC replacement. This was strongly criticized by the NGOs but promoted participation by all countries. Later, in practice, the 3.1 per cent proved to be high and was reduced to 2.8 per cent in 1995 through an adjustment. Furthermore, most countries consumed significantly less than the allowed amount until at least 2000 and only one or two countries approached the limit in 2001.

A similar scenario was played out in the case of methyl bromide. A strongly reluctant set of developing countries was exempted from controls in 1992, but was brought within controls partly in 1995 and fully in 1997. Every time controls were set, the consensus of all countries was preferred even if it resulted in weaker measures, despite criticism from NGOs. The measures were strengthened only to the extent of willing compliance of almost all the countries.

Multilateral negotiations and the leadership of UNEP

After the discovery of the link between CFCs and ozone depletion, UNEP held an expert meeting in Washington, DC, in March 1977. This meeting formulated a World Plan of Action for further study of the issues. UNEP formed a Coordinating Committee on the Ozone Layer to coordinate studies throughout the world. Due to pressure from NGOs and public opinion, some countries such as the USA, Sweden and Canada introduced regulations to control the use of CFCs in aerosols. These countries then took the initiative to persuade other industrialized countries to take similar steps and held three meetings in 1977, 1978 and 1980. However, these meetings led to no agreement because of deep differences of opinion. It was in 1981 that UNEP Governing Council formed an international negotiating group, which consolidated the results. The presence of a neutral but effective facilitator in UNEP, with a mandate to facilitate the protection of the environment, helped to achieve agreement.

Informal consultations

It was Mostafa Tolba, then Executive Director of UNEP, who made a very productive use of informal consultations from 1987. At every crucial stage, he would invite the key delegates from countries with strong opinions on the issues, who could make or mar the negotiations, to these meetings in their personal capacity to discuss the issues. He would also occasionally invite some experts. No records of these meetings were kept. All invitees were free to give their opinions in confidence without compromising their governments, and other invitees would probe on the extent to which each government may be prepared to compromise. At the end of the meeting, there was clarity on the issues and the contours of compromise were visible. Frequently, Tolba would follow up with his own proposals to the negotiating groups, based on his impression of what compromise would be acceptable. These proposals were very productive in leading to a compromise.

Informal groups during the meetings of the working groups

The reluctance of countries to compromise was also tackled in informal groups constituted during the meetings. All those with strong opinions on issues were included in these groups – fewer than 20 countries usually – and others were kept out, occasionally to their annoyance. The bargaining took place in these informal meetings behind closed doors and any individual dissenting countries were brought under tremendous pressure by their fellow delegates to agree to a compromise. The reports of the meetings rarely included the discussions in the informal groups and onlookers got the impression of easy harmony. The net result was that the amendments and adjustments were adopted by a consensus and there was never any voting on any subject.

Participatory process overrides legal hair-splitting

Legally, each amendment had to be ratified by each Party for its provisions to be applicable to that Party. For example, a Party that had not ratified the Copenhagen Amendment is not bound by the controls on HCFCs and methyl bromide. There are 49 such Parties to date. But there has always been an assumption that non-ratification is due only to internal procedural delays of countries and that it is only a matter of time before their ratification takes place. The Secretariat asks all Parties for data on all the substances, whether or not the Party has ratified the amendment(s) listing those substances. Of the Parties that have not ratified some amendments, some Parties do not submit their data but many others do. No Party has so far raised the legal issue of why they are asked to submit their data.

The consensus process has led to the positive result that, at every meeting, Parties explain why they have not so far ratified some amendments. Similarly, while some industrialized countries hold the view that their contributions to the Multilateral Fund are only voluntary, they provide an explanation in the meetings of the Executive Committee if they have arrears to the Fund. They have not, so far, taken the plea that it is only a voluntary fund and they are not obliged to pay.

PARTNERSHIP LED BY SCIENCE AND TECHNOLOGY

The initial period

A pattern of fruitful collaboration has been established between scientists, governments, NGOs, media and UN organizations – science leading to awareness, awareness leading to policy, policy leading to implementation. It is, however, noteworthy that this process took more than ten years to become established. The 1974 discovery by Molina and Rowland of the link between CFCs and ozone depletion led to immediate action in North America and Nordic countries to curb use of CFCs in cosmetic aerosols, which were considered a frivolous use. The consumption of CFCs decreased for some time.

But by 1981, other uses of CFCs increased and overall consumption of CFCs started increasing. The public concern over CFCs declined. After 1978, the UNEP-run Coordinating Committee on the Ozone Layer was the only forum in which scientists and governments met regularly to assess the situation and thus kept the issue alive. From 1981, the CCOL and the UNEP Working Group of Experts were the only forums in which there was action. The estimates of predicted ozone depletion kept declining from the original 15 per cent in 1978 to 3 per cent in 1984. The Vienna Convention of 1985 was only a statement of intent to protect the ozone layer from all dangers. CFCs were mentioned almost in passing as one of the many potential dangers. The advance of scientific knowledge, in the meantime, did little to impress governments.

International assessments from 1986

The discovery of the Antarctic ozone 'hole' by Japanese and British scientists, and Joe Farman and his British Antarctic Survey team boldly linking the ozone hole with CFCs, created the momentum for the Montreal Protocol. The assessment of 1986, spearheaded by NASA and with the participation of WMO and UNEP, was the first international and comprehensive assessment, and it influenced the negotiations. The assessment effectively ruled out the possibility that CFC consumption could grow without controls. It turned the industry from vociferous opposition to reluctant support of mild controls.

The expeditions organized in 1986 and 1987 to Antarctica, and the measurements by aircraft and ground instruments, allowed reasonable certainty that it was the CFCs and halons which caused the ozone depletion and created atmospheric conditions by which the depletion was much worse than predicted by earlier models. The report of the Ozone Trends Panel in 1988, in which WMO, UNEP and NASA collaborated, provided the unassailable arguments to support the phase-out of CFCs. The downward trend in ozone in the higher latitudes was clearly demonstrated. The correlation between the decline in ozone and increase in UV-B radiation was also firmly stated. There was enough evidence already on the many adverse effects of higher UV-B, including estimates of additional skin cancers; every 1 per cent decline in ozone would increase the skin cancers by nearly 2 per cent. The assessment was the critical factor in promoting the complete phase-out of CFCs.

Standing assessment panels of experts with wide participation

Article 6 established the dominance of science and technology in the Protocol by providing for the formation of Assessment Panels for Science, Environmental Effects, and Technology and Economics to assist the Parties to assess the control measures at least once in four years. These panels were established informally in October 1988 and were confirmed in May 1989 by the first Meeting of the Parties to the Montreal Protocol. The panels each involved hundreds of experts from around the world to synthesize the latest knowledge on the issues. Their reports replaced national or industry assessments and are regarded as objective. Occasionally, there were criticisms on the conclusions of the panels. For instance, the environmental NGOs felt that the reports overestimated the importance of HCFCs in replacing CFCs. The methyl bromide industry and some users of the chemical complained that the panels overestimated the ozone-depletion potential of the chemical or overstated the availability of alternatives. Such complaints were aired by the doubters in the Meetings of the Parties, allowing observers and the delegates to make their own judgements.

Freedom for scientists and technologists to recruit the best
The chairpersons of the panels, appointed by the Meetings of the Parties, had considerable freedom in recruiting members of their technical committees that helped them in preparing the reports. While the Meetings of the Parties emphasized that there should be equitable geographic distribution, it was subject to expertise available in the regions. Nominations made by governments for membership were scrutinized by the chairpersons and only those whose services were needed were recruited to the committees. All countries recognized that the expertise may not be equitably distributed due to several factors and that the cause is better served by experts than political representatives who in any event had the power of final decision. Most of the expertise was located in the industrialized countries since most of the use was in these countries. Only about one-third of the members of all the committees were from developing countries, but there were no objections to this.

Standing rather than ad hoc panels
While Article 6, under which the panels are constituted, stated that there shall be an assessment at least once in four years and that the Parties shall convene panels of experts at least one year before each assessment, the panel established in 1989 has continued until now with little change, except for change of members due to natural causes. So, instead of being ad hoc panels, as envisaged in Article 6, these panels became standing panels. The panels did more work than merely assessing the control measures. Every year the Meetings of the Parties took advantage of the expertise of the panels and gave the panels many tasks such as recommendations on essential-use exemptions and the replenishment of the Multilateral Fund. Every Meeting of the Parties started with presentations of the latest knowledge by the panels on the tasks entrusted to them. Thereafter, the merit of any proposal was almost always justified by delegates of governments on the basis of the scientific and technical facts given by the panels. The discussions were based on facts and to the point, with very little rhetoric on environment or the usual North versus South issues.

Development of an ozone community
The standing nature of the panels has developed, over the past 15 years, a community of scientists and technologists who thought in harmony, which spread to the delegates to the Meetings of the Parties. The membership of the panels from nearly 50 countries resulted in the benefit of the latest knowledge to the countries through these members. This spread of knowledge was reflected even in the political discussions.

World community supplies expertise to the panels at no charge
An important fact about the assessment panels is that the members are not paid any remuneration but gave their services voluntarily. In a few cases, some governments supported some members but, usually, the employers of the members spared their time for this global service and paid for their travel. It was true that the employers expected that the experts sent by them would represent their point of view in the panels. Indeed, it is the presentation of such wide varieties of opinion which made the reports of the panels realistic. The Trust Funds of the Protocol met the travel expenditure for members from developing countries and the countries with their economies in transition. Valuable reports emerged from the panels at a relatively small cost. The reports were not individual findings but represented a global technical consensus due to the panels' wide membership. This technical consensus easily translated itself into political consensus.

Variety of expertise
The members of the panels and the technical committees are drawn from government, industry, academia, consultancies and NGOs. They are bound by a code of conduct to ensure that they do not promote their employers' agenda. It is considered proper that they would contribute their knowledge gained during their employment for the purpose of protecting the ozone layer. The panels enabled the extraordinary situation of, for instance, a representative of ODS manufacturers working out scenarios for the phase-out of ODSs.

WHY DID INDUSTRY COOPERATE: REGULATION OR AVAILABILITY OF ALTERNATIVES TO ODSS?

The worldwide cooperation of industry in the phase-out of ODSs after 1988 has already been noted. The Montreal Protocol to phase out the ODSs according to a schedule could not have been successful without the industry stakeholders cooperating wholeheartedly. Why did they cooperate, and not use their considerable influence to defeat the implementation of the Protocol? Is it because they had developed the alternatives, HCFCs and HFCs, and wanted to make greater profits on these costlier substances? As detailed in Chapter 5, the facts are that the manufacturers of CFCs were looking into alternatives from 1980, prompted by the discovery of the links between CFCs and ozone depletion, and had a good idea that the alternative chemicals would be costlier. The alternatives, if developed then, would not have found a market if the cheaper CFCs continued to be marketed. Hence the development of the alternatives was abandoned, since before 1986 nobody expected that the CFCs would be destined for a phase-out. By 1988, the dangers from CFCs were articulated clearly by scientists, the Montreal Protocol was in place and scientific opinion declared that a total phase-out was clearly imperative.

Industry was under threat of litigation from NGOs in the USA. Did it also fear liability litigation from those whose health would be adversely affected by ozone depletion? The CFC manufacturing industry knew that alternatives to

CFCs, however costly, would have a market. They therefore reversed their earlier stand and threw their support behind total phase-out; they intensified research on alternatives. A usually competitive industry took up the mandatory testing of the alternatives as a cooperative venture without losing time and came out with the alternatives in 1990. The CFC manufacturers thus smoothly moved from CFCs to alternatives, thanks to regulations. This is a case of regulations forcing the development of technology. Signals of predictable controls were given to the industry with fair lead times, and the industry picked up the signals. Sensing the goodwill to be earned by cooperation, many companies, such as Northern Telecom, took the lead in spreading alternative solvent technologies throughout the world.

The case of the thousands of industries using ODSs, such as those involving refrigeration, foams, solvents and fire-fighting equipment, is even clearer. Once they knew that ODSs would not be available, they innovated alternative chemicals and processes quickly. This innovation not only served the transition to non-ODS technologies but also resulted in general improvement of products and reduction of costs.

CAUTION FOR THE FUTURE

The positive developments so far lead to confidence that the Montreal Protocol will succeed in its objective of protecting the ozone layer. However, this should not lead to complacency that the problem has been solved. There have so far been no signs of the reversal of the depletion that has already occurred. The Antarctic Ozone 'hole' continues to be an annual feature in the Antarctic spring, September–October, with an area of about 22 million square kilometres. While the scientists assure us that the ozone layer will recover if the Montreal Protocol is fully implemented, caution and vigilance are needed to avoid creating new problems for the ozone layer. We must remember to navigate carefully around the potholes on the road ahead.

Climate change and other effects on the atmosphere

Major volcanic eruptions such as the one in Mount Pinatubo in 1996 could temporarily increase ozone depletion. Cold, protracted winters in the Arctic could make the ozone depletion worse. While the studies so far have not revealed any alarming feature about the impact of air transportation on the ozone layer, studies are still continuing and there is no finality on the issue yet. There are also several connections between the ozone layer and the human-induced greenhouse effect responsible for global warming and other climate change. CFCs and other ozone-depleting substances are also greenhouse gases, contributing to global warming. On the one hand, the depletion of the ozone layer more than offsets the global warming impact of CFCs, perhaps offsetting up to about 30 per cent of the global warming since the late 1970s. Global warming due to greenhouse gases will therefore be even more serious with the recovery of the ozone layer. On the other hand, increases in various greenhouse

Box 10.1 Lessons from the ozone-layer experience

Mostafa Tolba, Mario Molina and Elizabeth Cook

'What did we learn from the ozone negotiations that could influence the negotiations of other environmental regimes in the future? I would like to cite a number of key ingredients that made a difference time and again during the negotiations:

- A core group of countries with similar objectives pushing for action (essentially the Toronto Group).
- The very strong role of science and technology, and the way this information was integrated into the negotiation and renegotiation process. Bob Watson, Dan Albritton, Stephen Andersen, Vic Buxton, Rumen Bojkov and their colleagues certainly built and are continuing to build the strong scientific foundation upon which the regime was established.
- A willingness to compromise and move forward one step at a time. This feature was evident throughout the entire process of negotiating the ozone regime, and I am sure it will continue.
- Several strong government personalities at the negotiating table: Winfried Lang, Richard Benedick, Patrick Szell, Willem Kakebeeke, L J Brinkhorst, Eileen Claussen, Per Bakken, Vic Buxton, Ilkka Ristimaki, Maneka Gandhi, David Trippier, Bill Reilly and many others.'

Mostafa Tolba, former Executive Director, UNEP[3]

'One very important lesson we learned from the CFC-ozone depletion phenomenon is that society is capable of successfully addressing such global problems. It illustrates how basic research, motivated by scientific curiosity to understand nature, can lead to practical results that benefit society. The CFC problem is to a large extent under control, thanks to the Montreal Protocol. The Protocol not only provides an example of science in the service of public interest, but sets a very important precedent that demonstrates how the different sectors of society (scientists, industry people, policymakers and environmentalists) can work together and be very productive by functioning in a collaborative rather than in an adversarial mode. It emphasises that global problems cannot be solved without the active participation of all countries of the world (developing and developed). It has also established a new way of addressing environmental problems by applying the "precautionary principle": the original agreement was negotiated on the basis of the CFC–ozone depletion theory (which predicts that human-made CFCs would deplete the stratospheric ozone layer), rather than direct unambiguous observation of noticeable damage to the ozone layer from CFCs. Another important precedent established by the Protocol is the inclusion of procedures for periodic revision of its terms: the international agreement was strengthened first in London in 1990, and then in Copenhagen in 1992, leading to a complete ban on industrialised countries' manufacture of CFCs by the end of 1995. And, of course, the society still enjoys the benefits of refrigeration, air-conditioning, plastic foams, and aerosol spray *cans – now with new, CFC-free technologies. My own experience with the CFC–ozone depletion phenomenon offers me hope that as we face new environmental challenges (such as global warming, increased tropospheric ozone, loss of biodiversity, deforestation and degradation of land), society will again rely on scientific understanding to provide the foundation for responsible action. The quality of life of future generations will be based to a large extent on our ability to deal intelligently with global-scale environmental problems.'*

Mario J Molina, Professor, Massachusetts Institute of Technology[4]

'To put these success stories into perspective, consider how great the challenge of phasing out CFCs seemed less than a decade ago. In 1987, the year governments signed the Montreal Protocol on Substances That Deplete the Ozone Layer, Americans were literally surrounded by CFCs; these "wonder chemicals" helped cool their cars and offices, preserve and package their food, and manufacture their new high-tech computer and electronic gear. US industries were responsible for using one-third of all CFCs produced worldwide – and US companies sold more than $500 million worth of the chemicals every year. American goods and services involving CFCs were worth $28 billion annually, and installed equipment worth more than $128 billion relied on CFCs. Not surprisingly, industry uniformly argued that moving away from CFCs would be a prohibitively costly, slow process that would harm the quality of products and services.

'But the worst never happened. Today, industry has developed alternatives for virtually all CFC applications. The public has not been denied popular products, and the economy has not been seriously disrupted. In case after case, firms have eliminated CFCs faster, at lower cost, or with greater technological improvements than ever imagined. In fact, the whole phase-out is proving less costly than the Environmental Protection Agency originally estimated.'

Elizabeth Cook, Director, Management Institute for Environment and Business, World Resources Institute[5]

gases in the atmosphere will have impacts on the recovery of the ozone layer. Increases in methane will shorten the recovery period while increase in nitrous oxides, HFCs, PFCs and SF_6 will extend the recovery period. The greenhouse gases cool the stratosphere, increase the destruction of ozone by the ODSs and extend the recovery period.

Ozone depletion will reach its peak in a few years; the ozone layer will recover gradually over the next few years only if the Montreal Protocol is fully implemented by both developed and developing countries and only if climate change does not adversely affect recovery of the ozone layer. While scientists have not yet determined all the connections between climate change and ozone-layer depletion and the ultimate net effects, some connections are clear. For example, climate change will cool the stratosphere, leading to ice-particle formation that will increase ozone depletion, with relatively lower absolute ozone abundance and prolonged effects. Emissions of greenhouse gases could have an adverse impact on the ozone layer. Despite the Framework Convention on Climate Change signed ten years ago, Parties are not yet bound to specific and meaningful targets to reduce the greenhouse gas emissions. The Kyoto Protocol of 1997, which is a first step in reducing the emissions, has not yet attracted sufficient ratifications to enter into force. This could delay the recovery of the ozone layer.

Future ratification and implementation of existing agreements

Non-ratification of the Copenhagen Amendment by some countries

The Copenhagen Amendment adopted in November 1992 entered in to force after 20 ratifications only in June 1994. Now, after nearly ten years, only 141 of the 183 Parties to the Montreal Protocol have ratified the Amendment. The 49 Parties that have not ratified the amendment are not bound by the controls for HCFCs and methyl bromide and include the most populous country in the world, China. This is a cause for concern. China informed the Ozone Secretariat in March 2002 that they would ratify the Copenhagen Amendment before the end of 2002.

Completion of the phase-out by developing countries

The developing countries will have to fulfil their part of the bargain and complete the phase-out. The indications so far are positive in that the total ODS consumption of developing countries has started to decline, thanks to the assistance of the Multilateral Fund. However, the developing countries are yet only in the initial stage of ODS phase-out. Thanks to the conversion of large and middle-sized CFC-using industries with support from the Multilateral Fund, most developing countries will be able to meet the CFC freeze requirements and probably also the 50 per cent reduction in 2005. But in order to achieve the required 85 per cent CFC reduction by 2007 and the full elimination by 2010, the widespread CFC use for servicing of existing equipment in the refrigeration and air-conditioning sectors will also have to be managed. In many countries, big and small, this consumption accounts for more than half of the total CFC consumption. To tackle this challenge, countries require coherent and comprehensive policies, using a variety of tools, and it will take some years before the actions can yield results in terms of required reductions. Large-scale import of second-hand CFC equipment by these countries would also make the required reductions of CFC use difficult. Other sectors where the ODS consumption is spread over many small users – as is for instance the case with methyl bromide in the agricultural sectors – will confront governments with similar challenges.

The Multilateral Fund will have to be replenished with new contributions until the task is completed. 'Donor fatigue' or complacency should not divert funds. The funds will have to be used in a way that effectively assists developing-country governments to put policies in place which eliminate ODS consumption in a sustainable manner.

Illegal trade in ODSs

Illegal trading in ODSs – boosted by the continued demand for these substances and currently especially relevant for CFC-12 for servicing in the refrigeration/air-conditioning sector – counteracts the phase-out efforts by governments in both developed and developing countries. Apart from making policy measures less effective, illegal trade also maintains the risk of losing control over how much ODSs are really used – what is illegally imported or produced will not show up in official reports.

HCFC phase-out

While HCFCs are being gradually phased out in the developed countries, there is now a fast-growing use in developing countries, as replacement for CFCs. The Protocol allows continued growth up to 2015, with no ceiling attached. Funding for HCFC installations in developing countries has been granted on the condition that no assistance from the Multilateral Fund will be requested to phase out use of HCFCs. Under the Protocol, no reductions are required until 2040, the date when HCFCs should be totally eliminated. Will the world community in forty years' time be vigilant enough to ensure that this will really happen? What tools will the international community have at its disposal with no financial assistance available as a carrot?

New ozone-depleting chemicals

Five new ozone-depleting chemicals, not controlled by the Montreal Protocol, have been introduced into the market in the recent years. Since they have not been listed in the Protocol, they are not amenable to control by the Protocol unless it is amended to include them. The amendments, if adopted, have to be ratified by a specified number of Parties for entry into force. Even when the amendments enter into force, they will bind only the Parties that ratify the amendments. As already noted, only 134 of the 183 Parties to the Montreal Protocol have so far ratified the Copenhagen Amendment of 1992. The 49 Parties that have not ratified this amendment are not bound by the controls for HCFCs and methyl bromide. Does this bode well for the control of new ODSs in future years when the ozone issue will no longer be in the spotlight? Who is responsible for testing the ozone-depletion potential of new chemicals introduced in the market? In 20 years, will the world community be watchful for the emergence of new ozone-depleting chemicals? Can a simplified procedure control new ODSs? The Parties have discussed these issues since 1997, but with no resolution. A simpler procedure for listing these chemicals, other than the procedure of amendment, has eluded the Parties so far. The importance of this issue to the protection of the ozone layer is clear. Let us hope that the Parties will continue their discussions and find a solution soon.

On the occasion of International Ozone Day, 16 September 2001, a message by Klaus Töpfer, Executive Director of UNEP, said:

> *'But we must remain vigilant if our success story is to ultimately have a happy ending. Some of these new, replacement chemicals may prove to be no threat at all to the ozone layer although they may pose threats to human health, wildlife*

and the environment generally. Others, however, may have the potential to cause significant damage to stratospheric ozone, undermining our efforts to date. I would urge countries to carry out immediate scientific assessments of these new chemicals and to ban those that are shown to have real ozone-depleting potential. Finally governments, industry and organizations like UNEP must, based on sound science, work together to devise a long-term strategy so that we know in advance the ozone-depleting potential of future chemicals before they appear on the market.'

Mario Molina, who with Sherwood Rowland discovered the link between CFCs and ozone depletion in 1974, echoed Töpfer's views on the same occasion:

'It is important to keep the issue of new, emerging, chemicals in perspective. Firstly, we must be worried about those substances with a large potential for depleting the ozone layer like the old CFCs. But even these new, short-lived, substances need to be watched...

'Until recently it was thought that these new substances could not damage the ozone layer. It was thought they did not live long enough to reach the stratosphere. However, new research is changing this view. There is evidence that under the right conditions, such substances and their breakdown products can travel far enough to reach the Earth's protective shield.

'At the moment I believe we do not have a big problem with these new substances. But we cannot be complacent. If enough of them are manufactured and emitted, we will delay the recovery of the ozone layer quite significantly.'

He called for more scientific studies to assess fully the ozone-damaging potential of the new chemicals:

'We need to know which of them are safe and which may be a worry in the future. We have enough experience to know that we cannot take the ozone layer, which shields life on Earth from levels of ultraviolet light which can cause cancers and eye cataracts, for granted.'

Closing note: eternal vigilance

It is clear that humanity should be vigilant forever to ensure that ozone-depleting substances are banished from the world. New technologies must be screened continuously to ensure that they are ozone-safe. The extraordinary spirit of cooperation of the governments, scientists, technologists, NGOs, media and UN organizations must continue until the objective is achieved. Eternal vigilance is the price of a safe world.

'If we have injured the space, earth or heaven
or if we have offended mother or father
from that, May Agni, fire of the house, absolve us
and guide us safely to the world of goodness.'

Prayer from *Atharva Veda*[6]

'The tumult and shouting dies
The captains and the kings depart
Still stands thine ancient sacrifice
A humble and contrite heart
Lord God of Hosts, be with us yet
Lest we forget, lest we forget.'

Rudyard Kipling, 1865–1936,[7] *Recessional Hymn*

'Yet this we ask ere you leave us, that you speak to us and give us of your
truth.
And we will give it unto our children, and they unto their children, and it
shall not perish.'

Kahlil Gibran, 1883–1931[8]

'Human history becomes more and more a race between education and
catastrophe.'[9]

H G Wells, 1866–1946

'Now is not the end. It is not even the beginning of the end. But it is,
perhaps, the end of the beginning.'

Winston Churchill, 1874–1965[10]

Ozone layer timelines: 4500 million years ago to present*

---

	c4500 million years ago
Science	The Earth, planets and moons are formed.

	c400 million years ago
Science	The ozone layer becomes thick enough to allow life on land.

	c250 million years ago
Science	Photosynthesis from blue-green bacteria produces oxygen.

	c900 BC
Science	Homer notes the smell of ozone after lightning storms.

	1500s
Science	Leonardo da Vinci determines that air contains a constituent that supports combustion.

	1700s
Science	**1773** Joseph Priestley isolates gaseous oxygen **1785** Martinus Van Marum creates ozone in a laboratory by sparking electricity in oxygen.
NGOs	**1798** Thomas Malthus publishes *An Essay on the Principle of Population*.

	1800–1829
Science	**1812** Benjamin Rush notices chemical pollution.
Technology	**1803** Thomas Moore coins the word 'refrigerator' to describe his ice-cooled, rabbit-fur-insulated wooden butter-transport box. **1824** Nicholas Leonhard and Sadi Carnot invent the Carnot Refrigeration Cycle.

	1830s
Science	**1839–1840** Christian Friedrich Schönbein discovers and names 'ozone'.
Technology	**1834** Jacob Perkins receives British patent 6662 for closed-cycle refrigeration. **1839** Carbon tetrachloride discovered.

	1860s
National	**1863** British chemical industry pollution leads to Alkali Act, allowing government to question industry and suggest voluntary actions.

Science	**1860** Surface ozone measured at hundreds of locations.
	1860 French Academy of Sciences appoints special commission to review ozone science.
	1864 J L Soret identifies ozone as an unstable form of oxygen composed of three oxygen atoms (O₃).
Technology	**1861** US Civil War stops the import of natural ice to the South; first mechanical ice machines imported from France.

1870s

Science	**1878–1879** Marie-Alfred Cornu theorizes that a gas in the atmosphere filters UV radiation.

1880s

Science	**1880–1881** Walter Noel Hartley identifies ozone as the gas filtering UV.
Technology	**1884** William Whiteley invents air conditioning for horse carriages consisting of a wheel-driven fan blowing air across blocks of ice.
NGOs	**1886** First local Audubon Society formed by George Bird Grinnell.

1890s

Technology	**1890s** Carbon tetrachloride is a popular metal cleaning solvent.
	Late 1890s F Swarts pioneers fluorocarbon chemistry.
NGOs	**1892** Conservationist John Muir organizes the Sierra Club.

Late 1800s

Technology	Warm winters, industrial pollution, mining waste and sewage force ice collection further from markets and favour 'artificial' manufactured ice.
NGOs	The first death attributed to any ozone-depleting substance (methyl bromide).

1900–1919

International	**1919** League of Nations founded at the Paris Peace Conference.
Science	**1906** Erich Regener first to study the decomposition of ozone using ultraviolet light.
	1908 L Teisserenc de Bort names 'stratosphere'.
	1913 M Charles Fabry and M H Buisson use UV measurements to prove that most ozone is in the stratosphere.
Technology	**1900** Methyl bromide and carbon tetrachloride introduced as fire-extinguishing agents, solvents, plastic ingredients and other uses.
	17 December 1903 Orville and Wilbur Wright fly at Kitty Hawk, North Carolina.
	1915 Sears Roebuck markets the 'status icebox' with mahogany cabinet and a brass spigot that dispenses chilled water through the door.
NGOs	**5 January 1905** National Association of Audubon Societies for the Protection of Wild Birds and Animals formed in US.
	1913 Passenger pigeon becomes extinct.
	1919 First two deaths attributed to carbon tetrachloride used as a fire-extinguishing agent.

1920s

International	**1924** Members of the League of Nations adopt a protocol aimed at resolving all international disputes through arbitration.
Science	**1920** M Charles Fabry and M H Buisson take quantitative measurements of total column ozone in Marseille.

1924 Gordon M B Dobson and D N Harrison invent a prism spectrophotometer (Dobson meter) to monitor total atmospheric column ozone.

1925 R O Griffith and A M McKeown discover that bromine greatly speeds the decomposition of ozone.

1928 George H Findlay discovers that ultraviolet radiation causes skin cancer.

1929 F W P Götz invents the Umkehr method for measuring the vertical distribution of ozone.

May 1929 First International Ozone Conference in Paris.

Technology **1923** Compressed gas propellants first used as a means of dispersing insecticides.

1928 Thomas Midgley, working with Albert Henne and Robert McNary, invents CFCs.

31 December 1928 Frigidaire (owned by General Motors) receives the first CFC patent.

1930s

Science **1930** Sydney Chapman establishes the photochemical theory of stratospheric ozone.

27–28 May 1931 Auguste Piccard and Charles Kipfer reach the stratosphere in a pressurized aluminium ball below a hydrogen balloon.

25 June 1931 Sydney Chapman defines the chemical reactions whereby the ozone layer maintains a 'steady-state' equilibrium.

1933–1934 Dorothy Fisk and Charles Abbot promote the importance of the ozone layer in protecting life on Earth from harmful UV radiation.

1936 Second International Ozone Conference in Oxford.

Technology **1930** Wills Carrier meets with Thomas Midgley and recommends CFC-11 for air conditioning.

27 August 1930 General Motors and DuPont form a joint stock company – the Kinetic Chemical Company – to manufacture and market CFC refrigerants.

1939 Packard Motor Company produces the first car with ODS vehicle air conditioner (HCFC-22).

1939 Lyle Goodhue and William Sullivan invent aerosol products propelled by liquefied compressed gas, recommending CFC-12 as the best propellant.

1939 DDT first compounded.

1940s

International **1 January 1942** The name 'United Nations' was first used in the *Declaration by United Nations*.

5 May–26 June 1945 *United Nations Charter* drawn up.

24 October 1945 The United Nations officially came into existence.

Science **1948** International Ozone Commission organized at the IUGG General Assembly in Oslo.

Technology **1940s** Carbon dioxide used in the manufacture of refrigerators as a drop-in replacement for CFC-12, in short supply during World War II.

1942 Westinghouse markets the first CFC-12 aerosol pesticide 'bug bombs' for use by the US military in World War II.

1950s

National **1950** Los Angeles, California, begins effort to control pollution from automobiles.

Science   **1950** David R Bates and Marcel Nicolet propose photochemical theory of catalytic ozone destruction by hydrogen radicals.
1952 London's 'Killer Smog' kills 4,000 in a weekend, leading to air pollution path-breaking regulations.
1955 International Ozone Commission (IOC) and World Meteorological Organization (WMO) propose a global ozone station network.
1957–1958 International Geophysical Year encourages scientific cooperation and integrates Earth-monitoring stations. WMO with IOC establishes the Global Ozone Observing System and develops standard data-collection and reporting procedures.
4 October 1957 USSR launches Sputnik 1, the first artificial Earth satellite.
Technology **1956** First air-conditioned shopping mall opens in Edina, Minnesota.

1960s

International **1961** Antarctic Treaty signed, allowing only scientific, tourist and other limited use of the continent.
National   **1961** International Clean Air Congress, London.
Science   **1960** A W Brewer and D R Milford construct the Brewer-Mast electrochemical ozonesonde.
1960 World's first weather satellite, Television and Infrared Observation Satellite, functions for 78 days transmitting 20,000 photographs to Earth.
12 April 1961 Russian Yuri Gagarin is first man in space riding Vostok 1.
5 May 1961 Alan Shepherd, America's first astronaut, describes the atmosphere's fragile ozone layer.
1963 USSR scientist G P Gustin describes the first filter ozonometer.
1966 First publication of UV backscatter measurements on board satellites by R D Rawcliffe and D D Elliot in the US and V A Krasnopolskij in the USSR.
1966 US National Academy of Sciences (NAS) asks James McDonald to investigate the possibility that global climate could be affected by supersonic transport aircraft (SST) exhaust.
1963–1965 Gordon Dobson publishes the first paper indicating the anomalous Antarctic ozone behaviour for the period 1956–1963.
1960s J Hampson postulates a series of reactions involving hydrogen compounds produced from water vapour that are capable of rapidly destroying ozone.
1969 Buckminster Fuller publishes the *Operating Manual for Spaceship Earth*.
Technology **29 November 1962** French and British agree to build an SST for commercial passenger service.
1964 DuPont introduces CFC-113 for dry-cleaning of clothes.
31 December 1968 First flight of Russian Tupolev-144 SST.
2 March 1969 First flight of Anglo-French Concorde SST.
20 July 1969 Americans Neil Armstrong and Buzz Aldrin land on the moon in Apollo 11.
NGOs   **1962** Rachel Carson publishes *Silent Spring*, alerting the public to the harmful effects of pesticides on natural ecosystems and human survival.
24 February 1967 William Shurcliff organizes the Citizens League Against the Sonic Boom.
July 1967 John E Gibson publishes 'The Case Against the Supersonic Transport' in *Harpers* magazine.
6 October 1967 Environmental Defense Fund (EDF) is founded to stop the use of DDT.
1969 Friends of the Earth founded by David R Brower.

1970	
Science	**1970s** Damaging effects of UV-B radiation on plants reported and quantified. **1970s** James Lovelock measures CFCs in the atmosphere. **April 1970** Nimbus 4 satellite with Backscatter Ultraviolet Spectrometer begins first ozone observations from space. **1970** Paul Crutzen hypothesizes that nitrogen oxides, possibly from fertilizers, will deplete the ozone layer. **3 October 1970** US National Oceanic and Atmospheric Administration created. **1971** Paul Crutzen and Harold Johnston discover an ozone-destroying catalytic cycle involving nitrogen compounds.
NGOs	**1970** Natural Resources Defense Council founded. **22 April 1970** First 'Earth Day'.

1971	
National	**24 March** US Congress halts funding of the American SST.
Science	James McDonald suggests that exhaust of SST might deplete stratospheric ozone levels and makes the first predictions of the increase of skin cancer from depletion of the ozone layer. NAS forms its Climatic Impact Committee to investigate the threat to the ozone layer from nitrogen oxides and other emissions from the exhausts of SSTs.
NGOs	Greenpeace founded.

1972	
International	**29 February** Signing of the Convention on the Prevention of Marine Pollution by Dumping of Wastes and Other Matter. **5–16 June** United Nations Conference on the Human Environment in Stockholm ranks stratospheric ozone depletion among priority issues; recommends that a global WMO network of 110 stations monitor the atmosphere, including the ozone layer; and proposes creation of the United Nations Environment Programme (UNEP) with Nairobi as its headquarters.
Science	Michael Clyne finds that at stratospheric temperatures, chlorine atoms would destroy ozone six times more efficiently than the nitrogen oxides associated with SSTs. Lester Machta (NOAA) reports on James Lovelock's unpublished measurement of CFCs in the atmosphere. F Sherwood Rowland is in the audience. Richard Stolarski and Ralph J Cicerone determine that exhaust from rockets slightly depletes the ozone layer.
Technology	DuPont organizes Seminar on the Ecology of Fluorocarbons for the world's CFC producers.
NGOs	DDT banned by US EPA. *The Limits to Growth* published by the Club of Rome.

1973	
International	**June** UNEP Executive Director Maurice Strong's address to the first meeting of the GC warns that damage to the ozone layer may endanger human life on the planet. Eighty nations sign the Convention on International Trade in Endangered Species of Wild Fauna and Flora (CITES).
Science	**December** Mario Molina and F Sherwood Rowland complete development of hypothesis that CFCs will release chlorine in the stratosphere and deplete the ozone layer.

374   *Protecting the Ozone Layer*

Michael McElroy and Steven Wofsy independently discover the chlorine chain but conclude that volcanoes are not a significant source of chlorine in the stratosphere.

James Lovelock publishes his report finding that CFCs are ubiquitous in the atmosphere.

Chuck Kolb identifies CFCs as a potentially significant source of chlorine in the atmosphere.

Technology **3 June** Russian Tupolev supersonic transport crashes at Paris Air Show.

Pan American Airways, Trans World Airlines, Quantas and Japan Air cancel their orders for the Concorde SST while British Airways and Air France proceed.

NGOs E F Schumacher publishes *Small Is Beautiful*.

1974

National Autumn City Council of Ann Arbor, Michigan, enacts a voluntary ban on CFC aerosol cosmetic products.

Science **January** NASA sponsors scientific assessment meeting that concludes that 50 space-shuttle launches could result in 'small but significant' ozone depletion.

28 June Mario Molina and F Sherwood Rowland publish CFC ozone-depletion hypothesis and estimate 7–13% ozone depletion at current levels of CFC production.

10 September Rowland and Molina present theory at the American Chemical Society.

September Manufacturing Chemists Association announces financing of research into the ozone-depletion theory.

James Lovelock and Lester Machta determine that CFCs must have a very long atmospheric lifetime.

Philip Krey presents data showing a falloff in CFC-11 concentrations at higher altitudes.

Technology **11 December** 'If credible scientific data, developed in this experimental program (of research), show that any chlorofluorocarbons cannot be used without a threat to health, DuPont will stop production of these compounds' – Raymond L McCarthy, DuPont.

NGOs **July** NRDC biochemist Karim Ahmed and NRDC co-founder Thomas Stoel learn of the Molina and Roland theory and initiate ozone-aerosol campaign.

September NRDC scientist Karim Ahmed arranges presentation of the Molina–Rowland theory at the meeting of the American Chemical Society.

November NRDC petitions US Consumer Product Safety Commission to ban CFC aerosol products.

Worldwatch Institute is founded.

1975

International **April** UNEP GC backs a programme on ozone protection proposed by Maurice Strong.

30 August The Convention on the Prevention of Marine Pollution by Dumping of Wastes and Other Matter enters into force.

National **January** US Council on Environmental Quality creates inter-agency task force on the Inadvertent Modification of the Stratosphere (IMOS).

March NAS appoints a Panel on Atmospheric Chemistry to report to its Climatic Impact Committee.

13 June IMOS reports that the ozone-depletion theory is plausible though not proven and recommends that CFC propellants be banned by January 1978.

15 June Oregon bans aerosol products containing CFCs, effective 1 March 1977.

9 August New York requires labels on CFC aerosol products warning of possible harm to the environment and prescribing a ban in 1978 if they are found to pose a hazard.

14 October US EPA Administrator Russell Train raises the CFC/ozone-depletion issue at the North Atlantic Treaty Organization.

Canada and the US ask the Environment Committee of the Organisation for Economic Co-operation and Development (OECD) to take action on ozone-layer protection.

Science   **21 January** US National Academy of Sciences and Department of Transportation conclude that atmospheric levels of CFCs would deplete the ozone layer.

February Michael McElroy identifies methyl bromide as potential ozone-depleting substance.

3 October Veerabhadran Ramanathan determines that CFCs are highly potent greenhouse gases.

WMO conducts the first international assessment of global ozone: 'Modification of the Ozone Layer due to Human Activities and some Possible Geophysical Consequences'.

Arthur Schmeltekopf and NOAA colleagues publish the first observations of CFCs in the stratosphere, conclusively demonstrating that hypothesized tropospheric sinks for these chemicals are negligible.

Technology   **18 June** S C Johnson, Inc announces it will phase out CFCs as aerosol product propellants.

16 July The Sherwin-Williams Co announces phase-out of CFC aerosol products.

Bristol Meyers and Mennen launch advertising campaigns promoting non-aerosol products as substitutes for CFC products. 'Get off the Can and on the Stick!'

NGOs   **February** An episode of the popular US television drama *All in the Family* declares that CFC aerosol hairspray will destroy the ozone layer and 'kill us all'.

April Tom Stoel (NRDC) meets with UNEP official Rames Mikhail to request that UNEP takes the lead on the ozone issue.

1976

International   **April** UNEP GC requests the Executive Director to convene a meeting 'to review all aspects of the Ozone Layer, identify ongoing activities and future plans, and agree on a division of labour and a co-ordinating mechanism for: inter alia, the compilation of research activities and future plans; and the collection of related and commercial information'.

Mostafa K Tolba appointed Executive Director of UNEP.

National   **February** US EPA publishes the first comprehensive assessment of alternatives to selected chlorofluorocarbon uses.

February The Canadian Minister of Environment announces intention to ban non-essential uses of CFCs.

May The UK Department of the Environment concludes that scientific uncertainties make any discussion of CFC controls premature.

15–17 September Russell Peterson, Chairman of the US Council for Environmental Quality and former DuPont executive, declares that, 'unlike criminal defendants, chemicals are not innocent until proven guilty'.

15 October US Food and Drug Administration (FDA) proposes an orderly phase-out of non-essential uses of CFC propellants in food, drug and cosmetic products, and an interim warning label on the same products in the meantime.

Science   **13 September** NAS concludes that CFCs affect the ozone layer and recommends 'selective regulation'.

15 September James Anderson reports finding chlorine and chlorine monoxide rising steadily in abundance up to an altitude of 42km – proving that chlorine was reacting with ozone.

November Atmospheric Environment Service of Canada (AESC) concludes that: 'the scientific evidence is sufficiently strong to warrant the government making a decision on regulation at this time'.

WMO initiates Global Ozone Research and Monitoring Project.

Technology   **21 January** First supersonic passenger service offered on British/French Concorde.

1977

International   **1–9 March** UNEP GC sponsors the first international conference on CFCs in Washington, DC.

25 April UNEP GC devises a 'World Plan of Action' and establishes the Coordinating Committee on the Ozone Layer (CCOL).

11 November First meeting of UNEP CCOL.

National   **26 April** Second intergovernmental conference on CFCs, Washington, DC.

August The Commission of the European Communities recommends a cap on CFC-11 and -12 production capacity.

15 December Sweden bans CFC aerosol products, effective 1 July 1979.

Science   British Antarctic Survey records ozone depletion above its Halley Bay research station, but does not discuss its findings with other scientists.

Technology   Hydrofluorocarbon (HFC) 134a developed by several companies.

Vehicle manufacturers reduce the quantity of CFCs used in air conditioners. German and Swiss aerosol industry associations agree to a 30% voluntary reduction from the 1976 baseline to be achieved by 1979.

1978

International   **28–30 November** Meeting of CCOL.

6–9 December A conference on CFCs in Munich, Germany, recommends scientific and regulatory research and a voluntary reduction in global CFC emissions.

National   **15 March** US bans most CFC food, drug, device and cosmetic aerosol products (except metered-dose inhalers).

15 December US halts product manufacturing with CFC propellants.

December Norway announces plans to ban CFC aerosol products from July 1981.

The OECD Environment Committee publishes its report on CFC substitutes, policy options and economic impacts.

Science   **October** Nimbus 7 satellite launched with Total Ozone Mapping Spectrometer (TOMS) and other instruments to study stratospheric chemistry.

Technology   **20 March** 'An additional study period of 3–5 years prior to any regulation of nonpropellant uses will not pose significant risk to the public health or the environment' – DuPont.

NGOs   Tom Stoel and Alan Miller (NRDC) author a study showing that Canada, The Netherlands, Sweden and the US had moved quickly to regulate CFCs but that other countries had done little or nothing.

	1979
International	**November** Third meeting of CCOL, Paris, France.
National	**1 June** Norway bans aerosol products. effective 1 September 1981.
	1 July Swedish ban on most CFC aerosol products enters into force.
Science	Second major US National Academy of Sciences Report predicts 16.5% ozone depletion, much-increased rates of skin cancer and serious damage to the marine food chain and agricultural crops. It recommended taxes, quotas, or a ban on uses of CFCs with alternatives.
	Stratospheric Research Advisory Committee to the UK Department of Environment concludes it is too early to confirm the ozone-depletion hypothesis.
Technology	**28 March** Three Mile Island nuclear power plant almost melts down.

	1980
International	**29 April** UNEP GC appeals to governments to reduce their CFC consumption and not to increase production capacity. (Resolution 8/7B).
	November The UNEP CCOL meets in Bilthoven, The Netherlands.
National	**March** The Council of European Communities asks its members to cap CFC manufacturing capacity and achieve a 30% reduction in aerosol products by the end of 1981.
	14–16 April International conference in Oslo, Norway recommends negotiations on an international convention for protection of the ozone layer.
	September Japan announces plans to reduce CFC consumption and cap production capacity.
	7 October The US issues an 'Advance Notice of Proposed Rulemaking' outlining a cap on production capacity to limit total CFC use.
	Canada prohibits use of CFCs in hairsprays, antiperspirants and deodorants.
Technology	**June** Having patented several HFC and HCFC chemical substitutes for CFCs, DuPont, ICI and Daikin Kongyo suspend their research.
	Alliance for Responsible CFC Policy is formed.

	1981
International	**26 May** UNEP Ninth GC passes a resolution establishing a Working Group of Legal and Technical Experts (Working Group) to elaborate the design of a Global Framework Convention for the Protection of the Ozone Layer. (Resolution 9/13B.)
	12–16 October The UNEP CCOL meets in Copenhagen, Denmark.
	28 November–6 December UNEP Working Group meets in Montevideo, Uruguay, to develop environmental law and select the protection of the ozone layer as one of three areas for the first legal guidelines.
National	**1 July** Norwegian ban on most CFC aerosol products enters into force.
	Sweden asks UNEP to form a working group to start negotiations on an international ozone agreement.
Science	**October** Low Antarctic ozone levels detected by the British Antarctic Survey at Halley Bay attributed to instrument malfunction.
	December Scientists from DuPont and elsewhere publish a two-dimensional model of stratospheric chemistry.
	First scientific assessment of the state of the ozone layer issued by WMO in collaboration with UNEP and national research agencies.
	NASA scientists reprogram computers to analyse only the ozone data falling between 180 and 650 Dobson units on the false assumption that readings outside that range would be the result of instrument malfunction.

1982

International	**20–28 January** Inaugural meeting of the UNEP Working Group for the Preparation of a Convention for the Protection of the Ozone Layer, Stockholm. **31 May** UN Conference on the Global Environment held in Nairobi on the occasion of the tenth anniversary of the Stockholm Conference. **June** UNEP GC adopts Resolution 10/17 outlining an environmental law programme – including top priority for ozone protection. **10–17 December** Meeting of Working Group, Geneva.
National	Denmark prohibits methyl bromide in open fields due to health concerns.
Science	**March** Third US NAS/National Resources Council Report lowers the previous estimates of ozone depletion to 5–9% and notes that the link between ozone depletion and skin cancer is not sufficiently established to allow quantitative prediction. **28 March** El Chichon volcano erupts in Mexico, contributing to ozone depletion and complicating ozone monitoring.
Technology	International Institute of Refrigeration (IIR) Congress in Paris focuses on ozone depletion and global warming.

1983

International	**5–8 April** The UNEP CCOL meets in Geneva. **11–15 April** Working Group meets in Geneva. **24 May** Resolution 11/7 of UNEP Governing Council. **17–21 October** Working Group meets in Geneva.
National	Taiwan (a province of China) bans CFCs in cosmetic aerosol products.
Science	**October** A new Dobson meter detects low Antarctic ozone levels at Halley Bay, but British Antarctic Survey scientists do not report the data. Fourth NAS study: *Causes and Effects of Changes in Stratospheric Ozone: Update 1983* lowers ozone depletion estimates to 2–4%. Ozone-monitoring at Arosa, Switzerland, Hohenpeissenberg, Germany, and five stations in Canada record ozone depletion of 3–8%.
NGOs	**March** NRDC files intent to sue US EPA for failure to protect the ozone layer as required by the Clean Air Act. **10 June** 'Analysis of ground-based and satellite ozone measurements have failed to detect ozone depletion, either in the total amount of overhead ozone or at 40 kilometres altitude…since the risk is quite low, the Alliance advocates waiting for the results of further research' – Alliance for Responsible CFC Policy.

1984

International	**16–20 January** Working Group meets in Vienna. **28 May** Resolution 12/14 of UNEP GC requests more meetings to draft a Convention and possibly a Protocol. **June** WMO Workshop on Current Issues in our Understanding of the Stratosphere and Future of the Ozone Layer, Feldafing, Germany. **15–19 October** Seventh meeting of the UNEP CCOL, Geneva. **22–26 October** Working Group meets in Geneva.
National	**5–7 September** Canada hosts first meeting of the Toronto Group. **29 November** Denmark bans CFC aerosol products as of 1 January 1987.
Science	**28 June** Michael Prather, with Michael McElroy and Steven Wofsy, publishes a new model predicting 15% ozone depletion by 2050 at 3% annual growth in CFC use. **31 July** David Lee notices very low satellite measurements of Antarctic ozone, but supervising NASA scientists dismiss the readings as an instrument problem.

Fourth NAS report.

S Chubachi (Japan Meteorological Agency) is the first scientist publicly to report, at the Halkidiki Ozone Commission Symposium, extremely low Antarctic ozone during the 1982 Antarctic spring over the Syowa Research Station.

Technology   **23 December** Union Carbide pesticide plant leaks methyl isocyanate in Bhopal, India, killing thousands.

1985

International   **21–25 January** The CCOL and Working Group meet in Geneva.

18–22 March Conference of the Plenipotentiaries creates the Vienna Convention for the Protection of the Ozone Layer and adopts the Toronto-Group-sponsored Resolution on workshops and continued negotiations on a Protocol scheduled for adoption in 1987.

22 March Vienna Convention is signed by 21 countries and the European Community.

24 May UNEP GC requests that the Executive Director convene workshops on CFCs to reconcile the differences between countries, as specified by the Vienna Convention.

National   **4–5 March** The Toronto Group decides to abandon the proposal on CFC aerosol controls and prepares a draft resolution on workshops and continued negotiations to identify an alternative control methodology.

October A US Federal Court issues a consent order in the lawsuit between NRDC and US EPA imposing a timetable on EPA to conclude its consideration of regulations for non-aerosol CFC applications.

Science   **16 May** British scientists, led by Joseph Farman, announce abnormally low Antarctic ozone – 30–40% depletion observed since 1977.

Second Scientific Assessment of the State of the Ozone Layer is issued by WMO in collaboration with UNEP and national research agencies.

Technology   DuPont announces a major expansion of CFC production capacity in Japan.

NGOs   **10 July** French government agents sink the Greenpeace vessel *Rainbow Warrior* in New Zealand, killing one Greenpeace member.

26 December Dian Fossey murdered while working to protect mountain gorillas in Rwanda and Zaire.

1986

International   **24–28 February** Seventh meeting of the UNEP CCOL, Nairobi.

26–30 May UNEP Workshop on CFC Production and Consumption and CFC Controls, Rome, Italy.

June UNEP/US EPA International Conference on the Effects of Ozone Modifications and Climate Change on Health and Environment, Washington, DC.

8–12 September UNEP Second Workshop on Demand and Technical Controls and General Control Strategies, Leesburg, Virginia.

19–21 November The UNEP CCOL meets in Bilthoven, The Netherlands, to update environmental-effects science.

1–5 December First Session of the Negotiations Working Group, Geneva.

National   **10 January** US EPA publishes its *Stratospheric Ozone Protection Plan* (setting the deadline of 1 November 1987 for a Final Rule).

4 June Canada is the first country to ratify the Vienna Convention.

12 September Russian authorities Boris Gidaspov and Vyacheslav Khattatov and US EPA's Stephen Andersen propose placing US TOMS satellite in orbit aboard Russian spacecraft.

16 September Gidaspov and Khattatov agree to secure Russian agreement to place TOMS in orbit.

18 December USSR and US agree to pursue ozone and climate cooperation.

Canada proposes key elements of the Montreal Protocol including burden-sharing, the ODP-weighted 'basket' production and consumption controls, and the double majority for the Protocol to enter into force.

First meeting of the European Community Industry Group for the Protection of the Ozone Layer (IGPOL).

Science   **January** WMO/UNEP 1985 *Atmospheric Ozone Assessment* published.

9 February F Sherwood Rowland first to call Antarctic ozone depletion an 'ozone hole'.

28 January US space shuttle *Challenger* explodes during launch and its crew of seven astronauts killed.

28 August US NASA corroborates the ozone depletion measured by British and Japanese Antarctic Surveys by re-analysis of satellite data and concludes that there is a continental-scale, seasonally repeating 'ozone hole' phenomenon.

Susan Solomon leads first ground-based National Ozone Expedition (NOZE) to study stratospheric chemistry over Antarctica in order to determine the cause of the Antarctic ozone hole.

Susan Solomon, Rolando Garcia, F Sherwood Rowland and Don Wuebbles publish their theory that polar stratospheric clouds accelerate the destruction of ozone by chlorine/bromine compounds, thereby leading to the Antarctic ozone hole.

Technology   **16 September** The Alliance changes its original position, and calls on the US government to 'work in cooperation with the world community under the auspices of the United Nations Environment Programme to consider establishing a reasonable global limit on the future rate of growth of fully halogenated CFC production capacity'.

26 September 'We now conclude that it would be prudent to take further precautionary measures to limit the growth of CFCs world-wide…[and call for]…the development and adoption of a protocol under the United Nations Vienna Convention for the Protection of the Ozone Layer to limit worldwide CFC emissions' – DuPont.

Representatives of DuPont, Allied and Imperial Chemical Industries (ICI) separately report that between 1975 and 1980 they had identified compounds meeting environmental, safety and performance criteria for some CFC applications, but had terminated research and development because none were as inexpensive as CFCs.

DuPont restarts the research on substitutes that had been halted in 1980.

NGOs   **26 April** Chernobyl, Ukraine, nuclear power plant explodes, killing 31 quickly and thousands more from chronic effects.

May A court order accepts US EPA Ozone Protection Plan schedule in settlement of NRDC lawsuit.

8 September Seventy-nine European and American environmental NGOs urge the total phase-out of CFCs within ten years.

Citizens' Clearinghouse on Hazardous Wastes publicly challenges McDonald's use of CFCs in food packaging.

Greenpeace activists campaign against CFCs at Hoechst, and at a DuPont plant in Luxembourg.

1987

International **23–27 February** Second session of the Working Group, Vienna, Austria.
9–10 April UNEP/NASA meeting on ozone scientific models, Wuerzberg, Germany.
27–30 April Third session of the Negotiations Working Group in Geneva.
May Tolba issues first 'Executive Director's Paper' outlining desired outcome of the Montreal meeting.
10 June Venice Economic Summit declaration lists stratospheric ozone depletion first among environmental concerns.
June UNEP Governing Council instructs Working Group to address all potential ozone-depleting substances (clearing the way to add halon).
29–30 June Tolba holds informal consultations in Brussels among key delegation heads and trade experts.
6–8 July UNEP Ozone Protocol Legal Drafting Group, The Hague, The Netherlands.
8–11 September Preliminary Session to the Conference of Plenipotentiaries on the Protocol on Chlorofluorocarbons to the Vienna Conference for the Protection of the Ozone Layer.
14–16 September Conference of the Plenipotentiaries creates the Montreal Protocol on Substances that Deplete the Ozone Layer. Signed by 24 nations and the European Economic Community on 16 September 1987.
Tolba convenes first Informal Advisory Group.

National **1 January** Danish ban on CFC aerosol products enters into force.
27 February Prince of Wales bans CFCs from his home.
13–16 September Canada and US sponsor a trade exhibition at Montreal.
US President Ronald Reagan and USSR General Secretary Mikhail Gorbachev strongly endorse cooperation on stratospheric ozone and climate protection.

Science Airborne Antarctic Ozone Experiment (AAOE) conclusively demonstrates key role of chlorine in chemical reaction associated with ozone-hole formation; 'the smoking gun'.
Second National Ozone Expedition (NOZE II) continues to gather data, concluding that the cause of the Antarctic ozone hole is acceleration of chlorine- and bromine-catalysed cycles of ozone destruction by polar stratospheric clouds. The source of chlorine and bromine is emissions of ODSs.
First Report of the UK Stratospheric Ozone Research Group (SORG) published. Ozone depletion in the lower stratosphere documented to be 10% per decade.

Technology **19–21 February** A panel of experts from Germany, Japan, UK and the US form the 'US EPA Chlorofluorocarbon Chemical Substitutes International Committee' to identify environmentally superior chemical substitutes to ODSs.
March Washington, DC, Conference on CFCs and Ozone Protection Programs sponsored by the Alliance and the Center for Energy and Environmental Management.
8 May DuPont states: 'Our position during 1981–82 is the same as our position today; that is, that there are no commercially available substitutes for CFCs-11, 12 and 113 in spite of our substantial efforts.'
24 May Chlorofluorocarbon Chemical Substitutes International Committee concludes that the absence of a market – rather than technical or environmental issues – is the principal barrier to commercialization of CFC substitutes.
28 July DuPont reports: 'there is no scientific consensus to suggest imminent danger to man or the environment, so there is no need for severe global cutbacks in CFC production.'

NGOs   **March–April** Environment Action urges CFC-product boycotts, letter-writing campaigns, and leak-prevention maintenance of refrigeration and air conditioning.
8 September Fifty-nine environmental NGOs from 22 countries issue a policy statement: 'Ozone Depletion and Climate Change: A Statement on CFCs and Related Compounds by the International Environmental Community'.
USA Today features a list assembled by NRDC from US EPA data of America's largest emitters of CFCs: AT&T, General Electric Company, General Motors, IBM, US Air Force and United Technologies.

1988

International **9–11 March** Working Group for the Harmonization of Data meets in Nairobi.
19–21 October The Scientific, Environmental, Technology and Economics Assessment processes are informally initiated at a UNEP Conference on Science and Development, CFC Data, Legal Matters and Alternative Substances and Technologies, The Hague, The Netherlands.
24–26 October Second session of the Working Group for the Harmonization of Data, The Hague.
Paul Horwitz appointed Secretary, John Carstensen, first Deputy Secretary, and Megumi Seki, first Science Officer for the Ozone Meetings.
Workshop to establish the framework for the 'Scientific Assessment of Stratospheric Ozone 1989', London, UK.
A United Nations resolution characterizes climate as a 'common concern of mankind' and the Intergovernmental Panel on Climate Change (IPCC) is formed.

National   **13–15 January** The first International Conference on Alternatives to CFCs and Halons, Washington, DC, is sponsored by the Conservation Foundation, Environment Canada and US EPA.
February Three US Senators write to DuPont asking it to fulfil its 1974 promise that if additional research showed 'that any chlorofluorocarbons cannot be used without a threat to health, DuPont will stop production of these compounds'.
31 March Mexico is the first country to ratify the Montreal Protocol.
March Sweden releases its plan to cut CFC emissions by half by 1991, and eliminate them by 1995.
9 June Sweden prohibits the commercial use of CFCs on specific sector timetables.
23 June The Swedish EPA issues a Refrigerants Order to limit use and emissions of CFCs and HCFCs.
15 December The Swedish EPA introduces high fees for CFC exemptions; fees to increase sharply over time.
4 October The Council of Ministers invites the European Commission to conclude voluntary agreements with industry for the greatest possible reduction of CFCs and halons and the labelling of CFC-free products.
Brazilian Ministry of Health bans CFCs in household and cosmetic products.
Norwegian Action Plan to reduce CFC consumption by 50% by 1991, and by 90–100% by 1995.
Packaging industries agree to pay Massachusetts US$700,000 for illegal emissions of CFCs.
State legislation or executive orders controlling CFC products are either enacted or pending in California, Massachusetts, Maine, Minnesota, Rhode Island, and Vermont.

Science   **8 March** Canadian scientists report evidence of ozone 'hole' over the Arctic.
15 March Third Scientific Assessment of the State of the Ozone.

July John S Hoffman and Michael J Gibbs of the US conceive presentation of policy options in terms of chlorine and bromine abundance.
Second UK SORG Report.

Technology **5 January** Fourteen global ODS producers form the Programme for Alternative Fluorocarbon Toxicity Testing (PAFT) to expedite regulatory approval of fluorocarbon substitutes.

13 January AT&T and Petroferm announce a semi-aqueous solvent that cleans electronics as well as CFC-113.

18 February Provigo (Canadian food retailer) announces it will replace CFC egg cartons with pulp-based cartons and requires suppliers to reduce the CFC content in other foam packaging.

March ICI announces it will not market HCFC-22 for personal-care aerosol products due to lingering concerns about toxicity.

24 March Ten days after the Ozone Trends Panel Report is published, DuPont announces that it accepts the scientific evidence linking CFCs and ozone depletion and becomes the first ODS producer to commit to 'an orderly transition to phase-out of fully halogenated chlorofluorocarbon production' – DuPont.

24 March Pennwalt Corporation seconds DuPont's position and urges CFC production be 'discontinued as soon as practical'.

12 May US industry and environmental NGOs announce the world's first voluntary national CFC phase-out in food packaging.

16 May Dow Chemical Company announces it will stop using CFCs in its packaging and insulating foam products.

1 September Eight global CFC producers form second Programme for Alternative Fluorocarbon Toxicity Testing (PAFT) for HCFC-141b.

23 September Alliance for Responsible CFC Policy notifies US EPA that Parties to the Montreal Protocol should develop additional controls with the objective of phasing out production of fully halogenated CFCs.

October ICI urges strengthening of the Montreal Protocol, saying it 'aims to be the first chemical company in the world to make alternatives to CFCs commercially available'.

30 December The House of Fraser Group announces ban of all aerosol products containing ODSs.

The Body Shop calls for a ban on CFC aerosol and foam products.

First International Halon Conference, Lugano, Switzerland.

Institute of Interconnecting and Packaging Electronic Circuits announces a plan to protect the ozone layer.

Nortel and Seiko Epson announce corporate goals of a complete CFC-113 phase-out on accelerated schedules.

Toys-R-Us announces an end to sales of CFC foam party streamers and other CFC toys.

US Air Force restricts new applications of halons.

NGOs **10 March** Friends of the Earth (FOE) launches 1000 helium balloons at 7 Canadian locations to symbolize that CFCs and halons released in any location affect ozone concentrations elsewhere.

9 October Asia-Pacific People's Environmental Network runs a half-page advertisement in *The Sunday Tribune* advocating stronger ODS controls.

22 December Chico Mendes is murdered in Brazil because of his campaign to halt logging in the Amazon Basin.

FOE Netherlands declares campaign successful when four packaging manufacturers, representing 85% of the Dutch market, agree to switch from CFCs to HCFC-22 and pentane by March 1989.

FOE UK secures agreement of most restaurants, egg packers and supermarkets to halt the use of CFC foam packaging.

1989

International   **1 January** Vienna Convention and Montreal Protocol enter into force.
22–24 January Tolba holds informal consultations on financial mechanisms in Nairobi.
26–28 April COP 1 (first meeting of the Conference of the Parties to the Vienna Convention), Helsinki, Finland.
2–5 May MOP 1 (first Meeting of the Parties to the Montreal Protocol), Helsinki, Finland, officially establishes the Assessment Panels with Terms of Reference (TOR) according to Article 6 of the Montreal Protocol. Stephen O Andersen, John Hoffman and George Strongylis Co-Chairs of the Economic Panel; Jan van der Leun and Manfred Tevini Co-Chairs of the Environmental Effects Panel; Daniel Albritton and Robert Watson Co-Chairs of the Scientific Assessment Panel; and Stephen O Andersen and G Victor Buxton Co-Chairs of the Technology Review Panel. Meeting also establishes the Amendments and Financial Mechanism Working Groups and adopted UNEP as Secretariat.
3–5 July Senior Advisers meeting on funding investment in developing countries.
10–15 July Conference of the Environment Ministers convened by UNEP in Geneva; protocol among topics discussed.
26 July Assessment Panel Synthesis Report published.
11–14 August First meeting of the ad hoc working group of legal experts on non-compliance, Geneva.
14 August Technical Panel Report and five Technical Option Reports sent to UNEP.
22 March Basel Convention on the Control of Transboundary Movements of Hazardous Wastes and their Disposal.
21–25 August First session of the Open-Ended Working Group of the Parties to the Montreal Protocol (OEWG 1), Nairobi, Kenya.
28 August–5 September Second session of OEWG 1, Nairobi, Kenya.
18–22 September Third session of OEWG 1, Geneva, Switzerland.
13–17 November First session of OEWG 2, Geneva, Switzerland.
The Technology Review, Economic Assessment, Scientific Assessment and Environmental Effects Panels are formed, and publish assessment reports.
Tolba holds informal consultations on financial mechanism and alternatives.

National   **5–7 March** British Prime Minister Margaret Thatcher and the UK host 123 nations at the Conference on Saving the Ozone Layer in London to encourage developing countries to ratify and to promote strengthening of the protocol.
6 March Prince Charles calls for the total and immediate elimination of CFCs.
15–17 March US EPA hosts first international workshop on Integrating Case Studies Carried Out Under the Montreal Protocol.
May The Environment Protection Authority of the State of Victoria organizes the first international Conference on The Future of Halons.
10 May The Alhambra California City Council prohibits restaurants and retail food vendors from using plastic foam and non-biodegradable plastic packaging materials containing CFCs.
August Australian Government, with the leadership of the Association of Fluorocarbon Consumers and Manufacturers of Australia (AFCAM), decides to phase out halon for all but critical applications by 31 December 1995.
10–11 October International Conference on CFC and Halon Alternatives, Washington, DC.
November US Congress enacts tax on ODSs.
Brazilian Ministry of Health bans cosmetic, solvent and convenience aerosol

products, allowing only medical uses.

A European Commission public opinion poll shows that more than 60% of Europeans believe chemicals cause the ozone hole.

Mexican industry phases out CFC in flexible foams.

Science First Airborne Arctic Stratospheric Expedition (AASE).

Science Assessment Panel (SAP) explains the Antarctic ozone 'hole'.

WMO/UNEP Scientific Assessment published in cooperation with national scientific organizations includes a comprehensive review of the current understanding of environmental properties of alternative fluorocarbons.

Technology **January** Mobile Air Conditioning Society (MACS) and its partners develop a CFC recycling standard that is accepted worldwide under factory warranties.

1 July Australia Pipefitters and Gasfitters Union adapts policy to install no new halon systems from 1 July 1989, not to discharge test from 12 May 1989, and not to service halon systems after July 1991.

15 July IBM sets an internal goal to eliminate CFCs from manufacturing by 1993.

1 August General Motors announces recovery and recycling of CFC-12 at all its service centres worldwide.

August AT&T pledges to eliminate all CFC emissions from its manufacturing processes by 1994.

10 October Industry Cooperative for Ozone Layer Protection (ICOLP) is formed.

8 December Toyota announces recovery and recycling of CFC-12 at all its domestic service centres.

50%-reduced CFC foam systems for refrigerator manufacture first used in Germany and then quickly throughout Europe.

Halon Alternatives Research Corporation (HARC) is founded.

Industry forms the Alternative Fluorocarbons Environmental Acceptability Study (AFEAS) to promote research into environmental properties of HCFCs and HFCs.

Nissan commits to phase out CFCs in vehicle manufacturing.

Woolworths Australia sets a goal of halting the use of CFC refrigerants.

NGOs **January** 'Endangered Earth' gets *Time* magazine's 'Person of the Year Award'.

24 March *Exxon Valdez* oil tanker runs aground in Prince William Sound, Alaska.

2 May Ninety-three environmental NGOs from around the world demand a 'crash programme' to protect the ozone layer.

Greenpeace activists are arrested for defaming the Finnish Flag outside MOP 1 in Helsinki when they float a banner reading 'Our Lives Are in Your Hands' to the top of the flagpole.

Greenpeace hangs a banner at the DuPont facility in Maitland, Ontario, protesting the continued manufacture of CFCs.

Greenpeace hangs a banner in Luxembourg reading 'DuPont Ozone Killer No 1 in Luxembourg; and Worldwide – Greenpeace'.

Greenpeace ship *Trojan Horse* docks at the Hoechst CFC Plant in Frankfurt and posts a banner reading 'Skin Cancer Has a Name: Hoechst'.

1990

International **19–23 February** Regional workshop to encourage developing country participation in the Protocol, Penang, Malaysia.

26 February–5 March Second session of OEWG 2 on Amendments, Financial and HCFCs, Geneva, Switzerland.

6–7 March Tolba holds 'informal' consultations with representatives of selected countries 'to exchange views, not as representatives of their governments, but in personal capacity' (10 developed and 10 developing countries).

8–14 March First session of OEWG 3, Nairobi, Kenya.

12 March Tolba holds 'informal' consultations (8 developed and 4 developing countries).

7 April Tolba holds 'informal' consultations.

9–11 April Private consultations on negotiations in Moscow (UNEP, Canada, Norway, US, USSR).

26–28 April Montreal Protocol Bureau meeting, Geneva.

9–11 May Second session of OEWG 3, Geneva, Switzerland.

26–27 May Meeting of experts in cooperation with the World Intellectual Property Organization on the role of intellectual property rights in technology transfer.

18–19 June Tolba holds 'informal' consultations.

20–29 June OEWG 4, London, UK.

27–29 June MOP 2, London, amends the Protocol to create a financial mechanism including an Interim Multilateral Fund (MLF) of US$240 million.

3–5 December OEWG 5, Nairobi, Kenya.

First meeting of the Multilateral Fund Executive Committee.

Parties combine the Technology and Economic Panels to form the Technology and Economic Assessment Panel (TEAP) and appoint Stephen Lee-Bapty to replace G Victor Buxton as Co-Chair. HCFCs are reclassified as 'transitional substances'.

National **1 June** Regional Seminar on Ozone Depletion, Mexico City.

5 July Canada is first to ratify the London Amendment.

20 September Sweden bans imports of certain foam products made with or containing CFCs.

November International Conference on CFC and Halon Alternatives, Baltimore, MD.

28 November US EPA presents first Stratospheric Protection Awards.

Mexican government signs voluntary agreements with industrial associations to phase out CFCs and halons earlier than required by the Montreal Protocol.

Science Third UK SORG published.

Technology **5 April** Digital Equipment Corporation (DEC) donates patented aqueous cleaning technology to the public domain.

30 April German chemical makers Hoechst and Kali Chemie announce they will voluntarily stop production of CFCs by 1995.

18 October ICI first to commercialize HFC-134a with plant start-up at Runcorn, UK.

NGOs **22 April** Twentieth Earth Day celebrated by 200 million people in 141 countries.

27 June FOE demonstrators dress as penguins with signs reading: 'Don't blow our cover' and 'Ban all ozone destroyers'.

The environmental coalition CFC STOP petitions, protests and advertises for stronger German and European ozone-protection measures.

Greenpeace activists suspend a giant pair of inflatable sunglasses from roof of Sidney Opera House with the message 'Ozone Depletion: Stop It or We'll All Go Blind'.

Greenpeace activists block the entrance to Environment Canada, and hang a banner at the Kali Chemie plant in Hannover, Germany, reading 'Here in this Romantic Place Starts Ozone Damage'.

Paul McCartney and Capital Records launch world 'Rescue the Future' tour promoting the FOE ozone protection campaign.

NRDC releases *Who's Who of America's Ozone Depleters* listing the 3014 companies with the highest emissions.

1991

International **1 January** Start of the Interim Multilateral Fund.

April Secretariat for Vienna Convention and the Montreal Protocol (Ozone Secretariat) fully formed with K Madhava Sarma as Coordinator (later designated as Executive Secretary).

8–12 April Second meeting of the ad-hoc working group on non-compliance.

17–19 June COP 2, Nairobi, Kenya.

19–21 June MOP 3, Nairobi, approves Annex D list of Products for Trade Measures against non-Parties.

5–8 September Second meeting of the ad hoc working group of legal experts on non-compliance.

October Global Environment Facility launched.

5–8 November Third meeting of the ad hoc working group of legal experts on non-compliance, Geneva.

Reports of the Assessment Panels published with Synthesis Report find that more stringent control measures are environmentally necessary and technically and economically feasible.

UN Antarctica Treaty (Madrid Protocol) signed by 39 countries prohibiting mining, limiting pollution and protecting animal species.

UNEP establishes the OzonAction Programme under the Multilateral Fund (MLF) at its Industry and Environment centre in Paris to assist Article 5 Parties as a clearinghouse and for capacity building. OzonAction publishes its first newsletter and holds its first Regional Workshops in Thailand, Egypt and Venezuela.

National **8 May** Sweden prohibits the use of halons in new equipment by July 1991.

8–12 May Military and environmental authorities from Norway, Sweden, UK and US tour facilities in Norway and Sweden, concluding in new agreements to cooperate on stratospheric ozone protection.

1 July Norwegian regulations ban imports and most uses of CFCs and halons and products containing CFCs and halons.

11–13 September The First International Conference on the Role of the Military in Protecting the Ozone Layer, sponsored by NATO, US Air Force and US EPA, is held in Williamsburg, Virginia, US, with participants from Canada, Germany, Japan, The Netherlands, Norway, Union of Soviet Socialist Republics, UK and US.

9 October US EPA, the European Commission, the Japan Ministry of International Trade and Industry, the Japan Ministry of Health and Welfare, and the Programme for Alternative Fluorocarbon Toxicity Testing (PAFT) (an international consortium of chemical-producing companies) announce agreement on testing for safety of new HCFC and HFC substitutes for ODSs.

December Sweden bans halon in existing equipment (by 1 January 1998); 1,1,1-trichloroethane (by 1 January 1995); carbon tetrachloride (by 1 January 1998); and HCFCs except for rigid foam for insulation and refrigerants (by 1 January 1994). Import ban of CFC products expanded to include some HCFC products.

Camara Nacional de la Industria de la Transformacion, Nortel/Northern Telecom, ICOLP/ICEL and the US EPA form a partnership to phase out CFC solvents in Mexico by 2000.

Finland provides financing to UNEP to assist non-Party developing countries.

Mexico, with the help of US EPA, establishes the first Ozone Protection Unit.

Many developing countries, including China and India, ratify the Montreal Protocol.

Norwegian navy announces the first military acceptance of alternatives for halons used on combat vessels.

US Department of Defense (DoD) recommends that CFC solvents be phased out.

Science   **12 June** Mount Pinatubo volcano in the Philippines begins to erupt with largest blast on 15 June, contributing to ozone depletion and complicating ozone monitoring.

Fourth scientific assessment of the state of the ozone layer issued by WMO.

Fourth UK SORG Report.

NASA scientists, led by Richard Stolarski, publish satellite observations of global trends in the ozone above mid-latitudes, establishing that measurable ozone depletion is occurring in all areas of the globe except the tropics.

Second TOMS instrument launched on Russian Meteor 3 spacecraft, a joint Russian/American space project.

Technology   **10 September** Nortel donates patented 'tackiness tester' (vital to no-clean soldering) to the public domain.

Australia, Austria, Denmark, Germany, Mexico, New Zealand, UK and Venezuela industry complete the phase-out of aerosol cosmetic and convenience products.

Daikin, DuPont and Showa Denko commercialize HFC-134a. Asahi Glass builds the first HCFC-225 production plant.

HCFC-141b foam introduced in building products.

Hoechst distributes millions of leaflets in Germany stating that Greenpeace is endangering the lives of children by opposing HFC refrigerants.

Nissan and then Mercedes-Benz become world's first automobile manufacturers to introduce CFC-free air conditioning (HFC-134a).

Nortel becomes the first multinational telecommunications company to eliminate CFC-113 from its global manufacturing operations.

US–Japan–Russia Environmental Executive Leadership Workshop, Woods Hole Massachusetts, promotes technology cooperation.

NGOs   **2 April** US District Court allows DuPont to refuse to consider stockholder resolutions urging faster phase-out because CFC production is DuPont's 'ordinary business'. FOE appeals.

31 August–2 September NGOs from Indonesia, Malaysia, Thailand, Philippines and Papua New Guinea meet in Jakarta for Regional Forum on Global Climate Change Issues.

Greenpeace erects billboards in Hamburg, Germany, featuring photographs of Hoechst and Kali Chemie with the words 'Everybody is Discussing the Climate – We're Ruining It'. Hoechst threat to sue Greenpeace stimulates media coverage with photos of the billboard featured in newspapers and magazines and shown on TV. An 8-year judicial process concludes that CEOs are public figures and subject to such scrutiny.

1992

International   UNEP conducts four regional workshops with participants from 90 developing countries.

6–15 April OEWG 6, Nairobi, Kenya.

3–14 June Earth Summit, Rio de Janeiro, Brazil. The Framework Convention on Climate Change (FCCC) is signed. More than 100 heads of state attend.

UNEP conducts four regional workshops with participants from 90 developing countries.

25 June Interim Assessment Report on Methyl Bromide published.

8–17 July OEWG 7, Geneva, Switzerland.
August London Amendment enters into force.
17–20 November OEWG 8, Copenhagen, Denmark.
23–25 November MOP 4, Copenhagen, Denmark. Parties appoint Lambert Kuijpers to replace Stephen Lee-Bapty as Co-Chair of the TEAP and appoint new Co-Chairs Pieter Aucamp to the SAP, Andre P R Cvijak to the TEAP, and Xiaoyan Tang to the EAP.
MOP 4 approves the Copenhagen Amendment. HCFCs and methyl bromide included in the list of controlled substances.

National **1 January** The Netherlands completes phase-out of methyl bromide soil treatments.
11 February US President George Bush announces that the US will unilaterally accelerate the phase-out of ODSs and calls on other nations to ratify the Protocol and join faster phase-out.
23 February US Department of Defense directs its contracting officers to purchase new equipment that is manufactured and maintained without ODSs and directs technical experts to rewrite specifications to eliminate dependence on ODSs.
18 March The European Commission's *Fifth Environmental Action Programme* cites international action on the ozone layer as evidence that environmental problems can be globally solved.
1 July US Clean Air Act amendment makes it illegal to vent CFC or HCFC refrigerants.
11 August US Department of Defense limits use of ODSs and directs the Defense Logistics Agency to establish and manage a reserve for 'mission-critical' uses.
3 September Sweden expands Refrigeration Order to cover handling of HFCs.
Australia, Austria, Brazil, Canada, Denmark, Egypt, Finland, Germany, Hungary, Mexico, New Zealand, Sweden, Switzerland, Trinidad and Tobago, UK, US and Venezuela have banned most cosmetic and convenience aerosol products.
European Community (EC) and Norway announce plans to phase out CFCs and CTC by 1995.
Mexico commits itself to an accelerated ODS phase-out with target year 2000 and establishes a computerized system to monitor and control ODS imports and exports.
NATO supports accelerated phase-out and encourages global technology cooperation.
Sweden provides financing to UNEP IE OzonAction Programme to establish the first Regional Network of ODS Officers in the Southeast Asia & the Pacific Region.
Switzerland prohibits CFC, HCFC, 1,1,1-trichloroethane, carbon tetrachloride solvents and new halon installations.
US military orders rapid elimination of ODSs from all weapons acquisition and creates ODS banks for mission-critical uses.

Science Extremely low ozone during Antarctic spring observed, covering very large area and with the lowest ever ozone during the northern winter–spring.
Second Airborne Arctic Stratospheric Expedition (AASE II) documents high levels of the chemically active forms of chlorine over North America – indicating the potential for significant ozone depletion.
A team of international scientists, led by V Ramaswamy, shows that stratospheric ozone loss causes a cooling of the lower stratosphere, which can give rise to a cooling of the surface–troposphere climate system.

Technology   **11 February** Alliance for Responsible CFC Policy petitions US EPA to accelerate reduction steps and further reduce CFC production and consumption below levels allowed by the Montreal Protocol and US Clean Air Act, and to also accelerate the phase-out of HCFC-22, HCFC-141b and HCFC-142b.

12 February Ford Motor Company begins production of its first vehicles with CFC-free air conditioning.

21 February Alaska North Slope oil and gas producers agree to shift to recycled halon for explosion and fire protection.

May Chrysler is first to implement full-scale production of vehicles with CFC-free air conditioners at its Jeep Grand Cherokee plant.

19 November Elf Atochem announces HCFC-142b and HFC-134a production.

Brazil, Egypt, Finland, Hungary, Norway, Switzerland and Trinidad and Tobago industry competes the phase-out of aerosol cosmetic and convenience products.

The Coca-Cola Company halts the purchase of CFC refrigerated equipment, rapidly motivating manufacturers in developed and developing countries to meet global demands.

General Dynamics is first to eliminate ODS solvents from aircraft manufacture; Lufthansa is first to halt the use of most ODS solvents in aircraft maintenance.

Greenpeace, the German Dortmund Institute and refrigerator manufacturer DKK Scharfenstein form a partnership to develop a hydrocarbon refrigerator. Within one year, the 'Greenfreeze' is commercialized.

Japanese refrigerator companies in Thailand announce plans to phase out CFCs in 1997.

Seiko Epson is the first multinational manufacturer of precision instruments to eliminate CFC-113.

Spuma Pac is first in Brazil to phase out CFCs in food packaging in response to market demands by McDonald's. This leadership motivates a rapid voluntary global phase-out in food packaging.

A team of Brazilian soldering experts at Ford Brazil implements one of the first no-clean soldering technologies and transfers this advanced technology to developed and developing countries.

US–Japan Environmental Executive Leadership Workshop is held in Yountville, California.

NGOs   **11 February** National Toxics Campaign accuses the US government – including the Departments of Defense and Energy, NASA and their contractors – of emitting up to two-thirds of all CFC-113 used in America.

23 November Activists from Friends of the Earth International dress as penguins and hand out bottles of sunscreen to delegates to the MOP.

Greenpeace activists hang a banner reading 'Now Save the Ozone Layer' on House of Lords at the Palace of Westminster, UK; protest at DuPont, ICI, and Seagram stockholders' meetings; persuade more than 500 UK doctors to pledge not to prescribe ICI CFC drugs and begin boycott of ICI's brand of paint; and officially complain to the UK Advertising Standards Authority (ASA) for misleading ICI advertising.

US Securities and Exchange Commission (SEC) finds that DuPont was wrong to prevent a shareholder from offering a resolution at the company's 1991 annual meeting. The SEC emphasizes that: 'timing of the CFC phase-out has strategic, long-term implications for the company's business and raises other significant policy issues'.

1993

International **1 January** Interim Fund transitions to Multilateral Fund.
7–10 June First workshop for the ODS Officers Network for Southeast Asia and the Pacific (ODSONET/SEAP) in Bangkok, Thailand.
30 August–1 September OEWG 9, Geneva, Switzerland.
17–19 November MOP 5, Bangkok, Thailand, replenishes the Fund with US$455 million. Parties appoint Suely Carvalho to replace Andre P R Cvijak as Co-Chair of the TEAP.
22 November COP 3, Bangkok, Thailand.
Tolba awarded the 'Sasakawa Environment Prize' and the US EPA 'Stratospheric Ozone Protection Award'.
UNEP initiates the International Halon Bank Management Information Clearinghouse.

National **7 January** US Air Force bans purchase of ODSs.
1 February US requires labels on products made with or containing ODSs, causing suppliers to US markets to phase out rapidly to avoid consumer backlash.
1 March Saudi Arabia is the first country to ratify the Copenhagen Amendment.
21 April US President William J Clinton directs US federal agencies to maximize the use of safe alternatives to ODSs.
21 May US DoD declares that no contract after 1 June 1993 may include a specification or standard requiring the use of an ODS and orders re-evaluation of existing contracts dependent on ODS.
20–23 October International Conference on CFC and Halon Alternatives, Washington, DC.
Australia establishes the first national halon bank.
Austria, Belgium, Botswana, Canada, Denmark, Finland, Germany, Israel, Italy, Liechtenstein, The Netherlands, Sweden, Switzerland, UK, US and Zimbabwe sign a declaration to phase out methyl bromide consumption.
Mexico freezes its national CFC and halon consumption at 1989 levels.
Nordic Council of Ministers publishes a catalogue on CFC-free foam.
Norway bans most new installations of halon and announces plan to remove non-essential systems by 2000.
UK sets up the halon bank: Halon Users' National Consortium.
Uruguay launches 'ozone friendly' product labelling.
US plans to phase out methyl bromide by 2001; Denmark issues a draft regulation proposing to phase it out by 1998; and Sweden completes phase-out of methyl bromide soil treatments.

Science **6 October** Scientists observed the lowest Antarctic ozone value recorded from 1957 to 2000.
Fifth UK SORG report.
NOAA scientist A R Ravishankara and colleagues show that perfluorocarbons (PFCs) are highly potent greenhouse gases.

Technology **1 January** US production of halon ceases.
April Minebea, the company previously consuming the largest quantities of ODSs in Thailand, completely phases them out and donates to the public domain its patented deoxidized water-washing and vacuum-degassing technology.
1 June US Air Force halts purchase of newly produced halon and prohibits the purchase of commercial vehicles and equipment requiring CFCs.
Cadbury, Sainsbury's and Woolworths Australia achieve complete CFC phase-out.

CSIRO Australia announces carbonyl sulphide as a replacement for commodity fumigation with methyl bromide.

Hortico, Zimbabwe, introduces alternatives to methyl bromide for strawberry production.

With Greenpeace as a facilitator, GTZ, Liebherr and Haier help start up the production of Greenfreeze in China.

Matsushita is the first multinational company to manufacture kitchen appliances and entertainment equipment without CFCs.

NGOs   FOE initiates a 'Save Our Skins (SOS)' global campaign; presents a suit to the Norwegian Environment Minister that can be washed in water rather than ODS solvents; demonstrates in Belgium, Cyprus, Germany, Ireland, Malta, Netherlands, Poland, Spain and Sweden; and launches a campaign against methyl bromide through its youth network called Earth Action.

A Greenpeace protest at Dow Chemicals Norrkoping Sweden foam plant gets wide media coverage when the CEO holds up CFC product and says, 'This is the problem, I mean the product'.

Methyl Bromide Alternatives Network (MBAN), a coalition of 16 environmental, farmer, labour and consumer NGOs, is created to promote the phase-out of methyl bromide.

UK Advertising Standards Authority (ASA) rules for Greenpeace in four out of five complaints that ICI advertising of the environmental acceptability of HFCs is misleading.

Ozone Action founded.

1994

International   UN Convention on climate change enters into force.

1 January Halon production and consumption halted in developed countries. UN Convention on Climate Change enters into force. Thirty-four nations pledge $2 billion in support of GEF's activities.

16 May First workshop for the ODS Officers Network for South America (ODSONET-LA-S) in Montevideo, Uruguay.

5–8 July OEWG 10, Nairobi, Kenya.

6–7 October MOP 6, Nairobi, Kenya.

28 November and 2 December First workshop for ODS Officers in Central America and Spanish-speaking Caribbean (ODSONET/LA-C) in Antigua.

19 December United Nations General Assembly designates September 16 as the 'International Day for the Preservation of the Ozone Layer'.

Reports of the assessment panels published.

UNEP IE OzonAction and the Ozone Secretariat hold the first Workshop for Countries with Economies in Transition (CEIT) in Minsk, Belarus.

National   **24–25 January** The Second International Conference on the Role of the Military in Protecting the Ozone Layer is held in Brussels, sponsored by NATO, the US Department of Defense, US EPA, ICOLP, Aerospace Industries Association, American Electronics Association, Electronic Industries Association and the University of Maryland Center for Global Change.

22 December Sweden establishes a phase-out schedule for use of HCFCs in rigid foams for insulation (ending 1 January 1998) and expands the ban on imports of CFC and HCFC to include some rigid foam products. The import regulations are revised to become combined sales and import regulations due to Sweden's new position as member of the European Union by 1 January 1995.

Indonesia, Denmark and Sweden issue regulations to phase out methyl bromide by 1998.

Malaysia prohibits the use of CFCs as aerosol propellants.

State Fire Service of Poland conference for fire authorities in Poland and other countries with economies in transition (CEIT). Halon alternatives and halon-banking options are identified that allow Poland to withdraw its pending essential-use nomination for halon.

Taiwan prohibits CFC air conditioners in manufactured or imported automobiles and mandates recycling in the servicing and disposal of automobile air conditioners.

The US EPA's Significant New Alternatives Program (SNAP) approves the first environmentally acceptable alternatives to ODSs. SNAP prohibits PFCs in most applications.

Science   Extremely low ozone during Antarctic spring covers very large area, lowest ever ozone during the northern winter–spring.

Fifth WMO/UNEP assessment concludes that 'the atmospheric growth rates of several major ozone-depleting substances have slowed, demonstrating the expected impact of the Montreal Protocol and its Amendments.

NOAA scientists, led by A R Ravishankara, show that HFC-134a is a suitable 'ozone-friendly' substitute.

Technology   **10 January** Coca-Cola donates technical and training manuals on ODS phase-out for use by food service enterprises in developing countries.

April Hoechst shuts down production in Brazil, becoming the first CFC producer to halt manufacture worldwide.

August Electrolux introduces its first hydrocarbon refrigerator and pledges to make all European models with hydrocarbons by 1996.

Almost every new automobile air-conditioning system worldwide uses HFC-134a.

Greenpeace agrees to place labels on Bosch Denmark's hydrocarbon refrigerators stating 'Greenfreeze technology as recommended by Greenpeace'.

US–Japan Environmental Executive Leadership Workshop held in Osaka, Japan to promote international cooperation and actions by multinational companies to speed the phase-out in Article 5 countries.

NGOs   Center for International Environmental Law (CIEL) publishes *The Industry Cooperative for Ozone Layer Protection: A New Spirit at Work*, praising the aerospace and electronics industry for global collaboration to eliminate ODS solvents.

Greenpeace seizes canisters containing ozone-destroying substances from Dehon, and protests at an Elf Atochem meeting in Argentina and the World Bank's 50th anniversary meeting in Spain.

1995

International   **8–12 May** OEWG 11, Nairobi, Kenya.

15–18 May First workshop for ODS Officers in English-speaking Africa (ODSONET/AF-E) in Nairobi, Kenya.

28 August–1 September OEWG 12, Geneva, Switzerland.

18–21 September First workshop for ODS Officers in French-speaking Africa (ODSONET-AF-F) in Dakar, Senegal.

27 November–4 December Preparatory meeting in Vienna for the seventh meeting of the Parties to the Montreal Protocol.

5–7 December MOP 7, Vienna, Austria.

Assessment Reports of the SAP, the EEP and the TEAP are published. A Synthesis Report integrates the findings of the four Panels.

Tenth anniversary of the Vienna Convention is celebrated.

Tolba receives the 'Decoration for Distinct Services for the Global Environment' from Mexico and Venezuela.

National   **1 January** Austria, Finland and Sweden become members of the European
Union implementing EU restrictions on HCFC use, unless already covered by
stricter national regulations.
13 March EU and US agree to coordinate negotiating positions on major
environmental issues, including ozone depletion.
11 May Sweden introduces a phase-out schedule for all remaining use of
CFC and HCFC refrigerants.
Twenty-one countries sign a declaration to limit use of methyl bromide to
strictly necessary applications and to phase out as soon as possible.
Australian farming sectors agree to methyl bromide reduction targets and
finance alternatives trials with a voluntary levy on sales.
Austria finalizes regulations to phase out use of methyl bromide for storage
facilities by 1998.
European Union halts CFC production for all but export and essential uses,
reduces the HCFC 'cap' from 3.1% to 2.6%, and introduces HCFC use
controls.
EU regulation enters into force, mandating 25% cut in methyl bromide in
1998, and HCFC use controls.
Poland hosts the International Seminar for Countries with Economies in
Transition.
Thailand proposes prohibition of manufacture and import of CFC
refrigerators.

Science   **10 October** Paul Crutzen, Mario Molina and F Sherwood Rowland receive
Nobel Prize in Chemistry for the 'pioneering contributions to explaining how
ozone is formed and decomposes' which have 'contributed to our salvation
from a global environmental problem that could have catastrophic
consequences'.
Scientists observe record low ozone during January to March over Siberia
and a large part of Europe.

Technology   3M markets the world's first CFC-free metered-dose inhaler (MDI).
Calor Gas markets a range of hydrocarbon refrigerants.
European tobacco companies inform suppliers in Zimbabwe and Asia that
they will no longer accept tobacco treated with methyl bromide.
Introduction of cyclopentane/isopentane insulating foam systems.
Japanese companies in Thailand phase out CFCs for household refrigerators
– the first sectoral phase-out in an Article 5 country.
More than 40 multinational companies from seven countries pledge to help
the Government of Vietnam protect the ozone layer by investing only in
modern, environmentally acceptable technology in their Vietnam projects.
Sainsbury's opens the first UK supermarket to use Greenfreeze technology in
Horsham, Sussex, UK.
Walmart opens the first ODS-free shopping mall.

NGOs   Greenpeace erects a banner reading 'Stop Ozone Killers' at MOP 7 in Vienna.
International NGO Alliance to Protect the Ozone Layer is formed with
hundreds of NGOs across the globe coordinated by the Pesticide Action
Network. Activities included letter campaigns in 17 countries urging bans of
methyl bromide.
Methyl Bromide Alternatives Network (MBAN) contacts 1000 scientists, policy-
makers and journalists about the dangers of methyl bromide use and the
availability of environmentally sustainable alternatives.
Ozone Action and the Environmental Law Foundation sue Amana, General
Electric and Whirlpool and several retailers for misleading advertising, arguing
that 'CFC-free' labels imply that the product doesn't deplete the ozone layer.

1996

International **1 January** Production of CFCs, carbon tetrachloride and 1,1,1-trichloroethane halted in developed countries.

May World Health Organization (WHO) calls for full implementation of the Montreal Protocol and new ODS controls as the only way to prevent the emergence of serious health problems in Europe.

26–29 August OEWG 13, Geneva, Switzerland.

19–26 November OEWG 14, San Jose, Costa Rica (also referred to as the 'Preparatory meeting for the fourth COP and eighth MOP').

25–27 November MOP 8, San Jose, Costa Rica, decides a US$465 million replenishment of the Multilateral Fund.

Assessment reports of the TEAP and its Task Force on Countries with Economies in Transition (CEIT) are published.

Executive Secretary K Madhava Sarma (Ozone Secretariat) earns the US EPA 'Stratospheric Protection Award'.

National **1 January** Denmark phases out use of methyl bromide on tomatoes.

23 January UK concludes voluntary agreements with industry to limit use and emissions of HFCs from aerosol, air conditioning and refrigeration, fire protection and foam industries.

12 June Bruce Burrell becomes the first person to be extradited to the US for an environmental crime when Costa Rican authorities return him to the US to face charges on CFC smuggling.

24 September A jury convicts Sun-Diamond Growers on eight counts of providing US$9000 in illegal gratuities to a US Department of Agriculture official. The indictment states that the company 'had an interest' in delaying the ban of methyl bromide use.

Australia introduces substantial fees for methyl bromide.

Colombia bans methyl bromide with limited exemption for quarantine treatments.

Danish EPA and UNEP IE issue a joint study of Strategic Options for Accelerated ODS Phase-Out in CEITs.

EU 'Eco-Label' authorized for hydrocarbon, but not fluorocarbon, refrigerators.

Ghana reduces CFC use by 85%, mainly through training in refrigeration servicing.

The State of Sao Paulo, Brazil, prohibits the government purchases of new products containing ODSs.

Taiwan bans manufacture or import of CFC refrigerators.

Science **29 June** Third TOMS instrument and US Earth Probe satellite launched.

Japanese Advanced Earth Observing Satellite (ADEOS) launched.

Atmospheric concentrations of CFC-11, CTC and methyl chloroform decrease, and growth rate of CFC-12 concentrates slows.

Sixth SORG Report shows clear evidence of a decline in effective tropospheric chlorine equivalent, and prediction of continued decline through the next century.

NOAA scientists, led by Steve Montzka, show that the combined abundance of ozone-depleting compounds in the lower atmosphere peaked in about 1994, and have since begun to show a decline.

Technology **2 February** Elstar (UK) becomes the world's first commercial refrigeration manufacturer to switch entirely to hydrocarbon refrigerants.

May Poly-Urethane Industria e Comercio in Brazil installs the first polyurethane foam blown with castor oil.

NASA ODS-elimination effort experiences a serious setback when hot exhaust gases penetrate into all three field joints on both solid rocket motors during launch of the space shuttle *Columbia*. The most likely cause was the use of a new ODS-free adhesive and a new ODS-free cleaning agent in the assembly of the rockets. The launch of space shuttle *Atlantis* to the Russian Space Station *Mir* was delayed six weeks to allow retrofit with rocket motors that were assembled using the original ODS products.

World Semiconductor Council pledges voluntary reductions in emissions of PFCs.

US–Japan Environmental Executive Leadership Workshop is held in Nara, Japan, to study lessons from ozone layer protection that will be important to climate protection.

Whirlpool agrees to introduce Greenfreeze in Argentina by 2000.

NGOs   **25 November** Japan's Save the Ozone Network (JASON) activists – dressed as the sun, plants and children – perform an ozone dance for delegates to the MOP accompanied by the music of a song called *What's the Ozone Layer?*

21 December A settlement is reached in the suit by Ozone Action and the Environmental Law Foundation accusing Amana, General Electric and Whirlpool and several retailers of misleading advertising (see 1995 above). Manufacturers agree to explain HCFC content and to provide a $100,000 grant to research alternatives to ODSs.

Canada, Switzerland and US governments fund FOE study 'The Technical and Economic Feasibility of Replacing Methyl Bromide: Case Studies in Zimbabwe, Thailand and Chile'.

Greenpeace activists scale Whirlpool billboard in Buenos Aires and change the messages from 'Whirlpool Brings Quality to Your Life' to 'Whirlpool Destroys the Ozone Layer'.

1997

International   **12 May** First workshop for English-speaking ODS Officers in Caribbean and Barbados.

19 May First workshop for ODS Officers in West Asia in Bahrain.

3–6 June OEWG 15, Nairobi, Kenya.

9–12 September OEWG 16, Montreal, Canada.

15–17 September, MOP 9, Montreal, Canada.

16 September Gala celebration of the tenth anniversary of the Montreal Protocol held in Montreal. 'Best-of-the-Best Stratospheric Ozone Protection Awards' presented to 71 individuals and organizations.

1–11 December Representatives from more than 160 countries meet in Kyoto, Japan, and 122 nations sign the Kyoto Protocol on Climate Change. Assessment Report of the TEAP is published, including a *Handbook on Essential Use Nominations*.

National   **26–27 February** Life After Halon Conference in Melbourne highlights Australian and international leadership in environmental fire protection.

24 March Thailand Ministry of Industry bans the import and manufacture of refrigerators with CFCs: first environmental trade barrier enacted by a developing country.

August US EPA publishes *Champions of the World* elaborating the contributions of Stratospheric Protection Award winners.

6–7 November Third International Workshop on the Role of the Military in Implementing the Montreal Protocol and the First International Workshop on the Military Role in Climate Protection are held in Herndon, Virginia, US, with participants from 12 countries.

Brazil prohibits the use of new ODS equipment in all government-owned enterprises and institutions.

Denmark phases out all uses of methyl bromide, including quarantine treatments. Sweden completes methyl bromide phase-out except for wood exported to Australia.

Guyana hosts workshop for business and media to promote its national phase-out programme.

Hungary launches its Halon Recycling Station.

Norway requires recovery and destruction of ODSs from foams and refrigerants taken out of service.

Philippines prohibits halon imports.

Vietnam plans to phase out methyl bromide by 2006, nine years faster than the Protocol schedule.

Science   Photochemistry of Ozone Loss in the Arctic Region in Summer (POLARIS) A/C campaign conducted on Earth Resources-2 (ER-2) aircraft.

Space shuttle makes first limb-scatter measurements of ozone throughout the stratosphere.

Technology   **13 May** US airlines begin new installation of halon 1301 systems in cargo bays of passenger planes as a response to the 13 May 1996 Value Jet crash that killed all 109 persons on board.

Dutch flower auction houses initiate the Floriculture Environment Project to pay higher prices for crops grown without methyl bromide. Flower farms in Kenya and Israel qualify.

The Co-op food company in the UK bans the use of methyl bromide as a soil fumigant on its own farms and encourages suppliers to its retail supermarkets to phase out methyl bromide.

Japan's Ministry of Trade and Industry (MITI) and the Semiconductor Industry Association of Japan (SIAJ) announce formation of a voluntary PFC emission reduction partnership.

At the invitation of UNEP IE OzonAction Programme, 23 multinational corporations voluntarily pledge to support the 1999 freeze goals for developing countries, including a commitment not to transfer new CFC-using technology to CEIT and Article 5 countries.

All major manufacturers of commercial refrigeration in Mexico eliminate the use of CFCs.

Brazilian industry halts CFC uses in motor-vehicle manufacturing.

NGOs   **1 February** Russian tanker *Nakhodka* spills oil off Japan's western coast.

PANNA and Health and Environment Watch organize a technical workshop called Existing and Potential Alternatives to Methyl Bromide in Cut Flower Production in Kenya with technical experts from Colombia's cut flower industry, farmers, NGOs, UNEP officials and government representatives.

Comite Nacional Pro Defensa de la Fauna y Flora (CODEFF) in Chile works closely with the government of Chile to organize a seminar on methyl bromide, a critical step in Chile's successful efforts to replace methyl bromide.

International Alliance for Ozone Layer Protection sends two letters to the European Commission, signed by 78 NGOs, urging the Commission to phase out methyl bromide quickly.

NGOs demonstrate and use street theatre to pressure governments to phase out methyl bromide at MOP 9. An international alliance sends letters to 17 countries (including Mexico, Japan, Australia, Zimbabwe, Kenya, China and Malaysia) urging governments to ban methyl bromide.

NGOs publish several issues of *ECO* urging governments to take action on methyl bromide, illegal ODS trade and other issues.

FOE protests attempts by industry to open methyl bromide production facilities in Africa.

	1998

International **14–15 May** First workshop for ODS Officers in South and East Asia in New Delhi, India.
7–9 July OEWG 17, Geneva, Switzerland.
November Multilateral Fund has financed over US$850 million for the phase-out in Article 5 countries of more than 140,000 ODP tonnes of ODSs.
18–20 November OEWG 18, Cairo, Egypt.
23 November UNEP honours Tolba with a Global Ozone Award, 'in celebration of his outstanding services to the protection of the ozone layer'.
23 November Children's Painting Competition for Protecting the Ozone Layer, organized by UNEP IE OzonAction and the Egyptian Environmental Affairs Agency (EEAA).
23–24 November MOP 10, Cairo, Egypt, requests the Assessment Panels to study the effect of the Kyoto Protocol on the implementation of the Montreal Protocol and also asks for a study of the impact of new, uncontrolled, ODSs then being marketed.
Assessment reports are published, including separate reports from the Technical Options Committees. A Synthesis Report integrates the findings of the four Panels.
Thirty-six nations pledge $2.75 billion to the GEF for protection of the global environment (including the phase-out of ODSs in countries with economies in transition) and to promote sustainable development.

National **3 February** US EPA proposes ban of HFCs in 'self-cooled beverage containers'. EPA had calculated that a 5% market penetration would release 96 million metric tonnes of carbon equivalent (MtCE).
27 March Canada is first country to ratify the Montreal Amendment.
European Commission adopts proposal for the phase-out of all ODSs in the European Community, and a strategy to phase out CFCs in metered-dose inhalers (MDIs) in the Community.
Mexico approves legislation banning production and imports of air conditioning and refrigeration equipment containing CFCs.

Science Scientific Assessment Panel finds that the Antarctic ozone hole 'continues unabated', but the rate of ozone decline at mid-latitudes has slowed; the level of ozone-depleting compounds in the lower atmosphere peaked in 1994 and is declining. 'The Montreal Protocol is working.'
Upper Atmospheric Research Satellite (UARS) measures peaking chlorine concentrations in the stratosphere.

Technology DuPont announces a plan to shut down ODS production in South America by the end of 1999.
Greenpeace endorses Iceland Frozen Foods Company 'Kyoto' models of Greenfreeze refrigerators and freezers, and authorizes the use of the Greenpeace logo on advertising material.

NGOs **August** Sustainable Tomatoes Campaign is launched in Florida by a coalition of labour and environmental NGOs to raise awareness about methyl bromide and alternatives and to encourage consumers to buy tomatoes produced without methyl bromide.
1998–1999 NGO 'contact groups' established in Mexico, Chile, Senegal and Malaysia (PAN and FOE groups) to monitor methyl-bromide-alternatives projects being carried out under the Montreal Protocol's Multilateral Fund and to promote sustainable alternatives.
FOE activists deliver ozone-friendly, methyl-bromide-free tomatoes from Florida to all members of the US Congress to demonstrate that methyl bromide alternatives exist.

1999

International **May** Joint IPCC/TEAP Expert meeting on HFCs and PFCs in Petten, The Netherlands.

15–18 June OEWG 19, Geneva, Switzerland.

October TEAP publishes *The Implications to the Montreal Protocol of the Inclusion of HFCs and PFCs in the Kyoto Protocol*, finding that the Treaties can be compatibly administered.

29 November–3 December COP 5, Beijing, China.

29 November–3 December MOP 11, Beijing, China.

National **14 October** Sweden relaxes ban on CFC refrigerants, extending the grace period for small CFC refrigeration equipment to end of 2004.

Science Intergovernmental Panel on Climate Change (IPCC) report finds that 'there is no direct evidence that aircraft emissions have altered ozone'.

Technology **April and October** Test failures of a Titan solid rocket motor are attributed to the poor performance of proposed alternative to methyl chloroform, resulting in a setback in development of new process and rejection of production hardware manufactured prior to the test.

8 December State Environmental Protection Administration (SEPA) of China bans the use of ozone-depleting substances in new auto air conditioners from 1 January 2002.

28 December Matsushita announces plans to launch Greenfreeze in Japan by the end of 2002.

DuPont Brazil closes CFC production facility.

Greenpeace organizes the visit of Cuban scientists to Canada, leading to procurement of Canadian technology to purify gas for use as refrigerants for Cuba under support from the governments of Canada and Germany.

NGOs **1999–2000** Centre for Environmental Education in India educates schoolchildren about ozone depletion with an 'Educator's Kit and School Kit' that is widely disseminated. Four regional workshops with teachers integrate ozone protection in the environmental curriculum.

Greenpeace submits a petition with more than 10,000 signatures to the president of Matsushita refrigeration company demanding an early launch of Matsushita's hydrocarbon models in Japan.

Red de Accion sobre Plaguicidas y Alternativas en Mexico (RAPAM) releases a report on the use of methyl bromide in Mexico and recommends ways to eliminate its use.

Pesticide Action Network chapters from Africa, Chile, Colombia, Philippines and the US conduct a global survey for UNEP on methyl bromide laws and regulations in 100 countries.

Pesticide Action Network North America and UNEP publish an *Inventory of Technical and Institutional Resources for Promoting Methyl Bromide Alternatives*.

As a result of a lawsuit filed by Environmental Working Group, Friends of the Earth, Pesticide Action Network and Pesticide Watch, a Superior Court judge rules that the State of California must adopt regulations to protect the public from exposure to methyl bromide.

2000

International **11–13 July** OEWG 20, Geneva, Switzerland.

28 July Beijing Adjustments enter into force.

1 August Michael Graber becomes the acting Executive Secretary of the Ozone Secretariat, on retirement of K Madhava Sarma.

11–14 December MOP 12, Ouagadougou, Burkina Faso.

	Global Environment Facility (GEF) has funded more than US$160 million for the phase-out of ODSs in countries with economies in transition.
National	**3 May** Chile is the first country to ratify the Beijing Amendment.
	June EU adopts stricter regulations regarding phase-out of HCFCs, lower use of methyl bromide prior to the phase-out and controls to improve control of illegal trade.
Science	Antarctic ozone hole is at its largest ever recorded, covering 28 million square kilometres and for the first time extending over populated areas of Chile.
Technology	**28 June** Coca-Cola commits to expand its innovative research and development programme and, by the time of the Athens Olympic Games in 2004, to cease purchase of cold-drink equipment with HFC refrigerants or insulation materials, where cost-efficient alternatives are commercially available.
	9 August Olympic sponsor Foster's Brewing Group announces plans to phase out use of HFCs.
	5 September Olympic sponsor Unilever Foods announces plans to phase out the purchase of HFC ice-cream display cases worldwide by 2005.
	11 September Olympic sponsor McDonald's announces plans to find 'safe and viable HFC-free alternatives by 2004'.
NGOs	Comite Nacional Pro Defensa de la Fauna y Flora (CODEFF) with UNEP DTIE OzonAction Programme and the Government of Chile organize the Regional Policy Development Workshop to Assist Methyl Bromide Phase Out in Latin America.
	Fourteen NGOs from five countries send a letter to the Government of China urging China to ratify the Copenhagen Amendment and to help farmers adopt environmentally sustainable alternatives to methyl bromide.
	Greenpeace activists campaign against the use of HFCs by major Olympic sponsors Coca-Cola and McDonald's.
	Greenpeace activists place 'Enjoy Climate Change' stickers on Coke vending machines in Australian cities.

2001

International	**24–26 July** OEWG 21, Montreal, Canada.
	16–19 October MOP 13, Colombo, Sri Lanka. The issue of new ozone-depleting substances is discussed. Five such unlisted substances reported by the Parties.
National	**6–8 February** Fourth ozone (and second climate) workshop on The Importance of Military Organizations in Stratospheric Ozone and Climate Protection, Brussels, Belgium, with 160 military officers, environmental authorities, technical experts and environmental NGOs from 33 countries.
	1 March Denmark begins an eco-tax on high-GWP substitutes for ODSs (HFCs, PFCs and SF6).
	Sanya announces the goal of becoming China's first ODS-free city.
	Mauritius announces the goal of becoming the world's first ODS-free country.
Science	**17 October** NASA and NOAA scientists announce that the Antarctic ozone hole peaked at about 26 million square miles, similar in size to those of the past three years.
Technology	**8 November** Japanese refrigerator manufacturers Matsushita, Toshiba and Hitachi announce that they will produce and sell refrigerators using hydrocarbon refrigerants and insulating foam in the Japanese market by January 2002.
NGOs	**10–11 May** Consumer Unity and Trust Society (CUTS), SAWTEE and UNEP organize the South Asian Consultation on Atmospheric Issues in New Delhi, India. This is the first time a meeting is held in this region to educate and involve parliamentarians in ozone and climate protection.

1 October The Nestle Company reaffirms its commitment to 'natural refrigerants' and pledges to accelerate global phase-out of CFC, HCFC and HFC refrigerants.

Florida Consumer Action Network, Farmworker Association of Florida and FOE announce test results showing that high levels of methyl bromide have drifted from farmers onto the properties of two churches in Homestead, Florida, US.

A partnership of ten agricultural and environmental NGOs with UNEP DTIE OzonAction Programme raises awareness about the environmental dangers of methyl bromide and available alternatives in Chile, Costa Rica, Dominican Republic, Ethiopia, Kenya, Malawi, the Philippines, Thailand, Zambia and Zimbabwe. This is the first Multilateral Fund project to utilize environmental NGO experts.

2002

International   **1 January** All but ten countries that belong to the United Nations have ratified the Vienna Convention and all but eleven have ratified the Montreal Protocol.
25 February Beijing Amendment enters into force.
22–26 July OEWG, Geneva, Switzerland.
25–29 November MOP 14, Rome, Italy.

2010–

International   **2010** Global CFC production scheduled to end.
2040 Global HCFC production scheduled to end.
Science   **2050+?** Antarctic ozone hole predicted to disappear.
2500+? Ozone layer no longer affected by manufactured ODSs.

---

* This timeline was developed by Stephen O Andersen and K Madhava Sarma over the course of four years with the assistance of Daniel Albritton (US NOAA), Penelope Canan (University of Denver), Suely Carvalho (UNDP), Elizabeth Cook (World Resources Institute), Jorge Corona (National Chamber of Industry Mexico), Jim Curlin (UNEP DTIE), Superintendent Rolf Diamant (US National Park Service), David Doniger (NRDC), Yuichi Fujimoto (Japan Industrial Conference for Ozone Layer Protection), Corinna Gilfillan (London School of Economics), Michael Graber (Ozone Secretariat), Caley Johnson (US EPA), Ingrid Kökeritz (Swedish Environment Institute), Lambert Kuijpers (Technical University Eindhoven), János John Maté (Greenpeace), Mack McFarland (DuPont), Cecilia Mercado (UNEP DTIE), Alan Miller (GEF), Nora Mitchell (US National Park Service), E Thomas Morehouse (Institute for Defense Analyses), Gerald Mutisya (Ozone Secretariat), Nelson Sabogal (Ozone Secretariat), Stephen Seidel (US EPA), Rajendra Shende (UNEP DTIE), Mark W Robertson (Williams College), Lindsey Roke (Fisher Paykel), Robert Weyeneth (University of South Carolina) and many others. Thank you all.

World Plan of Action, April 1977

THE NATURAL OZONE LAYER AND ITS MODIFICATION
BY HUMAN ACTIVITIES

The natural stratosphere and its ozone layer have been the subject of extensive research over the last fifty years. The need for jet aircraft to fly at high altitudes has long spurred the interest of the world's meteorological community. The curiosity of atmospheric physicists and chemists has provided the main thrust to illuminate the dynamics and photochemistry of the ozone layer.

In recent years research, associated with man's accelerating activities in the stratosphere and aerospace, has greatly modified our perception of ozone photo-chemistry. It is widely accepted that natural ozone photochemistry is controlled to a large extent by the action of nitric oxide derived from nitrous oxide, which emanates from the Earth's surface as a result of denitrification processes. This has added a new dimension to the ozone problem and new interest in studies of the Earth's nitrogen cycle. It is now thought that stratospheric ozone is influenced by a variety of naturally occurring substances such as methane and methyl chloride. A continuing long-term programme is essential to our further understanding of these natural phenomena and as a basis for evaluating new effects.

Quite recently the impact of other substances such as the nitrogen oxides from aircraft exhausts and the man-made chlorofluoromethanes (CFMs) has been receiving intense scrutiny. As these are all trace substances with atmospheric concentrations as low as a fraction of a part per billion, the problems of measurement analysis and modelling have changed by an order of magnitude from those encountered when our understanding was based on the assumption of the simple Chapman (pure oxygen) system. Our knowledge and understanding have been greatly increased by a number of intensive national programmes, and there is a large measure of agreement on the model predictions that current aircraft emissions have minimal effects on the ozone layer but that the effects of continuing emissions of CFMs at the 1973 or higher levels are a matter of concern.

RESEARCH AND MONITORING

There are many known gaps in our knowledge of factors affecting the ozone layer, and there may be factors that are as yet unrecognized. An intensive and well-coordinated monitoring and research programme related to the occurrence of trace substances in the atmosphere, to test the model predictions and narrow their range of uncertainty, is particularly important.

Recommendations

The coordinated research and monitoring programme already initiated by the World Meteorological Organization (WMO), to clarify the basic dynamical photochemical and radiative aspects of the ozone layer and to evaluate the impact of man's activities on the ozone balance, should be encouraged and supported.

Specifically this should include action to:

- *Monitor ozone (WMO):* design, develop, and operate an improved system (including appropriate, ground-based, airborne and satellite subsystems) for monitoring and prompt reporting of global ozone, its vertical, spatial and temporal variations, with sufficient accuracy to detect small but statistically significant long-term trends.
- *Monitor solar radiation (WMO):* design, develop, and implement a system (including appropriate ground-based, airborne and satellite subsystems) for monitoring the spectral distribution of solar radiation extra-terrestrially to better understand the formation of ozone, to determine variations of solar flux and to clarify the influence of such variations on the Earth's radiation budget and on the accuracy of ground-based measurements.
- *Simultaneous species measurements (WMO):* design, develop, and implement an intensive short-term programme of simultaneous measurements of selected species in the odd-nitrogen, odd-hydrogen and odd-chlorine families to provide a better understanding of the ozone balance and to test the model predictions. This should be supported by a longer-term programme on all reactive species including their temporal and spatial variations and base line inventories.
- *Chemical reactions (WMO):* design, develop, and implement a broad programme to accurately determine chemical and photochemical reaction rates and quantum yields at appropriate temperatures and pressures; to provide accurate atomic and molecular line strength and location data for use in laboratory and remote sensing measurements of trace species; and to assess the sensitivity of chemical reaction schemes used in atmospheric models.
- *Development of computational modelling (WMO):* develop a hierarchy of improved (1-D, 2-D, 3-D) computational models by which the interrelationships of chemical, radiative, hydrodynamic and thermodynamic processes controlling the troposphere, and the stratosphere, may be established.
- *Large-scale atmospheric transport (WMO):* undertake studies to better describe and quantify large-scale atmospheric circulation and transport. Particular emphasis should be given to representation of vertical transport and transfer between the troposphere and the stratosphere.
- *Global constituent budgets (WMO):* develop a much improved understanding of the budgets of atmospheric constituents which affect the ozone balance, including their sources and sinks. At present this should give particular emphasis to the chlorine and nitrogen systems.

THE IMPACT OF CHANGES IN THE OZONE LAYER ON HUMANS AND THE BIOSPHERE

A prediction of depletion of ozone due to man's activities would have little meaning unless it could be shown, at least qualitatively, that it would have significant effects on people and their environment.

It has been demonstrated that excessive exposure to ultraviolet radiation at 254nm (radiation of a shorter wavelength than the UV-B which reaches the ground) causes tumors in laboratory animals and has deleterious effects on certain plants. Extensions of the studies to the UV-B band (from 290–320nm) have been made by theoretical and experimental work and by a series of epidemiological studies. There is some evidence that increased UV-B would be associated with an increase in skin cancer and possibly in eye damage in susceptible sections of the human population. It is also likely that large increases in UV-B would damage nucleic acids and proteins and thereby have deleterious effects on plant and aquatic communities, but the effect of smaller changes in UV-B is highly uncertain. Although it is clear that significant progress has been made in a wide range of related research activities in the biological and medical sciences, there are still many gaps in our knowledge of the effects of increased UV-B.

Recommendations

A wide variety of investigations of the impact of ozone depletion and increased ultraviolet radiation (UV-B) on humans and the biosphere should be encouraged, supported and coordinated. Specifically it is proposed that action be undertaken on the following aspects.

UV radiation

- *Monitor UV-B radiation (WMO/WHO, FAO):* monitor as far as possible with currently available instrumentation the spectral distribution of the UV-B radiation and the erythemally weighted integrated radiation intensities at the Earth's surface. This should be done for at least a complete solar cycle at globally distributed sites, and where possible at stations where ozone is being measured and/or skin-cancer data being collected and/or plant effects being studied.
- *Develop UV-B instrumentation (WMO/UNEP, WHO, FAO):* develop improved instrumentation and methods for measurements and providing precise levels of UV-B irradiance (both broad- and narrow-band) including satellite techniques.
- *Promote UV-B research (WMO, WHO, FAO):* develop standardized methodolo-gies for conducting UV-B research. Promote investigations to enable a better understanding of the spectral distribution of UV radiation, effects of flux, and effects of factors other than ozone such as atmospheric conditions and ground albedo, in determining the amount and wavelength of UV reaching the ground.

Human health

- *Statistics on skin cancer (WHO):* obtain improved worldwide statistics of skin cancer incidence and analyse such data in relation to latitude (with particular emphasis on locations where UV and total ozone can be measured simultaneously), type of cancer, age of onset, morbidity rates, location of cancer on body, sex, skin type, genetic and ethnic background, occupation and lifestyle. Develop internationally agreed upon protocols for design, collection and analysis of skin cancer data, and develop improved data storage, retrieval and dissemination mechanisms.
- *Research on induction mechanisms (WHO):* conduct experimental and theoretical studies at the molecular, cellular and tissue levels of the mechanisms of induction of skin cancer, skin-aging and eye damage by UV-B, as functions of dose, dose rate, and wavelength, with attention to possible synergistic factors including other wavelengths and possible effects of UV-B on DNA.

- *Other health aspects (WHO):* study potential impact of increased UV-B on aspects of human health other than skin cancer; develop a model for predicting human photobiological response for various skin types, genetic and ethnic background and environmental factors; and conduct research and modelling on the mechanisms of Vitamin D production and effects.

Other biological effects

- *Responses to UV-B (FAO):* conduct studies on the physiological, biochemical, and structural responses to increased UV-B radiation of selected wild and agricultural plants, animal species, and microorganisms. These should emphasize dose-response, reciprocity, synergisms, antagonisms, action spectra and interactions of stress factors on these organisms.
- *Plant communities (FAO):* study the effect of UV-B radiation on terrestrial plant communities (both agricultural and natural) inter alia by means of modelling studies of the impacts of UV-B radiation.
- *Aquatic systems (FAO):* study the effect of UV-B radiation on aquatic plants and animals emphasizing those related to primary productivity in the oceans; develop improved measurements of UV-B penetration into aquatic environment and study its effects on plankton.

Effects on climate

- *Development of computational modelling (WMO):* further develop modelling capability to consider the effects of ozone changes and also those of CFMs and similar compounds on the Earth's radiation balance and its climate.
- *Regional climate (WMO/FAO):* promote research on climate and its variability to provide better evaluations of the effects on critical climatic regions.

SOCIO-ECONOMIC ASPECTS

To provide a useful contribution to policy formulation, the assessment of a potential environmental hazard must include an evaluation of the costs and benefits to society that would result from a reduction of the hazard. Neither the methodology nor the available data are adequate to properly assess the socio-economic effects of stratospheric pollution or of measures taken to control it. Much work has been done on the costs of alternative courses of action, particularly with respect to possible limitations to total CFM emissions. However, further elaboration is needed before a complete evaluation can be made with confidence.

Recommendations

Studies of the socio-economic impact of predicted ozone-layer depletions and of alternative courses of action to limit or control identified ozone-depleting emissions to the atmosphere should be supported at the national and international level. Specifically it is proposed that action be undertaken to:

- *Production and emission data (UNEP, ICC, OECD, ICAO):* obtain more detailed data on global production, emission and use of substances that have the potential to affect stratospheric ozone. These data are required to model stratospheric changes as well as to evaluate socio-economic alternatives.

- *Methodology (UNESA, OECD):* develop improved methodologies for assessing the socio-economic factors related to CFM use, aircraft emissions, nitrogen fertilizers and other potential modifiers of the stratosphere.

INSTITUTIONAL ARRANGEMENTS

1   The Action Plan will be implemented by UN bodies, specialized agencies, international, national, intergovernmental and non-governmental organizations and scientific institutions.
2   UNEP should exercise a broad coordinating and catalytic role aimed at the integration and coordination of research efforts by arranging for:
    – collation of information on ongoing and planned research activities;
    – presentation and review of the results of research;
    – identification of further research needs.
3   In order for UNEP to fulfil that responsibility, it should establish a Coordinating Committee on Ozone Matters composed of representatives of the agencies and non-governmental organizations participating in implementing the Action Plan as well as representatives of countries which have major scientific programmes contributing to the Action Plan. The committee should be provided with adequate secretariat services. The committee should make recommendations to the Executive Director relevant to the continuing development and coordination of the Action Plan.
4   While much of the work included in the Action Plan is being and will be undertaken at the national level, and is the financial responsibility of countries, there is a continuing need for coordination of the planning and execution of monitoring and research related to particular segments of the Action Plan. This need can most effectively be met by the specialized agencies as indicated in the recommendations above.
5   Each agency should arrange for the provision of scientific advice relevant to its needs and those of the Coordinating Committee on Ozone Matters. In addition, the Executive Director of UNEP may from time to time convene a multi-disciplinary panel of experts to provide broadly-based scientific advice on the Action Plan.
6   UNEP should consider the need for and feasibility of establishing special co-ordinating mechanisms or procedures for certain areas of interdisciplinary work, such as photobiology, which presently lack such coordinating facilities.

Controlled substances under the Montreal Protocol

Annex A: Controlled substances

Substance		Ozone-depleting potential*
Group I		
CFCl3	(CFC-11)	1.0
CF2Cl2	(CFC-12)	1.0
C2F3Cl3	(CFC-113)	0.8
C2F4Cl2	(CFC-114)	1.0
C2F5Cl	(CFC-115)	0.6
Group II		
CF2BrCl	(halon-1211)	3.0
CF3Br	(halon-1301)	10.0
C2F4Br2	(halon-2402)	6.0

* These ozone-depleting potentials are estimates based on existing knowledge and will be reviewed and revised periodically.

Annex B: Controlled substances

Substance		Ozone-depleting potential
Group I		
CF3Cl	(CFC-13)	1.0
C2FCl5	(CFC-111)	1.0
C2F2Cl4	(CFC-112)	1.0
C3FCl7	(CFC-211)	1.0
C3F2Cl6	(CFC-212)	1.0
C3F3Cl5	(CFC-213)	1.0
C3F4Cl4	(CFC-214)	1.0
C3F5Cl3	(CFC-215)	1.0
C3F6Cl2	(CFC-216)	1.0
C3F7Cl	(CFC-217)	1.0
Group II		
CCl4	carbon tetrachloride	1.1
Group III		
C2H3Cl3*	1,1,1-trichloroethane* (methyl chloroform)	0.1

* This formula does not refer to 1,1,2–trichloroethane.

ANNEX C: CONTROLLED SUBSTANCES

Substance		Number of isomers	Ozone-depleting potential
Group I			
CHFCl2	(HCFC-21)**	1	0.04
CHF2Cl	(HCFC-22)**	1	0.055
CH2FCl	(HCFC-31)	1	0.02
C2HFCl4	(HCFC-121)	2	0.01–0.04
C2HF2Cl3	(HCFC-122)	3	0.02–0.08
C2HF3Cl2	(HCFC-123)	3	0.02–0.06
CHCl2CF3	(HCFC-123)**	–	0.02
C2HF4Cl	(HCFC-124)	2	0.02–0.04
CHFClCF3	(HCFC-124)**	–	0.022
C2H2FCl3	(HCFC-131)	3	0.007–0.05
C2H2F2Cl2	(HCFC-132)	4	0.008–0.05
C2H2F3Cl	(HCFC-133)	3	0.02–0.06
C2H3FCl2	(HCFC-141)	3	0.005–0.07
CH3CFCl2	(HCFC-141b)**	–	0.11
C2H3F2Cl	(HCFC-142)	3	0.008–0.07
CH3CF2Cl	(HCFC-142b)**	–	0.065
C2H4FCl	(HCFC-151)	2	0.003–0.005
C3HFCl6	(HCFC-221)	5	0.015–0.07
C3HF2Cl5	(HCFC-222)	9	0.01–0.09
C3HF3Cl4	(HCFC-223)	12	0.01–0.08
C3HF4Cl3	(HCFC-224)	12	0.01–0.09
C3HF5Cl2	(HCFC-225)	9	0.02–0.07
CF3CF2CHCl2	(HCFC-225ca)**	–	0.025
CF2ClCF2CHClF	(HCFC-225cb)**	–	0.033
C3HF6Cl	(HCFC-226)	5	0.02–0.10
C3H2FCl5	(HCFC-231)	9	0.05–0.09
C3H2F2Cl4	(HCFC-232)	16	0.008–0.10
C3H2F3Cl3	(HCFC-233)	18	0.007–0.23
C3H2F4Cl2	(HCFC-234)	16	0.01–0.28
C3H2F5Cl	(HCFC-235)	9	0.03–0.52
C3H3FCl4	(HCFC-241)	12	0.004–0.09
C3H3F2Cl3	(HCFC-242)	18	0.005–0.13
C3H3F3Cl2	(HCFC-243)	18	0.007–0.12
C3H3F4Cl	(HCFC-244)	12	0.009–0.14
C3H4FCl3	(HCFC-251)	12	0.001–0.01
C3H4F2Cl2	(HCFC-252)	16	0.005–0.04
C3H4F3Cl	(HCFC-253)	12	0.003–0.03
C3H5FCl2	(HCFC-261)	9	0.002–0.02
C3H5F2Cl	(HCFC-262)	9	0.002–0.02
C3H6FCl	(HCFC-271)	5	0.001–0.03
Group II			
CHFBr2		1	1.00
CHF2Br	(HBFC-22B1)	1	0.74
CH2FBr		1	0.73
C2HFBr4		2	0.3–0.8
C2HF2Br3		3	0.5–1.8
C2HF3Br2		3	0.4–1.6
C2HF4Br		2	0.7–1.2

C2H2FBr3		3	0.1–1.1
C2H2F2Br2		4	0.2–1.5
C2H2F3Br		3	0.7–1.6
C2H3FBr2		3	0.1–1.7
C2H3F2Br		3	0.2–1.1
C2H4FBr		2	0.07–0.1
C3HFBr6		5	0.3–1.5
C3HF2Br5		9	0.2–1.9
C3HF3Br4		12	0.3–1.8
C3HF4Br3		12	0.5–2.2
C3HF5Br2		9	0.9–2.0
C3HF6Br		5	0.7–3.3
C3H2FBr5		9	0.1–1.9
C3H2F2Br4		16	0.2–2.1
C3H2F3Br3		18	0.2–5.6
C3H2F4Br2		16	0.3–7.5
C3H2F5Br		8	0.9–14.0
C3H3FBr4		12	0.08–1.9
C3H3F2Br3		18	0.1–3.1
C3H3F3Br2		18	0.1–2.5
C3H3F4Br		12	0.3–4.4
C3H4FBr3		12	0.03–0.3
C3H4F2Br2		16	0.1–1.0
C3H4F3Br		12	0.07–0.8
C3H5FBr2		9	0.04–0.4
C3H5F2Br		9	0.07–0.8
C3H6FBr		5	0.02–0.7
Group III			
CH2BrCl	bromochloromethane	1	0.12

* Where a range of ODPs is indicated, the highest value in that range shall be used for the purpose of the Protocol. The ODPs listed as a single value have been determined from calculations based on laboratory measurements. Those listed as a range are based on estimates and are less certain. The range pertains to an isomeric group. The upper value is the estimate of the ODP of the isomer with the highest ODP, and the lower value is the estimate of the ODP of the isomer with the lowest ODP.
** Identifies the most commercially viable substances with ODP values listed against them to be used for the purposes of the Protocol.

ANNEX E: CONTROLLED SUBSTANCE

Substance		Ozone-depleting potential
Group I		
CH3Br	methyl bromide	0.6

Control measures of the Montreal Protocol

The continuous strengthening of the Montreal Protocol from 1990 through the adjustments in 1990, 1992, 1995, 1997 and 1999, and amendments in 1990, 1992, 1997 and 1999, ended in specific schedules of phase-outs for all the ozone-depleting substances (ODSs) identified.

The lists of controlled substances (see Appendix 3 above) are annexed to the Protocol as:

- Annex A (Group I – CFCs, Group II – halons);
- Annex B (Group I – other CFCs, Group II – carbon tetrachloride, Group III – methyl chloroform);
- Annex C (Group I – HCFCs, Group II – HBFCs, Group III – bromo-chloromethane);
- Annex E (methyl bromide).

The following is a summary of the phase-out schedules as of 1 July 2001.

SCHEDULES FOR NON-ARTICLE 5 PARTIES (PARTIES OTHER THAN DEVELOPING COUNTRIES OPERATING UNDER ARTICLE 5)

Annex A, Group I; Annex B, Groups I, II, III; Annex C, Group II

Production and consumption to be phased out by the end of 1995, but for possible essential-use exemptions granted from year to year by Meetings of the Parties.

For meeting the basic domestic needs (BDN) of Article 5 Parties, the following quantities were permitted.

Annex A, Group I (CFCs)

- Until the end of 2002: annual average of its production to meet the BDN for the period of 1995 to 1997 inclusive (base).
- Until the end of 2004: 80 per cent of the base.
- Until the end of 2006: 50 per cent of the base.
- Until the end of 2009: 15 per cent of the base.
- From 1 January 2010: zero.

Annex B, Group I (other CFCs)

- Until the end of 2002: 15 per cent of production in 1989.
- Until the end of 2006: 80 per cent of the (base) production for meeting the BDN during 1998–2000.
- Until the end of 2009: 15 per cent of the base.
- From 1 January 2010: zero.

Annex B, Groups II and III (carbon tetrachloride and methyl chloroform)

15 per cent of the production in 1989.

Annex C, Group II (HBFCs)

None.

Annex A, Group II (halons)

Production and consumption to be phased out by the end of 1993 but for possible essential-use exemptions. The additional production permitted to meet the basic domestic needs (BDN) was:

- Until the end of 2001: 15 per cent of the production in 1986.
- Until the end of 2004: annual average of production to meet the BDN in 1995–1997 (base).
- Until the end of 2009: 50 per cent of the base.
- From 1 January 2010: zero.

Annex C, Group I (HCFCs)

Consumption frozen at the base level (1989 HCFC consumption +2.8 per cent of 1989 CFC consumption) in 1996; 35 per cent reduction from 1 January 2004; 65 per cent reduction from 1 January 2010; 90 per cent reduction from 1 January 2015; 99.5 per cent reduction from 1 January 2020 and consumption restricted to servicing; and 100 per cent phase-out from 1 January 2030. Production frozen at the base level (1989 HCFC production +2.8 per cent of the 1989 HCFC production) from 1 January 2004; 15 per cent additional production allowed to meet the BDN.

Annex C, Group III (bromochloromethane)

Production and consumption phase-out from 1 January 2002. No exemptions.

Annex E (methyl bromide)

Production and consumption frozen at the base level of 1991 until the end of 1998; 25 per cent reduction until the end of 2000; 50 per cent until the end of 2002; 70 per cent until the end of 2004; complete phase-out from 1 January 2005 with possible critical-use exemptions.

Production to meet the BDN is as follows:

- Until the end of 2001: 15 per cent of the base level production.
- Until the end of 2004: 80 per cent of production in 1995–1998 to meet the BDN.
- From 1 January 2005: zero.

SCHEDULES FOR ARTICLE 5 PARTIES
(DEVELOPING COUNTRIES)

Annex A, Group I (CFCs)

Production and consumption frozen at the level of average during 1995–1997 (base) from 1 July 1999; 50 per cent reduction from 1 January 2005; 85 per cent reduction from 1 January 2007; and 100 per cent phase-out from 1 January 2010 with possible essential-use exemptions; 10 per cent base level production permitted to meet BDN until the end of 2009.

Annex A, Group II (halons)

Production and consumption frozen at the average 1995–1997 level (base) from 1 January 2002; 50 per cent reduction from 1 January 2005; 100 per cent phase-out from 2010 with possible essential-use exemptions; 10 per cent of base production allowed to meet BDN until the end of 2009.

Annex B, Group I (other CFCs)

Production and consumption reduction of 20 per cent from the level of 1998–2000 (base) from 1 January 2003; 85 per cent reduction from 1 January 2007; 100 per cent phase-out from 1 January 2010 with possible essential-use exemptions; 10 per cent of base level production allowed to meet the BDN until the end of 2009.

Annex B, Group II (carbon tetrachloride)

Production and consumption reduction of 85 per cent from 1 January 2005 from the level of 1998–2000 (base); 100 per cent phase-out by 2010 with possible essential-use exemptions; 10 per cent additional production allowed to meet the BDN until the end of 2009.

Annex B, Group III (methyl chloroform)

Freeze of production and consumption at the 1998–2000 level (base) from 1 January 2003; 30 per cent reduction from 1 January 2005; 70 per cent reduction from 1 January 2010; 100 per cent phase-out from 1 January 2015 with possible essential-use exemptions; 10 per cent additional production allowed to meet the BDN until the end of 2009.

Annex C, Group I (HCFCs)

Freeze of production and consumption from 1 January 2016 at 2015 level; phase-out of consumption from 1 January 1940; 15 per cent of base level allowed until the end of 2039.

Annex C, Group II (HBFCs)

Phase-out of production and consumption from 1 January 1996 with possible essential-use exemptions.

Annex C, Group III (bromochloromethane)

Phase-out of production and consumption from 1 January 2002 with possible essential-use exemptions.

Annex E (methyl bromide)

Freeze of production and consumption at 1995–1998 level (base) from 1 January 2002; 20 per cent reduction from 1 January 2005; 100 per cent phase-out from 1 January 2015 with possible essential-use exemptions; amounts used for quarantine and pre-shipment applications exempted at all stages.

Indicative list of categories of incremental costs: Annex VIII of the fourth Meeting of the Parties to the Montreal Protocol, 1992

The evaluation of requests for financing incremental costs of a given project shall take into account the following general principles:

(a) The most cost-effective and efficient option should be chosen, taking into account the national industrial strategy of the recipient party. It should be considered carefully to what extent the infrastructure at present used for production of the controlled substances could be put to alternative uses, thus resulting in decreased capital abandonment, and how to avoid de-industrialization and loss of export revenues;

(b) Consideration of project proposals for funding should involve the careful scrutiny of cost items listed in an effort to ensure that there is no double-counting;

(c) Savings or benefits that will be gained at both the strategic and project levels during the transition process should be taken into account on a case-by-case basis, according to criteria decided by the Parties and as elaborated in the guidelines of the Executive Committee;

(d) The funding of incremental costs is intended as an incentive for early adoption of ozone protecting technologies. In this respect the Executive Committee shall agree which time scales for payment of incremental costs are appropriate in each sector.

Incremental costs that once agreed are to be met by the financial mechanism include those listed below. If incremental costs other than those mentioned below are identified and quantified, a decision as to whether they are to be met by the financial mechanism shall be taken by the Executive Committee consistent with any criteria decided by the Parties and elaborated in the guidelines of the Executive Committee. The incremental recurring costs apply only for a transition period to be defined. The following list is indicated:

(a) Supply of substitutes:

(i) Cost of conversion of existing production facilities; cost of patents and designs and incremental cost of royalties; capital cost of conversion; and cost of retraining of personnel and cost of research to adapt technology to local circumstances;

(ii) Costs arising from premature retirement or enforced idleness, taking into account any guidance of the Executive Committee on appropriate cut-off dates; of productive capacity previously used to produce substances controlled by existing and/or amended or adjusted Protocol provisions; and where such capacity is not replaced by converted or new capacity to produce alternatives;

(iii) Cost of establishing new production facilities for substitutes of capacity equivalent to capacity lost when plants are converted or scrapped, including: cost of patents and designs and incremental cost of royalties; capital cost; and cost of training and cost of research to adapt technology to local circumstances;

(iv) Net operational cost, including the cost of raw materials; and

(v) Cost of import of substitutes.

(b) Use in manufacturing as an intermediate good:

(i) Cost of conversion of existing equipment and product manufacturing facilities;

(ii) Cost of patents and designs and incremental cost of royalties;

(iii) Capital cost;

(iv) Cost of retraining;

(v) Cost of research and development; and

(vi) Operational cost, including the cost of raw materials except where otherwise provided for.

(c) End use:

(i) Cost of premature modification or replacement of user equipment;

(ii) Cost of collection, management, recycling, and, if cost-effective, destruction of ozone-depleting substances; and

(iii) Cost of providing technical assistance to reduce consumption and unintended emission of ozone-depleting substances.

Awards for ozone-layer protection: Nobel Prize, United Nations and others

CHAMPIONS OF STRATOSPHERIC OZONE-LAYER PROTECTION

The Montreal Protocol is a democratic, scientific, diplomatic, regulatory, business and technology cooperation success. Environmental activists sounded the alarm and built public demands for action. Scientists developed the evidence that persuaded governments and companies to take precautionary action. Diplomats demonstrated that countries can work together to resolve global environmental security threats. Manufacturers of ozone-depleting substances (ODSs) who had long questioned the science, made an about-face and produced chemical substitutes. Leadership companies and military organizations that consumed ODSs announced plans to phase them out and formed partnerships to accelerate technical innovation. Governments and industry built technology cooperation and financed ODS phase-outs in countries with economies in transition and in developing countries. All the successes in protecting the stratospheric ozone layer depended on the work of individuals and organizations. One measure of the most significant contributions is the awards presented by the United Nations, the Nobel Committee, national environmental authorities and other organizations. Individuals and organizations from dozens of countries have earned these prestigious awards.

Awards recognize exceptional leadership, personal dedication and technical achievements in eliminating ozone-depleting substances. They distinguish those who have accomplished the extraordinary, the brilliant and the successful. They also serve as reminders to stand up for beliefs, as encouragement to take risks and as inspiration to protect the global environment.

WINNERS OF OZONE PROTECTION AWARDS COME FROM MANY COUNTRIES

Argentina, Australia, Belgium, Brazil, Bulgaria, Canada, Chile, Cuba, China, Denmark, the Dominican Republic, Egypt, Finland, France, Georgia, Germany, Greece, Hong Kong, Hungary, India, Indonesia, Iran, Ireland, Italy, Japan, Kenya, Malaysia, Mali, Malta, Mauritius, Mexico, The Netherlands, New Zealand, Niger, Norway, the Philippines, Poland, Romania, Singapore, Spain, Sri Lanka, Sweden, Switzerland, Syria, Thailand, the UK, the USA, Uruguay, Venezuela and Vietnam.

'Tending to world environmental problems – such as protecting the ozone layer, preventing global warming, and developing countermeasures for acid rain – is the common responsibility of all of us on earth, and is a task which should be confronted by concentrating the wisdom of mankind. For the purpose of solving the intrinsic problem, two concepts become important: 1) that this challenge now exceeds national borders, and 2) that countries, business or even individuals behave as though the problem is their own. There are many, many problems which mankind must overcome, but I think that the kind of global technological cooperation experienced during the ozone layer protection movement can serve as a valuable model for similar activities in the future.'

Tsuneya Nakamura, President, Seiko Epson Corporation, Japan
1993 EPA Award, 1997 Best-of-the-Best Award, 1998 Japan Nikkan
Kogyo Shimbun Award

'Protection of the ozone layer was judged technically impossible until industry rolled up its sleeves and made it a priority. The Awards are a testimony to those who worked the hardest and accomplished the impossible. When you read about the extraordinary people and organizations, you will begin to understand why and how business and environmental strategy can be merged. You will be inspired to join efforts to protect the global environment, including its fragile climate.'

Margaret Kerr, Vice President, Nortel/Northern Telecom, Canada
1990 EPA Award, 1997 Best-of-the-Best Award

'Through the 20+ year history of the stratospheric ozone issue, DuPont has been on the receiving end from many organizations due to our significant involvement in the manufacture and sale of ozone-depleting substances. DuPont's involvement has not always been praised, so we were very proud to have been an active leader in the CFC industry effort to phase out CFCs in a time frame that most of industry believed was unattainable.'

F A (Tony) Vogelsberg, DuPont Fluoroproducts, USA
1993 EPA Award

'Because AT&T was the first to take a leadership role in the elimination of CFC solvents in its manufacturing processes, I had the special honor of being among the initial winners of this (EPA) prestigious award. The Award was for leadership that was built on our confidence in the path-breaking work of engineers in the manufacturing facilities and at AT&T's Bell Laboratories. There are teams of unsung heroes around the world who helped protect the global environment. It gave me enormous pleasure to have been a part of this important work.'

David Chittick, Vice President, AT&T
1990 EPA Award, 1997 Best-of-the-Best Award

'I began my career as an environmentalist in the 1970s fighting with companies making and using CFCs. It was not a pleasant experience. In contrast, the spirit of cooperation engendered by the Montreal Protocol has been extremely rewarding and provides many lasting lessons. When people stop fighting and accept a common challenge, there is nothing we cannot accomplish. The greatest barriers are not technical or economic, but our beliefs and attitudes.'

Alan Miller, Global Environment Facility
1992 EPA Award, 1997 Best-of-the-Best Award

'The CFC phase-out effort remains one of the Foodservice & Packaging Institute's proudest moments and will have a lasting, positive legacy.'
Joseph W Bow, President, Foodservice & Packaging Institute
EPA Award 1990

'The US Department of Defense is a remarkably effective, mission-oriented organization. Once eliminating halons and CFCs became part of the mission, the results were astounding.'
Gary D Vest, US Principal Assistant Deputy Under Secretary of
Defense (Environmental Security)
1993 EPA Award, 1997 Best-of-the-Best Award

UNITED NATIONS AWARDS FOR STRATOSPHERIC OZONE PROTECTION, 1990–2001

UNEP Sasakawa Environment Prize

$200,000 prize for Outstanding Achievement
1993 Mostafa Tolba, Egypt
1999 Mario J Molina, USA[**]

UNEP Global 500 Awards

1988 F Sherwood Rowland, USA,[**] Joe Farman, UK[**]
1989 Mario Molina, USA,[**] Robert Watson, USA[**]
1996 Paul Crutzen, Germany[**]
1997 Jan van der Leun, The Netherlands[**]
1998 Stephen O Andersen, USA[*]

UNEP Global Ozone Award

1998 Mostafa Tolba, Egypt

1995 UNEP Ozone Awards

UNEP Awards for Science
Daniel L Albritton, USA[**]
Rumen D Bojkov, Canada/Bulgaria[**]
Paul Crutzen, Germany[**]
Joe Farman, UK[**]
Mario J Molina, USA[**]
F Sherwood Rowland, USA[**]
Manfred Tevini, Germany[**]
Jan C van der Leun, The Netherlands[*]
Robert T Watson, USA[**]

UNEP Awards for Technology
Stephen O Andersen, USA[*]
Gary Taylor, Canada[*]

---

[*] Member UNEP TEAP or its TOCs, Working Groups, or Task Forces.
[**] Member of UNEP Science or Environmental Effects Assessment Panels.

UNEP Awards for Policy and Implementation
Victor Buxton, Canada[*]
Eileen Claussen, USA
Tan Meng Leng, Malaysia
Juan Antonio Mateos, Mexico
Patrick Szell, UK
John Whitelaw, Australia

UNEP Awards for Non-Governmental Organizations
The Alternative Fluorocarbons Environmental Acceptability Study (AFEAS)
Programme for Alternative Fluorocarbon Toxicity Testing (PAFT)
Friends of the Earth (FOE)
The Japan Electrical Manufacturers' Association (JEMA)

1997 UNEP Ozone Awards

UNEP Awards for Science
James G Anderson, USA[**]
Ralph J Cicerone, USA[**]
Edward C DeFabo, USA[**]
Susan Solomon, USA[**]
Richard S Stolarki, USA[**]
Robert C Worrest, USA[**]
Christos S Zerefos, Greece[**]

UNEP Awards for Technology
Jonathan Banks, Australia[*]
Suely Machado Carvalho, Brazil[*]
Barbara Kucnerowicz-Polak, Poland[*]
Lambert Kuijpers, The Netherlands[*]
Melanie Miller, New Zealand[*]

UNEP Awards for Policy and Implementation
Richard E Benedick, USA
John Carstensen, Denmark
Willem J Kakebeeke, The Netherlands
Paul S Horwitz, USA[*]
Sateeaved Seebaluck, Mauritius[*]
Ilkka Ristimaki, Finland
The Department of Environment – Malaysia

UNEP Awards for Non-Governmental Organizations
Elizabeth Cook, USA
The Alliance for Responsible CFC Policy
Greenpeace International
Northern Telecom (NORTEL)

---

[*] Member UNEP TEAP or its TOCs, Working Groups, or Task Forces.
[**] Member of UNEP Science or Environmental Effects Assessment Panels.

UNEP 1998 Children's Ozone Painting Award (UNEP OzonAction and the Egyptian Environmental Affairs Agency)

Najla Husein Eid, Egypt
Laila Nuri, Indonesia
Rosa Kallontarpour, Iran
Maria Popescu, Romania
Bachari Saidou, Niger
Nan Qu, China (Jury Prize)

UNEP DTIE OzonAction Programme 2001 Global Video Competition on Ozone Layer Protection

A global video competition was held for professional and amateur participants in developing countries. A jury consisting of members from the private sector, television broadcasting and intergovernmental organizations selected the awards.

Global Winners
Pablo Massip Ginestà, Cuba
Nodar Begiashvili, Georgia
Mohammed Karesly, Syria

Regional Winners
Kiki Taher, Indonesia
Beijing Huadu Film and Television and Culture and Art, China
G S Algama, Sri Lanka
Nabil Maleh, Syria
Gabriel Rocca, Claudia Bustamante and Gabriel Stekolschik, Argentina
Omyma Khalil Kamel, Egypt
Diarra N'Golo, Mali

1997 EPA BEST-OF-THE-BEST STRATOSPHERIC PROTECTION AWARDS

Best-of-the-Best Corporate & Military Awards

Asahi Glass Company
The Coca-Cola Company
DuPont Company
Hitachi
IBM Corporation
ICI
Lockheed Martin Corporation
Lufthansa German Airlines
Department of Environment Malaysia
Minebea Group Companies
Mitsubishi Electric Corporation
Nissan
Nortel

Raytheon/Texas Instruments Systems
Seiko Epson Corporation
Thiokol/National Aeronautics and Space Administration
3M Corporation
US Air Force Space Launch Programs
US Army Acquisition Pollution Prevention Support Office
US Department of Defense (DoD)
US Naval Research Laboratory
US Naval Surface Warfare Center

Best-of-the-Best Association Awards

Air Conditioning and Refrigeration Institute (ARI)
Alliance for Responsible Atmospheric Policy (ARAP)
Australian Fluorocarbon Consumers and Manufacturers (AFCAM)
Halons Alternatives Research Corporation (HARC)
Institute for Interconnecting and Packaging Electronics Circuits (IPC)
CFC Benchmarking Team
International Cooperative for Environmental Leadership (ICEL)
Japan Electrical Manufacturers' Association (JEMA)
Japan Industrial Conference for Ozone Layer Protection (JICOP)
Mobile Air Conditioning Society (MACS)

Best-of-the-Best Individual Awards

Ward Atkinson, Sun Test Engineering[*]
James Baker, Delphi Harrison Thermal Systems[*]
Jay Baker, Ford Motor Company[*]
Jonathan Banks, CSIRO[*]
Walter Brunner, envico[*]
Suely Machado Carvalho, UNDP[*]
David Catchpole, British Petroleum[*]
David Chittick, AT&T
Jorge Corona de la Vega, Secretaria de Desarollo Urbano y Ecologia[*]
Philip J DiNenno, Hughes Associates[*]
Stephen P Evanoff, Lockheed Martin Corporation[*]
Kevin Fay, Alliance for Responsible Atmospheric Policy[*]
Joe Felty, Raytheon/Texas Instruments Systems[*]
Arthur FitzGerald, Nortel[*]
Yuichi Fujimoto, JEMA[*]
Kaichi Hasegawa, Seiko Epson Corporation
Andrea Hinwood, Environment Protection Authority Victoria[*]
Michael Jeffs, ICI Polyurethanes[*]
Margaret Kerr, Nortel
Joel Krinsky, US Navy
Lambert Kuijpers, Eindhoven Technical University[*]
Colin Lea, National Physical Laboratory
Eduardo Lopez, FONDOIN
Mohinder Malik, Lufthansa German Airlines[*]

[*] Member UNEP TEAP or its TOCs, Working Groups, or Task Forces.

Melanie Miller, Consultant*
John Minsker, Dow*
Mario Molina, Massachusetts Institute of Technology**
E Thomas Morehouse, Jr, US Department of Defense*
David Mueller, Fumigation Service & Supply*
Tsuneya Nakamura, Seiko Epson Corporation
Richard Nusbaum, Pennsylvania Engineering*
Simon Oulouhojian, MACS
Jose Pons Pons, SPRAY QUIMICA*
Sherwood Rowland, University of California**
Ronald W Sibley, US Defense Logistics Agency*
Gary Taylor, Taylor/Wagner*
Daniel Verdonik, US Army*
Gary Vest, US Department of Defense
Masaaki Yamabe, Asahi Glass Company*
Hideaki Yasukawa, Seiko Epson Corporation

US EPA STRATOSPHERIC OZONE PROTECTION AWARD WINNERS, 1990–2002

1990 EPA Award Winners

Corporate Awards
Digital Equipment Corporation
Dolco Packaging
DuPont Company

Association Awards
National Fire Protection Association (NFPA)
Foodservice & Packaging Institute (FPI)
The Institute for Interconnecting and Packaging of Electronic Circuits (IPC)
The Alliance for Responsible CFC Policy
Mobile Air Conditioning Society (MACS)

Laboratory Award
Underwriters Laboratories

Individual Awards
Ward Atkinson, Sun Test Engineering*
James A Baker, General Motors*
Jay Baker, Ford Electronics*
David Bergman, Institute for Interconnecting and Packaging of Electronic Circuits
James R Beyreis, Underwriters Laboratories
David Chittick, AT&T
Joe Felty, Texas Instruments*
Art FitzGerald, Nortel/Northern Telecom (Canada)*
Donald Grob, Underwriters Laboratories

* Member UNEP TEAP or its TOCs, Working Groups, or Task Forces.
** Member of UNEP Science or Environmental Effects Assessment Panels.

Leslie Guth, AT&T*
Kathi Johnson, China Lake Navy Weapons Center Electronics Manufacturing
  Productivity Facility
William Kenyon, Global Centre for Process Change*
Margaret Kerr, Nortel/Northern Telecom (Canada)
Simon Oulouhojian, Mobile Air Conditioning Society*
Robin Sellers, Naval Avionics Center
Gary Taylor, Taylor/Wagner*

1991 EPA Award Winners

Corporate Awards
Hitachi (Japan)
Motorola
Nissan Motor Company (Japan)
Nortel/Northern Telecom (Canada)
TEAM Aer Lingus (Ireland)
3M

Association Awards
Air-Conditioning and Refrigeration Institute (ARI)
The Industry Cooperative for Ozone Layer Protection (ICOLP)
The Japan Electrical Manufacturers' Association
Secretaria de Desarollo Urbano y Ecologia (Mexico)

Individual Awards
Elizabeth Cook, Friends of the Earth*
Jorge Corona, Camara Nacional de la Industria de la Transformacion (Mexico)*
Thomas E Daum, US Defense Reutilization & Marketing Service
David Doniger, Natural Resources Defense Council*
William Kopko, York International*
Colin Lea, UK National Physical Laboratory
E Thomas Morehouse, Jr, US Air Force*
Richard Nusbaum, Pennsylvania Engineering*
Robert C Pfahl, Motorola
Richard Stolarski, NASA Goddard Space Flight Center**
Dennis Tober, Florida Department of Environmental Regulation
Kjell Wetterlin, Astra ll-Draco (Sweden)

1992 EPA Award Winners

Corporate Awards
AT&T
The Boeing Company
British Aerospace Airbus
Chrysler Corporation
Ford Motor Company
General Dynamics
  Space Systems Division

---

* Member UNEP TEAP or its TOCs, Working Groups, or Task Forces.
** Member of UNEP Science or Environmental Effects Assessment Panels.

Fort Worth Division
IBM
  Endicott, New York
  Rochester, Minnesota
ICI Chemicals and Polymers (UK)
McQuay International
Mercedes-Benz (Germany)
Naval Air Warfare Center, Lakehurst
RECTICEL International
Seiko Epson (Japan)
The Trane Company
York International

Association Awards

Camara Nacional de la Industria de la Transformacion (Mexico)
Halon Alternative Research Corporation
Halon Essential Use Panel – EPA, Victoria, Australia
The Swedish Institute of Production Engineering Research (IVF)
Royal Norwegian Navy Materiel Command
US Army Acquisition Pollution Prevention Support Office

Individual Awards

Bryan H Baxter, British Aerospace[*]
Philip J DiNenno, Hughes Associates[*]
Stephen Peter Evanoff III, General Dynamics[*]
Yoshiyuki Ishii, Hitachi (Japan)[*]
Colin Lewis, UK Ministry of Defence
Milton Lubraico, Ford Motor Company, Brazil[*]
Shigeo Matsui, Toshiba (Japan)[*]
Alan S Miller, Center for Global Change[*]
Geno Nardini, Instituto Mexicano del Aerosol[*]
Tony L Phillips, General Dynamics
Laura J Turbini, Georgia Institute of Technology
Henry J Weltman, General Dynamics

1993 EPA Award Winners

Corporate Awards

AlliedSignal
Boeing Commercial Airplane Group
Cadbury (UK)
Charles County Board of Education
The Coca-Cola Company
Compaq Computer
Copeland
Defense Electronics Supply Center
Defense Logistics Agency
Department of the Navy – US Chief of Naval Operations
GEC-Marconi, Hirst Research Centre (UK)

---

* Member UNEP TEAP or its TOCs, Working Groups, or Task Forces.

General Services Administration
Hill Air Force Base
Hughes Aircraft
IBM – Austin, Texas
Kelly Air Force Base, Texas
Lufthansa German Airlines
Martin Marietta Astronautics
Matsushita Electric Industrial (Japan)
Minebea Group Companies (Thailand and Japan)
Motorola – Malaysian Project (Malaysia)
National Refrigerants
Naval Aviation Depot, Cherry Point
Naval Aviation Depot, Norfolk
New York State Energy Research and Development Authority – HFC Supermarket
  Refrigeration Demonstration Team
Nippondenso (Japan)
Rockwell International/US Army Air-To-Ground Missile Systems Project Office
J Sainsbury's (UK)
Shaw's Supermarkets
Texas Instruments, Missile Systems Division
Thiokol, Space Operations
Union Carbide/EKCO Housewares/Nordson
Unitor Ships Service (Singapore)
US Air Force, Air Base Fire Protection and Crash Rescue Systems Branch
Volvo Cars of North America
Woolworths Limited (Australia)

Association Awards
Center for Emissions Control
Heating, Refrigerating and Air-Conditioning Institute of Canada
Industry Cooperative for Ozone Layer Protection (ICOLP/ICEL)
Japan Industrial Conference for Ozone Layer Protection (JICOP)
Polyisocyanurate Insulation Manufacturers Association

Individual Awards
Robert Carter, Waste Reduction Resource Center for the Southeast
Nicholas T Castellucci, Northrop Grumman
David V Catchpole, BP Exploration (Alaska)*
Eileen Claussen, US Department of State
Timothy Crawford, Electronic Manufacturing Production Facility
Michael Earl Dillon, Dillon Consulting Engineers
Carl Eckersley, Compaq Computer
Carole K Ellenberger, Texas Instruments
Kevin Fay, The Alliance for Responsible CFC Policy*
John Fischer, Naval Air Warfare Center
Yuichi Fujimoto, Japan Electrical Manufacturers' Association*
Michael Hayes, Petroferm
Andrea Hinwood, Environment Protection Authority Victoria (Australia)*
Arthur G Hobbs, Jr, Four Seasons Division of Standard Motor Products

---

* Member UNEP TEAP or its TOCs, Working Groups, or Task Forces.

Mike Jeffs, ICI Polyurethanes (UK)[*]
Nancy Ketcham-Colwill, US Environmental Protection Agency
Lambert Kuijpers, UNEP Technology and Economic Assessment Panel
   (The Netherlands)[*]
Steve Lee-Bapty, UK Department of the Environment[*]
Kenneth W Manz, Robinair Division, SPX[*]
Thomas J Mathews, Hannaford Brothers
Yasuo Mitsugi, Seiko Epson (Japan)
Tsuneya Nakamura, Seiko Epson (Japan)
Sergio Oxman, KIEN Consultants (Chile)[*]
Cynthia Pruett, IBM Asia Pacific[*]
F Sherwood Rowland, University of California[**]
Terry Schaumberg, San Antonio Air Logistics Center
Angie Criser Schurig, Texas Instruments
Yoshihide Shibano, S&C (Japan)
John R Stemniski, The Charles Stark Draper Laboratory[*]
Robert E Tapscott, New Mexico Engineering Research Institute[*]
Steven D Taylor, BP Exploration (Alaska)
Mostafa Tolba, International Centre for Environment and Development (Egypt)
Gary D Vest, Principal Assistant Deputy Undersecretary of Defense (Environmental
   Security)
Clare Vinton, National Center for Manufacturing Sciences
F A (Tony) Vogelsberg, DuPont[*]
Carmen C Waschek, The Coca-Cola Company
Udo G Wenning, Bosch-Siemens (Germany)
Masaaki Yamabe, Asahi Glass (Japan)[*]

1994 EPA Award Winners

Corporate Awards
Aeronautical Systems Center, Wright Laboratory, Aircraft Halon Replacement Team,
   Wright-Patterson Air Force Base
The Aerospace Guidance and Metrology Center, Newark Air Force Base
US Army Communications-Electronics Command/Tobyhanna Depot
Asahi Glass (Japan)
Carrier
Falcon Halon Team, Wright-Patterson Air Force Base
Ford Motor Company
General Motors
Hewlett-Packard
Honeywell
Hussmann
ICI Polyurethanes (UK)
Lockheed
Martin Marietta
Mitsubishi Electric (Japan)
Norsk Forsvarsteknologi (Norway)
Northrop Grumman
Saab-Scania (Sweden)

---

[*] Member UNEP TEAP or its TOCs, Working Groups, or Task Forces.
[**] Member of UNEP Science or Environmental Effects Assessment Panels.

Separation Technologists
Tecumseh Products
Toyota Motor (Japan)

Association Awards
Alternative Refrigerants Evaluation Program, Air-Conditioning and Refridgeration
  Institute
National Association of Fire Equipment Distributors

Individual Awards
Daniel Albritton, National Oceanographic and Atmospheric Administration* **
Walter Brunner, envico (Switzerland)*
Brian Ellis, Protonique (Switzerland)*
James A Fain, Jr, Aeronautical Systems Center, Wright-Patterson Air Force Base
Mary Beth Fennell, Naval Aviation Depot, Cherry Point
Tetsuro Fukushima, Hitachi, Environmental Policy Office (Japan)
Victor Gatt, Department of Industry (Malta)
Charles Hancock, MDT*
John Hathaway, Arizona Department of Environmental Quality
John Hoffman, US Environmental Protection Agency*
Robert Holcomb, Motorola
Joel Krinsky, US Navy
Barbara Kucnerowicz-Polak, State Fire Service (Poland)*
Eduardo Lopez, FONDOIN (Venezuela)
Mohinder Malik, Deutsche Lufthansa (Germany)*
Marion McQuaide, UK Ministry of Defence*
Alvin Miller, National Weather Service
Tsutomu Odagiri, Japan Industrial Conference for Ozone Layer Protection
Steven Rasmussen, Hill Air Force Base
Wallace Rubin, Multicore
Franklin Sheppard, Jr, Office of the Chief of Naval Operations, US Navy
Steven Shimberg, US Senate Committee on Environment and Public Works
Ronald Sibley, Defense Logistics Agency*
Jack Swindle, Texas Instruments
James Vincent, US Army Aviation and Troop Command
Robert Watson, National Aeronautics and Space Administration* **
Hideaki Yasukawa, Seiko Epson (Japan)

1995 EPA Award Winners

Corporate Awards
Aberdeen Test Center, US Army
Advanced Cruise Missile DSO, US Air Force
AGM-130 Systems Program Office, US Air Force
Annapolis Detachment, Carderock Division, Naval Surface Warfare Center, US Navy
Australian Department of Administrative Services Centre for Environmental
  Management
Beverage-Air
Defence Institute of Fire Research (India)

---

* Member UNEP TEAP or its TOCs, Working Groups, or Task Forces.
** Member of UNEP Science or Environmental Effects Assessment Panels.

Dixie-Narco
Electrical & Mechanical Services Department, Hong Kong Government
Epson Group (Hong Kong)
GEO-CENTERS
H B Fuller
Low-Residue Soldering Task Force
Navy Technology Center for Safety & Survivability, US Naval Research Laboratory
Ontario Hydro (Canada)
SANYO Electric (Japan)
Sea-Land Service
Sharp (Japan)
Singapore Institute of Standards and Industrial Research
Texas Instruments
Titan IV Program ODS Reduction Team
Toshiba (Japan)
USBI
Xerox

Association Awards
Air-Conditioning & Refrigeration Institute
CFC Destruction Plasma Project, Clean Japan Center
Industrial Technology Research Institute (Taiwan)
Industry Technician Certification Team
Refrigerant Reclaim Australia

Individual Awards
Neil Antin, US Naval Sea Systems Command
Craig Barkhouse, Foamex Canada*
David Breslin, US Naval Sea Systems Command
Robert V Burress, SEHO USA
Denis Clodic, Ecole des Mines de Paris, Centre d'Energetique (France)*
Bjorn Egeland, Consolve
Robert Gay, US Defense Logistics Agency
Herbert T Gilkey, Engineering Consultants*
Casey Grant, National Fire Protection Association
Michael C Grieco, ICBM Systems Program Office, US Air Force
John Grupenhoff, National Association of Physicians for the Environment
Joop Van Haasteren, Ministry of Housing, Spatial Planning and the Environment
  (The Netherlands)*
Masatoshi Kinoshita, Japan Industrial Conference on Cleaning
Hiroshi Kurita, Japan Association for Hygiene of Chlorinated Solvents*
Michael J Leake, Texas Instruments
Cynthia Lingg, US Air Force
Hitoshi Mamiya, Honda Motors (Japan)
C K Marfatia, Real Value Appliances (India)
James A Mertens, Dow Chemical*
John Minsker, Dow Chemical*
David Mueller, Fumigation Service and Supply*
Goro Ogino, Minebea (Japan)

---

* Member UNEP TEAP or its TOCs, Working Groups, or Task Forces.

Douglas O Pauls, Contamination Studies Laboratories
Jose Pons Pons, Spray Quimica (Venezuela)*
A R Ravishankara, NOAA Aeronomy Laboratory **
Yasuomi Tanaka, Weyerhaeuser, Timberlands Nursery Team
Daniel P Verdonik, Hughes Associates*
Hans U Wäckerlig, Swiss Institute for the Promotion of Safety & Security
Peyton Weary, University of Virginia
George H White, Special Agent, US Customs Service, Miami, Florida
James Wolf, American Standard

1996 EPA Award Winners

Corporate Awards
Advanced Amphibious Assault Vehicle, US Marine Corps
Allied Signal/Carrier
Baxter Limited
Center for Technical Excellence for ODC Solvents
Draper Laboratory
F/A-18 Program Office and V-22 Program Office, US Navy
Tank-Automotive Research, Development, Engineering Center, US Army
ICBM System Program Office, US Air Force
Lockheed Martin Aeronautical Systems
Lockheed Martin Skunk Works
Ministry of Science, Technology and the Environment (Malaysia)
The Nikkan Kogyo Shimbun (Japan)
Operation Cool Breeze Enforcement Team
Philadelphia Detachment of the Carderock Division of the Naval Warfare Center,
  US Navy
Sanden
3M Pharmaceuticals
Tyler Refrigeration

Association Awards
Association of Fluorocarbon Consumers and Manufacturers of Australia (AFCAM)
Halon Recycling & Banking Support Committee (Japan)
International Institute of Refrigeration (IIR)
International Mobile Air Conditioning Association (IMACA)
Japan Industrial Conference on Cleaning
The Refrigerant Import Committee of the Alliance for Responsible Atmospheric Policy
Swedish Refrigeration Foundation

Individual Awards
Jonathan Banks, CSIRO Division of Entomology (Australia)*
Thomas A Bush, US Army
G Victor Buxton, Environment Canada*
Frank Cala, Church & Dwight
Suely M Carvalho, Companhia de Tecnologia de Saneamento Ambiental (Brazil)*
Stephen DeCanio, University of California*
Kaichi Hasegawa, Seiko Epson (Japan)

---

* Member UNEP TEAP or its TOCs, Working Groups, or Task Forces.
** Member of UNEP Science or Environmental Effects Assessment Panels.

Barbara Kanegsberg, BFK Solutions
Takeshi Kawano, Dai-Ichi Kogyo Seiyaku (Japan)
John King, US Air Force
Jean M Lupinacci, US Environmental Protection Agency*
Trish MacQuarrie, Environment Canada*
Melanie Miller, Consultant*
Peter Mullenhard, US Navy Clearinghouse
Larry Novak, Texas Instruments
John O'Sullivan, British Airways*
K Madhava Sarma, UNEP Montreal Protocol Secretariat*
Stephen Seidel, US Council for Environmental Quality*
Ronald S Sheinson, US Naval Research Laboratory*
Susan Solomon, National Oceanic and Atmospheric Administration**
Tsuyoshi Takaichi, Showa Denko (Japan)
Yuji Yamazaki, Seiko Epson (Japan)
Kiyoshige Yokoi, Matsushita (Japan)

1997 EPA Award Winners

Corporate Awards
Epson Portland
Lockheed Martin Michoud Space Systems
National Institute of Materials and Chemical Research (NIMC), New Energy and
  Industrial Technology Development Organization (NEDO), Research Institute of
  Innovative Technology for the Earth (RITE)
Naval Sea Systems Command Navy Strategic Systems Programs Fire Control &
  Guidance Branch
Operation Frio Tejas
Ozone Layer Protection Office (Japan)
Ozone Layer Protection Unit, Hazardous Substances Control Division (Thailand)
Solvent Sector ODS Elimination Team (Philippines)
RAH-66 Comanche Program Manager's Office
Refrigeration Systems Division, Appliance Systems Group, Sharp Corporation
Refrigerator Corporate Team in Thailand
ST-Microelectronics (Malta)
US Army, Pacific

Individual Awards
Paul Ashford, Caleb Management Services (UK)*
Antonio Bello Perez, Centro de Cinencias Medio Ambientales (Spain)*
Nick Campbell, ICI Klea (UK)*
Compressor Technical Support Team (Thailand)
Sheila Daar, Bio-Integral Resource Center*
Robert L Darwin, Naval Sea Systems Command*
Sukumar Devotta, National Chemical Laboratory (India)*
László Dobó, Ministry for Environment and Regulatory Policy (Hungary)*
Linda L Dunn, Agriculture and Agri-Food Canada*
Kiyoshi Hara, Japan Industrial Conference for Ozone Layer Protection
Richard L Helmick, Naval Surface Warfare Center

---

* Member UNEP TEAP or its TOCs, Working Groups, or Task Forces.
** Member of UNEP Science or Environmental Effects Assessment Panels.

Ismail Ithnin, Department of Environment (Malaysia)
Janusz Kozakiewicz, Industrial Chemistry Research Institute (Poland)
Jan-Karel B H Kwisthout, LLM, Ministry of Environment (The Netherlands)
Tan Meng Leng, Department of Environment (Malaysia)
John A Manzione, US Army CECOM
Tetsuo Nishide, Ministry of International Trade and Industry (Japan)
Maria U Nolan, Department of Environment (UK)[*]
Wiraphon Rajadanuraks, Department of Industrial Works (Thailand)
Rodrigo Rodríguez-Kábana, Auburn University[*]
Leroy E Sanderson, US Marine Corps
Ian P Tansey, 3M Health Care[*]
Helen Tope, Environment Protection Authority Victoria (Australia)[*]
Bruce G Unkel, Naval Sea Systems Command
Bert Veenendaal, RAPPA[*]
Rafael Veloz, Governmental Ozone Committee (Dominican Republic)
Viraj Vithoontien, UNEP
John Wilkinson, Vulcan Materials Company[*]
Ashley Woodcock, North West Lung Centre, Wythenshawe Hospital (UK)[*]
Wang Yangzu, National Environmental Protection Agency (China)
Hua Zhangxi, China Council of Light Industry[*]

1998 EPA Award Winners

Corporate & Military Awards
Atlas Roofing Corporation
C5F-Team, NIMC, Nippon Zeon (Japan)
General Headquarters of the State Fire Service (Poland)
INFRAS
The Joseph Company
Lockheed Martin Tactical Aircraft Systems
US Marine Corps Direct Reporting Program Manager, Advanced Amphibious Assault
LPD 17 Amphibious Transport Dock Ship Team
New Attack Submarine Program Office
Whirlpool Corporation

Association Awards
Hungarian Refrigeration and Air Conditioning Association
Halon Alternative Options Committee (India)

Individual Awards
J Godfrey Abbott, Dow Chemical (Switzerland)[*]
Radhey S Agarwal, Institute of Technology (India)[*]
Stephen O Andersen, US Environmental Protection Agency[*]
Dan O Chellemi, US Department of Agriculture
Michelle Maynard Collins, NASA[*]
Fran DuMelle, American Lung Association
Corinna C Gilfillan, Friends of the Earth[*]
Cheryl Hayden/American Academy of Dermatology
Fred J Keller, Carrier Corporation

---

[*] Member UNEP TEAP or its TOCs, Working Groups, or Task Forces.

David A Koehler, Ocean City Research
Thomas M Landy, US Army Tank-Automotive and Armaments Command
Doris Tan-Cheung Hui Lian, Singapore Productivity and Standards Board
Sally Rand, US Environmental Protection Agency[*]
Robert Sanders, VicOzone Monitoring Project (Australia)
Frédérique Sauer, DEHON Service[*]
Anne M Schonfield, Pesticide Action Network North America
Wim J M Sprong, The Netherlands Ministry of the Environment
Denis Taylor, British Airways (UK)
Gregory Toms, US Naval Sea Systems Command

1999 EPA Award Winners

Corporate & Military Awards
Canadian Forces Fire Marshal, Department of National Defence
The Cannon Group (Italy)
Idaho Army National Guard, Combined Support Maintenance Shop
Project Management Office for Bradley Fighting Vehicle Systems
Wei T'o Associates and the National Library of Canada

Individual Awards
Per M Bakken, Basel Convention (Switzerland)
Sheila Jones, Agriculture and Agri-food Canada
Ingrid Kokeritz, Stockholm Environment Institute (Sweden)[*]
Theresia Indrawanti Pudiyanto, Pertamina Petrochem Quality Control Center
  (Indonesia)
Robert T Wickham, Wickham Associates[*]
David John Clare, UK Department of Environment, Transport and the Regions[*]
Jacinthe Seguin, Environment Canada
James Shevlin, Environment Australia
Liu Yi, China Environmental Protection Administration

2000 EPA Award Winners

Corporate & Governmental Awards
Comisión Técnica Gubernamental de Ozono,
DINAMA/MVOTMA (Uruguay)
Quaker Oats Company of Canada Limited
US Air Force Research Laboratory

Individual Winners
David Ball, Kidde (UK)[*]
Tom Batchelor, European Community (Belgium)[*]
Penelope Canan, University of Denver[*]
Horst Kruse, University of Hannover (Germany)[*]
David J Liddy, Ministry of Defence (UK)[*]
Yasuko Matsumoto, Science University of Tokyo, Suwa College (Japan)
Steve McCormick, US Army Tank-Automotive Command[*]
Hideki Nishida, Hitachi (Japan)

---

* Member UNEP TEAP or its TOCs, Working Groups, or Task Forces.

Roberto de Aguiar Peixoto, Mauá Technological Institute (Brazil)[*]
Robert Van Slooten, Consultant (UK)[*]

2002 EPA Award Winners

Corporate & Governmental Awards

CO Keddy Nursery, Canada
Daikin Industries, Japan
Dow AgroSciences LLC
Fetzer Vineyards
Honeywell International
Kendall-Jackson Wine Estates
General Mills for the Recently Acquired Pillsbury Company
Raynor
USDA – ARS, Water Management Research Laboratory
Yoder Brothers

Association and Team Winners

Japan's Save the Ozone Network (JASON), Japan
Florida Telone Commercialization Team

Individual Winners

Sue Biniaz, US Department of State
Iwona Rummel-Bulska, World Meteorological Organization, Switzerland
Jim Cochran, Swanton Berry Farm
Kert Davies, Greenpeace
Paul Fraser, Commonwealth Industrial Scientific Research Organisation, Australia
Marcos Gonzalez, Costada Norte Parque, Costa Rica
Nikolai Kopylov, All-Russian Research Institute for Fire Protection, Russia[*]
James Frederick O'Bryon, US Department of Defense (retired)
John Okedi, National Environment Management Authority, Uganda
Nancy Reichman, PhD, University of Denver
Reva Rubenstein, US Environmental Protection Agency (retired)[*]
Darrel A Staley, The Boeing Company[*]
Sue Stendebach, National Science Foundation[*]
Patrick Széll, Department of the Environment, Transport and the Regions, UK
A Tcheknavorian-Asenbauer, United Nations Industrial Development Organization,
  Austria
Howard L Wesoky, National Aeronautics and Space Administration (NASA)

OTHER HIGH HONOURS AND STRATOSPHERIC OZONE PROTECTION AWARDS, 1990–2001

1995 Nobel Prize In Chemistry

Paul Crutzen, Germany[**]
Mario J Molina, USA/Mexico[**]
F Sherwood Rowland, USA[**]

---

[*] Member UNEP TEAP or its TOCs, Working Groups, or Task Forces.
[**] Member of UNEP Science or Environmental Effects Assessment Panels.

Better World Society Environment Medal

1988 Mostafa Tolba, Egypt

TEAP Certificates of Appreciation for Extraordinary Support from Parties

Australia, Canada, The Netherlands, Switzerland, the UK and the USA

Brazil Stratospheric Protection Awards

Stephen O Andersen, US Environmental Protection Agency[*]

1997 Japan Nikkann Kogyo Shimbun Awards

MITI Award
Hideaki Yasukawa, Seiko Epson

Director General of the Environment Award
Toshitami Kaihara, Hyougo Prefecture

Awards for Excellence
Shinya Ishizu, Asahi Glass
Akira Nakajima, Nakajima Auto Denso
Toshio Sugiura, Matsushita Air Conditioners

Special Awards for Excellence
Stephen O Andersen, US Environmental Protection Agency[*]
Hiromichi Seya, JICOP

1998 Japan Nikkann Kogyo Shimbun Awards

MITI Award
Hideaki Yasukawa, Seiko Epson

Environment Agency Award
Toshitami Kaihara, Hyougo Prefecture

Awards for Excellence
Toshio Sugiura, Matsushita Air Conditioners
Shinya Ishizu, Asahi Glass
Akira Nakajima, Nakajima Auto Denso

Special Awards for Excellence
Stephen O Andersen, USA
Hiromichi Seya, JICOP

** Member of UNEP Science or Environmental Effects Assessment Panels.

1999 Japan Nikkann Kogyo Shimbun Awards

MITI Award
Tsutomu Kanai, Japan Electrical Manufacturers' Association

Environment Agency Award
Seiji Sasuga, Kanagawa Prefecture

Awards for Excellence
Taku Morota, ICI Teijin
Masami Ushikubo, Sanden Corporation
Koji Nishigaki, NEC

Special Awards for Excellence
Shigeo Fukuchi, Asahi Breweries

2000 Japan Nikkann Kogyo Shimbun Awards

MITI Award
Hiromu Okabe, Denso

Environment Agency Award
Shunji Kono, Japan Halon Bank

Awards for Excellence
Naoki Hashimoto, JICC
Masuo Toba, Shizuoka Prefecture
Hiroshi Ikeda, Toshiba Carrier

Special Awards for Excellence
Shizuo Miyashita, Nisso Metallochemical
Masao Kobayashi, Nippon Paper Industries

2001 Japan Nikkann Kogyo Shimbun Awards

MITI Award
Tetsuo Shimaga, Maekawa Manufacturing

Ministry of Environment Award
Hiromi Nishizono, Japan's Save the Ozone Network

Awards for Excellence
Noriyuki Inoue, Daikin Industries
Shuhei Ishibe, Arakawa Chemical Industries
Tadayoshi Asada, Asada Company

Special Awards for Excellence
Akira Idemitsu, Idemitsu
Shozo Katayama, Toyo Tire & Rubber Company

US Department Of Defense
2001 Award for Excellence
Stephen O Andersen, Environmental Protection Agency (USA)*

Vietnam Ozone Award
1997
People's Army Newspaper
Department of News, Vietnam Television
Department of Industry, Hochiminh City
Department of Science, Technology & Environment of Hanoi
Saigon Cosmetic Company
DASO Company
Refrigeration Company (SEAREFICO)
Refrigeration & Engineering Enterprise (SEAREE)

1998
Vietnam Fumigation Company
Pioneer Youth Newspaper
Vietnam Aerological Observatory
Carrier Vietnam

1999
Stephen O Andersen, US Environmental Protection Agency*
Nguyen Duc Ngu, HMS
Department of International Organizations, Ministry of Foreign Affairs
Center for Assistance and Development of Small and Medium Enterprises (SMEDEC),
  Ministry of Science, Technology & Environment
Plant Quarantine Laboratory Center, Department of Plant Protection, Ministry of
  Agriculture and Rural Development
Refrigeration Electrical Engineering Corporation
Elf Atochem South East

2000
Department of International Cooperation, Ministry of Industry
Department of Science, Technology & Environment, Danang
Green Planet Programme, Vietnam Television 2
Southern Hydrology & Meteorology Center, HMS
Trane Vietnam
Le Quoc Khanh, Ministry of Industry
Pham Khoi Nguyen, Ministry of Science, Technology & Environment
Viraj Vithoontien, The World Bank

2001
Department of Foreign Economic Relations, Ministry of Planning & Investment
Department of Management of Technology & Product Quality, Ministry of Industry
UNIDO Field Office Vietnam
VIETSOVPETRO

---

* Member UNEP TEAP or its TOCs, Working Groups, or Task Forces.

1994 International Cooperative for Ozone Layer Protection (ICOLP) Global Achievement Award

Stephen O Andersen, US Environmental Protection Agency[*]

1996 World Resources Institute

US Representative Sherwood Boehlert
US Senator John Chafee
US Senator Joseph Lieberman
US Representative Henry Waxman

2000 Mobile Air Conditioning Society

Twentieth Century Award Environmental Leadership
Stephen O Andersen, USA[*]

Twentieth Century Award Engineering Leadership
Ward Atkinson, USA[*]

---

[*] Member UNEP TEAP or its TOCs, Working Groups, or Task Forces.

Appendix 7

Assessment Panels of the Montreal Protocol*

In 1974, when Mario Molina and F Sherwood Rowland published their thesis of atmospheric science linking chlorofluorocarbons (CFCs) to ozone depletion, there was already a comprehensive assessment underway by the US Department of Transportation to calculate the atmospheric, environmental and economic impacts on the ozone layer of supersonic flights.[1] In 1975, the United States Environmental Protection Agency (EPA) and the CFC and aerosol product industry published studies of the benefits and costs of the proposed CFC controls with opposite results. Later in 1975, the World Meteorological Organization (WMO) issued the first international science statement, 'Modification of the Ozone Layer Due to Human Activities and Some Possible Geophysical Consequences', considering both supersonic transport and CFC emissions.

In April 1977, the United Nations Environment Programme (UNEP) organized an international meeting of experts in Washington. This meeting recommended a 'World Plan of Action on the Ozone Layer' to coordinate further research on the depletion of the ozone layer and a Coordinating Committee on the Ozone Layer (CCOL) to undertake annual research reviews to be published as the *Ozone Layer Bulletin*. The Governing Council of UNEP accepted these recommendations. In 1981, a consortium of international organizations under WMO published a complete update (see Appendix 1 for more details).[2]

In May 1981, the UNEP Governing Council established an Ad Hoc Working Group of Legal and Technical Experts to elaborate a Global Framework Convention for the Protection of the Ozone Layer. This group met many times between 1982 and 1985 to arrive at the Framework Vienna Convention for the Protection of the Ozone Layer. The 1985 WMO/UNEP assessment chaired by Robert Watson was the precursor to the WMO/UNEP science assessments that were done under the auspices of the Montreal Protocol. Attempts to mandate specific controls on ozone-depleting substances (ODSs) did not succeed till 1987, in view of, in the opinion of many countries, scientific uncertainties on the extent of, and reasons for, ozone depletion. The scientific assessment, coordinated by many international organizations such as NAS and WMO and published at the end of 1985, and the 1986 UNEP workshops on ODS demand and technical options (held in Rome, Italy and Leesburg, Virginia), contributed enough certainty for the Montreal Protocol of September 1987.

Even at the time of signing of the Montreal Protocol, governments were aware of the likely inadequacy of the control measures. Hence, the 1987 Montreal Protocol included Article 6, Assessment and Review of Control Measures:

---

* This appendix was written by Stephen O Andersen, Suely Carvalho and Sally Rand, with the assistance of E Thomas Morehouse and Helen Tope.

'Beginning in 1990, and at least every four years thereafter, the Parties shall assess the control measures provided for in Article 2 on the basis of available scientific, environmental, technical, and economic information. At least one year before each assessment, the Parties shall convene appropriate panels of experts qualified in the fields mentioned and determine the composition and terms of reference of any such panels. Within one year of being convened, the panels will report their conclusions, through the Secretariat, to the Parties.'

The four Montreal Protocol panels – Panel for Scientific Assessment, Panel for Environmental Assessment, Panel for Technology Assessment, and Panel for Economic Assessment – were informally organized in The Hague at the October 1988 'UNEP Conference on Science and Development, CFC Data, Legal Matters, and Alternative Substances and Technologies' and formalized at the First Meeting of Parties, held in Helsinki in May 1989.[3]

After 1990, the technology and economic panels were united into the Technology and Economic Assessment Panel (TEAP), and the other panels re-labelled themselves the Science Assessment Panel (SAP) and the Environmental Effects Assessment Panel (EEAP).

The SAP and EEAP organized chapter committees, each with chapter co-chairs, on relevant topics while the TEAP organized Technical Options Committees (TOCs) on industry sectors, each with co-chairs.[4] The chairs had full freedom to select experts for these committees. The Chairs of each assessment panel meet jointly to 'synthesize' their results. It is important to note that the panels do not make recommendations of policy or action; members contribute their technical/scientific knowledge, and do not represent any government, company or organization position.

The Parties frequently outline terms of reference and timetables for completing assessments of available scientific, environmental, technology and economic information. Requests are clear but unelaborated. The 1990 request to TEAP was typical:

'to assess the earliest technically feasible dates and the costs for reductions and total phaseout of 1,1,1-trichloroethane (methyl chloroform); evaluation of the need for transitional substances (HCFCs) in specific applications; quantity of controlled substances needed by developing countries and availability; and comparison of toxicity, flammability, energy efficiency and safety considerations of chemical substitutes and their availability.'

The first assessment in 1989 was managed by an 'Inter-Governmental panel' (IGP) as the political supervisory body. The IGP was minimally involved in the drafting of the first Synthesis Report and mentioned at the Second Meeting of the Open-Ended Working Group in November 1989. The governments did not appoint a supervisory body for the subsequent assessment reports of the panels. The reports of the panels are published and distributed to Parties without any change or political review.

There are differences in the way the assessment panels function. The Science and Environmental Effects Assessment Panels had evolved from the WMO/NASA assessments, adopting the Chairs from that membership, honouring the scientific tradition of selecting worldwide government and academic members, and practising strict reliance on publications from peer-reviewed journals.

The Co-Chairs of the Scientific Assessment Panel – with input from an ad hoc international steering group of researchers – plan the scope, content, and authors of a forthcoming assessment report, having had input from the Parties regarding topics of interest. The Co-Chairs and the current set of lead authors meet to make further plans and coordinate the contents of the chapters and the preparation of first drafts. Only information that is based on peer-reviewed scientific journal articles is considered. The resulting drafts undergo a peer review by mail and a subsequent week-long panel review, at which the chapter conclusions are agreed upon, and the executive summary is finalized.

The Co-Chairs of the Environmental Effects Panel operate much like the Scientific Assessment Panel, but with simplified procedures and a small panel of 25 scientific members practising in photobiology and photochemistry, mainly in universities and research institutes. The appropriate panel members write individual chapters and review each others' chapters. The draft assessment is globally peer reviewed prior to publication.

When the panels were first established, UNEP Executive Director Mostafa Tolba, in consultation with Parties, allowed the Technology Panel to be organized in an entirely different framework that is based on a strong membership composed of respected and capable technical experts from industry.[5] This is because the newest technology developments are not documented in peer-reviewed journals,[6] and the fast-evolving changes caused by the Montreal Protocol in the market needed close and educated follow-up by experts with industry and market experience. Members of the Technical Options Committees are selected by the Committee Chairs from nominations of the Parties, and are supplemented by recruiting to assure the necessary balance of skills. Recruitment also takes into consideration geographical balance to the extent of expertise available but no experts are recruited merely to provide balance or to oblige governments. However, a deliberate search is made by the Chairs to increase developing countries' participation and this search has met with great success.

The members of the panel or committees are paid by their employers/sponsors and not by the Protocol. The Trust Fund for the Montreal Protocol, however, meets the travel expenses of experts of developing countries and the countries with their economies in transition for participating in the meetings of the panels and Committees. The members are essentially 'sworn in', agree to notify the TEAP of any potential conflicts of interest, and annually disclose the source of sponsorship. The three TEAP Co-Chairs manage a panel of 23 members, including senior experts and co-chairs of seven Technical Options Committees and occasional Task Forces. The committees each represent an industry sector dependent on ODSs. The TEAP consolidates findings from the committees and writes its own cross-cutting chapters.

The Co-Chairs of the three assessment panels meet, when necessary, to draft synthesis reports. These reports distil findings into 'policy-relevant technical information'. This information includes: policy options for further action; impact of each action on the ozone layer; and the technical and economic feasibility and impacts of the options.

Examples from the 1989 Assessments illustrate the simplicity and clarity of the synthesis:

> 'Even if the control measures of the Montreal Protocol were to be implemented by all nations, today's atmospheric abundance of chlorine (about 3 parts per billion by volume (ppbv)) will at least double to triple during the next century. If the atmospheric abundance of chlorine reaches 9 ppbv by about 2050, ozone depletion of 0–4 percent in the tropics and 4–12 percent at high latitudes would be predicted, even without including the effects of heterogeneous chemical processes known to occur in polar regions, which may further increase the magnitude of the predicted ozone depletion.
>
> 'Each 1 percent total column ozone depletion is expected to lead to... a worldwide increase of 100,000 blind persons due to UV-B induced cataracts... a 3 percent rise

of the incidence of non-melanoma skin cancer, (and)... reduced food yield (and food quality) by up to 1 percent.

'[I]t is technically feasible by the year 2000 to phase down by at least 95 percent the production and consumption of the five controlled CFCs, phase out totally the production and consumption of carbon tetrachloride..., (and) phase down by at least 90 percent the production and consumption of methyl chloroform.'

The synthesis report adopted the graphical policy display invented by John Hoffman and Michael Gibbs shown in Figure 10.3. The figure illustrates the past record of observed abundances of atmospheric 'equivalent' chlorine (which includes bromine, appropriately weighted) and the future abundances that would have been associated with each major 'decision step' of the Montreal Protocol. The area displayed under each of the curves represents the relative magnitude of the environmental effects of ozone depletion, the timing of peak chlorine concentration as the turning point in the rate of destruction, and the time when concentrations drop below 2ppbv (roughly the pre-hole level) representing the likely closing of the Antarctic ozone hole. This presentation dramatically portrays the opportunity to reduce ozone depletion with proposed control measures, and the reality that even the most stringent controls cannot close the Antarctic hole until 2050 or return the stratosphere to its natural condition until at least several hundred years from the present.

In addition to providing scientific and technical information as a basis for policy decisions, the assessment panels have also served as communication tools and catalysts for necessary change. The Scientific Assessment Panel helped to identify scientific uncertainties and had a constructive influence in focusing research efforts and securing necessary financing. The Environmental Effects Assessment Panel served a similar purpose and was also instrumental in building the global consensus that ozone depletion would be unacceptable to all countries – albeit that some people would suffer more. Participation in the TEAP helped industry experts to understand the importance of ozone-layer protection and identify new business opportunities, and helped government participants to appreciate the difficulties of implementing new technology.

TEAP members helped to remove barriers to change and inspired technical innovation. Experts from developed and developing countries worked together to identify new technologies applicable worldwide and helped develop a consensus of which technologies offered the best balance of technical efficiency, economy, and environmental acceptability. Perhaps most important, the TEAP process helped to build 'bottom-up' industry support for new technology in ways that governments, scientists, academics and others would not have been able to do.

Members of the TEAP are influential in technical standards organizations, industry associations, and private and public regulatory authorities. For example, a Co-Chair of the Halons Committee also chaired the international fire-protection organizations with authority to halt testing and training and to approve alternatives and substitutes. A Co-Chair of the Solvents Committee also chaired the US Department of Defense committee with authority to approve alternatives and substitutes. Members from environmental ministries use knowledge of emerging technology to time regulatory approval with commercialization. Industry experts are also influential in crafting regulatory incentives necessary to stimulate investment and rapidly achieve economies of scale.

TEAP experts can exercise considerable market clout. Industry and military members of the TEAP were typically responsible for the phase-out of ODSs from their own organizations. Experts working in world-class teams are more confident in selecting technology than when they work alone. Suppliers are more responsive when markets are primed to accept new technology.

For many technical experts, the TEAP was the first opportunity to team up with experts from competing companies. Motivated by regulatory, consumer and environmental leadership, the companies authorized extraordinary cooperation in identifying technologies and techniques to reduce and eliminate ODS emissions. Soon technical experts from industry were in high demand as national advisers and as speakers at industry conferences. Their experience and market knowledge represented an advantage, hard to be surpassed by government or academic sources. Technical optimism and case studies provided the confidence policy-makers needed to take even more decisive action. Technical reports pulled Montreal Protocol policy, and the increasingly stringent Protocol motivated the industry collaboration to achieve ultimate phase-out goals.

Appendix 8

Core readings on the history of ozone-layer protection

DESCRIPTIONS OF KEY WORKS

This section describes seminal and otherwise important readings on ozone-layer protection. The full citations are listed in the following section.

Mario Molina and F Sherwood Rowland, 'Stratospheric Sink for Chlorofluoromethanes: Chlorine Atom-Catalyzed Destruction of Ozone' (*Nature*, 1974) was first to hypothesize that CFCs would deplete the ozone layer. In a press conference at the American Chemical Society that autumn, Molina and Rowland were the first to call for a phase-out and persisted in building the global concern that led to the Vienna Convention and Montreal Protocol. Mario Molina, Sherwood Rowland, and collaborator Paul Crutzen won the 1995 Nobel Prize in Chemistry in recognition of their achievements.

Lydia Dotto and Harold Schiff, *The Ozone War* (1978) is the most comprehensive early account of conflict between scientists, citizens, industry and political activists. It was published during the decade when UNEP and national environmental ministries were created and environmental law was invented, but long before there was much hope of stratospheric ozone protection.

Paul Brodeur, 'Annals of Chemistry' *(New Yorker)* (1986) translated the atmospheric, environmental, and health science into language that persuaded the public and their policy-makers to take action.

Douglas G Cogan, *Stones in a Glass House: CFCs and Ozone Depletion* (1988); John Gribbin, *The Hole in the Sky: Man's Threat to the Ozone Layer* (1988); and J Nantze, *What Goes Up; The Global Assault on Our Atmosphere* (1991) are accounts of the triumph of science and diplomacy in securing the Montreal Protocol.

Richard Benedick, *Ozone Diplomacy: New Directions in Safeguarding the Planet* (1991) is a detailed account from the perspective of the senior US negotiator of the Montreal Protocol. Interpretations of the bargaining motivation and stratagems of other countries stimulated many others to tell the story from their own perspective.

Seth Cagin and Philip Dray, *Between Earth and Sky: How CFCs Changed Our World and Endangered the Ozone Layer* (1993) is the US story of CFCs and the ozone layer in a historical context. The authors show how chemical companies invented, defended and then abandoned tetraethyl lead and CFCs during an era of increased environmental awareness and growing scientific evidence.

Mostafa Tolba, *Global Environmental Diplomacy: Negotiating Environmental Agreements for the World, 1973–1992* (1998) shows that protecting the ozone layer was accomplished with a deliberate strategy. Mostafa Tolba is rightly credited with orchestrating scientists, diplomats, government experts and environmental and industry NGOs to agree on the Montreal Protocol and its most important amendments and adjustments.

Stephen O Andersen, Clayton Frech, and E Thomas Morehouse, *Champions of the World: Stratospheric Ozone Protection Awards* (1997); Stephen O Andersen, *Newest Champions of the World* (1997); and Stephen O Andersen, 'Industrial Responses to Stratospheric Ozone Depletion and Lessons for Global Climate Change' (1988), are detailed accounts of the individual and organization efforts that protected the ozone layer.

Penelope Canan and Nancy Reichman, *Ozone Connections: Expert Networks in Global Environmental Governance* (2002) is a sophisticated new analysis that reveals the extent and effectiveness of a small number of experts who were instrumental in protecting the ozone layer. Looking systematically at the connection between technology, global environmental policy, and the social connections among experts, the authors focus on the Technology and Economic Assessment Panel of the Montreal Protocol. By combining formal network analysis, biographical interviews and participant observation, they demonstrate that treaty implementation relies on social relations, trust and the formation of community among experts.

Reiner Grundmann, *Transnational Environmental Policy: Reconstructing Ozone* (2001) documents the advocacy role of science in the early politics of ozone layer protection based on lengthy interviews with key scientists in the ozone-layer debates. Grundmann shows that such advocacy is not typical in the scientific community but the willingness (and courage) of leading scientists to violate professional norms of policy detachment was instrumental in breaking political logjams and getting action on stratospheric ozone.

Well-documented histories of the science and diplomacy of UNEP, industry, and environmental non-governmental organizations (NGOs) are contained in: Duncan Brack, *International Trade and the Montreal Protocol* (1996); Elizabeth Cook, *Ozone Protection in the United States: Elements of Success* (1996); Philipe G Le Prestre, John D Reid and E Thomas Morehouse, *Protecting the Ozone Layer: Lessons, Models and Prospects* (1998); Alan Miller and Irving Mintzer, *The Sky Is the Limit* (1986); and Edward A Parson, *Protecting the Ozone Layer: Science and Strategy* (forthcoming).

UNEP, *Handbook for the International Treaties for the Protection of the Ozone Layer* (1996) and Ozone Secretariat, *Action on Ozone* (2000) are detailed and clear descriptions of the full spectrum of Montreal Protocol provisions.

Stephen J DeCanio, 'Managing the Transition: Lessons from Experience' (1991) and 'The Dynamics of the Phaseout Process Under the Montreal Protocol' (1994) explain how the global market transformation that phased out ODSs was undertaken pragmatically, economically and with minimal administration.

FULL CITATIONS OF WORKS DESCRIBED IN THIS APPENDIX

Andersen, Stephen O (1997) *Newest Champions of the World: Winners of the 1997 Stratospheric Ozone Protection Awards*, US EPA, Washington, DC

Andersen, Stephen O (1988) 'Industrial Responses to Stratospheric Ozone Depletion and Lessons for Global Climate Change, Volume 4: Responding to Global Environmental Change', in Tolba, Mostafa (ed), *Encyclopedia of Global Environmental Change*, John Wiley and Sons, New York

Andersen, Stephen O, Frech, Clayton and Morehouse, E Thomas (1997). *Champions of the World: Stratospheric Ozone Protection Awards*, US EPA, Washington, DC

Benedick, Richard (1991) *Ozone Diplomacy: New Directions in Safeguarding the Planet*, Harvard University Press, Cambridge, MA; and the updated edition, Benedick, Richard (1998) *Ozone Diplomacy: New Directions in Safeguarding the Planet, Enlarged Edition*, Harvard University Press, Cambridge, MA

Brack, Duncan (1996) *International Trade and the Montreal Protocol*, Royal Institute of International Affairs/Earthscan, London

Brodeur, Paul (1986) 'Annals of Chemistry: In the Face of Doubt', *New Yorker*, 9 June

Cagin, Seth and Dray, Philip (1993) *Between Earth and Sky: How CFCs Changed Our World and Endangered the Ozone Layer*, Pantheon, New York

Canan, Penelope and Reichman, Nancy (2002) *Ozone Connections: Expert Networks in Global Environmental Governance*, Greenleaf Publishing, Sheffield

Cogan, Douglas G (1988) *Stones in a Glass House: CFCs and Ozone Depletion*, Investor Responsibility Research Center, Washington, DC

Cook, Elizabeth (1996) *Ozone Protection in the United States: Elements of Success*, World Resources Institute, Washington, DC

Dotto, Lydia and Schiff, Harold (1978) *The Ozone War*, Doubleday & Co, Barden City, NY

DeCanio, Stephen J (1991) 'Managing the Transition: Lessons from Experience', Chapter 2 in *Report of the Economic Options Committee*, UNEP, Nairobi

DeCanio, Stephen J (1994) 'The Dynamics of the Phaseout Process Under the Montreal Protocol', Chapter 1 in *Report of the Economic Options Committee*, UNEP, Nairobi

Gribbin, John (1988) *The Hole in the Sky: Man's Threat to the Ozone Layer*, Bantam, New York

Grundmann, Reiner (2001) *Transnational Environmental Policy: Reconstructing Ozone*, Routledge, London

Le Prestre, Philipe G, Reid, John D and Morehouse, E Thomas (1998) *Protecting the Ozone Layer: Lessons, Models and Prospects*, Kluwer Academic Publishers, Boston

Miller, Alan and Mintzer, Irving (1986) *The Sky Is the Limit: Strategies for Protecting the Ozone Layer*, World Resources Institute, Washington, DC

Molina, Mario and Rowland, F Sherwood (1974) 'Stratospheric Sink for Chloro-fluoromethanes: Chlorine Atom-Catalyzed Destruction of Ozone', *Nature*, vol 249, 28 June

Nantze, J (1991) *What Goes Up: The Global Assault on Our Atmosphere*, William Morrow Publishing, New York

Ozone Secretariat (2000) *Action on Ozone: 2000 Edition*, UNEP, Nairobi

Parson, Edward A (forthcoming, 2003) *Protecting the Ozone Layer: Science and Strategy* Oxford University Press, New York

Tolba, Mostafa and Rummel-Bulska, Iwona (1998) *Global Environmental Diplomacy: Negotiating Environmental Agreements for the World, 1973–1992*, MIT Press, Cambridge, MA

UNEP (1996) *Handbook for the International Treaties for the Protection of the Ozone Layer*, UNEP, Nairobi

Appendix 9

Selected ozone websites

The websites listed in this appendix have links to many other relevant websites.

UNEP

UNEP DTIE OzonAction Programme: www.uneptie.org/ozonaction.html

UNEP Ozone Secretariat: www.unep.org/ozone or its mirror site www.unep.ch/ozone

UNEP Multilateral Fund: www.unmfs.org

UNEP TEAP: www.teap.org

SCIENCE

Atmospheric Chemistry Links (Jet Propulsion Laboratory, NASA): http://atmos.jpl.nasa.gov/links.htm#earth-l

British Antarctic Survey: www.antarctica.ac.uk/Corporate/The_British_Antarctic_Survey.html

John H Chafee Lecture on Science and the Environment, by F Sherwood Rowland and Mario Molina: http://cnie.org/NCSEconference/2000conference/Chafee

Research Center for Developing Fluorinated Greenhouse Gas Alternatives, National Institute of Advanced Industrial Science and Technology (AIST): http://unit.aist.go.jp/f-center/index_e.htm

Royal Swedish Academy of Sciences/Nobel Prize: www.nobel.se/announcement/1995/announcement95-chemistry.html

Socioeconomic Data and Applications Center (SEDAC): http://sedac.ciesin.org/ozone

US National Academies of Science: www.national-academies.org

US NASA/Antarctica ozone levels: www.toms.gsfc.nasa.gov

US National Oceanic and Atmospheric Administration ozone programs: www.al.noaa.gov/wwwhd/pubdocs/stratO3.html

World Meteorological Organization: www.wmo.ch/index-en.html

ASSOCIATIONS

Air Conditioning & Refrigeration Institute (ARI): www.ari.org

Alliance for Responsible Atmospheric Policy: www.arap.org

Alternative Fluorocarbons Environmental Acceptability Study (AFEAS):
www.afeas.org/index.html

American Society of Heating, Refrigeration, and Air Conditioning Engineers (ASHRAE): www.ashrae.org

Australian Fluorocarbon Council (AFC): www.afcweb.org

British Institute of Refrigeration (IOR): www.ior.org.uk

Halon Alternative Research Corporation (HARC): www.harc.org

Industrial Technology Research Institute (ITRI): www.itri.org.tw

International Institute of Refrigeration (IIR): www.iifiir.org

International Cooperative for Environmental Leadership (ICEL):
www.wri.org/wri/cpi/carbon/icel.htm

Japan Industrial Conference for Ozone Layer Protection (JICOP):
www.eco-web.com/cgi-local/sfc?a=index/index.html&b=register/01641.html

Japan Research Institute of Innovative Technology for the Earth (RITE):
www.rite.or.jp

Mobile Air Conditioning Society (MACS): www.macsw.org

National Academy of Engineering: www.nae.edu

National Fire Protection Research Foundation (NFPRF):
www.nfpa.org/Research/The_Research_Foundation/the_research_foundation.html

Programme for Alternative Fluorocarbon Toxicity Testing (PAFT):
www.afeas.org/paft

Swedish Refrigeration Foundation (KYS): www.kys.se

Society of Automotive Engineers (SAE): www.sae.org

NON-GOVERNMENTAL ORGANIZATIONS

Friends of the Earth (FOE): www.foe.org/ptp/atmosphere/index.html

Greenpeace: www.greenpeace.org/~ozone

Japan's Save the Ozone Network (JASON): www.jason-web.org (in Japanese)

World Resources Institute (WRI): www.igc.org/wri/climate/ozone/oz-net.htm

GOVERNMENT

Defense Environmental Network and Information Exchange (DENIX) Air Site: www.denix.osd.mil/denix/Public/Library/library.html#air

Environment Australia: www.ea.gov.au/atmosphere/ozone

Environment Canada: www.ec.gc.ca/ozone

European Commission: http://europa.eu.int/comm/environment/ozone/ozone_layer.htm

New Zealand Ministry for the Environment: www.mfe.govt.nz/issues/ozone.htm

US EPA: www.epa.gov/air/ozone

US EPA history office: www.epa.gov/history

US Navy Shipboard Environmental Information Clearinghouse (SEIC): http://navyseic.dt.navy.mil/ozone/ozone.htm

Notes

CHAPTER 1

1   This chapter was peer-reviewed by Richard Stolarski, US NASA, and Mack McFarland, DuPont.

2   Because ozone-depleting substances were already scheduled for phase-out under the Montreal Protocol, they are not included in the Kyoto Protocol. However, SF6, PFCs and HFCs that are substitutes for ozone-depleting substances in some applications are controlled by the Kyoto Protocol. The HFCs controlled under the Kyoto Protocol have global warming potentials (GWPs) that are generally lower than the CFCs they replaced. For example, the GWP of HFC-134a is about one-sixth the GWP of the CFC-12 it replaces as a refrigerant. See *The Implications to the Montreal Protocol of the Inclusion of HFCs and PFCs in the Kyoto Protocol*, Report of the HFC and PFC Task Force of the Technology and Economic Assessment Panel (Ozone Secretariat, United Nations Environment Programme, October 1999).

3   Stolarski, R (1999) 'History of the study of atmospheric ozone' (online). Available: http://hyperion.gsfc.nasa.gov/Personnel/people/Stolarski_Richard_S./history.html.

4   Leeds, A R (1880) 'Lines of Discovery in the History of Ozone', *Annals of the New York Academy of Sciences*, vol I, no 3, pp363–391 (as cited in Stolarski, 1999).

5   Cornu, M A (1879) 'Sur la limite ultra-violette du spectre solaire', *Comptes Rendus*, vol 88, pp1101–1108 (as cited in Stolarski, 1999).

6   Hartley, W N (1880) 'On the Probable Absorption of the Solar Ray by Atmospheric Ozone', *Chemical News*, vol 268, November 26 (as cited in Stolarski, 1999).

7   Fowler, A and Strutt, R J (1917) *Proceedings of the Royal Society of London*, Series A, vol 93, p577 (as cited in Stolarski, 1999).

8   Strutt, R J (1918) 'Ultra-violet Transparency of the Lower Atmosphere, and its Relative Poverty in Ozone', *Proceedings of the Royal Society of London*, Series A, vol 94, pp260–268 (as cited in Stolarski, 1999).

9   Dobson, G M B and Harrison, D N (1926) 'Measurements of the Amount of Ozone in the Earth's Atmosphere and Its Relation to Other Geophysical Conditions', *Proceedings of the Royal Society of London*, Series A, vol 110, pp660–693 (as cited in Stolarski, 1999).

10   'Ozone Milestones' (2000) US National Aeronautics and Space Administration/Jet Propulsion Laboratory (online). Available: http://atmos.jpl.nasa.gov/milestones.htm.

11   Chapman, S (1931) 'Some Phenomena of the Upper Atmosphere', *Proceedings of the Royal Society of London*, Series A 132 (as cited in Cagin, S and Dray, P (1993) *Between Earth and Sky: How CFCs Changed our World and Endangered the Ozone Layer*, Pantheon, New York).

12   Cagin and Dray (1993), op cit.

13   Götz, F W P, Meetham, A R and Dobson, G M B (1934) 'The Vertical Distribution Of Ozone In The Atmosphere', *Proceedings of the Royal Society of London*, Series A, vol 145, pp416–446 (as cited in Stolarski, 1999).

14   Bates, D R and Nicolet, M (1950) 'The Photochemistry of Atmospheric Water Vapor', *Journal of Geophysical Research*, vol 55, pp301–327 (as cited in Stolarski, 1999).

15   Nicolet, M (1955) 'The Aeronomic Problem of Nitrogen Oxides', *Journal of Atmospheric and Terrestrial Physics*, vol 7, pp152–169 (as cited in Stolarski, 1999).

16   Crutzen, P (1970) 'The Influence of Nitrogen Oxides on the Atmospheric Ozone Content', *Quarterly Journal of the Royal Meteorological Society*, vol 96, no 408, pp320–325.

17   Lovelock, J E, Maggi, R J and Wade, R J (1973) 'Halogenated Hydrocarbons in and over the Atlantic', *Nature*, vol 241, pp195 (as cited in Kowalok, M (1993) 'Common Threads: Research Lessons from Acid Rain, Ozone Depletion, and Global Warming', *Environment*, vol 35, no 6, pp12–20, 35–38 (online). Available: www.ciesin.org/docs/011-464/011-464.html.)

18   Harrison, H (1970) 'Stratospheric Ozone with Added Water Vapour: Influence of High-Altitude Aircraft', *Science*, November 13.

19   Johnston, H S (1971) 'Reduction of Stratospheric Ozone by Nitrogen Oxide Catalysts from Supersonic Transport Exhaust', *Science*, vol 173, pp517–522 (as cited in Stolarski, 1999).

20   Crutzen, P (1972) 'SST's – A Threat to the Earth's Ozone Shield', *Ambio*, vol I, no 2, pp41–51.

21   Glas, J P (1989) 'Protecting the Ozone Layer: A Perspective from Industry', National Academy of Sciences (online). Available: www.nap.edu/openbook/030900426X/html/137.html.

22   Wofsy, S C and McElroy, M B (1974) 'Hox, Nox, and ClOx: Their Role in Atmospheric Photochemistry', *Canadian Journal of Chemistry*, vol 52, pp1582–1591.

23   Stolarski, R and Cicerone, R (1974) 'Stratospheric Chlorine: A Possible Sink for Ozone', *Canadian Journal of Chemistry*, vol 52, pp1610–1615 (as cited in Benedick, Ozone Diplomacy, 1991, 1998).

24   Molina, M and Rowland, F S (1974) 'Stratospheric Sink For Chlorofluoromethanes: Chlorine Atom-Catalyzed Destruction Of Ozone', *Nature*, vol 249 (5460), pp810–812.

25   Cagin and Dray (1993), op cit.

26   Brodeur, P (1986) 'Annals of Chemistry', *The New Yorker*, 9 June.

27   Freon™ is the trademark of CFCs marketed by DuPont Corporation.

28   Wofsy, S C, McElroy, M B and Sze, N D (1975) 'Freon Consumption: Implications for Atmospheric Ozone', *Science*, vol 187, February 14.

29   Sullivan, W (1975) 'Ozone Depletion Seen as a War Tool', *New York Times*, February 28, pA20.

30   National Research Council (1976) 'Halocarbons: Effects on Stratospheric Ozone', National Academy of Sciences.

31   National Research Council (1979) 'Protection Against Depletion of Stratospheric Ozone by Chlorofluorocarbons', National Academy of Sciences.

32   Cagin and Dray (1993), op cit.

33   Gribbin, J (1988) *The Hole in the Sky: Man's Threat to the Ozone Layer*, Bantam, New York.

34   National Research Council (1982) 'Causes and Effects of Stratospheric Ozone Reduction: An Update', National Academy of Sciences.

35   National Research Council (1983) 'Causes and Effects of Changes in Stratospheric Ozone: Update 1983', National Academy of Sciences.

36   Chubachi, S (1984) 'Preliminary Result of Ozone Observations at Syoma Station from February 1982 to January 1983', *Memoirs of National Institute of Polar Research Special Issue No. 34, Proceedings of the Sixth Symposium on Polar Meteorology and Glaciology*.

37   Prather, M J, McElroy, M B and Wofsy, S C (1984) 'Reductions in Ozone at High Concentrations of Stratospheric Halogens', *Nature*, vol 312, pp227–231.

38   Farman, J S, Gardiner, B G and Shanklin, J D (1985) 'Large Losses of Total Ozone in Antarctic Reveal Seasonal ClO_x/NO_x Interaction', *Nature*, vol 315, pp207–210.

39   Solomon, S, Garcia, R R, Rowland, F S and Wuebbles, D J (1986) 'On the Depletion of Antarctic Ozone', *Nature*, vol 321, pp755–758.

40   'Atmospheric Ozone 1985: Assessment of our Understanding of the Processes
     Controlling its Present Distribution and Change' (1985) National Aeronautics and Space
     Administration, Federal Aviation Administration, US National Oceanic and Atmospheric
     Administration, United Nations Environment Programme, World Meteorological
     Organization, Commission of the European Communities, and Bundesministerium Für
     Forschung Und Technologie.
41   Goddard Space Flight Center, NASA (1996) 'The Airborne Antarctic Ozone Experiment'
     (online). Available: http://hyperion.gsfc.nasa.gov/Analysis/aircraft/aaoe.html.
42   International Ozone Trends Panel (1988) 'Report of the International Ozone Trends
     Panel 1988'. National Aeronautics and Space Administration, National Oceanic and
     Atmospheric Administration, Federal Aviation Administration, World Meteorological
     Organization, and United Nations Environment Programme.
43   Cagin and Dray (1993), op cit.
44   Cagin and Dray (1993), op cit.
45   NASA. 'Airborne Antarctic Ozone Experiment' (online). Available:
     http://geo.arc.nasa.gov/esdstaff/jskiles/fliers/all_flier_prose/antarcti.../
     antarcticO3_condon.htm.
46   Weisskopf, M (1987) 'Ozone Depletion Worsens, Is Linked to Man-Made Gas', *Washington
     Post*, 1 October, pA23.
47   Hoffman, J S and Gibbs, M J (1988) 'Future Concentrations of Stratospheric Chlorine
     and Bromine', EPA 400/1-88/005.
48   Saltus, R (1989) 'Ozone Loss is Seen as Threat to Body's Immune System', *Boston Globe*, 6
     July, p3.
49   OEWG (1989) 'Synthesis Report: Integration of the Four Assessment Panels Reports by
     the Open-Ended Working Group of the Parties to the Montreal Protocol', United
     Nations Environment Programme, Nairobi.
50   NASA (1992) 'End of Mission Statement: Second Airborne Arctic Stratospheric
     Expedition' (online). Available:
     http://hyperion.gsfc.nasa.gov/Analysis/aircraft/aase2/eom_aaseII.txt.
51   'Scientific Assessment of Ozone Depletion: 1991' (1991) National Aeronautics and Space
     Administration, National Oceanic and Atmospheric Administration, UK Department of
     the Environment, United Nations Environment Programme and World Meteorological
     Organization.
52   'Synthesis of the Reports of the Ozone Scientific Assessment Panel, Environmental
     Effects Assessment Panel, Technology and Economic Assessment Panel, Prepared by the
     Assessment Chairs for the Parties to the Montreal Protocol' (1991) United Nations
     Environment Programme.
53   NASA (1992) 'US Study Enhances Concern for Northern Ozone Depletion', 30 April
     (online). Available: www.alternatives.com/library/env/envozone/nasa30.txt.
54   'Scientific Assessment of Ozone Depletion: 1994; Executive Summary' (1995) National
     Aeronautics and Space Administration, National Oceanic and Atmospheric
     Administration, United Nations Environment Programme, and World Meteorological
     Organization.
55   'The 1994 Science, Environmental Effects, and Technology and Economic Assessments:
     Synthesis Report' (1994) United Nations Environment Programme.
56   The Royal Swedish Academy of Sciences (1995) 'The Ozone Layer – The Achilles Heel
     of the Biosphere' (online). Available:
     www.nobel.se/announcement/1995/announcement95-chemistry.html.
57   Ozone Secretariat (1999) 'Synthesis of the Reports of the Scientific, Environmental
     Effects, and Technology and Economic Assessment Panels of the Montreal Protocol',
     United Nations Environment Programme, Nairobi.

58   'Environmental Effects of Ozone Depletion: 1998 Assessment' (1998) United Nations Environment Programme.

59   'Synthesis of the Reports of the Scientific, Environmental Effects, and Technology and Economic Assessment Panels of the Montreal Protocol' (1999) United Nations Environment Programme. Available: www.unep.org/ozone/oewg/19oewg/ synthesis-main.pdf.

60   Intergovernmental Panel on Climate Change (1999) *Aviation and the Global Atmosphere*, Cambridge University Press, Cambridge, UK.

61   Solomon, S (1999) 'Stratospheric Ozone Depletion: A Review of Concepts and History'. *Reviews of Geophysics*, vol 37, no 3 (August 1999), pp275–316.

62   The European Commission (5 April 2000) 'Severe Stratospheric Ozone Depletion in the Arctic; Recovery of Ozone Layer Delayed Due to Climate Change?' (online). Available: http://europa.eu.int/comm/research/press/2000/pr0504en.html.

63   Steitz, D E, O'Carroll, C and Viets, P (2001) 'Ozone Hole About the Same Size as Past Three Years', Release 01-198, 16 October, NASA/NOAA, Washington, DC/Suitland, MD

CHAPTER 2

1   UNEP (1992) *The World Environment 1972–1992: Two Decades of Challenges*, Chapman and Hall, London (for UNEP), pp742–745.

2   Decision 29(III) of Governing Council of April 1975.

3   Decision 65(IV) of Governing Council of April 1976.

4   'Some Economic and Social Implications of a Possible Ban on Fluorocarbons', paper presented by the Department of Economic and Social Affairs, UK, to the Meeting of Experts on the Ozone Layer organized by UNEP on 1–9 March 1977 in Washington, DC, and printed in the *Ozone Layer*, edited by Asit K Biswas and published by Pergamon Press, Oxford, UK, for UNEP, 1979.)

5   See Appendix 2, Biswas, Asit K (ed) (1979) 'World Plan of Action', *Ozone Layer* Pergamon Press, Oxford, UK, for UNEP.

6   Stoel, Thomas, Miller, Alan and Milroy, Breck (1980) *Fluorocarbon Regulation: An International Comparison*, Lexington Books, Lanham, MD, USA.

7   Decision 84 C (v) of the Governing Council session of May 1977.

8   Australia, Canada, Denmark, France, Federal Republic of Germany, India, Italy, Japan, Kenya, The Netherlands, Norway, Sweden, Soviet Union, United Kingdom, United States, Venezuela, United Nations, WHO, ICAO, WMO and UNEP.

9   Australia, Belgium, Canada, Denmark, France, Federal Republic of Germany, Italy, The Netherlands, Norway, Sweden, Switzerland, UK and US.

10   Note by UNEP on the International Conference on CFMs in the Bulletin of January 1979.

11   Decision 8/7(B), 29 April 1980 of the Governing Council.

12   Decision 80/372/EEC, in March 1980.

13   Decision 9/13B, 26 May 1981 of the Governing Council.

14   Tolba, Mostafa with Rummel-Bulska, Iwona (1998) Global Environmental Diplomacy: Negotiating Environmental Agreements for the World, 1973–1992, MIT Press, Cambridge, MA, p6.

15   UNEP WG 69/10, 28 January 1982.

16   Decision 10/17, 31 May 1982 of the Governing Council.

17   UNEP/WG78/8, 5 January 1983.

18   Decision 82/795/EEC, November 1982.

19   UNEP/WG78/13, 17 June 1983.

20   Decision 11/7, Part II B, section 1, 24 May 1983.
21   UNEP/WG94/5, 10 November 1983.
22   UNEP/WG94/10, 1 February 1984.
23   Decision 12/14, Section 1 of 28 May 1984 of the Governing Council.
24   WNEP/WG110/4, 9 November 1984.
25   UNEP IG 53/4, 28 January 1985.
26   Document UNEP/IG/53/3.
27   Decision 13/18, 24 May 1985 of the Governing Council.
28   The report of the meeting UNEP/WG 148/2 dated 12 July 1986.
29   UNEP/WG 148/3, 12 September 1986.
30   The report of the Working Group, UNEP WG 151/4, 5 December 1986.
31   UNEP/WG151/2.
32   Jurak, Pamela S et al (1986) 'Tending the global commons', *Chemical and Engineering News*,
      vol 64, no 47, 24 November.
33   UNEP/WG 167/2, 27 February 1987.
34   Files of UNEP.
35   Files of UNEP.
36   UNEP/WGI/67/Inf I, 24 April 1987.
37   Report of the Meeting UNEP/WG 172/2, 8 May 1987.
38   UNEP/IG 79/3.

CHAPTER 3

1   UNEP/WG.Data1/3, March 1988.
2   UNEP/WG.Data/2/3, 26 October 1988.
3   UNEP/ozl.alt.1/3, 20 October 1988.
4   UNEP/Ozl.Conv/1/5, 28 April 1989.
5   UNEP/Ozl.Pro 1/5, 6 May 1989.
6   The meetings outside Nairobi took place in Helsinki, London, Copenhagen, Bangkok,
      Vienna, San Jose, Montreal, Cairo, Beijing, Ouagadougou and Colombo.
7   The report of the informal group of experts on financial mechanism in
      UNEP/Ozl.Pro.Mech.1/Inf.1, 16 August 1989.
8   The report of the meeting in UNEPozl.pro/LG1/3, 14 July 1989.
9   The report is in UNEP Ozl.Pro/WGI(1)/3, 25 August 1989.
10   The report of the second session in UNEP/Ozl.Pro/WG I(2)/4 dated, 4 September 1989.
11   UNEP/ozl.pro/WGI (3)/3, 25 September 1989.
12   UNEP/ozl.pro/WGII (1)/7.
13   Report of the Legal Drafting Group, UNEP/Ozl.Pro.WGII (1)/5, 20 November 1989.
14   UNEP/Ozl.Pro/WGIII (1)/3, 14 November 1989.
15   UNEP/Ozl.Pro/WGIII (2)/3, 22 May 1990.
16   UNEP/Ozl.Pro.Wg.III(2)/Inf.1, 17 April 1990.
17   UNEP/Ozl.Pro/WGIV/8, 29 June 1990.
18   UNEP/Ozl.Pro 2/7, 29 June 1990.
19   UNEP/Ozl.Pro/WG1/5/3, 5 December 1990.
20   LG second meeting April 1991.
21   UNEP/Ozl.Pro.3/11, 21 June 1991.
22   UNEP/OZL.Conv.2/7, 19 June 1991.
23   UNEP/Ozl.Conv.2/7, 19 June 1991.
24   LG meeting November 1991.
25   UNEP/OZL.PRO/WG1/6/3, 15 April 1992.

*26   The report of the meeting UNEP/ozl.pro/WG/1/7/4, 18 July 1992.
27   UNEP/Ozol.Pro/WG1/8/2, 26 November 1992.
28   The report of MOP 4, UNEP./Ozl.Pro.4/5, 25 November 1992.
29   Annex VII to the report of the fifth Meeting of the Parties.

CHAPTER 4

1   The report of the OEWG, UNEP/Ozol.Pro.WG1/9/7 dated 2 September 1993 and the report of the 5th MOP UNEP/Ozl.Pro. 5/12 dated 19 November 1993.
2   Austria, Belgium, Botswana, Denmark, EEC, Finland, Germany, Iceland, Italy, Liechtenstein, Malta, Netherlands, Norway, Sweden, Switzerland, UK and Zimbabwe.
3   Austria, Belgium, Denmark, Finland, Germany, Iceland, Israel, Italy, Liechtenstein, Netherlands, Sweden, Switzerland, UK, USA and Zimbabwe.
4   UNEP/Ozl.Conv.3/6 dated 23 November 1993.
5   The report of the OEWG UNEP/Ozl.Pro/WG1/10/6 dated 11 July 1994 and the report of the Sixth MOP UNEP/Ozl.Pro.6/7 dated 10 October 1994.
6   Argentina, Brazil, Chile, China, Colombia, India, Malaysia, Peru, Philippines and Uruguay.
7   UNEP/Ozol.Pro/WG1/11/10 dated 13 June 1995 and 1/12/4 dated 18 September 1995.
8   Report of the Seventh MOP, UNEP/Ozl.Pro. 7/12 dated 27 December 1995.
9   The proceedings of this symposium were published in 1996 by the Federal Ministry for Foreign Affairs of Austria as *The Ozone Treaties and Their Influence on the Building of International Environmental Regimes* (edited by Winfried Lang).
10   Argentina, Austria, Belgium, Botswana, Czech Republic, Denmark, EEC, Finland, France, Georgia, Germany, Ghana, Greece, Iceland, Ireland, Italy, Latvia, Liechtenstein, Lithuania, Luxembourg, Namibia, Netherlands, Norway, Poland, Portugal, Romania, Slovakia, Spain, Sweden, Switzerland, Uganda and the UK.
11   Bolivia, Burundi, Canada, Chile, Denmark, Ghana, Iceland, Namibia, Netherlands, New Zealand, Romania, Switzerland, Uruguay and Venezuela.
12   UNEP/Ozl.Pro/WG1/13/6 dated 6 September 1996, UNEP/Ozl.Pro.8/Prep/2 dated 24 December1996 and UNEP/Ozl.Pro.8/12 dated 19 December 1996. It may be noted that the fourteenth meeting of the OEWG was the preparatory meeting for the eighth MOP in San Jose, Costa Rica.
13   The reports of the OEWG, UNEP/WG1/15/5 dated 12 June 1997 and 1/16/2 dated 15 September 1997 and the report of the ninth MOP UNEP/Ozl.Pro.9/12/ dated 25 September 1997.
14   The proceedings of this workshop have been printed as *Protecting the Ozone Layer: Lessons, Models and Prospects* by Kluwer Academic Publishers, New York.
15   Argentina, Austria, Belgium, Botswana, Czech Republic, Denmark, EEC, Finland, France, Georgia, Germany, Ghana, Greece, Iceland, Ireland, Italy, Latvia, Liechtenstein, Lithuania, Luxembourg, Namibia, Netherlands, Norway, Poland, Portugal, Romania, Slovakia, Spain, Sweden, Switzerland, Uganda and the UK.
16   Bolivia, Burundi, Canada, Chile, Denmark, Ghana, Iceland, Namibia, Netherlands, New Zealand, Romania, Switzerland, Uruguay and Venezuela.
17   Report of the OEWG, UNEP/Ozl.Pro/WG1/17/3 dated 15 July 1998 and 1/18/2 dated 21 November 1998 and the report of the 10th MOP, UNEP/Ozl.Pro10/9 dated 3 December 1998.
18   Austria, Azerbaijan, Belgium, Bolivia, Botswana, Bulgaria, Costa Rica, Croatia, Cuba, Czech Republic, Denmark, Estonia, EEC, Finland, France, Georgia, Germany, Ghana, Greece, Iceland, Ireland, Italy, Lao Peoples Democratic Republic, Latvia, Lesotho, Liechtenstein, Lithuania, Luxembourg, Madagascar, Netherlands, Norway, Poland,

Portugal, Romania, Slovakia, Slovenia, Spain, Sweden, Switzerland, the UK and Uzbekistan.

19   The report of the OEWG, UNEP/WG1/19/7 dated 18 June 1999.

20   UNEP/Ozl.Pro.11/10 dated 17 December 1999.

21   UNEP/Ozl.Conv.5/6 dated 20 December 1999.

22   The report of the OEWG, UNEP/1/20/3 dated 13 July 2000 and the report of the twelfth MOP UNEP/Ozl.Pro.12/9 dated 10 January 2001.

23   UNEP/OzL.Pro/WG1/21/4 dated 26 July 2001 and the report of the 13th MOP UNEP/OzL.Pro/13/10 dated 26 October 2001.

CHAPTER 5

1   The authors are grateful to contributors and peer-reviewers of this chapter including: Radhey Agarwal (India Institute of Technology), Ward Atkinson (Sun Test Engineering), Jonathan Banks (Grainsmith Consultants), Steven Bernhardt (DuPont), David Catchpole (Catchpole Associates), Nick Campbell (Atofina), Sukumar Devotta (National Chemical Laboratory, India), Bryan Jacob (The Coca-Cola Company), Mike Jeffs (Huntsman), Horst Kruse (University of Hannover Germany), Mack McFarland (DuPont), Caley Johnson (US EPA), E Thomas Morehouse (Institute for Defense Analyses), Gary Taylor (Taylor-Wagner), Lindsey Roke (Fisher Paykel), Madhava Sarma (History co-author), Helen Tope (EPA Victoria Australia) and Robert Wickham (Wickham Associates).

2   Carbon tetrachloride is also used in the production of plastics, semiconductors, and petrol additives and in recovery of tin from tin plating waste.

3   Methyl chloroform is used as a solvent ingredient in aerosol products and in miscellaneous solvent, coating, and adhesive applications.

4   Methyl bromide is used in organic synthesis as a methylating agent (Torkelson, T R and Rowe, V K (1994) 'Methyl Bromide', in R L Harris, L J Cralley and L V Cralley (eds) *Patty's Industrial Hygiene and Toxicology*, 3rd revised edition, March, John Wiley and Sons, New York, pp3442–3446) and as a low-boiling-point solvent to extract oils from nuts, seeds and flowers (Matheson Gas Products (1980) *Matheson Gas Data Book*, Matheson Gas Products, East Rutherford, New Jersey, vol 6, pp 361–363).

5   Methyl bromide was used as a medicine and anaesthetic in the early 1900s.

6   CFC-11 is used as a solvent for flushing CFC refrigerating systems during service, as a solvent ingredient in aerosol products using CFC-12 propellants, and in miscellaneous solvent and adhesives applications.

7   The cumulative use of CFC-11 for miscellaneous purposes was significant. Uses included tobacco expansion, wind-tunnel test gases, as a product ingredient, etc.

8   Minor quantities of CFC-12 were used as a propellant in hand-held aerosol fire-extinguishers.

9   The cumulative use of CFC-12 in miscellaneous uses was significant. Uses included contact food freezing, explosion inerting of ethylene oxide mixtures used in pest fumigation and sterilization, vessel leak testing, nuclear fuel processing, aluminium processing, insulating between double glazing, and as a dielectric medium in scientific and medical equipment, thermostats and solar window controls, and tyre inflators. For examples of recent papers questioning ozone science see: Singer, S Fred (1990) 'Environmental Strategies with Uncertain Science', *Regulation*, CATO Review of Business & Government, Winter, pp65–70; Singer, S Fred (1994) 'Ozone, skin cancer, and the SST', *Aerospace America*, July, pp22–26; Singer, S Fred (1994) 'The Ozone Layer and Human Health', *The World & I*, pp214–219; Singer, S Fred (1994) 'Stratospheric Ozone Doubts', *Environmental Conservation*, vol 21, number 1, Spring, p6; Lieberman, Ben (1994)

'The High Cost of Cool: The Economic Impact of the CFC Phase-out in the Untied States', *Competitive Enterprise Institute*, June; Singer, S Fred (1994) '(N)O3 Problem', *The National Interest*, Summer, pp73–76; Easterbrook, Gregg (1995) *A Moment on the Earth, the Coming Age of Environmental Optimism*, Viking Press, New York; Paul Craig Roberts, 'Quietly, Now, Let's Rethink the Ozone Apocalypse' (1995) *Business Week*, June 19, p26; Baliunas, Sallie (1994) 'Ozone & Global Warming: Are the Problems Real?', George C Marshall Institute, Washington, and The Claremont Institute, Claremont, California, December 13; and Baliunas, Sallie (1995) 'Ozone Variations and Accelerated Phase-out of CFCs', US House of Representatives Subcommittee on Energy and Environment, September 20. Also see the exchange of comments: Taubes, Gary (1993) 'The Ozone Backlash', News and comment, *Science*, 11 June, p1580; Singer, S Fred (1993) 'Ozone Depletion Theory', Letters, *Science*, 27 August; and Rowland, F Sherwood (1993) 'Response to Letters', *Science*, 27 August.

10   The authors are grateful to Gary Taylor (Co-chair of the TEAP Halons Technical Options Committee and Chair of the National Fire Protection Association Halons Technical Committee) for providing the historic review of halon use.

11   Carbon tetrachloride (halon-104), methyl bromide (halon-1001), bromochloromethane (halon-1011), dibromodifluoromethane (halon-1202), bromochlorodifluoromethane (halon-1211), bromotrifluoromethane (halon-1301) and dibromotetrafluoroethane (halon-2402).

12   US National Institute of Health http://toxnet.nlm.nig.gov; World Health Organization (1999) 'Carbon Tetrachloride', Environmental Health Criteria 208, WHO, Geneva; World Health Organization (1995) 'Methyl Bromide', Environmental Health Criteria 166, WHO, Geneva.

13   World Health Organization (1999), op cit, p16.

14   Midgley, T Jr and Henne, A L (1930) 'Organic Flourides as Refrigerants', *Industrial and Engineering Chemistry*, vol 22, no 5, pp542–545; McLinden, M O and Didion, D A (1987) 'Quest for Alternatives: A Molecular Approach Demonstrates Tradeoffs and Limitations are Inevitable in Seeking Refrigerants', *ASHRAE Journal*, December, pp32–42; Nagengast, B (1988) 'A Historical Look at CFC Refrigerants', *ASHRAE Journal*, November, pp37–39; Kauffman, G B (1989) 'Midgley: Saint or Serpent?', *Chemtech Magazine*, December, pp717–725; Manzer, L E (1990) 'The CFC-Ozone Issue: Progress on the Development of Alternatives to CFCs', *DuPont Articles*, 6 July, vol 249, pp31–35.

15   Other refrigerants included ammonia, butane, carbon bisulphide, carbon dioxide, dichlorethylene, ethane, ethylamine, ethyl bromide, gasoline, isobutane, methyl formate, methylene chloride, methylamine, methyl chloride, naptha, nitrous oxide, propane, sulphur dioxide, trichloroethylene, and trimethylamine (Nagengast, 1988, op cit).

16   Toxicity testing available at the time CFCs were introduced was only able to identify short-term effects. For example, CFC-21 was considered relatively safe until 1983 when the TLV was reduced from 1000 parts per million (ppm) to 10 ppm.

17   A complete history of the development of CFC refrigerants is found in: Cagin, S and Dray, P (1993) *Between Earth and Sky: How CFCs Changed Our World and Endangered the Ozone Layer*, Pantheon, New York; Nagengast (1988), op cit; Midgley, T Jr (1937) 'From the Periodic Table to Production', *Journal of Industrial Engineering Chemistry*, February; Midgley and Henne (1930), op cit.

18   US National Institute of Health http://toxnet.nlm.nig.gov; World Health Organization (1999), op cit.

19   US National Institute of Health http://toxnet.nlm.nig.gov; World Health Organization (1999), op cit; TEAP Process Agent Task Force 1997 and 2001.

20   See: Dotto, Lydia and Schiff, Harold (1978) *The Ozone War*, Doubleday & Co, Barden, New York; Brodeur, Paul (1996) 'Annals of Chemistry (The Ozone Layer)', *New Yorker*,

June 9; Cogan, Douglas G (1998) *Stones in a Glass House: CFCs and Ozone Depletion*, Investor Responsibility Research Center, Washington, DC; Gribbin, John (1998) *The Hole in the Sky: Man's Threat to the Ozone Layer*, and Nantze, J (1991) *What Goes Up: The Global Assault on Our Atmosphere*, William Morrow, New York; Cagin, Seth and Dray, Philip (1993) *Between Earth and Sky: How CFCs Changed Our World and Endangered the Ozone Layer*, Pantheon, New York; Du Pont's Freon Products Division (1978) 'Information Requested by the Food and Drug Administration on Non-Propellant Uses of Fully Halogenated Chlorofluoroalkanes', January 13; Heckert, Richard E, Du Pont, (1988) Letter to the US Senate Explaining Du Pont's Position on the CFC/Ozone Issue, March 4.

21 Boeing abandoned SST development when the United States government halted public funding and the USSR abandoned its Tupolev-144 SST after a crash at the 1973 Paris Air Show and second crash of a revamped Tupolev-144D in 1977 inside the Soviet Union. In 2001, NASA rehabilitated a Tupolev SST for research use.

22 The history of the evolution of scientific proof of ozone depletion from the US CFC industry perspective is presented in DuPont (1990) 'Searching the Stratosphere: Industry's contribution to scientific understanding of the ozone depletion issue', DuPont, Wilmington, DE.

23 See: Rowland, S and Molina, M (1994) 'Ozone Depletion: 20 Years after the Alarm', *Chemical and Engineering News*, vol 72, no 33, August 15, pp8–13.

24 By 1975 DuPont had financed studies by J E Lovelock (Reading University UK), J N Pitts and J A Taylor (University of California), C Sandorfy (University of Montreal) and R A Rasmussen (Washington State University) that had confirmed that CFCs were rapidly accumulating in the atmosphere. J W Swinnerton (US Naval Research Laboratory) reported similar results. See McCarthy, Ray and Schuyler, Roy L (for DuPont) (1975) 'Testimony to the Subcommittee on the Upper Atmosphere', US Senate, October.

25 See, for example, R L McCarthy's testimony of 11–12 December 1974 before House of Representatives Subcommittee on Public Health and Environment.

26 See for example, DuPont (1980) 'Fluorocarbon/Ozone Update' June.

27 See: Orfeo, R S (1986) 'Response to Questions to the Panel from Senator John H Chafee', *Ozone Depletion, the Greenhouse Effect, and Climate Change, Hearings before the Senate Subcommittee on Environmental Pollution of the Committee on Environment and Public Works*, US GPO, Washington, DC, pp189–192, June; Strobach, Donald (1986) 'A Search for Alternatives to the Current Commercial Chlorofluorocarbons', in *Protecting the Ozone Layer: Workshop on Demand and Control Technologies*, US EPA, Washington, DC.

28 Hoppe, Richard (1986) 'Ozone: Industry is Getting its Head Out of the Clouds', *Business Week*, October 13, p110.

29 'Dear Freon™ Customer', letter dated 26 September 1986.

30 See: Andersen, S O and Miller, A (1996) 'Ozone Layer, the Road Not Taken', Correspondence, *Nature*, vol 382, p390. Andersen and Miller contend that the cost of the phase-out would have been much lower if science could have provided an earlier warning or if industry had been more willing to accept earlier scientific results. They also conclude that costs of a more rapid phase-out with a later start would have been much higher and that the choice of available options would have been less environmentally acceptable. In 1985 one third of global CFC use was for aerosol products that had switched to hydrocarbon propellants or not-in-kind substitutes. An additional 15–20 per cent was emitted from applications that halted use within one year of the initiation of industry efforts – typically using existing technology (reduced-CFC insulating foam, methylene chloride flexible foam, recovery and recycling of CFC refrigerants, halting discharge testing, venting, training, and accidental discharge of halon systems).

31 See: EPA (1998) 'How Industry is Reducing Dependence on Ozone-Depleting Chemicals', US EPA, Washington, DC, June.

32   On March 24, 1988 DuPont notified customers that it was beginning an orderly phase-out of ozone-depleting substances and urged customers to seek new solutions.

33   Mary Lu Carnevale, 'DuPont plans to Phase Out CFC Output', The WSJ, March 25, 1988.

34   Stephen O Andersen, John Hoffman, and George Strongylis Co-Chair the Economic Panel; Jan van der Leun and Manfred Tevini Co-Chair the Environmental Effects Panel; Daniel Albritton and Robert Watson Co-Chair the Scientific Assessment Panel; and Stephen O Andersen and G Victor Buxton Co-Chair the Technology Review Panel.

35   See: Andersen, Stephen O, Frech, Clayton and Morehouse, E Thomas (1997) *Champions of the World: Stratospheric Ozone Protection Awards*, US EPA, Washington, DC; Andersen, Stephen O (1997) *Newest Champions of the World: Winners of the 1997 Stratospheric Ozone Protection Awards*, US EPA, Washington, DC; Le Prestre, Philipe G, Reid, John D and Morehouse, E Thomas (1998) *Protecting the Ozone Layer: Lessons, Models and Prospects*, Kluwer Academic Publishers, Boston; Cook, Elizabeth (1996) *Ozone Protection in the United States: Elements of Success*, World Resources Institute, Washington, DC.

36   See: Berkman, Barbara N (1999) 'AT&T's Big Push to Stay Ahead of Environmental Woes' and Kerr, John (1999) 'Smart Answers for Toxic Troubles' and 'Tektronix on Toxics: Looking Good by Using Less' in *Electronic Business*, 18 September.

37   See: Kerr, John (1999) 'Smart Answers for Toxic Troubles', op cit.

38   US law required public reporting of toxic emissions including CFCs and other ODSs and in some countries chemical manufacturers and customers voluntarily disclosed similar information.

39   Stephen O Andersen, et al (1988) 'Stratospheric Ozone and the US Air Force', *The Military Engineer*, vol 80, no 523, August. See also Stephen O Andersen et al (1994) 'The Military's Role in Protection of the Ozone Layer', *Environmental Science and Technology*, vol 28, no 13; Cook (1996), op cit.

40   See: Andersen, et al (1994), op cit; Andersen and Morehouse (1997), op cit.

41   See: Andersen, Stephen O 'Halons and the Stratospheric Ozone Issue', Taylor, Gary 'Achieving the Best Use of Halons' and Willey, A Elwood ,'The NFPA's Perspective on Halons and the Environment', *Fire Journal*, vol 81, no 3 (May–June 1987).

42   Experts frequently use a variety of definitions in describing technical choices under the Montreal Protocol. The TEAP generally used the following definitions. When one chemical replaces another with the same basic function it is a 'substitute'. When a product or process change replaces a chemical it is an 'alternative'. A fluorinated chemical (ie HCFC or HFC) is an 'in-kind chemical substitute' while a non-fluorinated chemical (eg hydrocarbon, ammonia, carbon dioxide, etc) is a 'not-in-kind chemical substitute'. Examples of in-kind substitutes include HFC-134a replacing CFC-12 in vehicle air conditioning, HCFC-123 replacing CFC in building chillers, and HFC-225 replacing halon-1211. A product alternative satisfies the same consumer need with a different product or service. Examples of product alternatives to CFC aerosol deodorants include dispensing with pumps, sprays and sticks. A process change can eliminate the need for an ODS. For example, no-clean solders eliminate solvent use.

43   Montreal Protocol technology literature defines chemicals that directly replace ODSs as 'alternatives', and process changes and not-in-kind products as 'substitutes'.

44   A CD-ROM available from the Ozone Secretariat includes a descriptive list of the most important such efforts.

45   France (Atofina), Germany (Solvay), Japan (Mitsui-DuPont Fluorochemicals, Showa-Denko), Italy (Ausimont), United Kingdom (Ineas), and United States (Atofina, DuPont, Honeywell).

46   For example, in October 2001 Thailand approved a Finance Ministry proposal to discourage further the use of CFCs, halon, methyl chloroform and carbon tetrachloride ODSs by a new excise tax of 15 per cent immediately rising to 30 per cent from 2003 and an increase in import duties from 1 per cent to 5 per cent.

47   See: McFarland, M and Kaye, J (1992) 'Yearly Review: Chlorofluorocarbons and Ozone', *Photochemistry and Photobiology*, Vol 55, No 6, pp911–929; Manzer, L E and Rao, V N M (1993) 'Catalytic Synthesis of Chlorofluorocarbon Alternatives', *Advances in Catalysis*, Vol 39, pp329–350.

48   See: UNEP (1994) 'Practical Guidelines for Industry for Managing the Phase-Out of Ozone Depleting Substances', United Nations Environment Programme, DTIE, Nairobi, November, for descriptions of leap-frogging to economically and environmentally superior technology.

49   Current members are Bayer, Dow, Elastogran, ICI/Huntsman Polyurethanes, Lyondell and Shell.

50   European Regulation EC 3093/94 and European Regulation EC 2037/2000 respectively.

51   TEAP (1999) *The Implications to the Montreal Protocol of the Inclusion of HFCs and PFCs in the Kyoto Protocol*, UNEP, Nairobi.

52   Simple economic analysis fails to account for the investment that would occur without the phase-out (overstating the net investment required) and the improvements in operating costs, reliability, and performance (understating the benefits). Furthermore, many calculations improperly count tax revenue as a cost when it is properly considered as a transfer payment. See Vogelsberg, F A (1999) 'An Industry Perspective – Lessons Learned and the Cost of CFC Phase-out'.

53   E I DuPont de Nemours & Company (1979) 'Position Paper: Chlorofluorocarbon/Ozone Depletion Issue', August.

54   Lee, E D (1987) 'Pending Treaty Worries Chlorofluorocarbon Industry', *Wall Street Journal*, 15 September.

55   Alliance for Responsible CFC Policy (1986) *The Montreal Protocol: A Briefing Book*, Alliance for Responsible CFC Policy, Rosslyn, VA, December.

56   Putnam, Hayes & Bartlett, Inc (1987) 'Economic Implications of Potential Chlorofluorocarbon Restrictions: Final Report, Prepared for the Alliance for Responsible CFC Policy, Washington, DC, 2 December.

57   ICOLP (various authors) (1994) 'Protecting the Ozone Layer: A Global Business Initiative', *Supplement to Machine Design*, April 23.

58   Vogelsberg, F A (Tony) (1996) 'Industry Perspective on Lessons Learned and Cost of CFC Phase-out', *International Conference on Ozone Protection Technologies, 21–23 October 1996* (ten pages with ten-page supplement); and Vogelsberg, F A and Smythe, K D (1987) 'Stratospheric Ozone Depletion: Industry's Contribution to Solution', AFEAS, October.

59   EPA (1987) 'Protection of Stratospheric Ozone: Proposed Rule', 40 CFR Part 82, US Environmental Protection Agency, December 14.

60   Environment Canada (1997) 'Global Benefits and Costs of the Montreal Protocol on Substances that Deplete the Ozone Layer'.

61   David Doniger, NRDC, presentation at conference on 'Substitutes and Alternatives to CFCs and Halons', sponsored by the Conservation Foundation and the US Environmental Protection Agency, Washington, DC, 14 January 1988.

62   'Economic Implication of Potential Chlorofluorocarbon Restrictions: Final Report', Putnam, Hayes & Bartlett Inc, December 2, 1987.

63   Andersen et al (1997), op cit; Andersen (1998), op cit; Le Prestre et al (1998), op cit; Cook (1996), op cit.

64   See: Andersen and Morehouse (ASHRAE); Cook (1996), op cit; and ICOLP...

65   See: Carey, J (1995) 'Why Business Doesn't Back the GOP backlash on the Ozone', *Business Week*, 24 July, p 47. Examples of industry openly countering ozone sceptics include: Kevin Fay, 'Testimony to the House Committee on Science, Subcommittee on Energy and Environment', 20 September 1995 and Dave Stirpe, 'Memo to Alliance Members', 2 October 1995.

CHAPTER 6

1   UNEP/Ozl.Pro/2/3, Annex IV, Appendix II.
2   The MLF website, www.unmfs.org, presents the complete history of projects from inception, proposal, approval, implementation and completion.
3   UNEP/Ozl.Pro/Excom/8/29, 21 October 1992.
4   UNEP/Ozl.Pro/Excom/13/47, 27 July 1994.
5   UNEP/Ozl.Pro/Excom/22/79/Rev 1, 30 May 1997.
6   UNEP/Ozl.Pro/Excom/16/20, 17 March 1995.
7   UNEP/Ozl.Pro/Excom/28/55, 16 July 1999.
8   UNEP/OZL.Pro/Excom/15/45, 16 December 1994.
9   UNEP/Ozl.Pro/Excom/17/60, 28 July 1995.
10   UNEP/Ozl.Pro/Excom/20/72, 18 October 1996.
11   COWIConsult and Goss Gilroy Inc (1995) *Study of the Financial Mechanism of the Montreal Protocol*, UNEP Ozone Secretariat, Nairobi.
12   The most comprehensive comparisons of ODS regulations and related measures are found in UNEP, MLF, SEI (1996) *Regulations to Control Ozone-Depleting Substances: A Guidebook*, and the updated 2000 edition, published by the Division of Trade, Industry and Economics (DTIE), UNEP, Paris. Additional descriptions of national programmes with some comparisons to other countries are found on national websites.
13   Countries with taxes or fees on ODSs include Australia, Belarus, Bulgaria, the Czech Republic, Denmark, Hungary, Republic of (South) Korea, the Seychelles, Singapore, South Africa, Sweden, Thailand, Vietnam and the USA.
14   Australia, Bulgaria, Canada, China, the Czech Republic, Gambia, Germany, Jamaica, Jordan, Malaysia, New Zealand, Romania, Spain, Sweden, Syria and the USA require special environmental labelling for products containing ODSs; and Malaysia, Syria and the USA require labels of products made with ODSs. Some countries also require or encourage labelling of products not containing ODSs.
15   An appendix included in a CD-ROM available from the Ozone Secretariat lists many of the most influential NGOs and explains their significant contributions to ozone-layer protection; also, see Chapter 9 and the timeline in Appendix 1.
16   The seminars were entitled 'Trilateral Meetings of Japan, US and the venue country' as follows: Bangkok, Thailand (11–13 March 1992; 2–4 February 1994; 11 September 1995), Kuala Lumpur, Malaysia (26–28 April 1993), Jakarta, Indonesia (7–9 February 1994; 4–5 November 1996), Hanoi/Ho Chi Minh City, Vietnam (14–20 September 1995), and Manila, the Philippines (11 November 1996).
17   Hsian (18–19 December 2000), Hong Kong (20–22 June 1994), Nanjing (1–6 December 1997 and 1999), Shanghai (16–18 December 1997), Shenzhen (20–21 June 1994; 8–12 December 1998; and 26–30 June 2000).
18   Companies making the pledge included: Asahi Glass, Asea Brown Boveri, AT&T, British Petroleum, British Petroleum Vietnam, Carrier, the Coca-Cola Company, Daihatsu, DuPont, Ford, Fuji Electric, Fuji Heavy Industries, Hewlett-Packard, Hino, Hitachi, Honda, Honeywell, ICI, Isuzu, Kawasaki Heavy Industries, Lufthansa, Matsushita Electric, Mazda, Meidensha, 3M, Mitsubishi Electric, Mitsubishi Heavy Industries, Mitsubishi Motors, Motorola, Nissan, Nissan Diesel, Nortel (Northern Telecom), Sanyo, Seiko Epson, Sharp, Suzuki, Taiwan Fertilizer Company, Toshiba, Toyota, Trane, Yamaha, Yaskawa, Vulcan Materials and UNISYS.
19   Algeria, Argentina, Bangladesh, Brazil, Chile, China, Columbia, Costa Rica, Cote d'Ivoire, Egypt, Ghana, Hong Kong, India, Indonesia, Iran, Kenya, Democratic Peoples Republic of (North) Korea, Lao PDR, Malaysia, Maldives, Mauritius, Mexico, Mongolia, Nepal, Nigeria, Pakistan, Palestine, Papua New Guinea, Peru, Philippines, Samoa, Saudi Arabia,

Senegal, Sri Lanka, Syria, Tanzania, Thailand, Turkey, Venezuela, Vietnam and Western
Samoa.

CHAPTER 7

1   UNEP/Ozl.Pro/Impcom/1/2, 11 December 1990; UNEP/Ozl.Pro/ImPcom/2/3 of 14
    April 1991.
2   UNEP/Ozl.Pro/Impcom/7/2, 16 November 1993.
3   UNEP/Ozl.Pro/Impcom/9/2, 5 October 1994.
4   UNEP/Ozl.Pro/Impcom/21/3, 7 December 1998.
5   UNEP/Ozl.Pro/Impcom/27/4, October 2001.
6   The countries of Eastern and Central Europe and the former Soviet Union classified as
    developing are: Albania, Bosnia and Herzegovina, Croatia, the Former Yugoslav Republic
    of Macedonia, Georgia, Kyrgyzstan, Moldova, Romania and Yugoslavia.
7   UNEP/Ozl.Pro/Impcom/11/1, 14 September 1995.
8   UNEP/Ozl.Pro/Impcom/12/3, 21 December 1995.
9   Geneva, March 1996, UNEP/Impcom/13/3, 28 March 1996.
10 UNEP/Ozl.Pro/Impcom/23/3, 10 July 1999.
11 UNEP/Ozl.Pro/Impcom/20/4, 9 July 1998.
12 UNEP/Ozl.Pro/Impcom/3/3, 15 April 1992.
13 UNEP/Ozl.Pro/Impcom/4/2, 6 October 1992.
14 UNEP/Ozl.Pro.Impcom/18/3, 4 June 1997.
15 Bangladesh, Chad, Comoros, Dominican Republic, Honduras, Kenya, Mongolia,
    Morocco, Niger, Nigeria, Oman, Papua New Guinea, Paraguay, Samoa, Solomon Islands.
16 UNEP/Ozl.Pro/Impcom/27/4, 13 October 2001.

CHAPTER 8

1   The authors are particularly grateful to K Madhava Sarma, who made particular
    contributions to the identification and analysis of significant events in ozone history; to
    Stephen O Andersen, who assisted in the statistical interpretation and presentation of
    statistical results; and to Lani Sinclair, who is a masterful editor. Nancy Reichman proved
    valuable in helping frame the questions and in proving scientific contest for evaluating the
    importance of media.
2   Sachsman, D (2000) 'The Role of Mass Media in Shaping Perceptions and Awareness of
    Environmental Issues', *Climate Change Communication Conference Proceedings*, University of
    Waterloo and Environment Canada, A2, p1.
3   For more discussion about the influence of news coverage on public policy generally and
    environmental issues in particular see Ader, C (1995) 'A Longitudinal Study of Agenda
    Setting for the Issue of Environmental Pollution', *Journalism & Mass Communications
    Quarterly*, vol 72, no 2, p300; Gunter, B (2000) *Media Research Methods: Measuring Audiences,
    Reactions and Impact*, Sage Publications, London, Thousand Oaks and New Delhi, p199;
    Lowe, P and Morrison, D (1984) 'Bad News or Good News: Environmental Politics and
    the Mass Media', *Sociological Review*, vol 32, p85; McCombs, M and Shaw, D (1977)
    '"Agenda Setting" and the Political Process', in Shaw, D and McCombs, M (eds) *The
    Emergence of American Political Issues: The Agenda-Setting Function of the Press*, West Publishing,
    Minneapolis, Minnesota, p149; Nelkin, D (1995) *Selling Science: How the Press Covers Science
    and Technology*, W H Freeman and Company, New York, p13, p73; Strodthoff, G, Hawkins,
    R and Schoenfeld, A (1985) 'Media Roles in a Social Movement: A Model of Ideology
    Diffusion', *Journal of Communication*, vol 75, p136.

4   Unfortunately, comparable public-interest polls of ozone concern are not available for a significant number of years. However, ten years after the signing of the Montreal Protocol – and after the 1994 phase-out of halons and the 1996 phase-out of other ODSs – the Roper Poll asked 1000 American adults to estimate the priority of ozone depletion on a scale of 1 to 10, in which 1 meant the problem was a low priority, and 10 meant the problem was a high priority. In each year, ozone ranked 7 or higher on the scale (1997 mean = 7.3; 1998 mean = 7.0; and 2000 mean = 7.4). Source: Roper Center at the University of Connecticut.
5   The Lexis®-Nexis® 'News Group File, 'All' database using the 'Terms and Connectors' search scheme was undertaken on 25 and 26 February 2001 to identify all stories about the ozone layer for the period 1970–2000. The specific search scheme that provided the most on-point 'hits' was searching for the word 'ozone' within three words (in front or back but not counting words such as 'the', 'and', and 'or') of 'layer'. The use of the phrase 'ozone layer' was not a satisfactory search scheme since it would miss any references to 'layer of ozone'.
6   Ungar, S (2000) 'Why Climate Change is not in the Air: Popular Culture and the Whirlwind Effect', in *Climate Change Communication Conference Proceedings*, University of Waterloo and Environment Canada, A2, p10.
7   Between 1981 and 2001, UNEP issued more than 100 press releases on scientific, diplomatic and technical aspects of stratospheric ozone-layer protection.
8   All NGOs mentioned are listed in Table 8.1, which is found on a CD-ROM available from the Ozone Secretariat.
9   Brodeur, P (1986) 'Annals of Chemistry: In the Face of Doubt', *New Yorker*, 9 June, p70.
10   'Ozone controversy' (1975) *Wall Street Journal*, 3 December, p1; 'Spray-Can Scare: The Latest Findings' (1975) *US News & World Report*, 29 September, p62.
11   'Why aerosols are under attack' (1975) *Business Week*, 17 February, p50.
12   'Ozone controversy casts global shadow' (1975) *Chemical Week*, 15 October, p47.
13   Malakoff, D and Phillips, S (1996) 'The Spray Can Ban', in Cook, E (ed) *Ozone Protection in the United States: Elements of Success*, World Resources Institute, Washington, DC, p18.
14   'FDA requires fluorocarbon warnings' (1977) *Facts on File World Digest*, 30 April, 323 E1.
15   Brodeur (1986) op cit, p80.
16   Brodeur (1986) ibid, p84.
17   Jansen, A, Osland, O and Hanf, K (1998) 'Environmental Challenges and Institutional Change', in Hanf, K and Jansen, A (eds) *Governance and Environment in Western Europe*, Longman, New York, p291.
18   Ibid.
19   Benedick, Richard E (1998) *Ozone Diplomacy: New Directions in Safeguarding the Planet*, Harvard University Press, Cambridge, MA, p5.
20   Extracts from the 17 September 1987 'day after signing' coverage by several international news organizations are found in Table 8.2 on the Ozone Secretariat CD-ROM.
21   Benedick (1998), op cit, p102.
22   Ibid, p123.
23   Table 8.3, on the Ozone Secretariat CD-ROM, contains extended extracts from key editorials.
24   Butler, P (1989) 'London Conference on the Ozone Layer; CFC use "definitely to blame for ozone hole"', *Financial Times*, 6 March, p12.
25   Stammer, L (1989) 'Global Talks on Ozone Described as Successful', *Los Angeles Times*, 8 March, Part 1, p6.
26   Table 8.4, on the Ozone Secretariat CD-ROM, contains extracts from editorials published in the aftermath of the conference.
27   For example, Dumanoski, D (1990) 'In shift, US to Aid World Fund on Ozone; Opponents, Led by Sununu, Yield to Global Pressure', *Boston Globe*, 16 June, p1.

28   Clover, C (1990) 'Thatcher Promises 1.5 Million [pounds] to Help Phase Out CFCs', *Daily Telegraph*, 28 June, p4.

29   Johnson, C (1990) 'India Wins Hard-Fought Pledge from Ozone Conference', UPI, 30 June.

30   Benedick (1998) op cit, p165.

31   Riddell, P and Thomas, D (1990) 'US Backs Global Fund To Help Protect Ozone Layer', *Financial Times*, 18 June; Inter Press Service (1990) 'Environment: Group Calls on North to Help South Stop CFC Use', 19 June.

32   See Table 8.5, on the Ozone Secretariat CD-ROM, for extracts from several editorials and opinion pieces.

33   Brown, P (1997) 'Illegal Trade in Banned CFC Gases is Exposed', *Guardian*, 4 September, p8.

CHAPTER 9

1   Content and editorial contributions from Stephen O Andersen (EPA), Fernando Bejarano (RAPAM), Brent Blackwelder (FOE USA), Larry Bohlen (FOE USA), Elizabeth Cook (FOE and WRI), David Doniger (NRDC), Arjun Dutta (CUTS), János (John) Maté (GP), Alan Miller (NRDC, WRI and CGC), Anne Schonfield (PANNA), Rajendra Shende (UNEP), and Miguel Stutzin (CODEFF). For a summary of the Greenpeace Ozone Campaign's activities and publications see: Maté, János (John) (2001), *Making a Difference: A Non-Governmental Organization's Campaign to Save the Ozone Layer: Case Study of the Greenpeace Ozone Campaign*, Greenpeace International, Amsterdam.

2   Hudock, A (1999) *NGOs and Civil Society: Democracy by Proxy?* Polity Press, Cambridge, p1.

3   Conca, Ken (1996) 'Greening the UN: Environmental Organizations and the UN System', in Weiss, T and Gordenker, L (eds), *NGOs, the UN, and Global Governance*, Lynne Rienner Publishers, Boulder, Colorado, p113. The legal significance of UNEP's recognition of NGOs is illustrated by the rules of procedure of the Executive Committee for the Multilateral Fund that provide for notification to qualified NGOs of meetings where they may participate as observers (without the right to vote, upon invitation of the Chairman, and if there is no objection from committee members present). See Kiss, A and Shelton, D (2000) (2nd edn) *International Environmental Law*, Transnational Publishers, Ardsley, NY, p135.

4   Annan, Kofi, *Report to the 53rd Session of the UN General Assembly* (Item 58 'Strengthening of the UN System') Document A/53/170, 10 July 1998.

5   Agricultural, environmental, health and labour NGOs involved in ozone protection issues included: American Federation of Labor Congress of Industrial Organizations (USA), American Lung Association (USA), Arab Office for Youth and Environment (Egypt), Asthma and Allergy Foundation of America (USA), Allergy and Asthma Network (USA), Australian Conservation Foundation (Australia), Bio-Integral Resource Center (USA), Californians for Alternatives to Toxics (USA), Canadian Environmental Network (Canada), Center for Environmental Education (India), Center for Environmental Friends (Qatar), Center for Global Change (USA), Center for International Environmental Law (USA), Center for Science and Environment (India), Climate Institute (USA), Climate Action Network (USA), Climate Network Europe (Belgium), Comite Nacional Pro Defensa de la Fauna y Flora (Chile), Consumer Information Network (Kenya), Consumer Unity and Trust Society (India), Coordination Unit for the Rehabilitation of the Environment (Malawi), Dansk Naturfredning (Denmark), Earthcare Africa (Kenya), ENDA-Ethiopia (Ethiopia), Environmental Conservation Association of Zambia (Zambia), Environmental Defense Fund (USA), Environmental Health Coalition (USA),

Environmental Investigation Agency (UK), Environmental Liaison Center International (Kenya), Environmental Working Group (USA), Farmworker Association of Florida (USA), Farmworker Self-Help (USA), Friends of the Earth International (Netherlands), Florida Consumer Action Network (USA), Fundación Agricultura y Medio Ambiente (Dominican Republic), Green Africa (Kenya), Greenpeace International (Netherlands), Health and Environment Watch (Kenya), International Brotherhood of Teamsters (USA), International Council of Environmental Law (Kenya), Investor Responsibility Research Center (USA), Instituto Regional de Estudios en Sustancias Tóxicas (Costa Rica), Israel Economic Forum on the Environment (Israel), Israel Union for Environmental Defense (Israel), Japan's Save the Ozone Network (Japan), Kenouz Sinai Environment Protection Sharkiya (Egypt), Lebanese Environment Forum (Lebanon), Mothers for Asthma (USA), National Association of Physicians for the Environment (USA), National Safety Council/Environmental Health & Safety Institute (USA), Natural Resources Defense Council (USA), Legal Environmental Assistance Foundation (USA), Ozone Action (USA), Pesticide Action Network North America (USA), Pesticide Action Network Africa (Senegal), Pesticide Action Network (Philippines), Pesticide Trust (UK), Polish Ecological Club (Poland), Political Ecology Group (USA), Proconsumers (Mozambique), RAPALMIRA (Colombia), Royal Institute of International Affairs (UK), Red de Accion sobre Plaguicidas y Alternativas en Mexico (Pesticide Action Network, Mexico), Sierra Club (USA), Sustainable Agriculture Food and Environment (UK), Transnational Resource and Action Center (USA), Uganda Consumers Protection Association (Uganda), United Farmworkers (USA), Worldwatch Institute (USA), World Resources Institute (USA) and WWF (Switzerland).

6   Cagin, Seth and Dray, Philip (1993) *Between Earth and Sky*, Pantheon Books, New York, explains how opposition was first organized against the effect of sonic booms. However, scientists and citizen activists quickly came to appreciate that noise would damage natural and cultural treasures and that high-altitude emissions would affect the Earth's climate and its fragile ozone layer.

7   For a discussion of the modern role of NGOs and how they have evolved over time, see Higgott, Richard A, Underhill, Geoffrey R D and Bieler, Andreas (eds) (2000) *Non-State Actors and Authority in the Global System*, Routledge, London. For a discussion of how environmental NGOs have influenced the World Bank see Cleary, Seamus (1996) 'The World Bank and NGOs' in Willetts, Peter (ed) *The Conscience of the World: The Influence of Non-Governmental Organizations in the UN System*, the Brookings Institution, Washington, DC, p92.

8   Alan Miller worked at NRDC 1978–1984, at the World Resources Institute 1984–1986, at the Center for Global Change 1989–1996, and is now at the Global Environment Facility.

9   Maté (2001), op cit.

10   Benedick, Richard E (1998) 'The Montreal Protocol as a New Approach to Diplomacy', in Le Prestre, Philippe G, Reid, John D and Morehouse, E Thomas Jr (eds) *Protecting the Ozone Layer: Lessons, Models and Prospects*, Kluwer Academic Publishers, Boston, p88.

11   Sand, Peter H (1990) *Lessons Learned in Global Environmental Governance*, World Resources Institute, Washington, DC.

12   See Morphet, S (1996) 'NGOs and the Environment', in Willetts, op cit, p141. Morphet divides environmental NGOs that participate in UN activities into four categories: scientific; high-status, high-expertise; major international 'political'; and influential national.

13   Princen, Thomas (1994) 'NGOs: Creating a Niche in Environmental Diplomacy', in Princen, Thomas and Finger, Matthias (eds) *Environmental NGOs in World Politics*, Routledge, London, pp35–36.

14   Hurrell, A and Kingsbury, B (1992) 'The International Politics of the Environment: An Introduction', in Hurrell, A and Kingsbury, B (eds) *The International Politics of the Environment: Actors, Interests, and Institutions*, Clarendon Press, Oxford, UK, p10.

15   Greenpeace has nearly 6 million members worldwide, affiliated offices in 30 countries and an annual budget of about $100 million; Friends of the Earth International has half a million members and national affiliate groups in 46 countries. A small number of environmental think-tanks such as the World Resources Institute (WRI) get their influence primarily from reputation and transnational connections. See Conca (1996), op cit, pp106–107; Santos, M (1999) *The Environmental Crisis*, Greenwood Press, New York, estimates the following memberships for 1995: Environmental Defense Fund had 250,000 members (p169); Friends of the Earth had 35,000 members (p170); Greenpeace had 1,500,000 members (p171); Natural Resources Defense Council had 170,000 members (p176); Sierra Club had 550,000 members (p178); WWF had 5,000,000 members (p180).

16   David Doniger worked at NRDC from 1978–1992; he then worked for eight years in environmental posts in the Clinton Administration, including as Director of Climate Change Policy for EPA; he is now the Policy Director of NRDC's Climate Center.

17   Stoel, Thomas B Jr, Miller, Alan S and Milroy, Breck (1980) *Fluorocarbon Regulation: An International Comparison*, Lexington Books, Lexington, MA.

18   Gupta, Joyeeta and Gagnon-Lebrun, Frédéric (2000) 'Non-State Actors in International Environmental Negotiations: Increasing transparency or creating confusion', in Gupta, J *'On Behalf of My Delegation,...' A Survival Guide for Developing Country Climate Negotiators*, Institute for Environmental Studies, Amsterdam.

19   Edelman Public Relations Worldwide, Press Release, 1 December 2000.

20   Brunnée, J (1988) *Acid Rain and Ozone Layer Depletion: International Law and Regulation*, Transnational Publishers Inc, New York, p257.

21   Wuori, M (1997) 'On the Formative Side of History: The Role of Non-Governmental Organizations', in Rolén, M, Sjöberg, H and Svedin, U (eds) *International Governance on Environmental Issues*, Kluwer Academic Publishers, New York, pp159–166.

22   Bramble, B and Porter, G 'Non-Governmental Organizations and the Making of US International Environmental Policy', in Hurrell and Kingsbury (1992), op cit, pp340–341.

23   Andersen, Stephen O and Frech, Clayton 'Champions of the World, US Environmental Protection Agency' (EPA430-R-97-023) August 1997; Andersen, Stephen O, 'Newest Champions of the World, US Environmental Protection Agency' (EPA430-K-98-003) August 1998; Cook, Elizabeth (ed) (1996) *Ozone Protection in the United States: Elements of Success*, World Resources Institute, Washington, DC; and French, H F (1997) 'Learning From the Ozone Experience', in L Brown et al *State of the World 1997: A Worldwatch Institute Report on Progress Towards a Sustainable Society*, Earthscan Publications, London, pp159–162.

24   Wuori (1997), op cit, pp159–169; Cook, Elizabeth and Kimes, Jeffrey (1996)'Dangling the Carrot' in Cook (1996), op cit, p59.

25   For an elaboration and analysis of the influence of Greenpeace on CFC producers and their employees, see Wapner, P (1996) *Environmental Activism and World Civic Politics*, State University of New York Press, New York, pp53–54.

26   Cagin and Dray (1993), op cit, pp193–194.

27   Wapner (1996), op cit, pp127–128.

28   Benedick, R E (1998) *Ozone Diplomacy: New Directions in Safeguarding the Planet*, Harvard University Press, Cambridge, MA, pp311–312.

29   European Commission (2001) Communication to the Council, the European Parliament, the Economic and Social Committee and the Committee of the Regions, 'Environment 2010: Our Future, our Choice', COM 31 Final, 24 January, p62.

CHAPTER 10

1   Buckminster Fuller, R (1969) *Operating Manual for Spaceship Earth*, Lars Müller Publishers, New York (reissued 2000).
2   GEF (1998) *Valuing the Environment*, Global Environment Facility, Washington, DC, pp134–154; World Bank (1997) *Partnerships for Global Eco-systems Management*, the World Bank, Washington, DC, pp 45–77; Tolba, Mostafa K with Rummel-Bulska, Iwona (1998) *Global Environmental Diplomacy*, MIT Press, Cambridge, MA, pp55–88; Benedick, Richard (1998) *Ozone Diplomacy: New Directions in Safeguarding the Planet*, Harvard University Press, Cambridge, MA, chapter 19.
3   Lang, Winfried (ed) (1996) *The Ozone Treaties and Their Influence on the Building of International Environmental Regimes*, Federal Ministry for Foreign Affairs, Vienna.
4   From the Foreword to *Global Environmental Diplomacy* by Mostafa K Tolba, op cit.
5   Le Prestre, Philippe G, Reid, John D and Morehouse, E Thomas Jr (eds) (1998) *Protecting the Ozone Layer: Lessons, Models and Prospects*, Kluwer Academic Publishers, Boston.
6   One of the four Vedas, religious scriptures of Hindus (around 2000 BC), from Panikkar, R (1994) *Vedic Experience*, Motilal Banarsidas Publishers, India, p636.
7   *Complete Verse*, First Anchor Books edition, 1989, Random House, New York.
8   *The Prophet*, 1957, Alfred A Knopf, New York.
9   *The Outline of History*, 1920, Garden City Doubleday, New York, 1961, Book Club Edition.
10   Speech at the Mansion House, London, 10 November 1942.

APPENDIX

1   US Department of Transportation (1975) 'Climate Impact Assessment Program: Environmental Impacts of Supersonic Flight: Biological and Climate Effects of Aircraft Emissions in the Stratosphere', USDT, Washington, DC.
2   Consortium members include the National Aeronautics and Space Administration (NASA), National Oceanic and Atmospheric Administration (NOAA), United Kingdom Department of the Environment, United Nations Environment Programme (UNEP) and World Meteorological Organization (WMO).
3   Daniel Albritton (USA) and Robert Watson (USA) co-chaired the first Panel for Scientific Assessment; Jan C van der Luen (Netherlands) and Manfred Tevini (Germany) co-chaired the first Panel for Environmental Assessment; Victor Buxton (Canada) chaired and Stephen O Andersen (USA) vice-chaired the Panel for Technical Assessment, and George Strongylis (EC Commission) chaired and Stephen O Andersen (USA) and John S Hoffman (USA) vice-chaired the Panel for Economic Assessment. In 1990, Stephen Lee-Bapty (UK) replaced G Victor Buxton as Co-Chair of the TEAP. In 1992 Lambert Kuijpers (The Netherlands) replaced Stephen Lee-Bapty as Co-Chair of the TEAP and Pieter Aucamp (South Africa) was added to the SAP, Andre P R Cvijak (Brazil) to the TEAP, and Xiaoyan Tang (China) to the EAP. In 1993, Suely Carvalho (Brazil) replaced Andre P R Cvijak as Co-Chair of the TEAP. In 1997, Parties appointed Gerard Megie (France) as the fourth Co-Chair of the SAP. In 2000, Ayite-Lo Nohende Ajavon (Togo) replaced Pieter Aucamp on the SAP.
4   The TEAP currently has six Technical Options Committees: Aerosols, Sterilants, Miscellaneous Uses and Carbon Tetrachloride; Flexible and Rigid Foams; Halons; Methyl Bromide; Refrigeration, Air Conditioning and Heat Pumps; and Solvents, Coatings and Adhesives. There are also occasional Working Groups and Task Forces that consider special technical topics.

5   For elaboration, see S M Carvalho 'Technology Assessment for the Montreal Protocol';
     S Rand and L Singh, 'Importance of the TEAP in Technology Cooperation'; L Kuijpers,
     H Tope, J Banks, W Brunner and A Woodcock, 'Scientific Objectivity, Industrial Integrity
     and the TEAP Process'; and R Van Slooten, 'TEAP Terms of Reference' in Le Prestre, P
     G, Reid, J D and Morehouse, E T (eds) (1998) *Protecting the Ozone Layer: Lessons, Models and
     Prospects*, Kluwer Academic Publishers, Boston, MA.

6   Of all TEAP, TOC, and Working Group members, 75 per cent are from industry, 10 per
     cent from environment ministries, 5 per cent from academic universities, and 10 per cent
     from technical institutes, industry and agriculture ministries, and non-governmental
     organizations. The 1989 Assessment explicitly excluded participation of experts from
     companies producing CFCs.

List of acronyms and abbreviations

AFC	Australian Fluorocarbon Council (was AFCAM)
AFCAM	Association of Fluorocarbon Consumers and Manufacturers (now AFC)
AFCFCP	Alliance for Responsible CFC Policy (now ARAP)
AFEAS	Alternative Fluorocarbons Environmental Acceptability Study
ARAP	Alliance for Responsible Atmospheric Policy (was AFCFCP)
ASHRAE	American Society of Heating, Refrigerating, and Air-Conditioning Engineers
ASTM	American Society of Testing and Materials
ATOC	Aerosols, Sterilants, Miscellaneous Uses and Carbon Tetrachloride Technical Options Committee (TEAP)
BBC	British Broadcasting Corporation
CCOL	Coordinating Committee on the Ozone Layer (UNEP)
CEIT	countries with economies in transition
CFC	chlorofluorocarbon
CFM	chlorofluoromethane
CMA	Chemical Manufacturers Association
COAS	Council of Atmospheric Sciences
COP	Conference of the Parties (to the Vienna Convention)
COP 1	first meeting of the Conference of the Parties (Helsinki, Finland, April 1989)
COP 2	second meeting of the Conference of the Parties (Nairobi, Kenya, June 1991)
COP 3	third meeting of the Conference of the Parties (Bangkok, Thailand, November 1993)
COP 4	fourth meeting of the Conference of the Parties (San Jose, Costa Rica, November 1996)
COP 5	fifth meeting of the Conference of the Parties (Beijing, China, November–December 1999)
COPD	chronic obstructive pulmonary disease
CTC	carbon tetrachloride
CUTS-CSPAC	Consumer Unity & Trust Society – Centre for Sustainable Production and Consumption
DLA	Defense Logistics Agency (US)
DME	dimethyl ether
EAP	Economic Assessment Panel
EC	European Commission
ECE	Economic Commission for Europe (UN)

ED	Executive Director
EDF	Environmental Defense Fund (US)
EEAP	Environmental Effects Assessment Panel
EEC	European Economic Community
EIA	Environmental Investigation Agency
ENGO	environmental non-governmental organization
EOC	Economic Options Committee (TEAP)
EPA	Environmental Protection Agency (US)
EU	European Union
FAA	Federal Aviation Administration (US)
FAO	Food and Agriculture Organization of the United Nations
FC	fluorocarbon
FCCC	Framework Convention on Climate Change (UN)
FDA	Food and Drug Administration (USA)
FOE	Friends of the Earth
FTOC	Foams Technical Options Committee (TEAP)
GAW	Global Atmosphere Watch (established by WMO)
GATT	General Agreement on Tariffs and Trade
GC	Governing Council (UNEP)
GEF	Global Environment Facility
GEMS	Global Environment Monitoring System
GHG	greenhouse gas
GTZ	Deutsche Gesellschaft für Technische Zusammenarbeit
GWP	global warming potential
HARC	Halon Alternatives Research Corporation (US)
HBFC	hydrobromofluorocarbon
HCFC	hydrochlorofluorocarbon
HFC	hydrofluorocarbon
HFE	hydrofluoroether
HTOC	Halons Technical Options Committee (TEAP)
ICAO	International Civil Aviation Organization (UN)
ICEL	International Cooperative for Environmental Leadership (was ICOLP)
ICI	Imperial Chemical Industries
ICOLP	International Cooperative for Ozone Layer Protection (now ICEL)
ICSU	International Council of Scientific Unions
IGPOL	Industry Group for the Protection of the Ozone Layer (European Community)
IIR	International Institute of Refrigeration
IMO	International Maritime Organization (UN)
INGO	industry non-governmental organization
IOC	International Ozone Commission
IPAC	International Pharmaceutical Aerosol Consortium
IPCC	Intergovernmental Panel on Climate Change (UN Framework Convention on Climate Change)
IPM	integrated pest management
ISO	International Organization for Standardization

JACET	Japan Association of Cleaning Engineering and Technology
JAHCS	Japan Association for Hygiene of Chlorinated Solvents
JASON	Japan's Save the Ozone Network
JEMA	Japan Electrical Manufacturers' Association
JICOP	Japan Industrial Conference for Ozone Layer Protection
K	kelvin (unit of thermodynamic temperature)
LCCP	Life-cycle climate performance
MACS	Mobile Air Conditioning Society Worldwide
MBAN	Methyl Bromide Alternatives Network
MBTOC	Methyl Bromide Technical Options Committee (TEAP)
MDI	metered-dose inhaler
MITI	Ministry of International Trade and Industry (Japan) (now Ministry of Economy, Trade and Industry (METI))
MLF	Multilateral Fund for the Implementation of the Montreal Protocol
MOP	Meeting of the Parties (to the Montreal Protocol)
MOP 1	first Meeting of the Parties (Helsinki, Finland, May 1989)
MOP 2	second Meeting of the Parties (London, UK, June 1990)
MOP 3	third Meeting of the Parties (Nairobi, Kenya, June 1991)
MOP 4	fourth Meeting of the Parties (Copenhagen, Denmark, 1992)
MOP 5	fifth Meeting of the Parties (Bangkok, Thailand, November 1993)
MOP 6	sixth Meeting of the Parties (Nairobi, Kenya, October 1994)
MOP 7	seventh Meeting of the Parties (Vienna, Austria, December 1995)
MOP 8	eighth Meeting of the Parties (San Jose, Costa Rica, December 1996)
MOP 9	ninth Meeting of the Parties (Montreal, Canada, September 1997)
MOP 10	tenth Meeting of the Parties (Cairo, Egypt, November 1998)
MOP 11	eleventh Meeting of the Parties (Beijing, China, December 1999)
MOP 12	twelfth Meeting of the Parties (Ouagadougou, Burkina Faso, December 2000)
MOP 13	thirteenth Meeting of the Parties (Colombo, Sri Lanka, October 2001)
MOPIA	Manitoba Ozone Protection Industry Association Inc
MP	Montreal Protocol
NASA	National Aeronautics and Space Administration (US)
NCAR	National Center for Atmospheric Research (US)
NDSC	Network for the Detection of Stratospheric Change
NFPA	National Fire Protection Association
NGO	non-governmental organization
NOAA	National Oceanic and Atmospheric Administration (US)
NOZE	National Ozone Expedition
nPB	n-propyl bromide
NRDC	Natural Resources Defense Council (US)
O_3	ozone
ODP	ozone-depletion potential
ODS	ozone-depleting substance
OECD	Organisation for Economic Co-operation and Development
OEWG	Open-Ended Working Group (of the Parties to the Montreal Protocol)
OORG	Ozone Operations Resource Group (World Bank)

PAFT	Program for Alternative Fluorocarbon Toxicity Testing
PAN	Pesticide Action Network
PANNA	Pesticide Action Network of North America
PATF	Process Agent Task Force (TEAP)
PFC	perfluorocarbon
ppbv	parts per billion volume
ppt	parts per trillion
QPS	quarantine and pre-shipment
R&D	research and development
RMP	refrigeration management plan
RTOC	Refrigeration, Air-Conditioning and Heat Pumps Technical Options Committee (TEAP)
SACEP	South Asia Cooperative Environment Programme
SAE	Society of Automobile Engineers
SAGE	Stratosphere Aerosol and Gas Experiment
SAP	Scientific Assessment Panel (UNEP)
SAWTEE	South Asia Watch on Trade, Economics and Environment
SBSTA	Subsidiary Body for Scientific and Technological Advice (UN Framework Convention on Climate Change)
SBUV	solar backscatter ultraviolet
SCOPE	Standing Committee on the Problems of the Environment
SEI	Stockholm Environment Institute
SEPA	State Environmental Protection Administration (China)
SESAME	Second European Stratospheric Arctic and Mid-altitude Experiment
SMEs	small and medium-sized enterprises
SNAP	Significant New Alternatives Program (US EPA)
SORG	Stratospheric Ozone Research Group (UK)
SST	supersonic transport aircraft
STOC	Solvents, Coatings and Adhesives Technical Options Committee (TEAP)
TCA	1,1,1-trichloroethane (also methyl chloroform)
TEAP	Technology and Economic Assessment Panel (UNEP Montreal Protocol)
TEWI	total equivalent warming impact
TOC	Technical Options Committee (of TEAP)
TOMS	total ozone-mapping spectrometer
TRP	Technology Review Panel (merged with Economic Panel to become TEAP)
UARS	upper atmosphere research satellite
UN	United Nations
UNCED	United Nations Conference on Environment and Development (known as the Earth Summit)
UNDP	United Nations Development Programme
UNEP	United Nations Environment Programme
UNESA	United Nations Department of Economic and Social Affairs
UNESCO	United Nations Educational, Scientific and Cultural Organization
UN ESCWA	United Nations Economic and Social Commission for Western Asia

UNFCCC	United Nations Framework Convention on Climate Change
UNIDO	United Nations Industrial Development Organization
UNOPS	United Nations Office for Project Services
US DoD	United States Department of Defense
US EPA	United Stated Environmental Protection Agency
UV	ultraviolet
UV-A	ultraviolet A radiation
UV-B	ultraviolet B radiation
UV-C	ultraviolet C radiation
WHO	World Health Organization
WMO	World Meteorological Organization
WRI	World Resources Institute
WWF	*formerly* World Wide Fund For Nature (World Wildlife Fund in Canada and USA)

Glossary

aerosol products Containers filled with ingredients and an ODS, hydrocarbon or HFC propellant to release the product in a fine spray.

Amendments and Adjustments 'Amendments' (Articles 9 and 10 of the Convention) must be ratified by individual Parties and are binding, after entry into force of the Amendment, only on those Parties that ratify the Amendment. 'Adjustments' (Article 2, Paragraph 9 of the Montreal Protocol) to the ozone depletion potential or control measures of substances already listed in the annexes to the Protocol, once duly approved by Meetings of the Parties and notified, will be binding on all the Parties, without any process of ratification, after the expiry of six months after the notification.

Annex A substances Eight chemicals: five CFCs as Group I and three halons as Group II.

Annex B substances Three groups: Group I, ten CFCs; Group II, carbon tetrachloride; and Group III, methyl chloroform.

Annex C substances Three groups: Group I, 40 HCFCs; Group II, 34 HBFCs; and Group III, bromochloromethane.

Annex E substance Methyl bromide.

aqueous cleaning Cleaning parts with water.

Article 5 Party A developing-country Party to the Montreal Protocol whose annual calculated level of consumption is less than the limits prescribed in Article 5, Paragraph 1 – 0.3kg per capita of the controlled substances in Annex A, and less than 0.2kg per capita of the controlled substances in Annex B, on the date of the entry into force of the Montreal Protocol for that country or any time thereafter. An Article 5 Party is entitled to delay its implementation of the control measures of Article 2 by a period specified in Article 5 of the Protocol, in order to meet its basic domestic needs.

atmospheric lifetime A measure of the time a chemical remains without breaking up, once released into the atmosphere.

basic domestic needs The Protocol does not define this term. The first and seventh Meetings of the Parties made clarificatory decisions on this term.

blowing agents Expanding gases that foam the plastic matrix, creating open or closed foam cells.

carbon tetrachloride (CCl_4) A solvent with an ODP of approximately 1.1 that is controlled under the Montreal Protocol. It is used primarily as a feedstock material for the production of other chemicals.

cataract Damage to the eye in which the lens is partly or completely clouded, impairing vision and sometimes causing blindness. Exposure to ultraviolet radiation can cause cataracts.

chlorofluorocarbons (CFCs) A family of organic chemicals composed of chlorine, fluorine and carbon atoms.

chlorofluoromethanes A subset of chlorofluorocarbons that contain only one carbon atom.

command and control To prescribe regulations to compel implementation of particular policies.

consumption As defined by the Montreal Protocol, production plus imports minus exports.

controlled substance Any ozone-depleting substance that is subject to control measures under the Montreal Protocol, such as a phase-out requirement. Specifically, it refers to a substance listed in Annexes A, B, C or E of the Protocol, whether alone or in a mixture.

Copenhagen Schedule The control measures prescribed by the Copenhagen Adjustments and Amendment by the fourth Meeting of the Parties to the Montreal Protocol in 1992.

countries with economies in transition (CEITs) States of the former Soviet Union and of Central and Eastern Europe that have been undergoing a process of major structural, economic and social change since Communism collapsed in about 1990, resulting in severe financial and administrative difficulties for both government and industry.

country programme A national strategy prepared by an Article 5 country to implement the Montreal Protocol and phase out ODSs

decommissioning The physical process of removing a halon system from service.

developing countries Countries listed as 'developing' by the first Meeting of the Parties to the Montreal Protocol in 1989 and subsequently modified by Meetings of Parties.

Dobson unit A unit to measure the total amount of ozone in a vertical column above the Earth's surface. A typical value for the amount of ozone in a column of the Earth's atmosphere is 300 Dobson units.

drop-in replacement The procedure of replacing CFC-refrigerants with non-CFC-refrigerants in existing refrigerating, air conditioning and heat pump plants without doing any plant modifications.

essential use Decision IV/25 of the Parties to the Montreal Protocol defined criteria for essential uses for which Parties could get exemptions for continued use of ODSs after the phase-out date.

Executive Committee (of the Multilateral Fund) Established by Article 10 of the Montreal Protocol to administer the Multilateral Fund.

feedstock A chemical that is entirely transformed or destroyed while being used for manufacture of another chemical.

financial mechanism Established by Article 10 of the Montreal Protocol for providing financial and technical cooperation, including the transfer of technologies, to Article 5 Parties. It may also include other means of multilateral, regional and bilateral cooperation.

Freon™ The trade name of CFC and HCFC products marketed by DuPont.

Global Environment Facility (GEF) Provides financing to the eligible developing countries and countries with economies in transition for programmes that achieve global environmental benefits in one or more of four focal areas: biological diversity; climate change; international waters; and the ozone layer.

global warming The warming of the Earth due to the heat-trapping action of natural and manufactured greenhouse gases, which leads to climate change.

global warming potential (GWP) The relative contribution of greenhouse gases, eg carbon dioxide, methane, CFCs, HCFCs and halons, to the global warming effect when the substances are released to the atmosphere. The standard measure of GWP is relative to carbon dioxide, whose GWP is 1.0.

greenhouse gas A gas, such as water vapour, carbon dioxide, methane, CFCs and HCFCs, that absorbs and re-emits infrared radiation, warming the Earth's surface and contributing to climate change.

Group of 77 A group of developing countries, formed in the United Nations, which originally numbered 77, but continuously increased its number with new countries joining.

halocarbon A compound derived from methane (CH_4) and ethane (C_2H_6), in which one or several of the hydrogen atoms are substituted with chlorine (Cl), fluorine (F) and/or bromine (Br). CFCs, HCFCs and HFCs are examples of halocarbons. If all the hydrogen atoms are substituted in a halocarbon, it is 'fully halogenated'. Otherwise it is 'partially halogenated'.

halon A halon is a bromochlorofluorocarbon, a chemical consisting of one or more carbon atoms surrounded by fluorine, chlorine and bromine. Halons are fully halogenated hydrocarbons that are used as fire-extinguishing agents and as explosion suppressants.

halon bank The total quantity of halon existing at a given moment in a facility, organization, country or region. The halon bank includes the halon in fire protection systems, in portable fire-extinguishers, in mobile fire-extinguishers and in storage containers.

halon bank management A method of managing a supply of banked halon. Bank management consists of keeping track of halon quantities at each stage: initial filling; installation; 'recycling'; and storage. A major goal of a halon bank is to avoid demand for new (virgin) halons by re-deploying halons from decommissioned systems or non-essential applications to essential uses.

Harmonized System A numbering system developed by the World Customs Organization to cover all imported or exported goods in international trade.

HCFC cap The non-Article 5 Parties must freeze their consumption of HCFCs from 1996 at the level of the HCFC cap, ie their 1989 consumption of HCFC plus 2.8 per cent of their 1989 CFC consumption.

hydrobromofluorocarbons (HBFCs) A family of hydrogenated chemicals related to halons consisting of one or more carbon atoms surrounded by fluorine, bromine, at least one hydrogen atom and sometimes chlorine.

hydrocarbons (HCs) Chemical compound consisting of one or more carbon atoms surrounded only by hydrogen atoms. Examples of hydrocarbons are propane (C_3H_8, HC-290), propylene (C_3H_6, HC-1270) and butane (C_4H_{10}, HC-600). HCs are commonly used as a substitute for CFCs in aerosol propellants and refrigerant blends and have an ODP of zero. Although they are used as refrigerants, their highly flammable properties normally restrict their use as low concentration components in refrigerant blends.

hydrochlorofluorocarbons (HCFCs) A family of chemicals related to CFCs that contain hydrogen, chlorine, fluorine and carbon atoms. HCFCs are partly halogenated and have much lower ODP than the CFCs.

hydrofluorocarbons (HFCs) A family of chemicals related to CFCs that contain one or more carbon atoms surrounded by fluorine and hydrogen atoms. Since no chlorine or bromine is present, HFCs do not deplete the ozone layer. HFCs are widely used as refrigerants. Examples of HFC refrigerants are HFC-134a (CF_3CH_2F) and HFC-152a (CHF_2CH_3). HFCs have a high global warming potential.

Implementation Committee A committee of the Parties to the Montreal Protocol, established by the non-compliance procedure of the Protocol to investigate and recommend action to the Meetings of the Parties on instances of non-compliance by Parties with the provisions of the Montreal Protocol.

implementing agencies United Nations Development Programme (UNDP), United Nations Environment Programme (UNEP), United Nations Industrial Development Organization (UNIDO) and the World Bank have been designated as the implementing agencies of the Multilateral Fund.

incremental costs The Multilateral Fund meets the agreed incremental costs of Article 5 Parties. An indicative list of such incremental costs was agreed to by the second and fourth Meetings of the Parties to the Montreal Protocol.

industrial rationalization An exchange of production quotas between Parties, provided total production is within the limit set by the Protocol.

Langmuir periodic table A variation of the periodic table that grouped all the known elements by their atomic weight.

London Schedule The control measures approved by the second Meeting of the Parties to the Montreal Protocol in London in 1990.

low volume ODS-consuming countries (LVC countries) Defined by the Multilateral Fund's Executive Committee as Article 5 Parties whose calculated level of ODS consumption is less than 360 ODP tonnes annually.

metered-dose inhalers (MDI) A method of dispensing inhaled pulmonary drugs.

methyl bromide A colourless, odourless, highly toxic gas composed of carbon, hydrogen and bromine, used as a broad spectrum fumigant in commodity, structural and soil fumigation. Methyl bromide has an ODP of approximately 0.6.

methyl chloroform see 1,1,1-trichloroethane.

Montreal Protocol An international agreement, under the Vienna Convention to Protect the Ozone Layer, to phase out the production and consumption of ozone-depleting chemicals according to a time schedule.

Multilateral Fund Part of the financial mechanism under the Montreal Protocol, established by the Parties to provide financial and technical assistance to Article 5 Parties.

national ozone unit The government unit in an Article 5 country that is responsible for managing the national ODS phase-out strategy as specified in the country programme. These units are responsible for, inter alia, fulfilling data reporting obligations under the Montreal Protocol.

no-clean soldering A method of soldering which leaves no residues to be cleaned up.

non-Article 5 countries Parties that do not operate under Article 5. They are obliged to implement the control measures of Article 2.

non-Party With respect to any controlled substance, the term 'state not party to this Protocol' (non-Party) includes a country that has not agreed to be bound by the control measures for that substance. For instance, a Party to the Montreal Protocol of 1987, which listed and defined the control measures for the Annex A substances, is a Party for the Annex A substances. If the same Party does not ratify the 1990 Amendment to the Protocol, which listed and defined the control measures for the Annex B substances, it is a non-Party for the Annex B substances.

not-in-kind alternatives/substitutes Approximately 80 per cent of ozone-depleting compounds that would be used today if there were no Montreal Protocol have been successfully phased out without the use of other fluorocarbons. ODS use was eliminated with a combination of 'not-in-kind' chemical substitutes, product alternatives, manufacturing process changes, conservation and doing without.

ODP tonnes The number of tonnes of a substance multiplied by its ozone depletion potential.

OzonAction Programme The programme implemented by UNEP with the assistance of the Multilateral Fund. It provides assistance to developing countries under the Montreal Protocol through information exchange, training, networking, country programmes and institutional strengthening projects.

ozone A gas consisting of three oxygen atoms, formed naturally in the atmosphere by the association of molecular oxygen (O_2) and atomic oxygen (O). It has the property of blocking the passage of dangerous wavelengths of ultraviolet radiation in the upper atmosphere.

ozone-depleting substance (ODS) Any substance that can deplete the stratospheric ozone layer.

ozone depletion Accelerated chemical destruction of the stratospheric ozone layer by the presence of ozone-depleting substances produced, for the most part, by human activities.

ozone depletion potential (ODP) A relative index indicating the extent to which a chemical product may cause ozone depletion. The reference level of 1 is the potential of CFC-11 and CFC-12 to cause ozone depletion. If a product has an ozone depletion potential of 0.5, a given weight of the product in the atmosphere would, in time, deplete half the ozone that the same weight of CFC-11 would deplete.

ozone layer An area of the stratosphere, approximately 15 to 60 kilometres (9 to 38 miles) above the Earth, where ozone is found as a trace gas (at higher concentrations than other parts of the atmosphere). This relatively high concentration of ozone filters most ultraviolet radiation, preventing it from reaching the Earth.

ozone-safe A substance which has zero ODP (example, HFCs, hydrocarbons) or a product which contains no ozone-depleting substance.

Ozone Secretariat The Secretariat to the Montreal Protocol and Vienna Convention, provided by UNEP and based in Nairobi, Kenya.

ozonesonde A lightweight, balloon-borne instrument that is mated to a conventional meteorological radiosonde. As the instruments ascend through the atmosphere, information on ozone and standard meteorological quantities such as pressure, temperature and humidity are transmitted to ground stations.

Party A country that ratifies an international legal instrument (eg, a protocol or an amendment to a protocol), indicating that it agrees to be bound by the provisions of that instrument. Parties to the Montreal Protocol are countries that have ratified the Protocol.

perfluorocarbons (PFCs) A group of synthetically produced compounds in which the hydrogen atoms of a hydrocarbon are replaced with fluorine atoms. The compounds are characterized by extreme stability, non-flammability, low toxicity, zero ozone-depleting potential and high global warming potential.

phase-out The ending of all production and consumption of a chemical controlled under the Montreal Protocol, consumption being defined as production plus imports minus exports.

production As defined by the Montreal Protocol, the amount produced minus the amount destroyed by technologies approved by the Parties minus the amount entirely used as feedstock in the manufacture of other chemicals.

products made with, but not containing, ODSs Products made with ODSs but that after their manufacture, do not contain ODSs. Examples include computers that are cleaned with solvent CFC-113.

quarantine and pre-shipment applications Quarantine applications are treatments to prevent the introduction, establishment or spread of quarantine pests, including diseases. Pre-shipment applications are those treatments applied directly preceding and in relation to exports to meet the sanitary or phytosanitary requirements of the importing or exporting country.

reclamation As defined by the Parties to the Montreal Protocol in their Decision IV/24, 'the re-processing and upgrading of a recovered controlled substance through such mechanisms as filtering, drying, distillation and chemical treatment in order to restore the substance to a specified standard of performance. It often involves processing "off-site" at a central facility'.

recovery As defined by the Parties to the Montreal Protocol in their Decision IV/24, 'the collection and storage of controlled substances from machinery, equipment, containment vessels, etc, during servicing or prior to disposal'.

recycling As defined by the Parties to the Montreal Protocol in their Decision IV/24, 'the re-use of a recovered controlled substance following a basic cleaning process such as filtering and drying'. Refrigerants are normally cleaned by the on-site recovery equipment and recharged back into the equipment.

refrigerant The chemical or mixture of chemicals used in refrigeration equipment.

refrigerant management plan (RMP) A strategy for cost-effective phase-out of ODS refrigerants, which considers and evaluates all alternative technical and policy options.

service tail After the phase-out date for ozone-depleting substances, permission to consume ODSs for the purpose of servicing existing equipment.

skin cancer There are three types of skin cancer: basal cell carcinoma; squamous cell carcinoma; and cutaneous melanoma.

stratosphere The part of the Earth's atmosphere above the troposphere, at about 15 to 60 kilometres (9 to 38 miles). The stratosphere contains the ozone layer.

Technology and Economic Assessment Panel (TEAP) The TEAP is one of the three assessment panels appointed by the Parties to the Montreal Protocol. The TEAP is responsible for reviewing and reporting to the Parties about the status of options to phase out the use of ODSs, recycling, re-use and destruction techniques, and their technological and economic viability. The TEAP is served by an Economic Options Committee and Technical Options Committees on Aerosols, Sterilants, Miscellaneous Uses and Carbon Tetrachloride; Flexible and Rigid Foams; Halons; Methyl Bromide; Refrigeration, Air Conditioning and Heat Pumps; and Solvents, Coatings and Adhesives. These committees consist of hundreds of experts from around the world.

transitional substance A chemical whose use is permitted as a replacement for ozone-depleting substances, but only temporarily due to the substance's ODP. The HCFCs were categorized as transitional by the second Meeting of the Parties in 1990, but were included in the list of controlled substances by the fourth Meeting of the Parties in 1992.

1,1,1-trichloroethane A hydrochlorocarbon commonly used as a blowing agent and as a solvent in a variety of metal, electronic and precision cleaning applications. It has an ODP of approximately 0.11. It is also known as methyl chloroform.

troposphere The lower part of the Earth's atmosphere, below 15 kilometres (9 miles). The troposphere is below the stratosphere.

ultraviolet radiation (UV radiation) Solar radiation at the top of the atmosphere contains radiation of wavelengths shorter than visible light. The shortest of these wavelengths (UV-C) are completely absorbed by oxygen and ozone in the atmosphere. Wavelengths in the middle (UV-B) range are partially absorbed by the ozone layer. The higher UV-A wavelengths are transmitted to the Earth's surface.

Vienna Convention The international agreement agreed to in 1985 to set a framework for global action to protect the stratospheric ozone layer. This Convention is implemented through its Montreal Protocol.

About the contributors

Daniel Albritton PhD (USA) is Director of the Aeronomy Research Laboratory of the US National Oceanographic and Atmospheric Administration (NOAA) which studies the global chemistry and dynamics of the Earth's atmosphere. He is a founding Co-Chair of the Science Assessment Panel and coordinating lead author of the Intergovernmental Panel on Climate Change (IPCC) report *Climate Change 2001: The Scientific Basis*. He has earned the 1994 US EPA Stratospheric Protection Award and the 1995 UNEP Ozone Award for his outstanding contribution to the protection of the ozone layer.

Richard E Benedick was the chief US negotiator of both the Vienna Convention and the 1987 Montreal Protocol. Author of *Ozone Diplomacy: New Directions in Safeguarding the Planet* (Cambridge, MA and London: Harvard University Press, enlarged edition, 1998), he is currently Senior Advisor of the Joint Global Change Research Institute of the Pacific Northwest National Laboratory (Battelle) and President of the National Council for Science and the Environment.

Fatma Can (Turkey) was a staff member of the National Ozone Unit in Turkey and has often been a member of the Turkish delegation to the Open-Ended Working Group and Meeting of the Parties.

Penelope Canan PhD (USA) is Professor of Sociology and Director of the International Institute for Environment and Enterprise at the University of Denver. She served on the Economic Options Committee of the Technology and Economic Assessment Panel of the Montreal Protocol. She is co-author, with Nancy Reichman, of *Ozone Connections: Expert Networks in Global Environmental Governance* (2002, Greenleaf Publishing, Sheffield, UK).

Suely Machado Carvalho PhD (UNDP) is Chief of the Montreal Protocol Unit of the United Nations Development Programme New York and a Co-Chair of the Montreal Protocol Technology and Economic Assessment Panel. Prior to joining UNDP she was Director of Environmental Quality at Companhia de Tecnologia de Saneamento Ambiental (CETSB) in Sao Paulo, Brazil.

Brigitta Dahl (Sweden) is presently the Speaker of the Parliament of Sweden. She, as the Minister for Environment of Sweden from March 1986, played a key role in driving the early phase-out of CFCs in Sweden and arriving at the

Montreal Protocol in 1987 as well as its crucial London Adjustment and Amendment to the Protocol in 1990.

David Doniger (USA) is Director of Climate Protection at the Natural Resources Defense Council (NRDC). Prior to rejoining NRDC he held political appointments at the White House and US EPA under President William J Clinton. Before that, he was a Senior Attorney at NRDC where he successfully sued EPA to agree to proceed with ozone layer protection.

Linda Dunn (Canada) is Team Leader, Trade Team Canada Environment at Industry Canada. Formerly, she was Senior Environmental Policy Advisor for Industry Canada and prior to that Senior Policy Advisor for Agriculture and Agri-Food Canada. She was responsible for demonstration of alternatives to methyl bromide and significantly influenced methyl bromide policies and controls in Canada. She was member of the Montreal Protocol Technology and Economic Assessment Panel Methyl Bromide Technical Options Committee.

Omar E El-Arini PhD (Egypt) has been the Chief Officer of the Secretariat of the Multilateral Fund for the Implementation of the Montreal Protocol since its establishment in 1991 and assisted the Executive Committee to successfully steer the Fund towards achieving its goals most effectively. He was previously Director of the Centre for Research and Conservation of the Egyptian Antiquities Organization, Research Scientist at BAYER AG, Germany, Scientific Affairs Specialist for the US based National Science Foundation in Egypt and Adjunct Professor in Cairo University and the American University in Cairo.

Mohamed El-Ashry PhD (GEF) is the Chief Executive Officer and Chairman, Global Environment Facility, Washington, DC. Previously, he worked as Chief Environmental Advisor to the President of the World Bank, Senior Vice-President of the World Resources Institute, Director of Environmental Quality at the Tennessee Valley Authority and as Special Advisor to the Secretary-General of the 1992 United Nations Conference on Environment and Development (the Earth Summit).

Yuichi Fujimoto (Japan) is Adviser of the Japan Industrial Conference for Ozone Layer Protection (JICOP). Before joining JICOP he was Director of the Japan Electrical Manufacturers' Association (JEMA) where he organized Japanese industry to protect the ozone layer. He is Senior Expert Member of the Montreal Protocol Technology and Economic Assessment Panel.

Maneka Gandhi (India) is presently a Minister of State in the Government of India. As the then Minister of State for Environment and Forests, she headed the Indian Delegation to the second Meeting of the Parties to the Montreal Protocol in London in 1990. She is deeply interested in animal welfare and has won international and national awards for her work in this area. She is author of numerous books on plants and animals.

Qu Geping (China) was formerly the Minister of Environment of China and presently is the Chairman of the Environmental Protection and Resources Conservation Committee of the National People's Congress of China. He was recipient of both the 1992 UNEP Sasakawa Prize and Japan's Blue Planet Prize for 1999.

Corinna Gilfillan (USA) is a graduate student at the London School of Economics. From 1998 to 2001 she was a consultant and then an associate programme officer in UNEP Division of Technology Industry and Economics OzonAction programme. From 1995 to 1998 she was Director of the Ozone Protection Campaign at Friends of the Earth, USA.

Michael Graber PhD (UNEP) is Deputy Executive Secretary, Ozone Secretariat, UNEP, Nairobi and has directed the Secretariat since the retirement of Madhava Sarma in August 2000.

Joop van Haasteren PhD (The Netherlands) has been active in almost all the Montreal Protocol meetings on behalf of the Ministry of Housing, Spatial Planning and the Environment of The Netherlands, where he worked until his 2001 retirement. He initiated and coordinated The Netherlands phase-out of methyl bromide that preceded the Montreal Protocol controls and spearheaded technology cooperation in support of the global phase-out.

Morio Higashino (Thailand/Japan) was Director of Environment for the Minebea Company, which is the largest miniature-bearing manufacturer in the world, with production in Japan, Thailand and elsewhere. He was responsible for the corporate leadership policy of Minebea to phase out ODS solvents worldwide by 1997.

Paul Horwitz (USA) is International Advisor, Stratospheric Protection Division, US EPA. He was closely associated with the evolution of the Montreal Protocol and its London Amendment, as a member of the Environmental Law Division of UNEP and as Secretary to the international meetings which negotiated these. He has represented the USA in all the ozone meetings since then. He was the Chairman of the Executive Committee of the Multilateral Fund and has led the US delegation in the Executive Committee since 1992.

Caley Johnson (USA) is the Project Manager for Strategic Climate Projects in the US EPA's Climate Protection Partnerships Division. He authored multiple appendices and also managed the collection and cataloguing of Montreal Protocol documents and the production of the book drafts.

Margaret Kerr MD (Canada) is an environmental consultant specializing in global sustainability. She was Vice-President of Environment, Health and Safety for Nortel and an organizer of the Industry Cooperative for Ozone Layer Protection (now called the International Cooperative for Environmental Leadership). Under her leadership, Nortel became the first multinational electronics company to announce a CFC phase-out goal and the first to achieve that goal.

Vyacheslav Khattatov PhD (Russia) is Chief of Laboratory, Central Aerological Observatory, Russia, and was a member of the USSR delegation to the 1987 Montreal Meeting and its preparatory meetings. He was one of the first scientists to conclude that ODSs were depleting the ozone layer and a successful advocate of Soviet participation in science assessments. He was a leader of the team securing agreement to place an American ozone-monitoring satellite on a Soviet rocket when the USA lost access to space as a result of the explosion of the Space Shuttle Challenger.

Naoki Kojima (Japan) headed the Ozone Protection Office in the Ministry of International Trade and Industry from 1989 to 1991 during the important negotiations of the London Amendments. He is now Executive Director of the Petroleum Energy Center, Tokyo.

Ingrid Kökeritz (Sweden) worked on ozone issues with the Swedish Environment Protection Agency since the early 1980s and, though retired recently, works as a consultant on ozone and other environmental issues, based in the Stockholm Environmental Institute. She played a key role in all the stages of evolution of the Montreal Protocol.

Geoffrey Lean (UK) is External Editor for UNEP Publications including *Our Planet*, and a frequent editor of books, magazines and articles. He has reported on environmental issues for more than 30 years, for the *Yorkshire Post*, the *Observer* and the *Independent on Sunday*, where he is now Environment Editor.

János (John) Maté (Greenpeace) is the Director of the Ozone and Greenfreeze Projects for Greenpeace International. He represents Greenpeace at all the key meetings of the Protocol. He is the author of several Greenpeace documents on technical and policy issues related to ozone-layer protection, and is the producer of two videos on ozone and climate-friendly technologies in refrigeration. He holds an MA in Sociology and has been with Greenpeace since 1989.

Yasuko Matsumoto PhD (Japan) is a Fellow of the National Institute for Environmental Studies, Tsukuba, Japan. Prior to becoming a Fellow, Yasuko was Professor at the Science University of Tokyo, Suwa, Japan, and before that directed Greenpeace Japan's Stratospheric Protection Programme and represented Greenpeace at international meetings.

Gordon McBean PhD (Canada) is Chair of the Board of Trustees of the Canadian Foundation for Climate and Atmospheric Sciences. He is a professor at the University of Western Ontario. For many years he was Assistant Deputy Minister of the Meteorological Service of Canada.

Alan Miller (GEF) is Team Leader for Climate Change and Ozone at the Global Environment Facility. Prior to joining GEF he was Executive Director of the University of Maryland Center for Global Change, a senior associate in the energy and climate programme at the World Resources Institute, and Senior Attorney at the Natural Resources Defense Council. He helped negotiate the voluntary phase-out of CFCs in food packaging and started the litigation forcing the US EPA to protect the ozone layer.

John Miller spent all his scientific career involved in the global/regional monitoring of atmospheric chemistry parameters such as ozone, greenhouse gases and long-range transport of acidic substances. He served as Director of Mauna Loa Observatory, Hawaii, and spent the last ten years at the WMO as Chief of the Environment Division where he was responsible for development of WMO's Global Atmosphere Watch. He is now a consultant at the Air Resources Laboratory, NOAA, USA.

Ryusuke Mizukami (Japan) played a major role in making the Minebea Group the first manufacturer in Thailand to completely eliminate CFCs and methyl chloroform from all production processes in 1993. He is currently leading the Minebea Group in environmental activities that have enabled their principal manufacturing facilities, which account for 95 per cent of total Minebea Group output in monetary terms, to achieve ISO 14001 certification.

E Thomas Morehouse (USA) is Research Associate at the Institute for Defense Analyses (IDA), Washington, DC, USA and Senior Expert Member of the Technology and Economic Assessment Panel. Prior to joining IDA, he was a career officer in the United States Air Force, responsible for fire protection research in 1987 when the Montreal Protocol was first signed.

Lawrence Musset (France) works in the Ministry of Environment, France and represents France in all the meetings related to the Montreal Protocol.

Tsuneya Nakamura (Japan) was President of Seiko Epson Corporation when the plan to phase out CFCs was announced. He guided the leadership and technology cooperation for Seiko Epson, including the decision to donate to the public domain valuable patented technology to eliminate CFC solvents. After methyl chloroform was added to the Montreal Protocol, he made the decision to phase it out on an accelerated basis.

Julian Newman (UK) works for the Environmental Investigation Agency, London and has done considerable work in investigating and exposing illegal trade in ozone-depleting substances.

Tetsuo Nishide (Japan) is Director of the Air Quality Management Division at the Japan Ministry of the Environment. Prior to his assignment at the Environment Agency, he directed chemical programmes at the Ministry of

International Trade and Industry, including programmes on ozone and climate protection from 1992 to 1994 and 1997 to 2000.

Tsutomu Odagiri (Japan) is Secretary-General for the Japan Industrial Conference on Cleaning (JICC). He was Deputy Secretary-General and Director of Science & Technology for the Japan Industrial Conference for Ozone Layer Protection (JICOP) from 1990 to 1997. He was responsible for national technology cooperation on the solvent phase-out, for technology cooperation with developing countries, and for participation at international conferences and workshops.

Sally Rand (USA) is Director of High-GWP Industry Partnerships in the US EPA Climate Protection Partnerships Division. She was Co-Chair of the Technology and Economic Assessment Panel (TEAP) Foams Technical Options Committee (FTOC) from 1996 to 1999, and member of the 1999 HFC and PFC Task Force of the TEAP.

Nelson Sabogal (UNEP) is Senior Scientific officer, Ozone Secretariat, UNEP, Nairobi. Prior to joining UNEP he worked as a scientist and project administrator in Columbia.

Lani Sinclair (USA) is a freelance writer and editor with government, academic and NGO clients. She previously worked on ozone layer protection at the World Resources Institute and was a communications director at the National Wildlife Federation. Before that she was on the staff of US Senator John Chafee.

Don C Smith (USA) is a senior researcher and fellow of the International Institute for Environment and Enterprise at the University of Denver, USA, where he also serves as an Adjunct Professor of Comparative Environmental Law. He is an attorney with American and British law degrees and has extensive experience in media, including work in book publishing, magazines and journals.

Yoshihiko Sumi (Japan) was Director of the Ozone Layer Protection Office at the Japan Ministry of International Trade and Industry from 1991 to 1992 where he directed ozone protection programmes.

Helen Tope PhD (Australia) is Senior Policy Officer, Environment Protection Authority, Victoria, Australia where she facilitated ODS phase-out programmes. She is a member of the Montreal Protocol Technology and Economic Assessment Panel and Co-Chair of its Aerosol Products Technical Options Committee.

Hideaki Yasukawa (Japan) was head of manufacturing at the Seiko Epson Company when then-president Tsuneya Nakamura announced the plan to phase out CFCs. He was President of Seiko Epson when the phase-out was achieved.

Index

Page numbers in *italics* refer to boxes, figures and tables

The names of all winners of awards for contributions to ozone-layer protection can be found in
Appendix 6.

Washington Workshops (1977/1986) 68,
200
Watson, Robert 22, 60, 76, 98, 107, 131,
202, 306, 315–16, *363*, *384*, 438
Waxman, Henry 77, 336
Wells, H G 368
Westinghouse *371*
Whirlpool *331*, 342, *394*, *396*
Whiteley, William *370*
Wickham, Robert xxvi
Wirth, David 69, 80
WMO 6, *16*, 19, 45–7, 49, 63–4, 66,
128–9, 234, *375*, *376*, *385*, *388*, *393*
Wofsy, Steven C xxvi, 8, 11, 16–17, *373*,
378
Woolworths *385*, *391*
World Bank 42, 99, 109, 119, 156, 234,
238, 248–9, 250, 283
World Customs Organization 99, 128,
234–5, 276, *277*
World Data Centres *27*
World Health Organization (WHO) xxii,
47, 235, *395*
World Meteorological Organization *see*
WMO
World Ozone Data Centre, Toronto *16*,
27

World Plan of Action on the Ozone
Layer (1977) 12, *16*, 46–7, 357
World Resources Institute 78, 109, 119,
292
World Semiconductor Council *396*
World Summit on Sustainable
Development (2002) 186
World Wide Fund For Nature (WWF)
120
Worldwatch Institute *374*
Wright, Orville *370*
Wright, Wilber *370*
Wuebbles, Donald J 17, *380*
Wuori, M 333
Wurzburg Workshop (1987) 77

Yalcindag, Seniz H xxv
Yasukawa, Hideaki xxv, 204, 488
Yemen *267*
Yugoslavia 103, 138, 147, *267*, 355

Zambia *267*, 336
Zakharov, Vladimir 101
Zhenhua, Xie 320
Zimbabwe *267*, 336, 339